D0722597

Dynamics of
POND
Aquaculture

Edited by

Hillary S. Egna

Oregon State University
Corvallis, Oregon

Claude E. Boyd

Auburn University
Auburn, Alabama

CRC Press
Boca Raton   New York

The contents of this book are primarily an outcome of the Pond Dynamics/Aquaculture Collaborative Research Support Program, sponsored in part by Grant No. DAN-4023-G-00-0031-00 of the United States Agency for International Development's Office of Agriculture and Food Security. The program is intended to support collaborative aquaculture research between the United States and developing countries' universities and research institutions. The opinions expressed herein are those of the authors and do not necessarily reflect the views of the United States Agency for International Development. The chapters in this book were written, submitted, reviewed, and revised over the period from 1992 to 1996.

Library of Congress Cataloging-in-Publication Data

Dynamics of pond aquaculture / edited by Hillary S. Egna and Claude E. Boyd.
     p. cm.
    Includes bibliographical references and index.
    ISBN 1-56670-274-7
    1. Pond aquaculture. 2. I. Egna, Hillary S. II. Boyd, Claude E.
SH137.4.D96 1997
639.3′1—dc21
for Library of Congress
                                                      96-52335
                                                      CIP

This book contains information obtained from authentic and highly regarded sources. Reprinted material is quoted with permission, and sources are indicated. A wide variety of references are listed. Reasonable efforts have been made to publish reliable data and information, but the author and the publisher cannot assume responsibility for the validity of all materials or for the consequences of their use.

Neither this book nor any part may be reproduced or transmitted in any form or by any means, electronic or mechanical, including photocopying, microfilming, and recording, or by any information storage or retrieval system, without prior permission in writing from the publisher.

All rights reserved. Authorization to photocopy items for internal or personal use, or the personal or internal use of specific clients, may be granted by CRC Press LLC, provided that $.50 per page photocopied is paid directly to Copyright Clearance Center, 27 Congress Street, Salem, MA 01970 USA. The fee code for users of the Transactional Reporting Service is ISBN 1-56670-274-7/97/$0.00+$.50. The fee is subject to change without notice. For organizations that have been granted a photocopy license by the CCC, a separate system of payment has been arranged.

The consent of CRC Press LLC does not extend to copying for general distribution, for promotion, for creating new works, or for resale. Specific permission must be obtained in writing from CRC Press LLC for such copying.

Direct all inquiries to CRC Press LLC, 2000 Corporate Blvd., N.W., Boca Raton, Florida 33431.

© 1997 by CRC Press LLC
Lewis Publishers in an imprint of CRC Press

No claim to original U.S. Government works
International Standard Book Number 1-56670-274-7
Library of Congress Card Number 96-52335
Printed in the United States of America 1 2 3 4 5 6 7 8 9 0
Printed on acid-free paper

DEDICATION

To Bucyanayandi Jean-Damascène, Gasana Félicien, Gatera Anaclet, Kajyibwami Emmanuel, Kamanzi Jean Bosco, Mukundwa Alvera, Murokore, Ndayisaba Jean, Ndoreyaho Valens, Nsanganiye Onesphore, Uwizeyimana Eugénie, and others too numerous to name but not too numerous to forget. We remember that war claims unfairly those we respect, love, admire, and care for. All of us in the Pond Dynamics/Aquaculture CRSP hope for peace in Rwanda.

CONTENTS

Chapter 3
WATER QUALITY IN PONDS
James S. Diana, James P. Szyper, Ted R. Batterson, Claude E. Boyd, and Raul H. Piedrahita

Chapter 4
FERTILIZATION REGIMES
C. Kwei Lin, David R. Teichert-Coddington, Bartholomew W. Green, and Karen L. Veverica

Chapter 5
CLIMATE, SITE, AND POND DESIGN
Anita M. Kelly and Christopher C. Kohler

Chapter 6
POND BOTTOM SOILS
Claude E. Boyd and James R. Bowman

Chapter 7
ENVIRONMENTAL CONSIDERATIONS
Wayne K. Seim, Claude E. Boyd, and James S. Diana

Chapter 8
ATTRIBUTES OF TROPICAL POND-CULTURED FISH
David R. Teichert-Coddington, Thomas J. Popma, and Leonard L. Lovshin

Chapter 9
FACTORS AFFECTING FISH GROWTH AND PRODUCTION
Richard W. Soderberg

Chapter 10
FRY AND FINGERLING PRODUCTION
Bartholomew W. Green, Karen L. Veverica, and Martin S. Fitzpatrick

Chapter 11
FEEDING STRATEGIES
James S. Diana

Chapter 12
DISEASES OF TILAPIA

Kamonporn Tonguthai and Supranee Chinabut

Chapter 13
COMPUTER APPLICATIONS IN POND AQUACULTURE —
MODELING AND DECISION SUPPORT SYSTEMS
Raul H. Piedrahita, Shree S. Nath, John Bolte, Steven D. Culberson,
Philip Giovannini, and Douglas H. Ernst

Chapter 14
EXPERIMENTAL DESIGN AND ANALYSIS IN AQUACULTURE
Christopher F. Knud-Hansen

Chapter 15
ECONOMIC CONSIDERATIONS
Carole R. Engle, Revathi Balakrishnan, Terry R. Hanson, and Joseph J. Molnar

Chapter 16
DEVELOPING AND EXTENDING AQUACULTURE TECHNOLOGY FOR PRODUCERS
Karen L. Veverica and Joseph J. Molnar

FOREWORD

In September 1980, a new concept of aquacultural research was undertaken. A state-of-the-art study of the principles and practices of aquaculture was conducted from 1980 to 1982, under a planning grant to Oregon State University, which culminated in the award of the first Pond Dynamics/Aquaculture Collaborative Research Support Program (PD/A CRSP) grant in 1982. The study reviewed and analyzed literature on pond aquaculture and conducted numerous site visits to determine the research needs and the extent of local support for the program in a number of developing countries. The initial evaluation indicated a need to (1) improve the technological reliability of pond production systems and (2) establish economic optimization strategies consistent with local conditions. The planning study concluded that maximum yields were being obtained from many ponds, but equivalent yields could be realized with lower input costs if the input requirements could be estimated. Thus, production efficiency became a focus.

With great foresight, the initiators of the research program decided to create a global experiment that could contribute to the understanding of pond aquaculture systems. Pond aquaculture is not new, having been practiced for centuries, but practices reported to succeed in one case have not been achievable in others; results have not been reproducible. The Global Experiment concept was conceived as a means of analyzing empirical data to evaluate efficiency of production. The development of the Global Experiment is described more fully in Chapter 2.

In 1982, a group of individuals representing Auburn University, the University of California at Davis, and a consortium comprised of Oregon State University, Michigan State University, the University of Michigan, the University of Hawaii, and the University of Arkansas at Pine Bluff teamed up with researchers from six host countries (Honduras, Indonesia, Panama, Philippines, Rwanda, and Thailand) to initiate a global experiment on pond aquaculture. The original experimental sites all rested within 15 degrees latitude of the equator and represented the three major regions of the tropics, Southeast Asia, Africa, and Latin America. Initial planning included projects for Sierra Leone and Jamaica, but reduced funding prevented the establishment of these research sites.

The primary, underlying objective of the research program was to increase the availability of food, particularly animal protein derived from aquaculture, in developing countries. Secondary yet important objectives were to foster greater employment opportunities and enhance the income of rural people and communities. The goals of the overall program were broader, focusing on strengthening research capabilities in the U.S. and developing countries (see discussion in Chapter 2). The program functions under the auspices of the Management Entity at Oregon State University, which is advised by three bodies: a Technical Committee, a Board of Directors, and an External Evaluation Panel. External Evaluation team members are selected for their range of disciplinary and professional experiences related to aquaculture. The Technical Committee includes representatives from both host countries and participating U.S. institutions. The Technical Committee provides the leadership for design of experiments, establishment of research protocol, and prescription of test methods, procedures, and equipment. The Board of Directors provides guidance to the Management Entity on matters of procedure and administration of the overall program.

Experimental design in the CRSP has involved monitoring fish production variables in 12 or more ponds at each of the geographical locations in accordance with standardized work plans established by the Technical Committee. The experimental design included the use of standardized experiments, methodology, equipment, and data collection. Uncertainties resulting during early trials conducted by the research team led to the establishment of rigorous experimental protocols that have provided statistically reliable results. The success of the program ultimately yielded numerous publications of individual studies in peer review journals, which are the foundation of this book.

A Pond Dynamics/Aquaculture (PD/A) CRSP Central Data Base, created through this program, has become a resource of importance to the CRSP and other aquaculture researchers. The data base has been used to develop pond dynamics models for in-depth analysis of the dynamics of pond systems; it has also been used to develop a decision support system, POND$^©$, that is intended to facilitate management decisions in technology transfer or production aquaculture. The pond dynamics modeling focuses on primary production and evaluation of the interactions between light, turbidity, depth, dissolved oxygen, and temperature gradients. Both the pond dynamics models and POND contribute significantly to the fundamental understanding of pond aquaculture, and both contribute significantly to establish state-of-the-art guidelines for effective management of pond aquaculture (see Chapter 13).

The program's initial focus on production (particularly primary productivity) has been broadened to encompass sustainability, with attention to environmental stewardship, social implications at both the farm and community levels, and profitability. Studies have been incorporated on water quality and interactions at the soil–water interface. In order to improve social and economic conditions of smallholders, a variety of outreach activities have encouraged local farmers to implement this new knowledge in their farming practices.

The results of the PD/A CRSP research and Global Experiment have significance around the world. Statistics provided by the United Nations Food and Agriculture Organization in 1990 indicated that enormous opportunities exist for development of pond aquaculture in developing countries. While 11 million metric tons of aquaculture production are provided by the Asian countries of China, Japan, South and North Korea, Taiwan, and the Philippines, other regions of the world are at least one order of magnitude less productive. For example, Europe and other Asian countries (Indonesia, Vietnam, Thailand, India, and Bangladesh) produce only about 1 million metric tons each. Latin America produces only about 0.2 million tons and Africa about 0.07. These statistics, taken together with the fact that pond aquaculture is a widespread culture system used for both warmwater finfish and shrimp production, clearly demonstrate the opportunity for the global impact of a more comprehensive understanding of the dynamics of aquaculture ponds.

In addition to providing valuable knowledge of pond dynamics and primary productivity, the PD/A CRSP has also contributed in a direct way to the advancement of aquaculture in those countries where the CRSP has remained active. Because of budgetary constraints, the CRSP reduced its host country participation in 1987 to Honduras, Rwanda, and Thailand. Additional activities in Egypt and the Philippines were initiated in the early 1990s. In all of these locations, the CRSP has worked aggressively to extend research information and transfer technology. A primary activity in each host country has been the training of aquaculture researchers and technicians who can then participate directly in technology transfer within their own countries. Projects have been conducted directly with extension personnel, funded either by host countries or other international agencies, and indirectly through extension activities that work with host country producers. In each country, pond aquaculture activities have expanded as a direct consequence of the CRSP program, making notable progress toward the fundamental goal of increasing the availability of animal protein derived from fish.

Robert B. Fridley
Hillary S. Egna
Philip Helfrich
Harry Rea
R. Oneal Smitherman
Lamarr Trott

PREFACE

"Die Löecher sind doch die Hauptsache an einem Sieb." (The holes are the main aspects of a sieve.)

Joachim Ringelnatz

"Reasoning should not form a chain which is no stronger than its weakest link, but a cable whose fibers may be ever so slender provided they are sufficiently numerous and intimately connected."

Charles Sanders Pierce

This book was written during a period of great change within the Pond Dynamics/Aquaculture Collaborative Research Support Program (CRSP). Budget shortfalls in foreign assistance caused a reframing of priorities. A new project in Egypt was initiated and completed. Most importantly, our long-time field site in Rwanda was lost, as were many of our friends and colleagues. The gestation of this book was influenced and attenuated by these factors, none of which could have been predicted when we embarked on this project.

This book is the culmination of over a decade's research on aquaculture by the CRSP. Most of the chapters of the book are based on data collected by the CRSP at its field research sites in East Africa, Southeast Asia, Central America, and North America. The research findings from CRSP fieldwork have yielded much technical information on the chemical, biological, and physical interactions occurring within a fish pond. These findings are particularly useful for gaining an appreciation for the complexity of variables involved in pond management. However, aquaculture is more than pond dynamics; thus, some chapters were added to provide a more complete picture of the pond system and the environment in which it exists. In particular, the chapters on fish diseases, fish reproduction, extension, social and economic considerations, and environmental effects were brought into this volume because of their critical importance in making aquaculture a feasible enterprise. Because these chapters rely heavily on the general aquaculture literature and on researchers not directly involved with the CRSP, they contain comparatively less actual CRSP field data than the chapters on pond dynamics.

This book expands on *Principles and Practices of Pond Aquaculture* (Lannan et al., 1986), an earlier state-of-the-art literature review by the CRSP, which set the framework for much of the biotechnological research undertaken by the CRSP in the subsequent decade. The purpose of the first book was to provide information "useful in moving pond aquaculture from a highly developed art form to an agricultural technology." The editors assumed that most of the information available in the earlier literature was descriptive as opposed to quantitative (see Chapter 2 of this book). They also indicated that a lack of standardization in experimental design, data collection, and analysis rendered the existing data base of limited utility in predicting the performance of pond culture systems.

The editors of *Principles and Practices of Pond Aquaculture* finally concluded that the unifying question was as follows: how do physical, chemical, and biological processes interact to regulate the productivity of these systems? The need for mathematical models was stressed. The editors also emphasized the need for an ecological systems model based on the principles of pond aquaculture, which could be used as a framework for generating other deterministic and stochastic models based on systematic quantitative field data. Hence, the first book set the research agenda for the early CRSP years, which were devoted primarily to generating data for models and data bases. Comparatively less attention was paid to the organism or the social and natural environment in which the fish pond exists. However, with changes in the technical knowledge base, the global political economy, and development assistance priorities, the CRSP began to modify its research agenda in the late 1980s to include new research on economics, sociology, fish reproduction, and nutrition. This second book goes well beyond the first CRSP book by presenting actual research findings and by drawing connections to the people who depend on their fish ponds for food security.

Aquaculture pond management techniques have improved markedly during the past decade, when much of the CRSP research was conducted. These improvements include better techniques in almost all areas, including seed fish production, pond preparation, fertilization, feed composition and manufacturing, aerator design, and harvesting. Additionally, increasing awareness of the need for greater quantification and standardization in research and recognition of environmental impacts of aquaculture have gained momentum in the past decade. Separating out CRSP results from the results of the aquaculture community at large is difficult at best. The CRSP has been responsible for conducting much of the research in the peer review literature on pond dynamics and for providing technical information to producers, as can be seen in the succeeding chapters. However, many other individuals and groups have also been conducting research on pond dynamics, and their efforts must be acknowledged when speaking of the overall advances in pond aquaculture since the early 1980s.

The authors of each of the chapters are responsible for the accuracy, reliability, and originality of the information presented within their chapters. As editors, we have attempted to allow for diversity of thought, ideas, and opinions, so there may be some disagreements among chapters. These disagreements reflect the state of knowledge in the field, as the understanding of aquaculture's effects on communities and on the natural environment continues to evolve. The reader is encouraged to read each chapter with a critical eye, rather than as a prescription for absolute success in pond aquaculture. Having said that, this book does provide rules of thumb on pond culture of tilapia that are not found elsewhere in a single volume. More importantly, it provides a synopsis of many methods, techniques, and ideas that have been explored by aquaculture practitioners and researchers over the years.

ACKNOWLEDGMENTS

Without the support and hard work of many people this book would not have been possible. Authors of each of the chapters are recognized for their efforts. In addition, many other people contributed in important ways. They include students, field workers, researchers, farmers, extension agents, technicians, government officials, and administrators in Rwanda, Honduras, Egypt, Philippines, U.S., Thailand, Panama, and Indonesia. The vast network of people involved in the conceptualization of the CRSP, in the collection and analysis of data, and in the implementation and logistics of each CRSP project is far too widespread and so we cannot name each one here, but we remember them and are indebted to them.

We are also grateful to the various CRSP advisory boards for their dedication to international development and for understanding how the CRSP could further contribute through the publication of this book. CRSP participants had talked about following up on the earlier CRSP book, *Principles and Practices of Pond Aquaculture*, which was written in 1983, with a new textbook, but the idea languished for years. Roger Pullin of ICLARM was instrumental in giving us a push to get this book started in 1992. Richard Neal, Gary Jensen, Homer Buck, Philip Helfrich, Robert Fridley, and Oneal Smitherman also assisted by motivating the CRSP Technical Committee to stand behind this effort. Raul Piedrahita, Wayne Seim, Marty Fitzpatrick, and Chris Knud-Hansen are highly commended for not allowing the Technical Committee to lose sight of the goal of completing this book.

Each chapter was sent out to external reviewers who offered suggestions toward the improvement of this book. They are named on a separate page. We appreciate their selfless efforts under exceedingly tight deadlines. We hope we have not burned too many bridges by sending them copious drafts of chapters, as future publications by the CRSP will need their diligence and insights.

The hardships of editing a multiauthored volume, such as one of this magnitude, were minimized by the hard work of several behind-the-scene individuals. Deb Walks, Deb Burke, and Sayea Jenabzadeh, all graduate students at Oregon State University, worked tirelessly to format chapters, check literature cited in sections, correct figures and tables, create the acronyms and reviewers lists, and do so many more tasks that made this book complete. Through all the strains of late and incomplete chapter submissions, they maintained a wonderful sense of humor and high spirits. Up to the final deadline, Penny Schumacher, an OSU student, and Sayea Jenabzadeh labored late nights over this book, and to them we are forever grateful. Brigitte Goetze is especially commended for providing valuable assistance and pithy quotations. Her conflict remediation skills proved a real asset during some of the rougher times during the gestation of this book.

We thank the editors at Lewis Publishers for their patience, understanding, and steadfast commitment to this project. We also acknowledge the support of the entire Program Management Office at Oregon State University. They had to bear many more responsibilities than usual when the CRSP Director volunteered to be an editor. The office itself became laden with more paper and bustle than would normally be acceptable in a university office. For not complaining, not even once, we thank Marion McNamara, Danielle Clair, Brad Herbison, Brigitte Goetze, and Ingvar Elle.

The book would not have been possible without the support and attention given to us by our families and friends during the lengthy period during which this book was written, edited, and rewritten. In all senses they, too, are part of the CRSP network. Finally, we recognize the U.S. Agency for International Development for providing us financial assistance to continue the important work of the CRSP and to see this book to fruition. With the recent year's, turmoil in foreign assistance, we were uncertain whether the program would outlast the development of the book. We are delighted to say that it has. We are grateful to Harry Rea, Lamarr Trott, and Harvey Hortik for their unfaltering championship of aquaculture research in international development.

EDITORS

Hillary Egna is Director of the PD/A CRSP. She has served the CRSP for 17 years in various research and administrative capacities at Oregon State University, the University of Michigan, and in Panama. She manages all faculty and staff who directly participate in the CRSP and is responsible for the fiscal, administrative, and technical oversight of the program. She received her B.S. in Natural Resources from the University of Michigan in 1980 and her M.Ag. in Aquaculture/Fisheries from OSU in 1985. Her Ph.D. in Resource Geography is expected in 1997 (presently ABD status) from OSU. She has worked professionally on international projects in Africa, Asia, and Latin America for over 12 years; her international experience extends to over 20 years. A Senior Faculty Research Assistant at OSU, she has authored a number of papers on subjects ranging from environmental impact assessment to remote sensing in aquaculture. She has been editor-in-chief of CRSP publications for eight years and has initiated most of mainstream CRSP serials now in circulation.

Claude E. Boyd is a Professor in the Department of Fisheries and Allied Aquacultures, Auburn University. He received his B.S. in entomology (1962) and M.S. in insect toxicology from Mississippi State University (1963), and his Ph.D. in water quality from Auburn University (1966). His research efforts have included ecology and utilization of aquatic weeds, pond liming and fertilization, water quality and aeration, pond bottom soils management, and hydrology of ponds. Both freshwater and brackish water aquaculture have been included in these efforts. Dr. Boyd's teaching has focused on graduate training in water quality management (three courses per year and major advisor for more than 75 graduate students) and continuing education for professionals. Dr. Boyd's publications have included research, review, and experience papers in scientific journals, five books, experiment station publications and articles in trade journals (totaling over 250). He has worked on projects of one week to six months duration in 21 countries including Thailand, Honduras, Philippines, Australia, Belize, Brazil, China, Canada, Colombia, Ecuador, France, Guatemala, India, Indonesia, Israel, Italy, Malaysia, Mexico, Panama, Taiwan, and Surinam.

CONTRIBUTORS

Revathi Balakrishnan
Women in Development Program
UNDP
Bangkok, Thailand

Ted R. Batterson
North Central Regional Aquaculture Center
Michigan State University
East Lansing, Michigan

John Bolte
Bioresource Engineering Department
Oregon State University
Corvallis, Oregon

James R. Bowman
Department of Fisheries and Wildlife
Oregon State University
Corvallis, Oregon

Claude E. Boyd
International Center for Aquaculture
 and Aquatic Environments
Department of Fisheries and Allied
 Aquacultures
Auburn University, Alabama

Deborah A. Burke
Pond Dynamics/Aquaculture Collaborative
 Research Support Program
Office of International Research and
 Development
Oregon State University
Corvallis, Oregon

Supranee Chinabut
Aquatic Animal Health Research Institute
 (AAHRI)
Kasetsart University Campus
Bangkok, Thailand

Steven D. Culberson
Department of Environmental Studies
University of California at Davis
Davis, California

James S. Diana
School of Natural Resources
and Environment
University of Michigan
Ann Arbor, Michigan

Hillary S. Egna
Pond Dynamics/Aquaculture Collaborative
 Research Support Program
Office of International Research
 and Development
Oregon State University
Corvallis, Oregon

Carole R. Engle
Department of Aquaculture and Fisheries
University of Arkansas at Pine Bluff
Pine Bluff, Arkansas

Douglas H. Ernst
Bioresource Engineering Department
Oregon State University
Corvallis, Oregon

Martin S. Fitzpatrick
Department of Fisheries and Wildlife
Oregon State University
Corvallis, Oregon

Robert B. Fridley
College of Agricultural and Environmental
 Sciences
University of Davis at California
Davis, California

Philip Giovannini
Department of Biological and Agricultural
  Engineering
University of California at Davis
Davis, California

Bartholomew W. Green
International Center for Aquaculture
  and Aquatic Environments
Department of Fisheries and Allied
  Aquacultures
Auburn University, Alabama

Terry R. Hanson
International Center for Aquaculture
  and Aquatic Environments
Department of Agricultural Economics and
  Rural Sociology
Auburn University, Alabama

Philip Helfrich
Hawaii Institute of Marine Biology
University of Hawaii at Manoa
Kaneohe, Hawaii

Anita M. Kelly
Fisheries Research Laboratory
Department of Zoology
Southern Illinois University at Carbondale
Carbondale, Illinois

Christopher F. Knud-Hansen
Department of Fisheries and Wildlife
Michigan State University
East Lansing, Michigan

Christopher C. Kohler
Fisheries Research Laboratory
Department of Zoology
Southern Illinois University at Carbondale
Carbondale, Illinois

C. Kwei Lin
Agricultural and Aquatic Systems Program
Asian Institute of Technology
Bangkok, Thailand

Leonard L. Loveshin
International Center for Aquaculture
  and Aquatic Environments
Department of Fisheries and Allied
  Aquacultures
Auburn University, Alabama

Joseph J. Molnar
International Center for Aquaculture
  and Aquatic Environments
Department of Agricultural Economics
  and Rural Sociology
Auburn University, Alabama

Shree Nath
Bioresource Engineering Department
Oregon State University
Corvallis, Oregon

Raul H. Piedrahita
Department of Biological and Agricultural
  Engineering
University of California at Davis
Davis, California

Thomas J. Popma
International Center for Aquaculture
  and Aquatic Environments
Department of Fisheries and Allied
  Aquacultures
Auburn University, Alabama

Harry Rea
Office of Agriculture and Food Security
U.S. Agency for International Development
Washington, D.C.

Wayne Seim
Department of Fisheries and Wildlife
Oregon State University
Corvallis, Oregon

R. Oneal Smitherman
Department of Fisheries and Allied
  Aquacultures
Auburn University, Alabama

James P. Szyper
Hawaii Institute of Marine Biology
University of Hawaii at Manoa
Kaneohe, Hawaii

David R. Teichert-Coddington
International Center for Aquaculture
  and Aquatic Environments
Department of Fisheries and Allied
  Aquacultures
Auburn University, Alabama

Kamonporn Tonguthai
Aquatic Animal Health Research Institute
 (AAHRI)
Kasetsart University Campus
Bangkok, Thailand

Lamarr Trott
Office of Environment and Natural Resources
U.S. Agency for International
 Development
Washington, D.C.

Karen L. Veverica
International Center for Aquaculture
 and Aquatic Environments
Department of Fisheries and Allied
 Aquacultures
Auburn University, Alabama

REVIEWERS

The editors extend their sincere appreciation for the efforts of the many scholars and researchers who participated in the prepublication review of these chapters. The individuals named below reviewed one or more chapters in this book.

David Acker
International Agriculture Programs
Iowa State University
Ames, Iowa

John Bardach
Office of Research and Development
East-West Center
Honolulu, Hawaii

Hilary Berkman
Department of Fisheries and Wildlife
Oregon State University
Corvallis, Oregon

James Bowman
Department of Fisheries and Wildlife
Oregon State University
Corvallis, Oregon

Homer Buck
Illinois Natural History Survey
Salem, Illinois

Gary Chapman
United States Environmental Protection
  Agency
Hatfield Marine Science Center
Newport, Oregon

Wilfrido Contreras-Sanchez
Division Academica de Ciencias Biologicas
  CICEA
Universidad Juarez Autonoma de Tabasco
Tabasco, Mexico

Maria Darmi
Rothamsted Experiment Station
The Institute of Arable Crops Research
Hertsfordshire, United Kingdom

James S. Diana
School of Natural Resources and Environment
University of Michigan
Ann Arbor, Michigan

Bryan Duncan
International Center for Aquaculture
  and Aquatic Environments
Department of Fisheries and Allied
  Aquacultures
Auburn University, Alabama

Carole R. Engle
Department of Aquaculture and Fisheries
University of Arkansas at Pine Bluff
Pine Bluff, Arkansas

Douglas H. Ernst
Bioresource Engineering Department
Oregon State University
Corvallis, Oregon

Arlo Fast
Hawaii Institute of Marine Biology
University of Hawaii at Manoa
Kaneohe, Hawaii

Grant Feist
Department of Fisheries and Wildlife
Oregon State University
Corvallis, Oregon

Martin S. Fitzpatrick
Department of Fisheries and Wildlife
Oregon State University
Corvallis, Oregon

Jere Gilles
Department of Rural Sociology
University of Missouri-Columbia
Columbia, Missouri

Bartholomew W. Green
International Center for Aquaculture
  and Aquatic Environments
Department of Fisheries and Allied
  Aquacultures
Auburn University, Alabama

John Hargreaves
Department of Wildlife and Fisheries
Mississippi State University
Mississippi State, Mississippi

Christopher F. Knud-Hansen
Department of Fisheries and Wildlife
Michigan State University
East Lansing, Michigan

Christopher C. Kohler
Fisheries Research Laboratory
Southern Illinois University at Carbondale
Carbondale, Illinois

Christopher Langdon
Department of Fisheries and Wildlife
Hatfield Marine Science Center
Oregon State University
Newport, Oregon

James Lannan
Department of Fisheries and Wildlife
Oregon State University
Corvallis, Oregon

David Little
Division of Agriculture and Food Technology
Asian Institute of Technology
Bangkok, Thailand

Don Moss
International Center for Aquaculture
  and Aquatic Environments
Department of Fisheries and Allied
  Aquacultures
Auburn University, Alabama

Shree S. Nath
Bioresource Engineering Department
Oregon State University
Corvallis, Oregon

Reynaldo Patino
Texas Cooperative Fish and Wildlife
  Research Unit
Texas Technical University
Lubbock, Texas

Peter Perschbacher
Department of Aquaculture and Fisheries
University of Arkansas at Pine Bluff
Pine Bluff, Arkansas

Ronald Phelps
International Center for Aquaculture
  and Aquatic Environments
Department of Fisheries and Allied
  Aquacultures
Auburn University, Alabama

Mark Prein
Life Sciences Division
International Center for Living Aquatic
Resources Management (ICLARM)
Manila, Philippines

James Rakocy
Agricultural Experiment Station
University of the Virgin Islands
Kingshill, Virgin Islands

Wayne Seim
Department of Fisheries and Wildlife
Oregon State University
Corvallis, Oregon

Sunantar Setboonsarng
Division of Agriculture and Food Engineering
Asian Institute of Technology
Bangkok, Thailand

William Shelton
Department of Zoology
University of Oklahoma
Norman, Oklahoma

Michael Skladany
Department of Sociology
Michigan State University
East Lansing, Michigan

Dick Soderberg
Fisheries Program
Mansfield University
Mansfield, Pennsylvania

Robert R. Springborn
Seattle, Washington

James P. Szyper
Hawaii Institute of Marine Biology
University of Hawaii at Manoa
Kaneohe, Hawaii

George Tchobanoglous
Department of Civil and Environmental
  Engineering

University of California at Davis
Davis, California

David R. Teichert-Coddington
International Center for Aquaculture
  and Aquatic Environments
Department of Fisheries and Allied
  Aquacultures
Auburn University, Alabama

Craig Tucker
Delta Research and Extension Center
Stoneville, Mississippi

Karen L. Veverica
International Center for Aquaculture
  and Aquatic Environments
Department of Fisheries and Allied
  Aquacultures
Auburn University, Alabama

Fred Wheaton
Agricultural Engineering Department
University of Maryland
Baltimore, Maryland

ACRONYM LIST

AB	Strain of IPNV
AEC	Anion exchange capacity
AIT	Asian Institute of Technology
ANCOVA	Analysis of covariance
ANOVA	Analysis of variance
APHIS	Animal and Plant Health Inspection Service
BCF	Bureau of Commercial Fisheries
BIFAD	Board for International Food and Agricultural Development
BOD	Biochemical oxygen demand
CDIE	Center for Development Information and Evaluation (USAID)
CEC	Cation exchange capacity
CGIAR	Consultative Group on International Agricultural Research
CIFAD	Consortium for International Fisheries and Aquaculture Development
COD	Chemical oxygen demand
CPU	Central processing unit
CRD	Completely randomized design
CRSP	Collaborative Research Support Program
CSC	Critical standing crop
CVM	Center for Veterinary Medicine
DAST	Data Analysis and Synthesis Team
DDT	Dichlorodiphenyl trichloroethane, a pesticide
DE	Digestible energy
DIN	Dissolved inorganic nitrogen
DIP	Dissolved inorganic phosphorus
DO	Dissolved oxygen
DOD	Deutsches Ozeanographisches Datenzentrum (German Oceanographic Data Center)
DOE	Department of Energy, U.S.
DSS	Decision Support System
EPA	Environmental Protection Agency, U.S.
FAC	Freshwater Aquaculture Center, Philippines
FAO	Food and Agriculture Organization of the United Nations
FBT	Fishborne trematode, fluke
FCR	Feed conversion ratio
FDA	U.S. Food and Drug Administration
FIFRA	Federal Insecticide, Fungicide, and Rodenticide Act, U.S.
FOA	Foreign Operations Administration
GIFT	Genetic Improvement of Farmed Tilapia

GPP	Gross primary productivity
HACCP	Hazard analysis and critical control point
HC	Host country
HCG	Human chorionic gonadotropin
HOD	Hypolimnetic oxygen deficit
IARC	International Agricultural Research Center
ICLARM	International Center for Living Aquatic Resources Management
IDRC	International Development and Research Centre of Canada
INAD	Investigational new animal drug
IPNV	Infectious pancreatic necrosis virus
JRC	Joint Research Committee
JSA	Joint Subcommittee on Aquaculture
LC50	Concentration resulting in 50% mortality in test organisms exposed for 48 or 96 hours
LR	Lime requirement
MS 222	Tricane methanosulfonate
MST	Mean square treatment
MT	Metric tons or 17α-methyltestosterone
NFY	Net fish yield
NMFS	National Marine Fisheries Service, U.S.
NOAA	National Oceanographic and Atmospheric Administration, U.S.
NPDES	National Pollution Discharge Elimination System
NPP	Net primary productivity
OC	Organic carbon
OSHA	U.S. Occupational Safety and Health Administration
PD/A CRSP	Pond Dynamics/Aquaculture Collaborative Research Support Program
PSTC	Program in Science and Technology, USAID
RAC	Regional Aquaculture Center
RCBD	Randomized complete block design
RDA	Resources Development Associates, Inc., California firm
SCS/USDA	Soil Conservation Service, U.S. Department of Agriculture
SD	Standard deviation
SDA	Specific dynamic action
SDD	Secchi disk depth
SE	Standard error
SEAFDEC	Southeast Asian Fisheries Development Center
SGR	Specific growth rate
SIFR	Study on International Fisheries Research
SRP	Soluble reactive phosphorus
STPP	Sodium tripolyphosphate
TAN	Total available nitrogen
TCA	Technical Cooperation Administration, U.S.
TKN	Total Kjaldahl nitrogen
TLC	Threshold lethal concentration
TMP	Triple monophosphate
TP	Total phosphorus
TS	Total solids
TSP	Triple superphosphate

TSPP	Tetrasodium pyrophosphate
URI	University of Rhode Island
USAID	U.S. Agency for International Development
USDA	U.S. Department of Agriculture
USDI	U.S. Department of Interior
USP	U.S. Pharmacopeia
VBGF	von Bertalanffy growth function
WBT	Waterborne trematode

1   INTRODUCTION

Hillary S. Egna, Claude E. Boyd, and Deborah A. Burke

Aquaculture is the cultivation of aquatic animals and plants. Its primary purpose is to produce aquatic food organisms for human consumption, but includes other purposes such as the cultivation of ornamental and aquarium fishes. Aquaculture may be done on many scales, ranging from small rainfed ponds to increase food production for rural families to large commercial farms to provide export products for international markets. Regardless of scale, aquaculture is an economic activity, and the value of aquacultural crops must exceed the cost of producing them. Knowledge of factors and interactions that determine success in aquaculture is not as well developed as in traditional agriculture; thus, aquaculture has tended to be risky and has suffered in some instances from unsustainability. But aquaculture technology is improving rapidly, and more reliable production systems are emerging.

Most aquaculture involves cultivating a species of interest under conditions that can be monitored and controlled. Sessile creatures such as mollusks can be cultivated by providing substrate for their attachment. However, fish, crustaceans, and other motile organisms are usually confined in order to cultivate them. The most common confinements used in aquaculture are ponds, raceways, cages, and pens. Ponds are by far the most widely used means of confining warmwater fish and crustaceans for cultivation. Therefore, it was quite appropriate for the U.S. Agency for International Development (USAID) to initiate a project on pond aquaculture as a Collaborative Research Support Program (CRSP). The purpose of this book is to summarize advances in pond aquaculture that have accrued from the CRSP effort. The intent of this introductory chapter is to summarize the role of aquaculture in world fisheries and to explain why research and development in the area of pond dynamics are critical to the sustainable development of aquaculture.

POND AQUACULTURE

The Role of Pond Aquaculture

According to FAO (1995), the total capture fisheries of the world peaked at about 90 million metric tons in 1989. Capture fisheries have provided around 85 million metric tons per annum since 1989. Aquaculture has been very important in supplementing capture fisheries. Since 1984, aquaculture has grown at annual rates of 8 to 14%, with an 89% increase in production between 1984 and 1992 (Anonymous, 1995). The total production of aquaculture was estimated at 19.3 million metric tons in 1992. Because of the contributions of aquaculture, world fisheries production has remained slightly above 100 million metric tons per year since 1988, despite the fact that capture fisheries peaked in 1989 and declined slightly in the following years (FAO, 1995).

In 1992, aquaculture produced 18.5% of the global fisheries output. Thirty-five countries produced 98% of the aquaculture products in 1992 (Anonymous, 1995). Of the total production of 19.2 million metric tons, 17 million metric tons were produced in Asia. In terms of total production, carps and other cyprinid fishes represented the greatest production of aquaculture products, with

0-56670-274-7/97/$0.00+$.50
© 1997 by CRC Press LLC

about 6.7 million metric tons. Other major contributors were tilapia and other cichlids, miscellaneous freshwater fishes, oysters, mussels, scallops, marine shrimp, trout and salmon, and seaweeds. Marine shrimp had a remarkably high monetary value, in spite of their relatively low contribution to total production.

Freshwater fish culture is particularly impressive in its contribution to the total freshwater fisheries production of the world. The production of freshwater aquaculture in 1992 was 8 million metric tons (FAO, 1995); this was 61% of the total freshwater fisheries output. Marine and brackish water aquaculture contributed 39.4% of the world mollusk production in 1992, and 25% of the world shrimp production in 1994, but it produced only 5.9% of seaweeds and 0.5% of fish in 1992.

It is not certain what percentage of total aquacultural production comes from ponds, but it is thought to be more than 50% of the total. Of course, if only fish and crustaceans are considered, well over 75% of total production is from ponds. Clearly, ponds are very important aquaculture production units.

Types of Pond Aquaculture

We may classify pond aquaculture along several lines, such as species, pond characteristics, and intensity of culture. In attempting to list all aquaculture species, Pillay (1990) named 95 finfish, 32 crustaceans, 35 mollusks, and 19 aquatic plants commonly used in aquaculture. Avault (1996) mentioned more than 100 aquaculture species in his discussion. More than 400 aquaculture species were listed by Jhingran and Gopalakrishnam (1974). Therefore, a classification of aquaculture by species would be unwieldy.

Culture techniques usually do not differ greatly for aquaculture products of similar characteristics. Thus, it is popular to speak of the culture of sport fish, food fish, ornamental fish, crustaceans, mollusks, frogs and other amphibians, alligators and other reptiles, vascular aquatic plants, seaweeds, etc. It also is useful to divide the culture categories according to temperature preference, i.e., coldwater, coolwater, warmwater, and tropical, and according to salinity preference, i.e., freshwater, brackish water, and marine.

The culture systems may be classified by type of grow-out units, i.e., ponds, cages, pens, raceways, tanks, silos, etc. This book emphasizes primarily pond culture. Ponds may be classified according to construction methods as watershed ponds, excavated ponds, and levee ponds (Yoo and Boyd, 1994; also see Chapter 5 for other pond types). Watershed ponds are formed by building a dam across a natural watercourse where topography permits water storage behind the dam. The dam is usually constructed between two hills that constrict the watershed. Watershed ponds may store only overland flow, or they may receive some combination of overland flow, stream flow, and groundwater inflow. Excavated ponds are formed by digging a hole in the ground. They may be filled by groundwater inflow where the water table is near the land surface, by overland flow if constructed in a low-lying area, or by well water. The water in levee ponds is impounded in an area surrounded by levees. Little runoff enters levee ponds, so they must be filled by water from wells, storage reservoirs, streams, or estuaries. Hydrologically, ponds may be classified in various ways — for example, as static ponds with little water exchange or as flow-through ponds where water exchange is used on a regular basis.

From the standpoint of pond dynamics, the most meaningful classification of ponds is the intensity of management inputs and the amount of production. In extensive aquaculture, there are few inputs of nutrients, and production is quite low. Larger nutrient inputs are provided and greater production is obtained in semi-intensive aquaculture. The greatest inputs of nutrients are provided in intensive aquaculture to achieve very high production. An attempt has been made in Table 1 to summarize the different levels of management and amounts of net production that are typical in the culture of marine shrimp, tilapia, and channel catfish. There are no standard definitions of

Table 1   Representative Net Production (kg/ha) for Channel Catfish, Tilapia, and Marine Shrimp with Different Levels of Management

Management	Channel catfish	Tilapia	Marine shrimp
Stocking only	50–100	200–500	100–200
Stocking and fertilization	200–300	1,000–3,000	400–600
Stocking and feeding	1,500–2,500	3,000–6,000	—
Stocking, feeding, and water exchange	—	—	1,000–2,000
Stocking, feeding, and aeration	2,500–5,000	6,000–10,000	—
Stocking, feeding, aeration, and water exchange	5,000–10,000	10,000–20,000	3,000–10,000

extensive, semi-intensive, and intensive aquaculture, but reasonable divisions between the three levels of management are as follows:

Extensive — production is enhanced only by manures or chemical fertilizers
Semi-intensive — feeds are used to increase production; manures and chemical fertilizers also may be used
Intensive — large amounts of feed are applied, manures and fertilizers may be used, and ponds are aerated by mechanical means

Water exchange is uncommon in freshwater pond aquaculture, but it may be used in very intensive systems. In brakishwater pond aquaculture, water exchange is used commonly at all levels of production. Water exchange rates are greatest in intensive, brakishwater aquaculture. As aquaculture gains momentum throughout the world, more and more farmers are leaning toward higher production strategies through increased intensification. For example, in rural, subsistence aquaculture, more farmers are beginning to use chemical fertilizers instead of manures and other organic fertilizers. In commercial aquaculture, there is a rapidly growing dependence upon feeds and mechanical aeration. Production is increasing in both subsistence and commercial aquaculture, but in developing countries, extensive and semi-intensive production techniques are still more common than intensive methods.

Pond Dynamics

The term *pond dynamics* was used by the CRSP to convey the physical, chemical, and biological factors that interact in pond systems. In pond aquaculture, ponds are constructed, filled with water, and stocked with fish, crustaceans, or other aquatic plants and animals. The amount of production in ponds depends heavily upon the quantity of food available to the culture organisms. In systems without supplemental feeding, primary productivity forms the base of the food web that culminates in fish and crustacean biomass. Natural levels of primary productivity are seldom high enough to provide sufficient natural food for high rates of aquacultural production. Some ponds have more nutrients in their bottom soils and waters than other ponds, and there is a large range in the natural productivity of ponds (Boyd, 1990). In order to enhance natural productivity, manures or chemical fertilizers are applied to stimulate primary productivity. Where watershed and pond soils are acidic, pond waters usually have low total alkalinity, and the response to fertilizers and manures will not be great. Such ponds are treated with liming materials to reduce acidity, increase total alkalinity, and enhance the response to fertilization. The amount of fertilizer or manure used in aquaculture ponds is highly variable, depending upon the resources and desires of the pond manager. Fertilization and manuring can result in 5- to 10-fold increases in fish production over natural productivity if the availability of phosphorus, nitrogen, and other limiting nutrients is increased to a sufficient level. Greater increases in production can be achieved through the use of manufactured feeds to supplement the availability of natural food organisms. Feeds also contain inorganic plant nutrients, and feed wastes fertilize the water to stimulate the primary productivity of ponds.

As the intensity of aquaculture is increased through the application of manures, fertilizers, and feeds, nutrient concentrations increase, plankton blooms occur, and dead organic matter settles to pond bottoms. Although phytoplankton produces dissolved oxygen through photosynthesis in the daytime, respiration also increases as the biomass of pond biota increases in response to nutrient inputs. This results in fluctuations in dissolved oxygen concentrations over a 24-h period, with highest concentrations in the afternoon and lowest concentrations at dawn. If phytoplankton abundance becomes too great, nighttime depletion of dissolved oxygen may stress or kill culture species. Accumulation of organic matter on pond bottoms can result in anaerobic conditions and the release of microbial metabolites, such as nitrite and hydrogen sulfide, into the water column. These substances are toxic to aquatic animals.

The major nitrogenous waste product of aquaculture species is ammonia, and as the intensity of production increases, ammonia concentrations may reach toxic levels. However, low concentration of dissolved oxygen is usually the first limiting factor that restricts production in ponds. If mechanical aeration is provided to enhance dissolved oxygen concentration, production can be increased until ammonia toxicity begins to limit production. The most reliable means of reducing ammonia concentration is water exchange. By exchanging water to flush ammonia from ponds, production levels may be increased. Production cannot be increased without limit through aeration and water exchange, because microbial decomposition of accumulated organic matter on pond bottoms will result in deterioration of bottom soil condition and high concentrations of toxic metabolites.

The discussion above reveals that many complex ecological factors and interactions are involved in the grow-out of aquaculture species in ponds. Aquacultural projects are even more complex, because they are conducted in a multiple-use environment where people live. Also, the value of aquaculture products must be greater than the cost of their production, and the value must be realized though direct use or by marketing. Thus, in addition to the ecological factors at play in aquaculture, many complex socioeconomic factors also are involved.

Tilapia Culture

Several species of tilapia are cultured in the world; this group of fishes is growing in importance both as a species for domestic consumption and for export. Estimated production in 1992 was almost 500,000 metric tons. Although there is some tilapia production in water recirculating systems in temperate climates and seasonal pond production in subtropical climates, the bulk of tilapia production is realized from tropical ponds. Tilapias are easy to culture because they are not commonly affected by diseases and water quality problems. They also grow very fast under crowded conditions, and high levels of production can be achieved. Because they feed on plankton and benthos, they are ideal for culture in manured or fertilized ponds. Nevertheless, they will feed upon pelleted, manufactured feed, and they can be raised under highly intensive conditions with aeration and water exchange.

The tilapias are popular species to culture in developing tropical countries. They can be important in local markets, and export possibilities are increasing because of the popularity of tilapia with consumers. Tilapias can be cultured at all levels of intensity by techniques common to most types of pond aquaculture. Therefore, tilapias are excellent for use in research on tropical aquaculture. Findings from research on tilapia aquaculture can also be transferable to other types of pond fish culture.

Pond Management

Pond fish culture consists of many more activities than just pond management. However, pond management is a key factor, because the major effort and expense occur in the grow-out of seed fish to marketable-sized animals in ponds. Mistakes in management also can greatly reduce yield and lead to the death of all or part of the fish in a pond.

Many factors are involved in pond management. They begin with site selection, design, and construction considerations (Hajek and Boyd, 1994). Unless a good site is selected and proper design and construction are used, pond aquaculture projects may have major problems from the beginning. In the grow-out phase, the manager must be concerned with climatic and meteorological factors such as temperature, rainfall, solar radiation, and wind. There is no control over such factors outside of planning. However, many other factors can be controlled to a greater or lesser extent by management. Culture species and stocking density can be selected. Plankton abundance can be increased by liming, fertilizing, and manuring at rates dictated by pond soil and water characteristics. When feeding is used, there usually is a choice about types of feed and feeding programs. Antibiotics and other chemicals are available to treat diseases and parasites of aquatic animals. Harvest times can be scheduled for optimum harvest size, ideal restocking time, or best market price. Pond bottom soils can be subjected to various treatments between crops to enhance organic matter decomposition and destroy pathogens and other unwanted organisms. In intensive culture, there are many possible options related to mechanical aeration, induced water circulation, water exchange, and chemical and biological treatments for improving pond water quality.

A major challenge in pond management is deciding upon an appropriate treatment for a given situation and determining the necessary rate at which the treatment should be applied. Most pond treatments have been developed by farmers through trial and error. Researchers have only recently investigated some of these treatments to determine efficacy and optimum application rates. Ponds are extremely complex ecological systems that require proper evaluation of benefits and potential problems that can result from some pond treatments. Unrealistic expectations about pond treatments can lead to disappointing yields. Applied research on pond management and effective extension programs for disseminating research findings are critical for the future development of aquaculture.

Sustainability and environmental effects are becoming important issues in all human endeavors because we are rapidly depleting many of the world's resources and damaging its environment in order to provide for the needs of an ever-growing population. Aquaculture must be sustainable as a business in order to survive. However, it also must be done in a way that does not waste valuable resources that could be better used for other purposes. Pond aquaculture can have negative social and environmental impacts if conducted in an irresponsible manner. The location of aquaculture projects in wetlands, water pollution by pond effluents, soil and water pollution by solid wastes from aquaculture, and social conflicts over resource and space allocations are examples of some of the problems that can result from poorly designed or badly managed aquaculture projects.

The subject of sustainability and environmental effects of aquaculture deserves further comment because aquaculture was considered to be an emerging "green industry" in the 1970s and 1980s. However, during the present decade, environmentalists have become alarmed at the rapid growth of aquaculture, and they are expressing many concerns. A number of conservation organizations, including both government and nongovernment agencies, recently made statements related to the sustainability of aquaculture before the United Nations Commission on Sustainable Development. Some of the more radical groups suggested boycotting some aquaculture products. However, more useful suggestions were made by the non-governmental organizations (NGOs). They urged the world governments to agree to the 16 items listed in Table 2 to ensure the sustainability of aquaculture. Aquaculture has indeed caused some social and environmental problems, and some

Table 2   List of Concerns about Environmental and Societal Effects of Aquaculture and Its Sustainability*

1. Ensure that artisanal fisheries and dependent coastal communities, and their access to community resources, are not adversely affected by aquaculture development or operations, including extensive and semi-intensive as well as intensive aquaculture methods
2. Ensure the use of environmental and social impact assessments prior to aquaculture development and the regular and continuous monitoring of the environmental and social impacts of aquaculture operations
3. Ensure the protection of mangrove forests, wetlands and other ecologically sensitive coastal areas
4. Prohibit the use of toxic and bioaccumulative compounds in aquaculture operations
5. Apply the precautionary approach to aquaculture development
6. Prohibit the pollution of surrounding areas resulting from the excessive discharge of organic wastes
7. Prohibit the development and use of genetically modified organisms
8. Prohibit the use of exotic/alien species
9. Prohibit the use or salinization of freshwater supplies, including groundwater important for drinking or agriculture
10. Prohibit the use of feeds in aquaculture operations consisting of fish that is or could be used as food for people
11. Prohibit the wholesale conversion of agricultural or cultivable land to aquaculture use
12. Ensure that the collection of larvae does not adversely affect species biodiversity
13. Ensure that abandoned or degraded aquaculture sites are ecologically rehabilitated and that the companies or industry responsible bear the cost of rehabilitation
14. Ensure that aquaculture and other coastal developments are addressed in integrated coastal management planning, which includes the meaningful participation of all coastal user groups
15. Ensure the development of aquaculture in a manner which is compatible with the social, cultural, and economic interests of coastal communities, and ensure that such developments are sustainable, socially equitable and ecologically sound
16. Ensure that multi-lateral development banks, bilateral aid agencies, the U.N. Food and Agriculture Organization and other relevant national and international organizations or institutions do not fund or otherwise promote aquaculture development inconsistent with the above criteria

* Anonymous. Comments presented by NGOs to the United Nations Commission on Sustainable Development, New York City, April 29, 1996.

projects have been abandoned. However, those problems are the exceptions rather than the rule. Commercial aquaculturists are aware of the environmental concerns listed in Table 2, and there is an obvious effort on their part to mitigate negative effects. Nevertheless, much more attention must be devoted to developing "environmentally-friendly" pond management systems, and the aquaculture industry also needs to be proactive in improving environmental stewardship. Most of the concerns in Table 2 can be addressed without great expense to the aquaculture industry, and environmental protection can be improved. Attention to environmental issues can improve water quality in ponds, protect the environment, enhance sustainability, and improve long-term profits. Therefore, it is very important that extension efforts deal with environmental issues and teach farmers that use of environmentally safe aquacultural methods is a win-win situation because they will benefit economically and environmental deterioration will not result.

RESEARCH NEEDS IN AQUACULTURE

Setting relevant priorities for global research in the area of aquaculture development is a difficult undertaking because of changing national and international policies and ill-defined needs for fishery products at different scales of development. Assuming that global demand for fish will continue to increase with population growth, and that capture fisheries, at least in the short-term, will not be able to meet this demand, aquaculture will play an increasingly important role (Figure 1). To date, however, fish production from aquaculture has not been sufficient to make up the deficit.

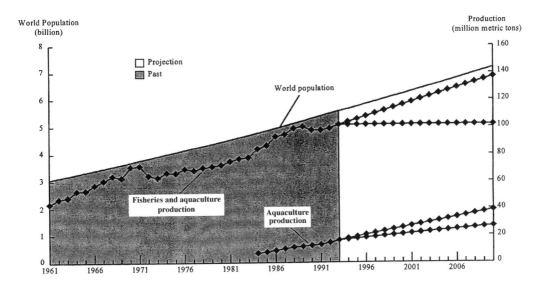

Figure 1.   World population and fisheries and aquaculture production, 1961–1993, with upper and lower projections to 2010. Projections are based on a major expansion in aquaculture (to 39 million metric tons) and a reversal in the decline of capture fisheries through good management and better use. (From Williams, M., Food, Agriculture, and the Environment, Discussion Paper 13, National Food Policy Research Institute, Washington, D.C., 1994, 4. With permission.)

Since the late 1980s, development assistance organizations (e.g., USAID, FAO) and some regional organizations have undertaken efforts to evaluate the potential for aquaculture to meet this growing demand for fish (Table 3). Constraints to the orderly development of sustainable aquaculture as they have been identified in the literature are summarized in Tables 4 and 5. In general, the literature notes that many of the major impediments to better aquaculture production are not research related but are rather issues of policy, institutional development, regulations, property and water rights, and human resources. However, there are key areas of aquaculture development that can benefit from a better understanding of aquaculture principles and practices.

Among the many constraints facing aquaculture, there is no one constraint that can be singled out for its global importance. Rather, most constraints are contextually defined through scale, system, and stage of development. Within this framework, research needs vary temporally and spatially. They also vary depending on the outlook of the national or international groups investing in research. For example, an African research agenda geared primarily toward small farmers with limited market access is intrinsically different from a Latin American agenda aimed at large-scale, export-oriented commercial producers. Therefore, assigning a priority order to world aquaculture research needs would not only be difficult but also not very useful. Having said this, there are

Table 3   Socioeconomic Characteristics: Annual Fish Production and Human Population by Regions of Developing Countries, 1988

	Asia/Pacific	LatinAmerica/ Caribbean	Sub- Saharan Africa	West Asia/North Africa	Total
Total production tonnes/year ($\times 10^3$)	29,264	16,206	4,163	1,947	51,580
Aquaculture tonnes/year ($\times 10^3$)	6,722	119	11	80	6,932
Population (1985) ($\times 10^4$)	2,575	404	421	245	3,645
Estimated population (2025) ($\times 10^4$)	4,379	761	1,296	609	7,045
Proportion of global population (2025) (%)	52	9	15	7	83

Note: FAO statistics for 1988.
Excerpted from ICLARM: A Strategic Plan for International Fisheries Research (1992).

certain general priorities that appear often in the literature for broad-based aquaculture development. Among the most frequently mentioned are

- Building national agricultural research systems by conducting collaborative research, thus providing a milieu for training and education
- Carrying out social sciences research, especially in cases where aquaculture is being extended to new areas
- Recognizing the importance of a systems approach, which would include an understanding of the interactions of technology and society, and the larger natural environment in which aquaculture occurs

Priorities for biotechnological research in aquaculture are varied and diverse. In general, research on topics related to broodstock and seed supply rank high for Asia, Africa, Latin America, and the United States. Feeds, nutrition, and environmental research (e.g., water quality, environmental impacts) related to aquaculture are also considered highly important. The U.S. was included in the survey because the CRSP model mandates that research address constraints common to both developing countries and the U.S. in order for research to be truly collaborative. Most of the chapters of this book are based on the CRSP model and reflect the research and development priorities inherent in the model (see Chapter 2).

In Africa, regional research priorities tend to focus on small aquaculture systems with limited market access. In these systems, research priority areas include broodstock and seed supply, culture of new or native species, aquaculture in small water bodies, and culture-based fisheries. Marketing is identified as the major priority research area within social sciences research. General development priorities for African aquaculture center on strengthening national agricultural research systems, improving the base of aquaculture information, and training and education. Again, as would be expected with such a large landmass, within the African continent aquaculture needs vary widely by region and specific locale.

Aquaculture has become big business in parts of Asia. In other parts, aquaculture is still a backyard operation. Typically, overall priorities for development cited in the literature vary considerably, depending on the importance of aquaculture within a given national context. For example, in 1995, shrimp farms in Asia harvested 558,000 metric tons, accounting for 78% of all shrimp produced from aquaculture (Anonymous, 1996). Thailand, currently the largest shrimp producer

Table 4 Aquaculture Research Priorities

	Global			Asia		Africa		Latin America		U.S.	
	Priority	Time period	Culture system	Priority	Culture system	Priority	Culture system	Priority	Culture system	Priority	Culture system
Biological and Technological Research											
I. Species-Driven Research Priorities:											
Broodstock & seed supply	P	all	all	P	all	P	E,S/FW	x	I/FW	P	S,I/FW, BW,M
Fry & fingerling production	x	***	***	P		x	S/FW	x	I/FW,M	x	
Controlled spawning, broodstock condition	x	***	***	x		x	E,S/FW			P	
Natural or local farm-made feeds, nutrition	x	all	E,S/FW	P	E,S,I/FW, BW	x	E,S/FW				
Genetics	x	LT	I/FW	x	S/FW	x	S,I/FW	x	I,FW	x	S,I/FW, BW,M
Diseases	x	LT	I/BW,M	x	I/FW,BW, M			P	I/M	x	S,I/FW, BW,M
General performance under culture conditions	x	***	all	x	all	x	all	x	all	P	S,I/FW, BW,M
Supplemental feeds & fish nutrition	x	MT, LT	S,I/FW, BW,M	P	S,I/FW, BW,M	x	S,I/FW, BW	x	I/FW,M	P	I/BW,M
Culture of new native or local species	x	MT, LT	E,S/FW	x		P	all	x	E,S,I/FW		

continues

Table 4 (continued)

	Global			Asia		Africa		Latin America		U.S.	
	Priority	Time period	Culture system	Priority	Culture system	Priority	Culture system	Priority	Culture system	Priority	Culture system
II. Systems-Driven Research:											
Integrated farming systems	x	all	E,S/FW			x	E,S/FW	x	E,S,I/FW		
Polyculture systems	x	all	E,S/FW, BW			x	E,S/FW				
Cage culture	x	all	E,S/FW, BW,M			x					
Integrating aquaculture in irrigation schemes	x	all	E,S/FW			x	E,S/FW				
Pond dynamics/ ecology	x	all	all	x	E,S/FW	x	E,S/FW	x	S,I/FW, BW,M		
Water quality	x	all	S,I	P	all			x	S,I/FW, BW,M	x	S,I/FW, BW,M
Marine aquaculture	x	MT, LT	S,I	x	S,I/BW,M	x	S,I/BW, M	P	S,I/M	P	S,I/BW,M
Temperate/cool water aquaculture	x	all	all			x	E,S/FW			x	S,I/FW
Organic fertilizers	x	ST,MT	E,S			x	E,S/FW				
Small water bodies & culture-based fisheries	x	all	E,S			P	E,S/FW	P	E,S,I/FW		

Environmental impact of aquaculture	x	all	S,I	P	S,I/FW, BW,M	x	S,I/FW, BW,M	x	I/BW	x	S,I/FW, BW,M
Regional analysis for siting farms	x	MT,LT	all	x	all	x	all	x	S,I/FW, BW,M		
Social Sciences Research											
Marketing	P			P	all	P	all	P	S,I/M	P	
Nutritional contribution of fish to human diet	x					x	E,S/FW				
Adoption and diffusion of technology	x			x	all	x	E,S/FW				
Target group identification	x			x	S/FW	x	E,S/FW				
Policy	x			P	all			x	***	P	***
Food security	x										
Management of aquaculture research	x			P	all	x	all	x	***	x	***
Regional analysis	x			x	all	x	all				
Diversification of production	x			x	S/FW						
Decision-making strategies: household level	x			P	S/FW	x	E,S,I/FW	x	all	x	***
Participatory rural appraisal (PRA)	x			x	S/FW	x	E,S/FW				
Economic feasibility studies & analysis	x			x	all			x	S,I/M	x	***

continues

Table 4   (continued)

Notes: (1) x and P indicate general priority areas. x means the area is considered important, but is not mentioned frequently enough to be considered a top priority. P means the research area appears frequently enough in the literature to be considered one of the top priorities in aquaculture R&D. Culture systems — E: extensive; S: semi-intensive; I: intensive / FW: freshwater; BW: brackish water; M: marine. Time period — ST (1–5 years); MT (5–10 years); LT (10–20 years). (2) In over 120 documents surveyed for compiling this table, social sciences research was generally considered more important than biotechnological research when extending aquaculture to new areas (CRSP Continuation Plan, 1996). (3) An attempt to standardize the various scopes and scales of documented studies was made. In some cases, research priorities were mentioned frequently enough to warrant individual consideration (e.g., fry production). Other research priorities also contained subpriorities, but these were not of sufficient concern to be included separately in this table. Some areas were so broad as to be inclusive of other subcomponents, e.g., integrated systems include other species-and systems–driven research areas. Broader research priorities are mentioned in this table to reflect their importance in the literature. Global/general priorities marked by a P are based on a summation of studies that focused on global significance. An x reflects the mention of a priority as a general concern. (4) *** indicates that data were unavailable or there was no clear trend for the system or time period. (5) These priorities were generated from a large body of literature PD/A CRSP, 1996). It should be noted that most of the authors and participants in the aquaculture evaluations contained in this literature were experts or a selected elite among the development community. Few farmers, women, North Americans, or host country representatives were included in the evaluations. Also, most of the evaluations were concerned with issues of equity, sustainability, and efficiency (following CGIAR recent thinking) rather than directly with export promotion or capital growth. Hence, bias is apparent in the approach and outcomes. Given this understanding, the priorities listed in this table can serve as a general indication of interest and perceived need for aquaculture research among donors, researchers, administrators, and bureaucrats.

Table 5 Aquaculture Development Priorities

	Global	Asia		Africa		Latin America		U.S.	
	Priority	Priority	Culture system	Priority	Culture system	Priority	Culture system	Priority	Culture system
General Development Priorities									
Strengthen national research systems	P	x	***	P	all	P	***		
Aquaculture information	P	P		P	all	P	***	P	***
Training and education	P		all	P	all	P	***	x	***
Production indicators	x	P		x	all	x	***	x	***
Extension	x	P	all	x	all			x	***
Post-harvest, processing	x	x	***	x	all	x	***	x	***
Access to credit, financing	x	x	I/M	x	E,S/FW			x	***
Land and water limitations		x	S,I/M					x	***
Technology transfer process	x	P	all	x	all			x	***
Women and youth	x			x	E,S/FW				
Policy/regulation implementation, enforcement	P	P	all	x	all	P	***	P	***

Notes: (1) x and P indicate general priority areas. x means the area is considered important, but is not mentioned frequently enough to be considered a top priority. P means the research area appears frequently enough in the literature to be considered one of the top priorities in aquaculture development. Culture systems — E: extensive; S: semi-intensive; I: intensive / FW: freshwater; BW: brackish water; M: marine. Time period — ST (1–5 years); MT (5–10 years); LT (10–20 years). Other — LAC: Latin America and Caribbean. (2) Priorities for aquaculture development that aim at efficiency, equity, and sustainability are many and diverse. Research will help to alleviate only some of these constraints. Only the most frequently mentioned priorities from the literature are included in this table. (3) *** indicates that data were unavailable or that there was no clear trend for the system or time period.

in Asia, harvested 220,000 metric tons from 20,000 farms; shrimp accounted for 32% of total 1995 export value of Thailand.* Much of the increase in the world's overall production from aquaculture in the past decade has thus come from Asia. During the past five years, the largest tilapia-producing country was Taiwan, followed by other Asian and Latin American countries. In Asian countries where aquaculture products have high export potential, government priorities for aquaculture tend to promote more intensive systems rather than the small-scale subsistence farms found in sub-Saharan Africa. Some donors and other organizations continue to promote efforts with extensive and semi-intensive systems in Asia, leading to great variety in the types of aquaculture practiced in the region.

Concurrent with the growth of the shrimp industry and industrial expansion in Asia, environmental impacts have emerged as a concern, especially in coastal areas. General biotechnological research priorities for Asia, therefore, include water quality and environmental impact as well as broodstock and seed supply, supplemental feeds and nutrition, and natural feeds.

As intensification of aquaculture occurs, research concerns shift toward stress reduction measures. In intensive systems, fish are held at higher densities and are usually fed formulated feeds. They are prone to disease, which is a relatively rare occurrence in extensive pond culture. Indeed, over the course of a decade, CRSP ponds have seen little mortality due to disease. Intensification also requires more energy and capital to overcome environmental obstacles related to water management and general site characteristics. Also, intensive systems are usually geared toward monocultures and are managed to meet market targets. For these systems, interest focuses on genetic engineering and selective breeding to ensure a reliable product. In Latin America, as in Asia, where aquaculture industries are rapidly expanding and systems are intensifying, research on disease prevention and genetics is gaining interest. However, in these regions, private sector funds generated from profitable commercial enterprises may be used for research in circumstances where the general public is not perceived to benefit directly.

The use of public sector funds to carry out private sector research has been a topic of interest for some time. Given shrinking public budgets for research, national governments and international donor organizations may increasingly identify specific research areas for their funds. For example, in many developing countries, effective environmental regulations often do not exist, and where they do exist, they are not strictly enforced. Transnational commercial farms typically have little incentive to curtail negative impacts resulting from their aquaculture operations. In such cases, governments have authorized the use of public funds to help develop aquacultural methods for mitigating negative environmental impacts. In other cases, a partnership of public and private research investment may be more effective. An example of this type of strategy is the CRSP work with a commercial shrimp farm in Honduras, where joint investigations were undertaken on feeding strategies that were effective for the grower yet resulted in lower discharge rates to the estuary.

Finally, research should be consistent with local and national development priorities for food security and economic growth. Much of biotechnological aquaculture research in developing countries to date has, however, tended to be donor-driven. Part of the reason for this is that local needs assessments are generally unavailable or unreliable, and national governments themselves have yet to articulate a strategy for aquaculture development. Other reasons center on the priorities of the international organizations funding research and the evolution of technical knowledge available on the status of aquaculture. Thus, the research constraints identified for the CRSP back in the late 1970s reflected a predominate U.S. perspective in spite of a concerted attempt to incorporate needs of the host countries. Even today, disaggregating external and internal input into the research agendas outlined in various governmental and non-governmental documents continues to be exceedingly challenging.

---

* In comparison, Honduras, which is one of the larger shrimp producing countries in Latin America, harvested 10,000 metric tons from 36 farms during the same time period.

PD/A CRSP CONTRIBUTIONS TO AQUACULTURE

The Goal of the PD/A CRSP

The PD/A CRSP, through the enhancement and development of sustainable aquaculture systems, endeavors to improve long-term food supplies and human nutrition. Implicit in this endeavor is the PD/A CRSP mission, which seeks to (1) raise small farmers' incomes and increase consumers' welfare through the enhancement of fish farm activity, (2) improve the well being of the rural poor, and (3) conserve or enhance the natural resource base.

Since 1982 the PD/A CRSP, through strategic and applied research and a multidisciplinary systems approach, has made progress toward removing the following initially identified constraints to the aquaculture sector:

- Lack of standardized baseline data on pond dynamics
- Lack of models describing principles of pond culture systems
- Inefficient aquaculture production practices
- Lack of fry availability
- Environmental degradation resulting from aquaculture operations
- Limited networking activities and lack of strong national research systems

Constraints and PD/A CRSP Contributions

The following provides a brief description of some PD/A CRSP contributions and accomplishments. The contributions are discussed within the context of the initial constraints identified by the PD/A CRSP in 1981 (see Chapter 2).

Constraint: Lack of Standardized Baseline Data on Pond Dynamics

CRSP Contribution: The PD/A CRSP Central Data Base, the world's largest standardized data base on tropical fish culture, contains over 1 million observations of pond variables from field sites around the world. The Central Data Base offers a unique contribution to the aquaculture literature, which lacks detailed standardized records. In addition, the data base design facilitates communications with other large data bases, thereby allowing opportunities for further collaboration.

Constraint: Lack of Models Describing Principles of Pond Culture Systems

CRSP Contribution: The CRSP Data Analysis and Synthesis Team (DAST) designed a model for simulating dissolved oxygen concentrations in shallow aquaculture ponds. In contrast with previous models, which assumed the water column to be homogeneous throughout its depth, this new model is able to predict oxygen concentrations in either homogeneous or stratified ponds. In addition, this model requires substantially less data and allows researchers to pose "what if" questions for the evaluation of different management strategies on dissolved oxygen content.

The DAST has also developed a water quality computer model to evaluate alternative responses to simulated conditions of oxygen depletion in ponds. Data collected from PD/A CRSP research were used to run simulations that tested alternative management strategies, such as nutrient enrichment to increase oxygen production, water exchange, and water level control. Simulation results indicated that nutrient enrichment was not as effective as water exchange and water level control. Water exchange or flushing with freshwater was a partially effective approach, and lowering the water level was the most effective approach for dealing with oxygen depletion.

A bioenergetic growth model was also developed by the DAST to simulate the interactive effects of fertilization, stocking density, and spawning on tilapia *(Oreochromis niloticus).* PD/A

CRSP data, indicating that growth is significantly higher in the absence of reproduction, were used to validate the bioenergetic simulation model. The simulation results fit the observed data relating growth to reproduction in ponds.

The DAST has developed two generations of decision support system (DSS) software. The first generation, PONDCLASS© (Versions 1.1 and 1.2), generates recommendations for fertilizing and liming individual ponds through the evaluation of climatic, water quality, and soil characteristics provided by the software user. An attribute of PONDCLASS© software is its ability to account for previous nutrient inputs of ponds. PONDCLASS© software has been tested by PD/A CRSP researchers in Thailand, the Philippines, and Honduras, and results indicate that PONDCLASS© software recommendations for fertilizer application are lower than traditionally used application rates of fertilizer. Lower fertilization rates suggested by PONDCLASS© may translate into lower fish production costs for the user.

The second generation of software, POND© Version 2.0, is able to perform simulation analyses and includes an economic analysis package for individual ponds and entire aquaculture facilities. POND©, a powerful analysis tool, includes the following features: (1) capabilities for defining a complete aquaculture facility in terms of geographic information, as well as multiple ponds, fish populations, and species characteristics; (2) complete functionality of PONDCLASS©; and (3) simulation models for fish bioenergetics and water quality (temperature, phytoplankton, zooplankton, and nutrients).

With the economic analysis package, POND© also provides capabilities for conducting multiple simulations to examine the effects of various pond management scenarios on fish yields and facility-level economics. Simulation models are automatically fed into the economics package, which is able to generate enterprise budgets for a pond facility. Fixed, variable, and depreciable costs, as well as income items pertinent to a particular facility, are calculated; economic returns can then be viewed in terms of the overall facility, per unit area or per unit of fish produced.

Constraint: Inefficient Aquaculture Production Practices

CRSP Contribution: The feasibility of combining the concepts of integrated fish culture and cage culture to improve overall yield was tested in Ayutthaya, Thailand. Walking catfish (*Clarias* sp.) were stocked and reared intensively in cages suspended in tilapia (*Oreochromis niloticus*) ponds. Supplemental feed, the only nutrient added to the ponds, was provided for the catfish; tilapia depended solely on plankton production derived from wastes of the supplemental feed and catfish feces. Research results demonstrated that wastes resulting from intensive catfish culture could be utilized by planktivorous species such as tilapia, with the added benefit of maintaining water quality at desirable levels. In the CRSP system, tilapia production averaged 8000 kg/ha/yr, which was greater than production levels reported for both a standard fish-livestock system and tilapia ponds optimally fertilized with chicken manure.

PD/A CRSP pond management strategies resulted in increased economic returns and lower economic risk in Abbassa, Egypt. The CRSP conducted a series of experiments to compare various management strategies for biological and economic efficiency. CRSP management guidelines resulted in the highest values of production per labor-hour, per kilogram of feed, and per Egyptian pound of variable cost.

Constraint: Lack of Fry Availability

CRSP Contribution: PD/A CRSP researchers studied the effects of stocking density on growth and survival of cultured hybrid tilapia (*O. niloticus* x *O. hornorum*) at the El Carao station in Honduras. Researchers found that daily growth rates of 0.2 to 0.4 g can be supported after initial stocking densities ranging from 50,000 to 83,000 fry/ha were established in ponds treated with chicken litter. In addition, research results indicated that increased amounts of chicken litter allow for similar growth rates at higher initial stocking densities. With PD/A CRSP assistance, production at the El Carao station increased nearly three-fold from 260,000 tilapia fingerlings in 1984 to 740,000 fingerlings in 1987.

Supplies of sand goby in Thailand in recent years have dwindled as a result of increased market demand and price of the fish. PD/A CRSP researchers conducted a one-year experiment in Thailand, which resulted in the production of nearly 150,000 fingerlings.

Constraint: Environmental Degradation Resulting from Aquaculture Operations

CRSP Contribution: The Universite Nationale du Rwanda (UNR) Rwasave Fish Culture Station did not have the capacity to analyze pond water quality at the onset of collaboration with the PD/A CRSP. The jointly financed construction of a water quality laboratory was one of the first efforts implemented by UNR and the PD/A CRSP. The laboratory regularly contracted for soil and water analyses, and was reputed as the premier analytical laboratory in Rwanda and one of the best in Central Africa.

Results of PD/A CRSP research in Honduras have been utilized by the Ministry of Natural Resources for the formulation of a series of ministerial decrees to govern the sustainable development of shrimp aquaculture. In 1992 PD/A CRSP researchers tested the effects of different levels of protein (20 and 40%) in diets offered to penaeid shrimp. The protein level did not significantly affect survival, yield, or average weight of the shrimp cultured in both dry and wet seasons. Results of this research suggested that diets of penaeid shrimp should contain no more than 20% crude protein, because higher protein levels would increase costs and add more waste nitrogen to the estuarine system without increasing production.

Constraint: Limited Networking Activities and Lack of Strong National Research Systems

CRSP Contribution: The PD/A CRSP project in Thailand is fully integrated into the international research network at the Asian Institute of Technology (AIT). PD/A CRSP research has attracted researchers from many countries in Southeast Asia and Europe involved with projects emphasizing integrated agriculture/aquaculture systems, utilization of nonconventional and low cost feed stuffs, development of fry production and nursing strategies, and regional aquaculture outreach. Asian Institute of Technology and PD/A CRSP collaboration has resulted in a connection with a large network of other research, government and donor agencies, including ICLARM, the Department of Fisheries of the Royal Thai Government, the DANIDA-funded Fisheries Sector Review of the Lower Mekong Basin, UNICEF's Family Food Production project in Cambodia, Dalhousie University of Canada, Sterling University and the Natural Resources Institute, United Kingdom.

PD/A CRSP has catalyzed linkages between various groups involved in aquaculture in Honduras. The PD/A CRSP's direct collaboration with the private sector and its facilitation of a public-private joint venture with the Ministry of Natural Resources, the National Association of Honduran Aquaculturists (ANDAH), the Panamerican Agriculture School (EAP), and the Federation of Producers and Exporters of Honduras (FPX) resulted in a newly furnished laboratory in 1993 and a study of water quality issues that affect shrimp production and the estuarine environment surrounding farms.

In Rwanda, the PD/A CRSP supported the establishment of a forum to facilitate communication and cooperation between women fish farmers, production scientists, non-government organization professionals, policy-makers, and program managers. A colloquium was held in 1990 to accomplish the following: (1) to begin a dialogue regarding the achievements and constraints involved with transferring fish production technology to Rwandan women farmers and (2) to develop recommendations for improving the relations between the various groups involved in the colloquium.

The PD/A CRSP has developed two technical publications series, which are an important aspect of the program's technology dissemination: *Collaborative Research Data Reports* and *CRSP Research Reports*. A broad domestic and international audience, approximately 300 people in 42 countries, receives the annual technical and program reports. PD/A CRSP publications are listed in the USDA National Agriculture Library data base and also appear in the peer-reviewed scientific literature.

REFERENCES

Anonymous, Comments presented by NGOs to the United Nations Commission on Sustainable Development, April 29, 1996, New York City, NY.

Anonymous, Status of World Aquaculture 1994, *Aquaculture Mag. Buyer's Guide '95,* 11–22, 1995.

Anonymous, *Fish Farming Int.,* 23 (1), 1, 1996.

Avault, J. W., Jr., *Fundamentals of Aquaculture,* AVA Publishing Company, Inc., Baton Rouge, LA, 1996.

Boyd, C. E., *Water Quality in Ponds for Aquaculture,* Alabama Agricultural Experiment Station, Auburn University, Auburn, AL, 1990.

FAO (Food and Agriculture Organization), *The State of World Fisheries and Aquaculture,* Food and Agriculture Organization of the United Nations, Fisheries Department, Rome, Italy, 1995.

Hajek, B. F. and Boyd, C. E., Rating soil and water information for aquaculture, *Aquacultural Eng.,* 13, 115–128, 1994.

Jhingran, V. G. and Gopalakrishnam, V., *Catalog of Cultivated Aquatic Organisms,* FAO Fisheries Technical Paper Number 130, Food and Agriculture Organization of the United Nations, Rome, Italy, 1974.

Pillay, T. V. R., *Aquaculture: Principles and Practice,* Fishing News Books, Inc., Surrey, England, 1990.

Pond Dynamics/Aquaculture Collaborative Research Support Program, *Continuation Plan: 1996–2001,* Oregon State University, Corvallis, 1996, 220 pp. with appendixes.

Williams, M., *Food, Agriculture, and the Environment,* Discussion Paper 13, National Food Policy Research Institute, Washington, D.C., 1994, 4.

Yoo, K. H. and Boyd, C. E., *Hydrology and Water Supply for Pond Aquaculture,* Chapman and Hall, New York, 1994.

2

HISTORY OF THE POND DYNAMICS/AQUACULTURE COLLABORATIVE RESEARCH SUPPORT PROGRAM

Hillary S. Egna

The research described in *Dynamics of Pond Aquaculture* resulted primarily from 14 years of work undertaken by the Pond Dynamics/Aquaculture Collaborative Research Support Program (CRSP). The long-term goal of the program has been to develop aquacultural technologies as a means of enhancing food security in developing countries and to strengthen the capacities of U.S. and international institutions to carry out aquacultural research. The Pond Dynamics/Aquaculture CRSP is one of a family of 11 CRSPs that were created to link the capabilities of U.S. agricultural universities with the needs of developing countries. Collaborative Research Support Programs are funded by the U.S. Agency for International Development (USAID) and by participating U.S. and host country institutions. They are among the longest lived international agricultural research efforts that exist within the U.S., and as such they have built vast networks through which research and development occur.

This chapter sets the context and frame of reference for subsequent chapters in *Dynamics of Pond Aquaculture*, many of which are based on the underlying premises of the CRSP model. A brief overview of the early history and philosophy of the Collaborative Research Support Programs is presented, beginning with its origins in Title XII legislation. Another full chapter could be devoted to the evolution of the CRSP as USAID responded to changes in foreign policy and development assistance in the U.S. and elsewhere; however, the focus on the early history of the CRSPs reflects the fundamental adherence of the various CRSPs to the structure and philosophy set forth by the framers of early CRSP legislation. A special section on the role of the social sciences in the CRSPs is presented to highlight how CRSP biotechnological emphases integrate with social and political concerns. Again, another chapter could be written on this subject; hence, the attempt is to set the context for understanding the last few chapters of this book, most of which report on combined efforts by the CRSP and others in making technologies available to farmers and researchers. The history of the Pond Dynamics/Aquaculture CRSP is explored from two angles — as an outgrowth of U.S. aquaculture and as a vehicle for alleviating hunger and poverty in developing countries. Finally, descriptions of Pond Dynamics/Aquaculture CRSP research sites, research priorities, research objectives, and program components are presented as a reference for Chapters 3 through 14.

HISTORY OF THE COLLABORATIVE RESEARCH SUPPORT PROGRAMS

Throughout the past century, the U.S. land grant university system has demonstrated its effectiveness in building human and institutional resources for agriculture in other countries. This recognition, along with the desire to bring U.S. universities more formally into U.S. foreign policy,

0-56670-274-7/97/$0.00+$.50
© 1997 by CRC Press LLC

eventually led to the creation of Title XII legislation in 1975, which contained provisions for the establishment of the CRSP framework. Prior to the passage of Title XII, *Famine Prevention and Freedom From Hunger*, the involvement of the U.S. university community in international agricultural research had been slowly declining. USAID was relying on universities for implementation rather than project design, and the universities themselves gave low priority to international activities of its faculty (Brady, 1984).

In the early 1970s, federal legislators recognized the vital role the universities played in the development of U.S. agriculture and that the academic community remained interested and capable of contributing to foreign assistance efforts. The successful 1887 Hatch Act, which gave rise to the U.S. system of state experiment stations, was viewed as a model for "mobilizing scientific and technical expertise of land grant institutions within a formal policy framework aimed at eliminating world hunger" (Lipner and Nolan, 1989, p. 20). In addition, during the late 1960s and early 1970s, concern about population growth and food crises gained popular attention. Grassroots groups within the U.S. university community responded with increasing interest to help resolve problems of food availability in developing countries (NRC, 1991). Thus, in 1975 the U.S. Congress, with participation from USAID and the university community, designed the Title XII legislation, still considered among the most significant amendments to the U.S. Foreign Assistance Act of 1961.

Title XII legislation specified that the President of the U.S. shall "provide program support for long-term collaborative university research on food production, distribution, storage, marketing, and consumption" (Title XII Sec. 297 a.3). Title XII legislation further recognized the connection between hunger and building an economic base for growth, particularly in the poorest countries. The legislation mandated that universities receiving funds through Title XII legislation provide matching funds "so as to maximize the contributions to the development of agriculture in the United States and in agriculturally developing nations" (BIFAD, 1985). This was the legislative foundation for the eventual CRSP structure (Yohe et al., 1990). Title XII legislation also authorized support for existing institutions such as the International Agricultural Research Centers (IARCs) to link with U.S. universities, the U.S. Department of Agriculture, and other public and private institutions. In 1971, USAID participated in the establishment of the Consultative Group on International Agricultural Research (CGIAR), which continues to act as the umbrella for the international centers.

The creation of Title XII was therefore dependent on the enrollment of many actors with their many interests in foreign assistance and policy. Besides the direct interests of the academic community in institution building, other forces in motion at the time Title XII legislation was passed included an increasing awareness of global interdependencies, a basic reassessment of the U.S. bilateral assistance strategy, the food crisis of the early 1970s and the resultant World Food Conference of 1974, and a growing awareness of the importance in investing in research (Thomas, 1977, p. 4). These forces attracted diverse interests and in so doing built a coalition of support for Title XII.

Congress, through Title XII, provided direct guidance for U.S. universities to participate "in innovative ways in all major aspects of the U.S. foreign assistance program in food and nutrition," and provided the infrastructural guidance or "machinery" for carrying out the mandate (Thomas, 1977, p. 9).

Title XII legislation thus established the U.S. Presidential Advisory Board for International Food and Agricultural Development (BIFAD) to assist USAID in the administration of the legislation. BIFAD advises USAID on university involvement in collaborative research, and assists in the planning, implementation, monitoring, and evaluation of activities funded under Title XII. At the time that collaborative research activity was defined within Title XII, it was a new programmatic concept in the international field. Between 1977 and 1982, BIFAD, through its Joint Research Committee (later the Joint Committee on Agricultural Research and Development), helped USAID design and implement the eight original collaborative research support programs considered by BIFAD in its Guidelines for CRSPs: Small Ruminant CRSP (1978); International Sorghum and Millet CRSP (1979); Bean/Cowpea CRSP (1980); Soils Management CRSP (1981); Nutrition CRSP (1981–1992); Peanut CRSP (1982); and the Pond Dynamics/Aquaculture CRSP (1982). Three other CRSPs were subsequently authorized: Fisheries Stock Assessment CRSP (1985–1994); Sustainable Agriculture and Natural Resources Management CRSP (1992); and Integrated Pest Management CRSP (1993).

Historical Overview of the Social Sciences in the Collaborative Research Support Programs

Concern over social issues in development assistance figured into the creation of the Title XII legislation. The formal recognition of the role of the social sciences in U.S. agricultural development projects began in the late 1960s after the first critical comments regarding unanticipated social impacts of the Green Revolution were levied (Lipner and Nolan, 1989). In the early 1970s, development workers became increasingly aware that previous research and development efforts had ignored the needs of small farmers and the rural poor. Development approaches that emphasized technology transfer and diffusion were found to have widened inequities between rich and poor. In response, Congress passed the New Directions mandate in 1973, an amendment to the 1961 Foreign Assistance Act, which stated that "United States bilateral assistance should give the highest priority to undertakings submitted by host governments which directly improve the lives of the poorest of their people and their capacity to participate in the development of their countries" (see Mickelwaite et al., 1979, p. 3). The land grant universities criticized the New Directions mandate for placing too much emphasis on capital transfer rather than on research and institution building. Their comments resounded within USAID, where there was a belief that basic research could expand the knowledge base regarding local conditions for solving food problems. The universities and other forces pushed toward new legislation, which was to become Title XII.

The general guidelines for Title XII programs such as the CRSPs developed from this historical context. With reference to social and economic concerns, Congress mandated that Title XII activities should "build and strengthen the institutional capacity and human resource skills of agriculturally developing countries so that these countries may participate more fully in the international agricultural problem-solving effort and to introduce and adapt new solutions to local circumstances" (Title XII, Sec. 297 a).

The emphasis on small farmers and the people they help to feed is evident in Title XII legislation and in the manner in which CRSPs have interpreted the BIFAD Guidelines. During the time when most CRSPs were being authorized, The Peasants' Charter, which challenged some of the traditional land grant approaches to agriculture, was defined (Axinn, 1984). The charter called for a reorientation of research, training, and extension priorities toward the adaptation and improvement of location-specific technologies suitable for small producers and cooperatives. One result of the charter was that some CRSPs developed strong social sciences components that focused on poverty alleviation through closely coordinated socioeconomic and technological research. The actual effort expended by the various CRSPs in integrating the social and technical spheres was often dictated by funding.

The BIFAD Guidelines for CRSPs, in interpreting Title XII, mandate the integration of social and natural sciences. BIFAD further recognized the challenge facing the CRSPs for providing a healthy forum for integration because interdisciplinary efforts in the university community were uncommon at the time. Not only must CRSP research focus on social and economic aspects of development, but social scientists must play an active role on the panels and advisory groups that comprise a CRSP. External Evaluation Panel members "are selected so that collectively they will cover the substance of the CRSP, including socio-economic factors that can influence research and adoption of technology generated from research" (BIFAD, 1985, p. 12).

CRSP Structure and Concept

All CRSPs were designed to recognize that agricultural research is most successfully carried out under the rubric of international collaboration. This internationalization of agricultural research is exemplified in the structure and organization of the CRSP model. Host country and U.S. researchers share in the identification of specific research needs, the design of experiments, and the analysis of results. Collaborative research is jointly planned, jointly implemented, and jointly evaluated. The concept of networking also extends to people and organizations not formally tied to a CRSP. The CRSPs provide formal and informal training through these networks and establish long-term researcher-to-researcher linkages. Since the inception of the first CRSP in 1978, these long-term collaborative linkages have survived institutional, organizational, and donor changes that so often are disruptive to established program directions (Yohe et al., 1995).

The CRSPs are communities of U.S. universities, developing country institutions, USAID Missions and Bureaus, other U.S. federal agencies, International Agricultural Research Centers, private agencies, industry, nongovernmental organizations, and other groups. Through shared resources, peer review, and institutional support, these communities give emphasis to the needs of small-scale producers, and to the rural and urban poor. Title XII provides for shared financing between state and federal organizations to "maximize the contributions to development of agriculture in the United States and in developing nations" (Title XII Sec. 297 b). The mandated cost sharing provided by the U.S. universities and institutions is further matched by contributions from host country institutions, so that USAID typically provides 40 to 60% of funds to a CRSP operation, and the participating institutions provide a nearly equal amount.

CRSP research integrates people, institutions, and disciplines. Research methods and approaches vary from CRSP to CRSP. However, all programs share fundamental elements. The BIFAD Guidelines for CRSPs note that the approach seeks to link U.S. and developing country institutions having common interests in organized programs of research on priority food, nutrition, and rural development problems. "The concept is to organize research on a program basis with [an] adequate number of scientific disciplines to solve the constraints. This is distinguished from the project, piece-meal approach to research, traditional in much technical assistance" (BIFAD, 1985, p. 8). Therefore, the CRSP research agenda is organized around constraints, which are then used to define priorities. A long-term approach is envisioned for each program; however, disciplines, universities, and other organizations collaborating in a CRSP may change as constraints are removed from the agenda. The CRSP structure allows for flexibility and re-orientation to new priorities.

A typical multitiered CRSP structure is presented in Figure 1. The two operational levels of organization are the Management Entity and its supporting advisory committees. The Management Entity is advised by a Board of Directors (composed of administrators), a Technical Committee (composed of leading scientists participating in the CRSPs), an External Evaluation Panel (composed of independent senior development specialists), and the USAID Global Bureau Project Officer (Yohe et al., 1995).

General outputs of a CRSP are increased production and improved consumption of food in developing countries. To bring this about, CRSP objectives primarily focus on generating technologies applicable to developing countries. A corollary is to improve institutional capacity in countries where research is being conducted "so that they can ultimately operate independently and play lead roles in spreading technology in their respective ecological zones and geographic regions" (BIFAD, 1985, p. 42). A spin-off is the benefit to U.S. agriculture, which provides the rationale for the contributions of resources by U.S. universities. While the CRSP effort itself is not the principal vehicle to extend research results, each program actively establishes scientific linkages and disseminates information. Further, each CRSP includes efforts to demonstrate the applicability of its research by conducting on-farm trials, field trials, and pilot tests. CRSP personnel are required to "manage research results until they can be passed on to an agency suitable for extending them" (BIFAD, 1985, p. 42).

Over the past decade, common themes important to international collaborative efforts have emerged from CRSP operations. Although the term "collaborative research" has become popularized in recent years, few groups actually have experience making collaboration a reality. Collaboration thus runs the risk of becoming another defeated buzzword. The CRSP model can offer perspective. CRSPs have learned that collegiality among persons of very different backgrounds and traditions does not necessarily occur naturally; it must be nurtured. CRSPs have learned how difficult interdisciplinary integration really is. The importance of establishing a legitimate role for each relevant discipline must be clear, and financial resources must be made available to make

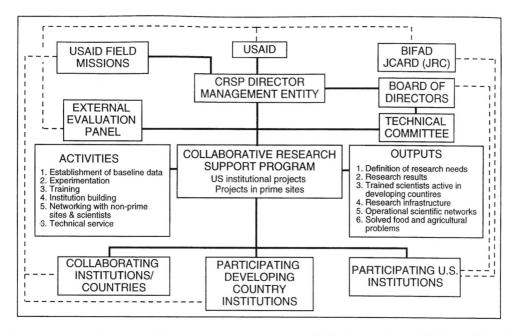

Figure 1. Organization, activities, and outputs of a typical CRSP. (Adapted from BIFAD, 1985, p. 5.)

collaboration happen. CRSPs have learned the importance of working effectively through institutional administrators. CRSPs have learned that integration occurs best when projects and people share not only objectives but also concern for similar populations in the same or related geographic regions. And CRSPs have learned that human resources have provided and will continue to provide the primary base upon which development can occur (Yohe et al., 1995).

OVERVIEW OF AQUACULTURE IN U.S. FOREIGN ASSISTANCE

Parallel Developments in Domestic and International Aquaculture

The difficulty of summarizing the U.S. involvement in international aquaculture comes in great measure from the vague and uncoordinated federal bureaucracy governing aquaculture in the U.S. Through the 1960s to the mid-1980s, the development of the nascent U.S. aquaculture industry and the emergence of aquaculture as a modern science brought about a response by many agencies and departments interested in aquaculture. Agencies became increasingly involved in aquaculture for the dual goals of protection of resources and investment opportunities through trade, which in some cases amounted to conflicting policies within a single agency and duplication at the state level. A common complaint voiced in 1978 was that "there were at least 42 government agencies and sub-agencies directly involved with the affairs of a typical aquaculture business" (Tiddens, 1990, p. 71).

Early federal involvement in aquaculture can be traced to 1871, when Spencer Fullerton Baird established the U.S. Fish Commission, which by 1887 had placed its major emphasis in hatcheries and fish culture (Tiddens, 1990; Meade, 1989, p. 4). Although the exact beginnings of aquaculture research are murky, it is thought to have emerged in response to declining wild fish populations in the Northeast as demand for hydroelectric utilities increased (Iversen and Hale, 1992). The "hatchery solution" was envisioned as a way to control nature, consistent with prevailing views of the time.

Aquaculture in the U.S. thus grew from the sports fishery rather than the food production industry and, accordingly, was never permanently located within one federal agency. Aquaculture was moved to the Department of Commerce in 1903, then to Interior in 1939, where the Fish and Wildlife Service opened laboratories that focused on science-related concerns of aquaculture, and then to many other agencies and departments. U.S. history is rich in examples of early fish culture;

Native Americans traditionally cultivated a number of freshwater and marine species, and trout were introduced to new environments as European settlers migrated across the continent. However, it was warmwater fish farming of baitfish, crappies, bass, paddlefish, buffalo fish, and catfish that drew the attention of the federal government. The 1960s saw the establishment of yet more laboratories for aquaculture on warmwater species, and the U.S. Fish and Wildlife Service coordinated its efforts with the experiment stations of USDA to extend laboratory research results to farmers (Brown, 1977, p. 14). By the late 1970s, Department of Interior work with freshwater fish had gone largely unnoticed, and with budget cutbacks and mounting rivalries with other agencies, aquaculture looked to private industry for support. Consequently, many of the federal fish culture activities were moved back to the Department of Commerce.

A national policy and plan on aquaculture was needed if U.S. aquaculture was to contribute substantially to the supply of high quality fishery products. With interests in ocean science gaining emphasis through the 1960s, aquaculture policy and support shifted toward the marine environment. President Johnson's 1965 Science Advisory Panel prepared a document stating national goals in marine policy. Among them were research on sea farming as a way to augment the food supply (Tiddens, 1990, p. 66). Shortly thereafter, other laws and measures were enacted that were directed primarily toward mariculture, notably, Public Law 89-454, "Marine Resources and Engineering Development Act," led by Senator Claiborne Pell from Rhode Island in 1966. The Stratton Commission was subsequently formed, and in its 1969 report related that most U.S. aquaculture production was actually in freshwater. "Finally, in contradictory language that all but ignored freshwater aquaculture programs within the federal government, the Commission recommended that a new independent agency, the National Oceanic and Atmospheric Administration (via its Bureau of Commercial Fisheries), be given the explicit mission to advance aquaculture, assist and encourage states through the Coastal Zone Authorities, and support more research (through the BCF and the newly established National Sea Grant College Program) on all aspects of aquaculture" (Tiddens, 1990, p. 67). The Sea Grant College and Program Act of 1966 brought to mind the Hatch Act of a century earlier, lending further substance to the notion that aquaculture was, at the time the CRSP programs were initiated, bureaucratically about 100 years behind agriculture. The Sea Grant Act was, however, not to have the same infrastructural significance as the Hatch Act, due to political and economic circumstances (Tiddens, 1990, p. 68).

Efforts to consolidate government organizations into one aquaculture department or to authorize one agency or department to be responsible for aquaculture development in the U.S. went largely unrealized through the mid-1980s. Even today, a "U.S. Department of Aquaculture" does not exist, nor does one agency control all aquaculture activities. Lack of a strong and broad-based constituency for aquaculture, major demographic movements from rural farm areas, and attitudes supporting reduced government involvement during the late 1970s and early 1980s relegated aquaculture to a complex existence in relation to government. Even passage of the 1980 National Aquaculture Act did little to consolidate government efforts in aquaculture.

The National Oceanographic and Atmospheric Administration (NOAA) was instrumental in setting forth ideas for the 1980 National Aquaculture Act. The 1977 NOAA National Aquaculture Plan, adding to previous government studies and reports, centered on biotechnological science but noted the confused state of bureaucratic interests involved in aquaculture. Also at that time, the passage of the Food and Agriculture Act of 1977 designated aquaculture as a mission of USDA. Congress considered these reports in their attempt to select a single lead agency for aquaculture in the late 1970s, but multiple interests in the end gave way to the responsibility for aquaculture being shared among agencies. Thus, the Joint Subcommittee on Aquaculture (JSA) was created, which now has 29 representatives from over 10 departments and agencies.* The Joint Subcommittee

* The 1995 JSA roster lists 29 representatives from the following agencies and departments: USDA, USAID, FDA, NOAA, EPA, NSF, DOE, DOD, SBA. In 1983 the JSA consisted of representatives from: USDA, USAID, SBA, EPA, NSF, DOE, Corps of Engineers, TVA, Farm Credit Administration (JSA, 1983).

assisted with the reauthorization of the Aquaculture Act, which in 1985 became part of the Farm Bill under the authority of the U.S. Department of Agriculture. The major impact of this legislation was the creation of the Regional Aquaculture Centers (RACs). At the time the RACs were established, the influence of the CRSP structure on the new entities was obvious, as leaders and promoters of the PD/A CRSP were involved in advising USDA on RAC structure and organization.* This is another manifestation of the overlap of domestic and international policy in the small arena of aquacultural research.

By the time a single department, Agriculture, was given primary leadership of aquaculture in the U.S., congressional funding had been curtailed, regulations had mounted, and the Departments of Commerce and Interior, which shared aquaculture funding responsibilities with Agriculture, had already begun reducing their financial involvement. Popular and scientific interest in aquaculture, however, continued to be strong. Interest in freshwater, brackish water, and marine aquaculture grew substantially after the late 1970s and was reflected in the explosive growth in the literature of aquaculture. Requests for information from USDA have increased sharply since the passage of the 1985 Farm Bill, from 250 requests in 1985 to 4200 in 1989, and the number of books on aquaculture has doubled every five years since 1961 (Iversen and Hale, 1992, p. 2). Iversen and Hale (1992, p. 1) write that: "[d]espite documentation that various types of aquaculture were practiced as early as 2000 BC in the Far East, interest in the Western World developed rapidly only about thirty years ago. Optimistic articles in the news media predicted how this 'newfound' protein source could literally feed the world. Based on these early reports, Western entrepreneurs had great expectations for profitable ventures." Successes were few at first in the Western World. The type of aquaculture practiced in Asia was primarily small-scale and subsistence oriented, and was not driven by a profit motive. Western entrepreneurs became big money losers, and aquaculture began to be seen as a high-risk venture. "They had not realized the high level of economic risk and many were unaware of how extremely limited biotechnological knowledge was for many species. Success required information" (Iversen and Hale, 1992, p. 1). Thus, the tone was set for the growth of aquaculture as a scientific endeavor in the U.S.

The 1995 Farm Bill, which was approved by Congress in 1996, did not contain the original language put forth by USDA for increasing its authority over aquaculture in the U.S. USDA had hoped to play a more central role in both the funding and accountability of aquaculture activities in the U.S. The draft bill called for reauthorization of the Regional Aquaculture Centers and the National Aquaculture Act of 1980. Aquaculture was specifically mentioned as an area deserving research investment. More importantly, the bill called for Congress to authorize USDA as the lead federal agency in the establishment of private aquaculture as a form of agriculture. Aquaculture would thus have been included in "all authorities for USDA research, education, and extension" (USDA, 1995, p. 85). The original USDA language also recognized the changing demographics of agriculture in that it "no longer describes the lion's share of the population as it did when Abraham Lincoln created the 'people's department'"; hence, the draft bill recommended changing the name of USDA to the Department of Food and Agriculture, to encompass both producers and consumers. This broadening would have helped to focus aquaculture as well, as the growth of aquaculture in the U.S. is clearly linked to the consumer.

Concurrent with these developments in domestic policy, an important report by the National Research Council of the National Academy of Sciences in 1978 ushered in explicit recommendations for foreign assistance in aquaculture research and development (NRC, 1978). *Aquaculture in the U.S.: Constraints and Opportunities* was prepared for the Senate Subcommittee on Foreign Agricultural Policy. Prior to that report, Congress had already written aquaculture specifically into Title XII legislation, but the selection of a CRSP to focus on aquatic issues had yet to be made. By 1978, however, the National Research Council report had coincided with the Planning Study

---

* The two entities — RACs and CRSP — share some organizational precepts but also have some noteworthy differences; in particular, U.S. industry plays a vital role in setting the RAC research agenda, and RACs tend to provide only seed money for a variety of applied research activities rather than for cohesive long-term programmatic activities.

for the PD/A CRSP, which contained many similar findings, including the need for long-term research in genetics, feeds, reproduction, water quality, and other aspects of aquaculture systems.

The bureaucratic linkages established with regard to domestic aquaculture factor into the Pond Dynamics/Aquaculture CRSP and make it a unique hybrid through which domestic and foreign assistance priorities are carried out. The Pond Dynamics/Aquaculture CRSP has been managed by fisheries experts, who were seconded by USAID from NOAA and USDA, almost from its inception. The U.S. Agency for International Development, which oversees the management of the CRSPs and is the vehicle through which foreign assistance funding from Congress reaches the CRSPs, has representation on the JSA. The JSA further involves many other agencies and departments that support international activities in aquaculture. Consequently, the diffuse coordination of federally supported aquaculture activities extends to the international arena as well.

Coordination of international efforts, although difficult, presents a small problem compared to the overall lack of attention paid by U.S. foreign assistance agencies to aquaculture and fisheries issues. Lamar Trott, former USAID/NOAA project director for the PD/A CRSP, writes (1992, p. 179), "[USAID] has traditionally paid minimal attention to fisheries issues, reflecting the same low recognition as given to the domestic fisheries sector." USAID, operating from 72 field missions and offices in 1992, had seven fisheries projects that were centrally funded and nearly a dozen more that were funded by the missions. Other donors have paid far greater attention to fisheries than USAID, with 13% of IDRC (International Development and Research Centre of Canada) and 14% of FAO's budgets, respectively, being devoted to fisheries (Trott, 1992). Although current USAID topics are relevant for fisheries and aquaculture (e.g., sustainability, agribusiness, biotechnology, biodiversity, income generation, natural resources, global warming, urban shift, role of women, and privatization), USAID has tended to pay little attention to fisheries and aquaculture in its overall development strategy.

Indeed, throughout the 1980s, a number of reports were commissioned to recommend a strategy for fisheries and aquaculture within USAID. A secondary motive was to make the Agency aware of its own activities in this arena and to draw attention to the importance of the fisheries sector. In 1982, the National Research Council recommended that USAID should be the lead agency for delivering U.S. assistance in fisheries and aquaculture to developing countries, and that it should foster joint projects and maintain close contacts with other organizations such as NMFS. Additional studies attempted to provide a rationale for fisheries and aquaculture within USAID (Donovan et al., 1984; NMFS, 1985; Neal, 1986, 1987; USAID, 1987; Campbell, 1987; Idyll, 1988; NRC, 1988). The National Research Council in 1988 again advised USAID of the importance of fisheries and aquaculture. But toward the end of the decade, key USAID aquaculture and fisheries programs were slated for cuts, and interest in aquaculture by field missions had fallen off. In this climate, the PD/A CRSP has struggled to survive and redefine itself.

Brief History of International Aquaculture Development Activities in USAID

Well before the creation of USAID, the U.S. was involved in foreign assistance activities in aquaculture (Table 1a). Among the earliest examples was a project in Technical Assistance in Agriculture in Mexico from 1942 to 1950. The fish culture portion of the project involved the participation of two biologists from the Fish and Wildlife Service of the Department of Interior and focused on improving the food and recreational resources of inland Mexico through hatchery and stocking efforts (Cremer, 1983). Throughout the 1950s, aquaculture activities were expanded through programs within the Technical Cooperation Administration (TCA), Foreign Operations Administration (FOA), and International Cooperation Administration (ICA). Current U.S. foreign aid programs for development assistance emerged from the 1938 program called the International Committee on Scientific and Cultural Cooperation (SCC). From small beginnings, the SCC came to provide the basis of major foreign aid programs during and after World War II, including the Lend-Lease program and the Marshall Plan. In 1949, President Truman followed these programs with his "Point Four" program of technical assistance to poorer countries. Truman's program

Table 1a Early U.S. Foreign Assistance Efforts in International Aquaculture or Related Activities Prior to the Establishment of USAID in 1961

Region/country	Project title or activity	Project code/grant no.	Project duration	Reference
Cambodia	Fisheries Development	n/a	1958–63	Cremer, 1983
Cambodia	Fish Development Training	4420230	1958–67	Idyll, 1988
Chile	Fishery Development	5130014	1953–63	Idyll, 1988
Dominican Republic	Fishery Conservation	n/a	1955–60	Cremer, 1983
East Pakistan	East Pakistan Fisheries	3910055	1955–62	Idyll, 1988
El Salvador	Inland Fisheries/ Fish Culture	n/a	1951–53	Cremer, 1983
El Salvador	Aquaculture	5190002	1958	Idyll, 1988
Ethiopia	Fishery Development	6630030	1957–60	Idyll, 1988
Iceland	Canning industry	1430030	1957–58	Idyll, 1988
Iceland	Byproduct utilization	1430040	1957–58	Idyll, 1988
Iceland	Fish control survey	1430084	1959–60	Idyll, 1988
India	Fisheries Development	n/a	1957–59	Cremer, 1983
Indonesia	Expansion of Inland Fisheries	4970042	1955–56	Idyll, 1988
Korea	Fisheries Development	n/a	1956–64	Cremer, 1983
Liberia	Cooperative Program in Agriculture, Forestry, and Fisheries	n/a	1951–67	Cremer, 1983
Liberia	Freshwater Fisheries	6690003	1951–67	Idyll, 1988
Mexico	Technical Assistance in Agriculture	n/a	1942–50	Cremer, 1983
Nigeria	Agriculture Advisory Service	6200212	1960–74	Idyll, 1988
Pakistan	Fisheries Development	n/a	1955–62	Cremer, 1983
Pakistan	West Pakistan Fisheries	3910054	1955–62	Idyll, 1988
Peru	Fish Development	5270924	1944–57	Idyll, 1988
Somali Republic	Fisheries Development	6490066	1957–66	Idyll, 1988
Taiwan	Fish Propagation	4840320	1956–58	Idyll, 1988
Thailand	Fisheries	4930012	1955–59	Idyll, 1988
Turkey	Fisheries Development	2770231	1954–56	Idyll, 1988
Vietnam	Fisheries	n/a	1958–75	Cremer, 1983
Vietnam	Inland Fisheries	5000053	1955–56	Idyll, 1988

Note: Projects were funded by ICA, TCA, and SCC. n/a = accurate information not available.

became the TCA, approved as PL 535 in June 1950. The TCA then merged in 1954 with the Mutual Security Act of 1953 to form the Foreign Operations Administration, which was to be reorganized a year later into the ICA and placed within the State Department by President Eisenhower. Seven years later, under the Foreign Assistance Act of 1961, the program was again reorganized to form the U.S. Agency for International Development.

During the 1960s, U.S. foreign assistance was redirected toward alleviating poverty among the poorest people in developing countries. The post–World War II fisheries programs that were aimed at commercial means for economic growth and increases in food supply (e.g., programs in India and Korea) were "curtailed in favor of ad hoc projects, generally tied to agricultural development programs.

AID attached a lowered priority to fisheries during this period and most of the agency's own expertise in fisheries was lost" (NRC, 1982). The failures of the program on fish protein concentrate, a joint effort with the Bureau of Commercial Fisheries (now NMFS), added to USAID's declining interest in supporting fisheries projects. However, a number of aquaculture activities were still carried out by USAID, some of which were notable for their influence on the university community. USAID continued many of the long-term programs begun under TCA and ICA in Asia. Short-term technical assistance in fact increased, and new long-term projects continued to be concentrated in the Asian region (Cremer, 1983). The NRC review of USAID (1982, p. 2) noted that much of the increased interest within USAID since the 1960s had been directed to aquaculture rather than fisheries.

In the late 1960s, USAID began efforts to strengthen U.S.-based technical capabilities in fisheries and aquaculture, through cooperative assistance programs with U.S. universities. Auburn University provides one of the earliest examples of institutional strengthening through its association with USAID. Building on funding from the Rockefeller Foundation in 1965 to carry out aquaculture development activities overseas, in 1967 Auburn began providing long-term technical assistance to USAID in establishing a program in inland fisheries and aquaculture (Cremer, 1983).

After a 1966 amendment to the Foreign Assistance Act of 1961, USAID became dedicated to providing long-term support to centers of excellence in fisheries and aquaculture at U.S. universities. Under Section 211(d), which allowed USAID "to fund U.S. educational and research institutions in order to strengthen their capability to develop and carry out programs concerned with economic and social development in less developed countries," Auburn University and the University of Rhode Island (URI) were provided institutional development grants. Support was used by URI to create the International Center for Marine Resource Development (ICMRD), with an emphasis on research and consultation services in fisheries and mariculture. At Auburn University, Section 211(d) support was used to establish the International Center for Aquaculture. Through the designation of these centers, USAID influenced how and which universities were to participate in development assistance. In the 1970s, Auburn University continued to provide technical assistance to USAID under a series of grants. Between 1970 and 1979, USAID used 3410 person-days of short-term technical services, and 54 person-years of long-term services overseas from Auburn's International Center for Aquaculture (Cremer, 1983). By the early 1990s, Auburn had accrued more than 125 person-years of experience in over 95 countries (ICAAE, n.d., p. 2). The NRC (1982, p. 34) report acknowledged that "lacking aquacultural expertise except through a seconding arrangement with NOAA's National Marine Fisheries Service, AID came to depend more and more heavily on Auburn to implement its aquaculture programs." Thus, Auburn University was supported through the U.S. bureaucracy to become one of the major players in the U.S. in international aquaculture.

Through the 211(d) mechanism, USAID tried to establish international linkages by maintaining close contacts with a few leading institutions in key fields. Although USAID wanted to make the creation of university consortia a condition of the grants, in practice this was often not the case (NRC, 1982). Rather, awards were given to replace or supplement existing USAID arrangements with universities. Problems of institutional isolation, lack of sustainability, and dependence on USAID for continuing support finally led to the repeal of Section 211(d) by the International Development and Food Assistance Act of 1978 (PL 95-424, 92 Stat. 942). Institution-strengthening grants awarded to Land and Sea Grant institutions under Title XII are now partially fulfilling the functions of Section 211(d). Of the universities active in the PD/A CRSP, itself a Title XII creation, Auburn University and The University of Michigan were each awarded separate strengthening awards by USAID under Title XII.

Another vehicle for international assistance in aquaculture has been the U.S. Peace Corps. USAID and other organizations have supported aquaculture activities since the first Peace Corps fish culture project in 1966 in Togo. The focus of Peace Corps fish culture activities has been on disseminating grassroots technologies in conjunction with extension, community participation, health, and nutrition efforts (Jensen, 1983). Although research is not an emphasis of the Peace

Corps per se, many volunteers gain training through their experience and later cycle in to U.S. university academic programs in aquaculture. Until the late 1980s, the Peace Corps had many volunteers active in aquaculture. After 1990, Peace Corps activities in aquaculture declined because many of the major projects (i.e., those in Zaire, Liberia, Philippines, and Guatemala) were closed for political reasons. Thus, the future pool of individuals who might have been advocates for aquaculture within USAID was reduced.

Through the 1970s and 1980s, USAID continued its development assistance in aquaculture (Table 1b). In addition to short- and long-term projects, USAID entered into a cooperative agreement with NOAA in 1975 to provide support to USAID's fisheries division. Tables 1a and 1b give an indication of the number and complexity of the various activities in aquaculture carried out by USAID. When one adds to this the other activities in international aquaculture in developing countries that are carried out through various federal agencies, it is no wonder that USAID has difficulty tracking and monitoring aquaculture activities. Although the information age offers some promise for improving the organization of the aquaculture portfolio, it remains to be seen whether any real changes will occur as informed fisheries staff leave USAID and are not replaced.

Finally, in the mid-1980s, as USAID and other donors became acutely aware of the mounting problems in the fisheries sector and the diminishing funding for research and development, a concerted effort was launched to enhance donor cooperation in fisheries and aquaculture. The Research Advisory Committee of USAID reported that donor assistance in fisheries and aquaculture had doubled during the ten-year period ending in 1984 but declined thereafter, with most support going to aquaculture as "it holds the promise for increases of the types popular with donors" (RAC, 1988, p. 4). The Study on International Fisheries Research (SIFR), commissioned by the World Bank and other groups, attempted to set the stage for the 1990s, whereby scarce financial resources could be best directed toward critical constraints facing aquaculture and fisheries.* Although it remains unclear how USAID will use the information generated by this multilateral report, it is quite clear that increased coordination of donor efforts and improved collaboration among institutions and people will become mandatory for all USAID fisheries and aquaculture projects.

The late 1980s brought about great upheavals in the underlying foundation for development assistance. With the end of the Cold War, many of the ideas that informed USAID strategies were quickly and immutably altered, causing a large chasm to open between actions in the field and policies from Washington, D.C. Within USAID, fisheries and aquaculture staff were reduced and those who remained continued to write strategy reports urging the agency to consider the importance of this sector in development (see Acker, 1989; USAID, 1990a, 1990b, 1993). A new administration under the first Democratic president in 12 years brought new people to power within USAID. The role of agriculture, not to speak of fisheries and aquaculture, was greatly diminished in the struggle for USAID to "re-engineer" itself. By the mid-1990s, the main efforts in fisheries and aquaculture had been substantially cut, ending support for the Fisheries/Stock Assessment CRSP, Oceanic Institute's Milkfish Project, and other smaller projects, in addition to the long-term technical assistance programs at Auburn University and the University of Rhode Island (USAID, 1993). Thus, among the larger efforts, the PD/A CRSP and ICLARM alone continue to be supported within the restructured Office of Agriculture and Food Security in the Global Bureau's Center for Economic Growth. (These programs were in the former Bureau of Science and Technology and the short-lived Bureau for Research and Development.) In the Center for Environment, another CRSP, Sustainable Agriculture and Natural Resources Management, supports an aquaculture component. In addition, a reduced number of smaller activities continue to be supported by USAID field missions, notably in Egypt, although very few missions have included aquaculture in their Mission Strategies (USAID, 1993).

---

* The SIFR strategy's original objective was to coordinate donor assistance in capture fisheries, but recognition of the inextricable links between fisheries and aquaculture prompted the inclusion of aquaculture in the study. Fisheries is thus defined in SIFR to include aquaculture.

Table 1b   A Partial List of Aquaculture Activities Funded by USAID from 1961 to 1995

Region/country	Project title or activity	Project code/grant no.	Project duration	Reference
Africa Regional	Improved Rural Technology	6980407	1979–90	CDIE, 1993
Africa Regional	Accelerated Impact Program	6980410	1976–87	CDIE, 1993
Asia Regional	Southeast Asia Fisheries Development Center	4980226	1969–73	CDIE, 1993
Brazil	Fish Production, Processing, and Marketing	5120247	1964–76	CDIE, 1993; Idyll, 1988
Brazil	Food Fortification	5120288	1971–75	Idyll, 1988
Brazil	Fishery Development	5122474	1966–74	Idyll, 1988
Brazil	Fisheries Training	9310042	1974–77	Idyll, 1988
Burundi	Highland Fisheries Development	6950102	1979–81	Idyll, 1988
Cameroon	Small Farmer Fish Production	6310022	1980–85	CDIE, 1993
Caribbean	Institutional Development	5380016	1978–82	Idyll, 1988
Caribbean	Fish Development	5380023	1978–82	Idyll, 1988
Caribbean	King Crab Culture	5380045	1978–82	Idyll, 1988
Central Africa Republic	Fish Culture	6760004	1977–80	Idyll, 1988; CDIE, 1993
Central Africa Republic	Rural Development	6760015	1982–88	CDIE, 1989
Chile	Agricultural Cooperative Development	5130277	1975–80	Idyll, 1988
Chile	Rural Development	5130296	1977–79	Idyll, 1988
Colombia	Fisheries Research	5140191	1975–81	Idyll, 1988; CDIE, 1993
Colombia	Research on Inland Fisheries	5140078	1975–78	Idyll, 1988
Costa Rica	Agricultural Development	5150038	1963–80	Idyll, 1988
Djibouti	Fisheries Development	6030003	1979–85	CDIE, 1993
Djibouti	Fisheries Development, Phase II	6030015	1984–88	CDIE, 1993
Dominican Republic	Inland Fisheries (Grant to Church World Services)	5170123	1978–82	Idyll, 1988; CDIE, 1993
Dominican Republic	Inland Fisheries (Grant to DR counterpart of Church World Services)	5170162	1982–85	Idyll, 1988; CDIE, 1993
East Africa	Freshwater Fisheries (to EAFFRO, Regional Organization)	6180649	1969–80	Idyll, 1988; CDIE, 1993
East Pakistan	Consultant Services to the East Pakistan Agricultural University at Mymensingh	n/a	1967–70	Cremer, 1983
Eastern Caribbean	Fish Sector Assessment	5380045	1978–82	Idyll, 1988
Egypt	Aquaculture Development	2630064	1978–87	CDIE, 1993

Table 1b   (continued)

Region/country	Project title or activity	Project code/grant no.	Project duration	Reference
Egypt	CRSP	n/a	1992–95	n/a
El Salvador	Inland Fisheries	n/a	1972–76	Cremer, 1983
El Salvador	Rural Community Development	5190094	1967–83	Idyll, 1988
El Salvador	Agribusiness Development	5190327	1987–94	CDIE, 1995; CDIE, 1995
Gambia	Market Feasibility	6350211	1980–81	Idyll, 1988
Guatemala	Family Fish Pond Development Program	5200290	1981–86	CDIE, 1993
Guatemala	Aquaculture Extension	5200351	1986–89	CDIE, 1993; CDIE, 1995
Honduras	Aquaculture	5220214	1976–79	Idyll, 1988; Cremer, 1983
Honduras	Core Services Rural Development	5220118	1973–79	Cremer, 1983; CDIE, 1989
Indonesia	University Development	n/a	1962–66	Cremer, 1983
Indonesia	Brackish Water Fishery Production	n/a	1976–81	Cremer, 1983
Indonesia	Expansion of Fisheries Facilities	4970001	1964–79	Idyll, 1988
Indonesia	Small Scale Fisheries Development (Government of Indonesia)	4970286	1980–86	CDIE, 1993; Idyll, 1988
Indonesia	Aquaculture	4970189	1969–81	Idyll, 1988
Indonesia	Assistance to Agriculture	4970236	1975–81	Idyll, 1988; CDIE, 1989
Indonesia	Resource Development	49702663	1979–83	Idyll, 1988; Cremer, 1983
Indonesia	Research and Training	49702664	1978–84	Idyll, 1988; Cremer, 1983
Indonesia	Assistance to Small Scale Fish Producers	n/a	1980–84	Kammerer, 1983
Indonesia	Brackishwater Fisheries Development	n/a	1980	Kammerer, 1983
Ivory Coast	Fish Training Center	6810005	1962–67	Idyll, 1988
Jamaica	Inland Fisheries Development	5320038	1976–80	Idyll, 1988; CDIE, 1993
Jamaica	Fish Production System Development	5320059	1979–85	CDIE, 1993; Idyll, 1988
Kenya	Inland Fisheries	6150130	1965–70	Idyll, 1988
Korea	Office of Fisheries	4890673	1971–76	Idyll, 1988
Korea	Rural Policy	489059402	1963–74	Idyll, 1988
Laos	Agriculture Development	439006506	1963–75	Idyll, 1988
Liberia	Agriculture Research and Extension	6690188	1984–86	Idyll, 1988
Mali	Mali-San Pilot Fisheries Production	6880220	1979–83	CDIE, 1993; Idyll, 1988

continues

Table 1b   (continued)

Region/country	Project title or activity	Project code/grant no.	Project duration	Reference
Nepal	Tribhvan University: Propagation of Fish in Reservoirs	9365542	1983–	Idyll, 1988
Nepal	Resource Conservation and Utilization	3670132	1980–88	CDIE, 1989
Nigeria	Fish Development	6200704	1962–68	Idyll, 1988
Palau	Mariculture of Tridachna (Giant Clam)	n/a	1982	Kammerer, 1983
Panama	Panama Aquaculture	n/a	1971–73	Cremer, 1983
Panama	Regional Development (Guaymi Area Development)	5250200	1979–83	Idyll, 1988; Cremer, 1983
Panama	Rural Development/Farm Ponds	5250186	1977	Idyll, 1988
Panama	Aquaculture	5250245	1984	Idyll, 1988
Panama	Managed Fish Production	5250216	1980–85	CDIE, 1993; Idyll, 1988; Cremer, 1983
Persian Gulf	Fisheries Development	2900164	1963–65	Idyll, 1988
Peru	Freshwater Fisheries Development	5270144	1977–81	CDIE, 1993; Idyll, 1988
Peru	Rural Enterprises II	n/a	1979–	Cremer, 1983
Philippines	Freshwater Fisheries Development	4920322	1979–84	CDIE, 1993; Cremer, 1983; Kammerer, 1983
Philippines	Inland Fisheries	4920234	1971–76	CDIE, 1993
Philippines	Bicol River Basin Development	4920260	1973–79	Idyll, 1988; Cremer, 1983
Philippines	Aquaculture Production	4920266	1974–80	Idyll, 1988; CDIE, 1993
Philippines	Resource Development/ Fish Management	4920366	1982–89	Idyll, 1988
Philippines	Coastal Zone Management	4920368	1983–87	Idyll, 1988
Portugal	Technical Consultants and Training, Institute of Azores	n/a	1977–79	Cremer, 1983
Portugal	National Fishing Study	1500002	1977	Idyll, 1988
Portugal	Fish Marketing, Extension, Training	1500001	1975–85	Idyll, 1988
Rwanda	Fish Culture	6960112	1981–89	CDIE, 1993; Cremer, 1983
Senegal	Lowland Fish Culture	6850240	1979–82	CDIE, 1993
Senegal	Bakel Crop Production	6850208	1977–85	Idyll, 1988; CDIE, 1989
Senegal	Fish Culture	6850240	1979–82	Idyll, 1988
Senegal	Fisheries Resource Assessment	6850254	1981–84	CDIE, 1993; Idyll, 1988

Table 1b (continued)

Region/country	Project title or activity	Project code/grant no.	Project duration	Reference
Sierra Leone	Improved Rural Technology	n/a	1979–	Cremer, 1983
Somali Republic	Fish Freezing Plan	6490006	1965–67	Idyll, 1988
South Pacific Region	PVO Integrated Rural Development	8790251	1981–86	Idyll, 1988
Thailand	Agriculture Development	4930180	1964–79	CDIE, 1989
Thailand	Applied Agriculture Research in Northeast	n/a	1968–71	Cremer, 1983
Thailand	Northeast Rainfed Agricultural Development	4930308	1981–89	CDIE, 1989
Thailand	Lam Nam Oon On-Farm Development	4930272	1977–85	Cremer, 1983; CDIE, 1989
Thailand	Village Fish Pond Development	4930303	1979–82	CDIE, 1993; Idyll, 1988
Thailand	Fisheries Development	4930179	1964–79	Idyll, 1988
Thailand	Fish Development	49301807	1964–79	Idyll, 1988
Thailand	Fish Farm Development	4930272	1977–78	Idyll, 1988
Thailand	Vocational Education	4930295	1980–83	Idyll, 1988
Thailand	Strengthening of SE Asian Aquaculture Institutions	9365543	1982	Idyll, 1988; Kammerer, 1983
Vietnam	Fish Training	5000242	1967–70	Idyll, 1988
West Indies-Eastern Caribbean	High Impact Agricultural Marketing and Production	5380140	1986–91	CDIE, 1989
Zaire	Fish Culture Expansion	6600080	1978–88	CDIE, 1993; Idyll, 1988
Zaire	Fishing Cooperatives	6600056	1976–80	Idyll, 1988
Zaire	Imeloko Integrated Rural Development (PVO-Zaire Church of Christ)	6600082	1978–81	CDIE, 1989; Idyll, 1988
Auburn University	International Aquaculture Development	9310120	1970–79	CDIE, 1993; Idyll, 1988
Auburn University	Fish Culture in LDC	9310787	1967–73	Idyll, 1988; CDIE, 1993, 1995
Auburn University	Aquaculture Tech Development	9311314	1977–88	Idyll, 1988; CDIE, 1993
Auburn University	Aquaculture Research and Support (follow on to 9311314)	9364180	1988–92	CID 1995
Auburn University	Fisheries Training Center	9310042	1974–77	CDIE, 1993
Auburn University	Water Harvesting/ Aquaculture: Joint PVO/University	9380240	1984–89	CDIE, 1993
University of Rhode Island	Small Scale Fisheries	9310113	1969–79	CDIE, 1993; Idyll, 1988

continues

Table 1b   (continued)

Region/country	Project title or activity	Project code/grant no.	Project duration	Reference
University of Rhode Island	SE Asia Fisheries Development Center (SEAFDEC)	9311155	1979	Idyll, 1988
University of Rhode Island	Workshop Stock Assessment	9311155	1980	Idyll, 1988
University of Rhode Island	Post Harvest Losses	9311157	1980	Idyll, 1988
University of Rhode Island	Fishery Development Support	9364020		Idyll, 1988
University of Rhode Island	Fishery Development Support Services	9364024	1982–92	CDIE, 1993
NOAA	Advisory Services	9310242	1976–88	Idyll, 1988
USDA	Agric Technology R&D	9364109	1981–	CID, 1995
S&T/USAID	New Protein Sources	9310459	1966–78	Idyll, 1988
S&T/USAID	Nutritional Value Protein Food	9310459	1967–75	Idyll, 1988
S&T/USAID	Evaluation of Fish Protein Concentrate	9310482	1967–78	Idyll, 1988
S&T/USAID	Evaluation of Fish Protein Concentrate	9310845	1969–74	Idyll, 1988
Oceanic Institute, Hawaii	Artifical Propagation of Milkfish	9310526	1976–78	CDIE, 1989; Idyll, 1988
Oceanic Institute, Hawaii	Reproductive Studies on Milkfish	9364161	1984–89	CDIE, 1989
ICLARM	Fish Development Conference	9311050	1979–85	Idyll, 1988; CDIE, 1993
SEAFDEC	SE Asia Fisheries Development Center (SEAFDEC)	9311156	1978–79	Idyll, 1988
University of California, Davis	Air Bladder Inflation	9311157	1980	Idyll, 1988
Fisheries & Aquaculture "CRSP"	Report Title XII Fish Research	9311306	1977–78	Idyll, 1988
PD/A CRSP	Pond Dynamics-CRSP	9364023	1982–88	CDIE, 1989
PD/A CRSP	Pond Dynamics-CRSP	n/a	1988–	n/a
Fisheries & Aquaculture "CRSP"	Fisheries Technical Assistance Service	9364024	1980–83	Idyll, 1988
Oregon State University/ CRSP	CRSP Planning-Fisheries/Aquaculture	9311306	1980–81	CDIE, 1993
University of Michigan	Thailand	9365543	1982	Idyll, 1988; Kammerer, 1983

Table 1b   (continued)

Note: This list is not inclusive of all USAID activities. Others (i.e., Cremer, 1983; Idyll, 1988) have commented on the difficulty of synthesizing data from all relevant USAID sources involved in funding aquaculture. The reader is referred to their studies for a detailed explanation of the difficulties in compiling a complete list. The reader should note that USAID field mission activities in aquaculture (and related areas) are very likely under-represented in this table. For example, USAID/Cairo has bilaterally funded several aquaculture projects in Egypt since the early 1990s, but a record of these does not appear in the CDIE data base. Also, USAID/Washington has funded projects under its competitive grants programs in Science & Technology Cooperation (i.e., PSTC project on spotted scat at the University of Hawaii, Grant no. PDC-5542-G-SS-5033-02,1988), and through the U.S.-Israel CDR program; these do not appear in the CDIE data base.

CDIE: The Center for Development Information and Evaluation in the Bureau for Policy and Program Coordination in USAID. Dates indicate CDIE output over a period of time. Data were requested in 1989, 1993, and 1995. Over 50 key words related to aquaculture were used to search the data base.

Data from USAID, 1981; Cremer, 1983; Kammerer, 1983; Idyll, 1988; and USAID (CDIE), 1989, 1993, 1995.

While this review focused on USAID's historical involvement in the area of aquaculture, one must remember that aquaculture and USAID are part of a broader picture, in which global forces and popular opinion influence development agendas directly. Thus, the summary presented here isolates aquaculture somewhat from its moorings but in so doing expresses the existence of complexities within complexities, the weaving of institutional linkages, and the emergence and re-emergence of various themes and advocates over time.

POND DYNAMICS/AQUACULTURE CRSP HISTORY AND EARLY IDENTIFICATION OF RESEARCH PRIORITIES

Title XII legislation clearly states that "the term 'agriculture' shall be considered to include aquaculture and fisheries," and "the term 'farmers' shall be considered to include fishermen and other persons employed in cultivating and harvesting food resources from salt and freshwaters" (Title XII, PL 94-161, Sec. 296 f and g). This language reflected the popular attention that was being paid at the time to aquaculture as a panacea for impending food crises, foretold in such writings as Hardin's "The Tragedy of the Commons" and Ehrlich's "The Population Bomb." The disappearing concepts of unrestricted fishing and freedom of the high seas, coupled with the idea of the ocean as one of the last great frontiers, drew the curious to aquaculture. Although aquaculture had been practiced for centuries, it was only just emerging as an organized field of scientific activity in the 1960s.* It gained recognition as an academic emphasis in a number of land grant universities at the time Title XII was being conceived. Tiddens (1990, p. 1) writes that "in America, aquaculture was one of the soldiers that could help this country maintain its technological leadership in the oceanic arena, and help in the battle against a huge foreign trade deficit. Consequently, aquaculture, in the view of many, was deserving of the same kind of federal support that had launched its terrestrial cousin agriculture a century earlier." Thus, aquatic resources and products became a focus for Congress as an eventual Title XII activity.

Information, interest, and advocacy for aquaculture coalesced, eventually leading to the establishment of the PD/A CRSP in 1982. In 1976, USAID reported on some of the thinking that later was to become part of the underlying assumptions and framework of the PD/A CRSP: "The art of [sic] science of managed fish production, as contrasted with agriculture, are in their infancy. For

---

* A testimony to this is that in the 1960s, one could not obtain an academic degree in aquaculture. Now, recent graduates of several universities hold degrees such as a Ph.D. in Aquaculture. Many of the older researchers in the CRSP, for example, come from other fields — none have degrees in aquaculture. They come from agricultural chemistry, limnology, biology, genetics, civil engineering, rural sociology, etc. Skladany (n.d.) emphasizes that they did not come for their love of fish. Rather they were academic entrepreneurs who saw that a new field with new possibilities might allow them to make a name for themselves. And many were right.

the most part, the world is still hunting for fish, rather than culturing them. Nevertheless opportunities exist for fairly dramatic advances in managed fish production with potential benefits in terms of income and employment for the A.I.D. target group, the world's poor majority. Similarly, advances in food processing and distribution technology applied to fishery products can contribute to improved nutrition and lower food costs for the consuming poor in less developed countries" (USAID, 1976, p. 1). This set the stage for planning a CRSP in fisheries and aquaculture. At the time the PD/A CRSP was authorized in 1982, another group, the World Bank, gave the CRSP its send off message when they reported that "for a given outlay of funds, more animal protein of high quality can be obtained from fish than from any type of meat" (see Donovan et al., 1984).

Following the BIFAD Guidelines for planning a CRSP, an independent consulting firm, Resources Development Associates, was selected in October 1977 to define fisheries and aquaculture research and development needs in developing countries (Craib and Ketler, 1978). The request for greater definition of needs grew out of recognition on the part of USAID, BIFAD, and JRC that fisheries and aquaculture represented a broad and complex sector, and that the range of constraints affecting fisheries and aquaculture in developing countries was diverse. The selection of an independent consulting firm to conduct the planning study presented a departure from the earlier established CRSPs, for which participating institutions conducted the planning study.

With encouragement from USAID, RDA began the planning study with a conference to allow input and comment from the diverse academic communities involved. The conference in December 1977 was attended by 110 scientists and professionals, representing 47 universities and institutes, and various government agencies including NOAA, the U.N. Food and Agriculture Organization, and USAID. The conference was divided into functional sectors covering aquaculture, fisheries, and processing and marketing.

At the outset of the conference it became clear that "there was substantial confusion in many sectors concerning the purpose of Title XII legislation, the real meaning of 'collaborative research support programs,' and other areas that might affect the [planning] study itself" (Ketler, 1977, p. i). One speaker mentioned that the university community's response to Title XII ran from excitement to disillusionment and even anger, and that some universities thought that Title XII would be an entitlement for each state at about $2 million a year (Whaley, 1977, p. 12). The first speaker, Dr. D. Woods Thomas of BIFAD, tried to involve the audience as stakeholders in the legislation. He reiterated how Title XII "was not the 'brainchild'. . . , of any individual or for that matter, any small group of individuals. It was really the product of deep-seated, common concerns in a number of different quarters. These concerns were relevant to the future course of a broad array of world events. They were concerns about the part that the United States could and should play in the shaping of this set of events" (Thomas, 1977, p. 4). The concerns he cited were prescient: increasing global interdependencies in politics, public policy, energy, economic growth, and food supply.

A spokesperson from USAID attempted to allay the fears of the universities by noting that Title XII also presented an institutional and bureaucratic innovation for USAID. The CRSP structure was to cut across bureaus, missions, and various levels of organization, prompting "from the top down" attention to the legislation. USAID had never had to deal with anything like the CRSPs before and was itself unsure of what changes would occur as a result of the Title XII legislation (Belcher, 1977). Thus, much of the uneasiness from the government and university community toward the unfolding CRSP model arose because institutional structures were evolving rapidly, absolute jurisdictions had yet not been claimed, and the stakes were high.

At the conference, USAID also commented on the shape of a future fisheries CRSP by noting that the planning study should focus on priority research areas that are consistent with the overall focus of USAID on small farmers and the rural poor. Consequently, USAID was less interested in supporting large-scale commercial deep sea fishing than fish farming, coastal fisheries, and brackish water fisheries. Even though Congress explicitly included fisheries and aquaculture in the language for Title XII, comments from USAID at the conference indicated at this early stage that the agency was having difficulty standing behind fisheries and aquaculture: "although the general area of fisheries and aquaculture is important, ranking somewhere between the first eight or twelve important areas [for a CRSP], its exact priority ranking has not yet been decided. We've acted on a hunch that some additional research in fisheries and aquaculture is so important that it is worth going ahead to plan it. But only when we see the results of the planning activity and see what a potential collaborative research program might look like will a decision be made as to whether to go ahead with a collaborative research program in this area" (Belcher, 1977, p. 22). USAID and JRC, in citing funding limitations, were additionally apprehensive about coordination between a fisheries CRSP and the newly established International Center for Living Aquatic Resources Management (ICLARM) to which they intended to provide support (Belcher, 1977; French, 1977). That two CRSPs were eventually established testifies to the perceived importance of the sector, but the cautious interest by USAID in fisheries and aquaculture served in the long run to undermine the overall capacities of these CRSPs to focus on the range of critical issues requiring research investment.

Findings from the Planning Study and State-of-the-Art Survey

The planning study furthered ideas generated at the conference and identified priority areas for research in fisheries and aquaculture (Table 2). Priorities for research areas were established by the following criteria:

Number of people in target groups benefiting from this research
Geographical area benefiting from this research
Sensitivity of research to geographical selection
Extent to which this research program matched the perceived needs of the developing
   countries
Time required to produce results
Cost of program
Probability of successful completion of research objectives
Probability that results obtained would be applied
Research capability in developing countries and potential for expansion
Research capability in the U.S.
Institutional preparedness in the developing countries
Institutional preparedness in the U.S. (Craib and Ketler, 1978, p. 113).

The planning study also prepared a detailed inventory of U.S. university capabilities in undertaking research consistent with the priority areas. USAID charged RDA to "take into consideration the depth and breadth of scientific talent in various institutions as well as the commitment of institutions to the collaborative research concept" (Jones, 1977, p. 50). An inventory of host country capabilities and needs was also developed. The final selection of the participants in the PD/A CRSP came in large part from the inventory generated by RDA and from the notion that USAID was endorsing the "consortium concept." Of the long-term entities participating in the PD/A CRSP, Oregon State University, the University of California, Michigan State University, and the University of Michigan were in the top category of meeting the criteria of capacity and institutional preparedness for undertaking program leadership. Auburn University and the University of Hawaii were in the second category, identified as those able to undertake project responsibilities. Other "eligible" universities were in the third category or not included in the inventory. Among the HC institutions that have formally participated in the PD/A CRSP, Institut Pertanian Bogor in Indonesia, Central Luzon State University in the Philippines, University of the Philippines, the Department of Fisheries in Thailand, and the University of Cairo in Egypt (where some Abbassa staff maintain linkages) received top rankings in the inventory.

**Table 2   Original Priority Research Areas Identified in 1978
for a Fisheries and Aquaculture CRSP**

Relative priority	Capture fisheries	Aquaculture	Product utilization
1. Principles and mechanisms of pond culture systems		X	
2. Resource assessment	X		
3. Feed and nutritional requirements		X	
4. Seed availability		X	
5. Causes and rates of spoilage of catch in tropical waters			X
6. New or expanded fisheries	X		
7. Loss of catch due to insects and pests			X
8. Marketing and distribution			X
9. Fisheries administration and the extended economic zone	X		
10. Control of spoilage in fishery products			X
11. Environmental analysis and habitat protection	X		
12. Culture systems for native species		X	
13. Morbidity and mortality causes and controls		X	
14. New product and processing technique development			X
15. Low energy preservation and processing techniques			X
16. Genetic improvement		X	

Adapted from Craib, K. B. and Ketler, W. R., 1978, p. 114.

One result of the planning study was to lead BIFAD and USAID to split fisheries and aquaculture into two separate CRSPs rather than the one program originally envisioned.* The JRC in 1977 hinted at the difficulties of combining these areas, in which the U.S. university role overseas appeared elusive and difficult to define. At the time fisheries and aquaculture were selected by JRC for planning studies, nutrition and sorghum-millet CRSPs were also being planned. In the case of sorghum-millet, a clear role for university involvement appeared obvious: "[A] limited number of researchers and universities are working in this area, and they can be identified through the USDA data retrieval system and through some of our national workshops" (Poponoe, 1977, p. 90). With fisheries and aquaculture this was not the case, and a better definition of boundaries could be attained through separation. As the planning study recommended research on pond culture systems as an appropriate area for a CRSP, BIFAD initiated further development of an aquaculture CRSP. By its approval of three main U.S. institutions in February 1980 — the Consortium for International Fisheries and Aquaculture Development (CIFAD), Auburn University, and the University of California at Davis — the Joint Research Committee set the organizational structure for what was to become the Pond Dynamics/Aquaculture CRSP. A second planning study was subsequently undertaken by these institutions, with the award in June 1980 of a Specific Support Grant to Oregon State University as Management Entity for an Aquaculture CRSP (Lannan et al., 1983). This study

* A third area, post-harvest loss and technologies, worthy of a CRSP in itself was never pursued, and when a CASP (Cooperative Agribusiness Support Program, as distinct from a CRSP) was created in 1993, it did not encompass fishery products in its mandate. Fish were among the top four areas identified for the CASP, but funding allowed only the first three areas to be included.

culminated in a book on pond dynamics and aquaculture — *Principles and Practices of Pond Aquaculture: A State of the Art Review* — upon which the present volume expands substantially.

The state-of-the-art survey concentrated on four aquaculture systems that were considered to have the greatest potential for contributing to the supply of low-cost animal protein in developing countries. These were, in order of importance,

1. Small, low-intensity tropical pond systems characterized by limited external inputs of feeds or fertilizers
2. Cooler water (15 to 25°C) tropical ponds at medium to high elevations
3. Brackish water and hypersaline ponds, including those in tropical mangrove zones
4. Higher intensity tropical pond systems characterized by high external inputs of feed or fertilizers (Lannan et al., 1986)

The findings from the state-of-the-art review were the following:

1. Most information on pond culture is descriptive and is of limited statistical and predictive value.
2. A lack of standardization among the limited quantitative data available renders the existing data base of limited utility in predicting the performance of pond culture systems. Much of the data lack statistical precision and are site specific, and the reproducibility of the results is unreliable. However, the existing data base can be used to formulate testable hypotheses about the performance of pond culture systems.
3. "Among the myriad technical questions about pond culture systems, there is a unifying question: how do physical, chemical, and biological processes interact to regulate the productivity of these systems?" (Lannan et al., 1983). The description of these processes involves the integration of disciplines: fish production, water chemistry, and physical and biological limnology.
4. "In considering pond culture practices, the number of correlations increases exponentially with the number of variables observed" (Lannan et al., 1983). A cost-effective approach to studying the principles of pond culture systems must initially address the limited subset of correlations which appear to be most relevant.
5. If the CRSP is to employ predictive mathematical models as management and research tools in improving the efficiency of pond culture systems, additional model development beyond what is presently available will be required.

Initial guidelines followed from the state-of-the-art review. These recommended that the CRSP

1. Involve existing host country infrastructure in the CRSP to the fullest extent possible
2. Quantitatively describe the physical, chemical, and biological principles of pond culture systems
3. Distinguish between site-specific and general considerations in understanding the principles of pond culture systems
4. Distinguish between general principles applicable to all production systems and principles that apply only to specific systems. (In addition to the four agroecosystems previously mentioned, the principles of fry production systems should be investigated concurrently whenever possible. The geographic application of various systems must also be addressed.)
5. Wherever possible, employ standardized experimental designs, methods, and data collection and analysis
6. Limit the number of variables to be investigated at any particular time. (Quality of data is more important than quantity.)
7. Make the output of the aquaculture CRSP applicable to improving the efficiency of pond culture systems in developing countries.

These guidelines subsequently formed the foundation of the first three cycles of CRSP research (Work Plans I through III), including the Global Experiment and the creation and maintenance of a standardized Central Data Base.

Research Sites and Institutional Linkages

The PD/A CRSP was formally initiated on 1 September 1982 as a Title XII program under the International Development and Food Assistance Act of 1975. The Consortium for Fisheries and Aquaculture Development (CIFAD), Auburn University, and the University of California at Davis were chosen by BIFAD and USAID to participate in the CRSP, with Oregon State University serving as the lead institution or Management Entity (Figure 2). CIFAD, no longer a functional entity, consisted of the University of Arkansas at Pine Bluff, the University of Hawaii, the University of Michigan, Michigan State University, and Oregon State University. Most of the CIFAD institutions continue to participate in the CRSP. As with any organization, changes in structure are apparent through the passage of time. However, the fundamental management and research units have been remarkably stable. This is partly due to the CRSP philosophy of long-term collaboration at each prime site and to the practical difficulties and costs related to initiating new projects (Figure 3).

At most of the host country (HC) research sites, several U.S. universities collaborate, with one U.S. and one HC university designated as lead (Figure 2). This structure provides for a more

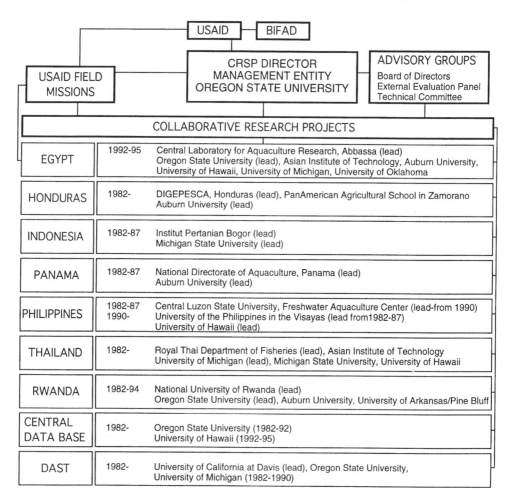

Figure 2. Organizational structure of the Pond Dynamics/Aquaculture CRSP, 1982–1995.

Figure 3. Two photographs showing the CRSP site in Butare, Rwanda at the initiation of the CRSP in 1982 (top) and after the station had expanded with the assistance of other donors and the National University of Rwanda (bottom). Before the war in 1994, the Rwasave Fish Culture Station was among the best equipped facilities in east and central Africa for aquaculture research.

equitable arrangement with host country institutions and helps in, the coordination of work plans, experiments and logistics. The mode of collaboration and the types of linkages differ from site to site, typically with formal linkages between CRSP institutions and a number of national institutions, non-governmental organizations, and international donor groups. Thus, the CRSP may become embedded within a host country institution and lose its autonomy but function as part of the whole. Consequently, benefits of CRSP research are rightfully ascribed not only to the CRSP but to all those working within the network at a particular place.

The Global Experiment

The findings from the state-of-the-art survey led to the establishment of a unifying strategy, called the Global Experiment, for optimizing the technological and economic efficiencies of pond production systems. The survey found that maximum yields were already being obtained from ponds (especially in Asia) with little scientific assistance, but methods for reproducing results were unreliable and input-output efficiencies were low. The Global Experiment was therefore conceived as a scientific method for examining variation in pond performance both spatially and temporally. Site-specific and regional variation in pond productivity, as measured by selected indicator parameters, was correlated with season. The global experiment further attempted to correlate climatological and aquatic characteristics over a broad geographic area in which CRSP sites were located (Figure 4).

The original Global Experiment was conducted from 1982 through 1989. The term Global Experiment was later used to convey that a designated experiment on pond dynamics was conducted following CRSP standardized protocols at CRSP research sites on three continents. The standardized Central Data Base incorporates all data collected from the various Global Experiments.

The statistical design for the original Global Experiment involved monitoring environmental and fish production variables at seven geographical locations. The different locations provided a spectrum of pond environments (Tables 3a and 3b). Observations specified in the biennial work plans (also called experimental cycles) were made on 12 or more ponds at each location, except at Gualaca, Panama, where 10 ponds were used (Egna et al., 1987). The pond parameters to be measured, frequency of observation, and materials and methods for determination were standardized in several CRSP documents: CRSP Instructions for Data Entry, CRSP Work Plans, and the CRSP Handbook of Analytical Methods.

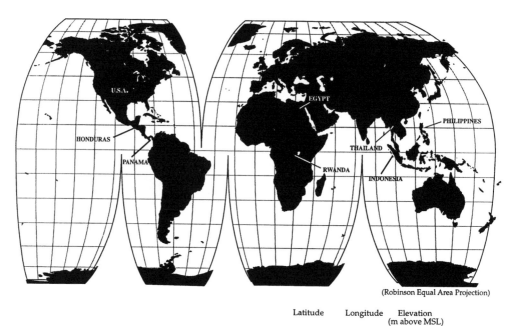

(Robinson Equal Area Projection)

	Latitude	Longitude	Elevation (m above MSL)
Egypt: Central Laboratory for Aquaculture Research, Abbassa	30°32'N	31°44'E	6
Honduras: El Carao Aquaculture Center, Comayagua	14°26'N	87°41'W	583
Indonesia: Institut Pertanian Bogor	6°6'S	106°7'E	220
Panama: Brackishwater Experiment Station, Aguadulce	8°15'N	80°29'W	0
Panama: Estacion Acuicola, Gualaca	8°31'N	82°19'W	100
Philippines: University of the Philippines in the Visayas, Iloilo	10°45'N	122°30'E	4
Philippines: Central Luzon State University	15°43'N	120°54'E	30
Rwanda: Rwasave Aquaculture Station, Butare	2°40'S	29°45'E	1625
Thailand: Royal Thai Department of Fisheries, Ayutthaya	14°11'N	100°30'E	5
Thailand: Asian Institute of Technology	14°20'N	100°35'E	5

Figure 4.   Map of Pond Dynamics/Aquaculture CRSP research sites, 1982–1995, with geographic coordinates and elevation for each site.

Table 3a   Selected Baseline Soils Data for CRSP Research Sites

	pH	Organic matter (%)	CEC meq/100 g	Clay (%)
Egypt (Abbassa)				
n (no. of ponds)	3	n/a	20	n/a
Mean ± SE	8.5 ± 0.2	n/a	41.6 ± 4.2	n/a
Median	8.3	n/a	37.1	n/a
Honduras (Comayagua)*				
n (no. of ponds)	12	11	12	12
Mean ± SE	8.1 ± 0.1	1.1 ± 0.1	31.1 ± 0.5	56.9 ± 0.5
Median	8.0	1.0	31.5	56.6
Honduras (Choluteca)				
n (no. of ponds)	[a]	[a]	[a]	[a]
Mean ± SE	7.1 ± 0.0	1.5 ± 0.2	20.9 ± 2.3	40.4 ± 2.8
Median	n/a	n/a	n/a	n/a
Indonesia (Bogor)*				
n (no. of ponds)	12	11	12	12
Mean ± SE	6.0 ± 0.1	2.7 ± 0.2	15.9 ± 0.3	67.0 ± 0.8
Median	6.1	2.9	15.9	67.2
Panama (Aguadulce)*				
n (no. of ponds)	12	12	n/a	13
Mean ± SE	7.0 ± 0.1	1.1 ± 0.1	n/a	37.0 ± 1.8
Median	6.9	1.2	n/a	34.0
Panama (Gualaca)*				
n (no. of ponds)	10	10	3	10
Mean ± SE	6.1 ± 0.3	2.9 ± 0.2	29.9 ± 0.7	36.0 ± 2.0
Median	6.2	2.4	30.0	30.0
Philippines (Iloilo)*				
n (no. of ponds)	n/a	18	n/a	n/a
Mean ± SE	n/a	2.8 ± 0.2	n/a	n/a
Median	n/a	2.8	n/a	n/a
Philippines (C. Luzon)				
n (no. of ponds)	2	2	2	2
Mean ± SE	7.6 ± 0.1	1.6 ± 0.1	38.8 ± 2.2	62.5 ± 3.0
Median	7.6	1.6	38.8	63.0
Rwanda (Butare)*				
n (no. of ponds)	12	12	12	12
Mean ± SE	4.2 ± 0.0	2.3 ± 0.3	21.0 ± 1.5	40.9 ± 1.6
Median	4.2	1.9	21.5	39.2
Thailand (Ayutthaya)*				
n (no. of ponds)	12	12	n/a	12
Mean ± SE	7.4 ± 0.1	0.8 ± 0.1	n/a	66.5 ± 0.7
Median	7.5	0.7	n/a	66.5
Thailand (AIT)				
n (no. of ponds)	20	20	n/a	20
Mean ± SE	Range 3.6–4.5	1.1 ± 0.5	n/a	51.0 ± 3.1
Median	3.7	1.0	n/a	48.0

Note: Most data were collected during the first few years of the CRSP Global Experiment, and were reported in CRSP Data Report Volume 1 (Egna et al., 1987). Additional detail on soils characteristics is presented in Chapter 6.

[a] Ponds at the Choluteca site are larger than ponds at the other sites and range in size from 0.7 to 1.0 hectares. n/a: summary data are not available.

* Indicates an original CRSP site.

Table 3b Selected Baseline Source Water Data for CRSP Research Sites

| | Freshwater sites | | | | | | | | Brackish water and marine sites | | |
	Egypt Abbassa	Honduras Comayagua	Indonesia Bogor	Panama Gualaca	Philippines C. Luzon	Rwanda Butare	Thailand Ayutthaya	Thailand AIT	Honduras Choluteca	Panama Aguadulce	Philippines Iloilo
Alkalinity (mg/L CaCO$_3$)	221.3	30.9	23.5	18.4	205.0	12.0–17.0	92.0	84.0	131.5	n/a	51.1–193.8
Ammonia (mg/L NH$_3$-N)	n/a	0.04	0.41	0.93	0.50	n/a	0.04	0.16	0.10	0.64	0–3.92
Boron (mg/L)	0.1	<1.0	0.0	***	n/a	n/a	n/a	n/a	n/a	10.5	n/a
Calcium hardness (mg/L CaCO$_3$)	121.2	15.6	***	3.4	n/a	n/a	n/a	n/a	n/a	713.0	n/a
Chloride (mg/L)	19.0	5.2	7.2	6.0	n/a	n/a	700.0	n/a	n/a	18700.0	n/a
Iron (mg/L)	0.2	1.2	7.4	0.1	n/a	n/a	7.2	n/a	n/a	1.9	n/a
Magnesium (mg/L)	16.4	1.9	2.5	0.9	n/a	n/a	217.3	n/a	n/a	1395.0	n/a
Nitrate (mg/l NO$_3$-N)	n/a	0.075	***	0.001	n/a	***	n/a	n/a	0.132	0.052	n/a
Nitrate-nitrite	n/a	0.075	0.280	n/a	n/a	***	0.033	0.010	0.166	n/a	n/a
pH	n/a	8.05	6.95	6.40	7.91	6.50	8.60	7.30	n/a	8.79	7.05–9.72
Potassium (mg/L)	6.70	4.25	1.97	0.43	n/a	n/a	3.00	n/a	n/a	365.00	n/a
Salinity (ppt)	n/a	***	***	n/a	n/a	n/a	n/a	n/a	18.5	40	n/a
Soluble orthophosphate (mg/L PO$_4$-P)	n/a	0.085	0.120	0.011	0.330	n/a	<0.005	(a)	0.120	n/a	n/a
Sulfate (mg/L)	n/a	<1.0	7.3	2.0	n/a	n/a	816.0	n/a	n/a	n/a	n/a
Total hardness (mg/L CaCO$_3$)	188.71	22.82	19.00	14.60	n/a	43.00	184.00	n/a	n/a	n/a	n/a
Total phosphorus (mg/L PO$_4$-P)	0.20	0.13	0.42	n/a	n/a	***	0.05	0.15	0.28	0.24	n/a
Year data collected	1994	1984/86	1986	1985/86	1993	1985–94	1982/83	1995	1993/94	1983	1983

Note: *** trace; ppt: parts per thousand; (a) Reactive phosphorus: 0.02 mg/l; n/a: summary data are not available.

Data Management, The Central Data Base, and Data Synthesis

The approach taken by the Pond Dynamics/Aquaculture CRSP toward developing quantitative expressions that can be used to improve production technology and facilitate economic analyses has been to create a standardized data base for evaluating pond performance over a broad range of environments (PD/A CRSP, 1986). Early in the development of the CRSP, it was determined that the use of a standardized research plan at the various CRSP project sites meant that the resulting data could be processed best using a centralized management system. By July 1984, the CRSP had an operational data management system, which has continued to evolve with changing technology and needs to the present system, now available electronically through the Internet. Further, data collection is being facilitated through the use of data loggers, in addition to the traditional *in situ* measurements.

Standardized data are tabulated at each research location for each experimental cycle in accordance with CRSP work plans. CRSP work plans specify which standard methods are to be used and which variables are to be measured (Table 4). The number of variables and methods have been modified through the years, depending on funding levels, and more importantly on research objectives. Also, some parameters have been substituted for other earlier parameters because CRSP researchers found that they served as better predictors and indicators of pond productivity.

Adherence to standardized methods has continued to be a cornerstone of all research conducted by the CRSP; this has led to the formation of the world's largest standardized data base on tropical pond aquaculture. The CRSP Central Data Base provides a great service through its collection of daily measurements of photosynthetically active radiation, rainfall, evaporation, air temperature, and wind speed concurrently with experimental data from ponds. Detailed, standardized, and verifiable records such as these are rare in the aquaculture literature. This is particularly true for photosynthetically active radiation and on-site rainfall, which are important features of water and nutrient budgets for ponds in the wet tropics. The data base was also designed to facilitate communications with other data bases, thereby increasing opportunities for collaboration and improving efficiency of resource use.

Consistent with its long-term goal, the CRSP began in 1985 to develop practical pond management models to improve the efficiency of pond culture systems. Integrated data management, standardization of data collection, and data synthesis were part of the program since its inception (Figure 5). The activity was designed to create continuity by functioning through the life of the CRSP.

A data analysis and synthesis component (now referred to as the *Data Analysis and Synthesis Team*, or DAST) was formally added in 1986 with the following objectives:

- To statistically analyze data from the field experiments to describe global and site-specific variations in pond culture systems
- To synthesize data from the Global Experiment and develop descriptive models of the physical, chemical, and biological processes that regulate the productivity of pond culture systems
- To develop conceptual frameworks for one or more pond management models and develop operating instructions consistent with each conceptual framework
- To compile a manual of operating instructions describing pond management procedures for optimizing yields, increasing the reliability and improving the efficiency of pond culture systems

Research Objectives

Research Objectives from 1982 Through 1987

The main research objectives for the first five years of the CRSP (*1982–1987 CRSP Grant*) were

- To compile a quantitative baseline of chemical, physical, and biological parameters for each work location, and to correlate responses of these parameters to various levels of

Table 4   A Selection of Variables Collected from 1982 to 1987 Following
Standardized Procedures

Physical Environment Measurements (required at all sites)
  Solar radiation ($E/m^2/d$)
  Rainfall (cm/d)
  Wind speed (km/h)
  Air temperature (C)
  Pond soil characteristics (e.g., pH, P, organic matter, extractable bases, cation exchange
    capacity)
  Pond temperature extremes (max C, min C)
  Morphometric characteristics (length and width in m, pond depth in m, area in m^2,
    volume in m^3)
  Hydrologic characteristics (e.g., surface inflow, precipitation, outflow, evaporation,
    seepage; in m^3/d)
Water Analyses (required during production experiments)
  Dissolved oxygen (mg/l)
  Temperature (C)
  pH
  Alkalinity (mg $CaCO_3/l$)
  Total hardness (mg $CaCO_3/l$)
  Water quality characteristics (e.g., chloride, sulfate, zinc, iron, magnesium)
  Total nitrogen (mg/l)
  Ammonia nitrogen (mg/l)
  Nitrate (mg/l)
  Total phosphorus (mg/l)
  Dissolved orthophosphate (filterable reactive phosphorus) (mg/l)
  Silicates (brackish water and marine sites only)
Growth and Yield Measurements (required during production experiments)
  Growth (kg/number of individuals, g for a sub-sample of individuals)
  Reproduction (number of individuals, g)
  Survival
Biological Limnology Measurements (required during production experiments)
  Secchi disk visibility (cm)
  Chlorophyll a (mg/m^3)
  Chlorophyll b, c (brackish water and marine sites only)
  Primary productivity (mg $C/m^3/d$)
  Qualitative identification of phytoplankton, zooplankton, and benthos (order, relative
    abundance)

Note: These data were collected at each CRSP research site and are recorded in the CRSP
        Central Data Base.
Adapted from Egna, H. S., Brown, N., and Leslie, M., 1987.

organic and inorganic fertilizer applications to pond culture systems (referred to as the
"Global Experiment")
• To compile a baseline of information on hydrology, locally available nutrient inputs,
  geography, and water quality in each participating country, utilizing available host coun-
  try resources
• To observe and document technical constraints limiting fry availability in each partici-
  pating host country, and to test alternative fry production methods where appropriate
• To develop models describing the principles of pond culture systems

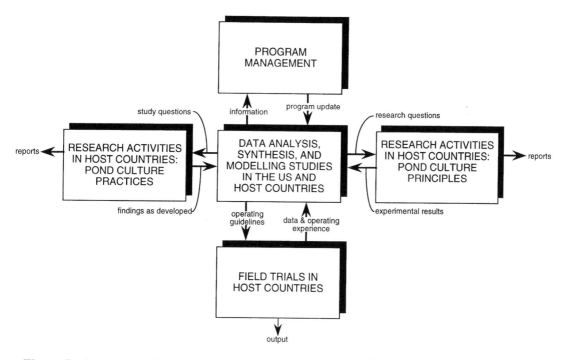

Figure 5. Interrelationships between and among activities in the Pond Dynamics/Aquaculture CRSP. (Adapted from Pond Dynamics/Aquaculture CRSP, 1986.)

Research Objectives from 1987 Through 1995

The *1987–1990 Continuation Plan* addressed the most important objectives of the original plan, with the goal of synthesizing the results of the first three work plans as a staged progression into a conceptual model of pond aquaculture systems. This model was used to identify research needs, which were prioritized and translated into objectives for field research projects specific for each host country (Figure 6).

The programmatic and operational objectives in the *1990–1995 Continuation Plan* were

- To continue to develop technology, through research, to overcome major problems and constraints affecting the efficiency of pond aquaculture in developing countries
- To maintain or improve environmental quality through proper management of aquaculture systems
- To stimulate and facilitate the processing and flow of new technologies and related information to researchers, to extension workers, and ultimately, to fish farmers in developing countries
- To promote activities that encourage faculty and researchers to build and maintain linkages
- To create opportunities for greater multidisciplinary research in aquaculture and to enhance the socioeconomic and ecological aspects of the CRSP
- To encourage informational and data exchange among international agricultural research centers, universities, the non-government research community, and USAID centrally funded and mission-funded projects
- To expand results derived from the site-specific research to regional recommendations through a global analysis of the data
- To use an ecosystem approach to prioritize the research agenda and integrate technologies

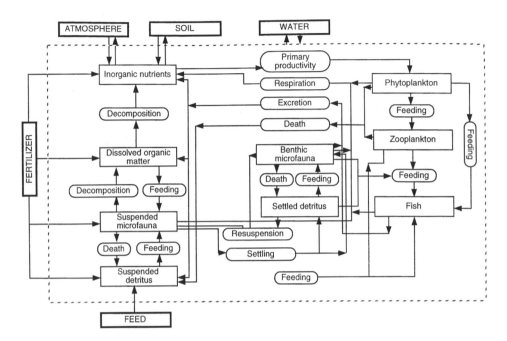

Figure 6. A CRSP conceptual model of an aquaculture pond. This conceptual model, which was revised for later CRSP work, served as a reference for the Global Experiment. The arrows connecting the model's components represent the paths for movement of mass in the system. The system includes both biological and nonbiological components. The strictly biological components are phytoplankton, zooplankton, and fish, whereas the nonbiological components are the inorganic nutrients that are considered likely to limit productivity. Sediments include decomposing organic matter that settles from the water, the parent soil material, and benthos. Particulate organic matter is a composite of dead particulate organics and bacteria that either coat the particles or are free in suspension. Most research on fertilized aquaculture ponds has been based on the premise that yields and production rates are determined by primary productivity. Nutrients added to a pond must undergo transformations that include fixation by phytoplankton before they are available to fish.

Evolving Directions in Pond Dynamics/Aquaculture Research, 1980 to 1995

The state-of-the-art survey, which was undertaken in 1980, set the direction for the PD/A CRSP for its first decade of research. The first and second CRSP grants (1982 through 1990) were largely devoted to collecting and analyzing baseline data to quantitatively describe the physical, chemical, and biological principles of pond culture systems. A global experiment and a standardized system of data collection and management eventually led to research results, which to date have appeared in a number of journals and program publications. The information from these experiments is presented in various chapters throughout the book, which concentrate on describing the natural pond environment and fish productivity.

The results themselves, however, have been slow to move off the shelf, and consequently the CRSP in recent years has devoted more resources to on-farm trials, field trials, and social sciences studies of adopters and non-adopters of CRSP technologies. A major impediment to the perceived lack of adoption of CRSP technologies has spawned a number of projects in the social sciences, including a Women in Development study, a global economics study, and extension and social sciences workshops. Results from these studies and efforts are presented in Chapters 15 and 16, which describe CRSP and non-CRSP activities to show impacts of technologies and methods for extending research to farmers.

Concurrent with these developments in the field, the DAST, after statistically summarizing fish growth rate variability across sites, has headed in two directions. They have developed a decision support system called POND© to provide more cost-effective and biologically efficient pond management strategies. They have also developed a number of water quality models, including deterministic and stochastic models for better defining the critical processes involving oxygen dynamics in ponds. The efforts of the DAST are presented in Chapter 13.

With respect to the non-research mandate of the CRSP, that is, to strengthen institutions and capacities, the program has increased its linkages with non-governmental organizations, national agricultural research systems, other international groups engaged in aquaculture development, and individuals from non-CRSP countries. As of 1994, 121 students had received training in academic degree programs and over 500 individuals had participated in informal training in short courses, workshops, and symposia. Improved systems for managing information and feedback to research have been instituted, and the CRSP organizational model continues to be emulated by various groups. These strengthened research and management connections have helped the CRSP weather the changes wrought upon any long-term program. Even though funding levels, political upheavals, and new directions in U.S. foreign policy have brought about modifications to the research portfolio from year to year, the underlying research and development agenda has remained true to the work plans and the tenets of Title XII legislation.

Indeed, as the CRSP has matured over the years, it has come to be recognized in some quarters less for its technologies than as "an institutional innovation, a means for fomenting technical change in national aquaculture industries. Although not organized to deliver technology on the village level, it is a mechanism for improving the lot of family farms and the villages they encompass, albeit indirectly, through each nation's institutional network for aquacultural development (Molnar et al., 1995)." Thus, the impacts of CRSP tend to be perceived at "the institutional level where individuals are training, procedures [are] put in place, and facilities are planned with expatriate technical assistance" (Molnar et al., 1995, pp. 1–6). Recent trends in the CRSP to consider not just the pond itself but also the natural and social system in which the pond exists will become a greater focus for the program in future years.

ACKNOWLEDGMENTS

I am grateful to Mr. Harry Rea, USAID project officer to the PD/A CRSP, for his assistance in collecting information on aquaculture projects funded by USAID since 1981 and for reviewing this chapter; to Ms. Hilary Berkman, former CRSP data base manager, for her review of the history of the CRSP data base; to Dr. Michael Skladany for his unique insights into the sociology of aquaculture; to Dr. Martin Fitzpatrick for a thorough editorial review and for technical assistance on the figures and tables; and to family and friends who encouraged me in this effort. Finally, I wish to pay a tribute to our CRSP project in Rwanda, which after the recent tragedies of the Rwandan Civil War serves as a somber reminder to us all of the importance of historical perspectives.

REFERENCES

Acker, D., A Need within AID: Movement in Animal Agriculture and Aquaculture, *Information Memorandum to the Administrator,* from the Assistant to the Administrator, U.S. Agency for International Development, May 30, 1989, 1.

Axinn, N. W., The family and the farm system: some thoughts on collaborative research, *Rural Sociol.* 4(4), 269, 1984.

Belcher, M., Title XII and Collaborative Research Support Programs: the USAID perspective, in *Proceedings of the Fisheries Research Planning Workshop,* Craib, K. B. and W. R. Ketler, Eds., December 14–15, 1977, prepared for USAID under Contract No. AID/afr-C-1135-12 by Resources Development Assoc., Los Altos, CA, 1977, 19.

Board for International Food and Agricultural Development, *Guidelines for the Collaborative Research Support Programs,* U.S. Agency for International Development, Washington, D.C., 1985.

Brady, N. C., Title XII in retrospect and prospect, *Rural Sociol.* 4(4), 269, 1984.

Brown, E. E., *World Fish Farming: Cultivation and Economics,* AVI Publishing, Westport, CT, 1977.

Campbell, F. R., *Fisheries Technology for Food and Income,* Office of Agriculture, Bureau for Science and Technology, U.S. Agency for International Development, 1987, 26 pp.

Collaborative Research Support Program Proposal, *Efficiency of Pond Culture Systems: Principles and Practices,* prepared by Auburn University, University of California at Davis, and the Consortium for International Fisheries and Aquaculture Development for the Joint Research Committee, Board for International Food and Agriculture Development, Agency for International Development, 1981, 63 pp.

Craib, K. B. and Ketler, W. R., Eds., *Fisheries and Aquaculture Collaborative Research in the Developing Countries: A Priority Planning Approach,* prepared for USAID and the Joint Research Committee of BIFAD by Resources Development Assoc., Los Altos, CA, 1978.

Cremer, M., History of international development activities by U.S. government agencies, in Lannan, J. E., Smitherman, R. O., and Tchobanoglous, G., Eds., *Principles and Practices of Pond Aquaculture: A State of the Art Review,* prepared for USAID under Specific Support Grant No. AID/DSAN-G-0264, Oregon State University, Newport, OR, 1983, chap. 1.

Donovan, G. A., Sutinen, J., Pollnac, R. B., and McCreight, D., World Fishery Development: Roles for USAID and US Universities, paper presented at the 66th Meeting of the Board for International Agricultural Development, Washington, D.C., October 11, 1984, 12 pp.

Egna, H. S., Brown, N., and Leslie, M., General reference: site descriptions, materials and methods for the Global Experiment, *Pond Dynamics/Aquaculture Collaborative Research Data Reports,* Vol. 1, Oregon State University, Corvallis, OR, 1987.

French, C. E., Comments at the fisheries research planning workshop: the Title XII program, in *Proceedings of the Fisheries Research Planning Workshop,* Craib, K. B. and Ketler, W. R., Eds., December 14–15, 1977, prepared for USAID under Contract No. AID/afr-C-1135-12 by Resources Development Assoc., Los Altos, CA, 1977, 79.

Idyll, C. P., *Fisheries Projects,* United States Agency for International Development, Washington, D.C., 1988, 33 pp.

International Center for Aquaculture and Aquatic Environments, *Brochure*, Auburn University, Auburn, AL, n.d., 15 pp.

Iversen, E. S. and Hale, K. K., *Aquaculture Sourcebook: A Guide to North American Species,* Van Nostrand Reinhold, New York, 1992, 308 pp.

Jensen, G. L., History of United States Peace Corps involvement in freshwater aquaculture, in Lannan, J. E., Smitherman, R. O., and Tchobanoglous, G., Eds., *Principles and Practices of Pond Aquaculture: A State of the Art Review,* prepared for USAID under Specific Support Grant No. AID/DSAN-G-0264. Oregon State University, Newport, OR, 1983, chap. 2.

Joint Subcommittee on Aquaculture, *National Aquaculture Development Plan,* Federal Coordinating Council on Science, Engineering, and Technology, Volume II, Washington, D.C., 1983, 196 pp.

Jones, D. M., The fisheries planning process: what to expect, in *Proceedings of the Fisheries Research Planning Workshop,* Craib, K. B. and Ketler, W. R., Eds., December 14–15, 1977, prepared for USAID under Contract No. AID/afr-C-1135-12 by Resources Development Assoc., Los Altos, CA, 1977, 47.

Kammerer, K. C., Correspondence with Mr. Wayne Smith, Staff Director for the Subcommittee on Fisheries and Wildlife Conservation and the Environment, Committee on Merchant Marine and Fisheries, House of Representatives, Washington, D.C., 7 April 1983, 4 pp.

Ketler, W. R., Welcome and introduction, in *Proceedings of the Fisheries Research Planning Workshop,* Craib, K. B. and Ketler, W. R., Eds., December 14–15, 1977, prepared for USAID under Contract No. AID/afr-C-1135-12 by Resources Development Assoc., Los Altos, CA, 1977, i.

Lannan, J. E., Smitherman, R. O., and Tchobanoglous, G., Eds., *Principles and Practices of Pond Aquaculture: A State of the Art Review,* prepared for USAID under Specific Support Grant No. AID/DSAN-G-0264, Oregon State University, Newport, OR, 1983, 240 pp.

Lannan, J. E., Smitherman, R. O., and Tchobanoglous, G., Eds., *Principles and Practices of Pond Aquaculture,* Oregon State University, Corvallis, OR, 1986, 252 pp.

Lipner, M. E. and Nolan, M. F., Dilemmas of opportunity: social sciences in CRSPs, in McCorkle, C. M., Ed., *The Social Sciences in International Agricultural Research: Lessons from the CRSPs,* Lynne Rienner Publishers, Boulder, CO, 1989, 20.

Meade, J. W., *Aquaculture Management,* Van Nostrand Reinhold, New York, 1989, 175 pp.

Mickelwaite, D. R., Sweet, C. F., and Morss, E. R., *New Directions in Development: A Study of U.S. AID,* Westview Press, Boulder, CO, 1979.

Molnar, J. J., Hanson, T. R., and Lovshin, L. L., *Socioeconomic Dimensions of Aquacultural Development: Impacts of The Pond Dynamics/Aquaculture CRSP in Rwanda, Honduras, The Philippines, and Thailand,* Department of Agricultural Economics and Rural Sociology, Auburn University, Auburn, AL, draft, September 1995, 1–6.

National Marine Fisheries Service, *AID Fisheries Strategy Statement,* Draft 1985, 20 pp.

National Research Council, *Aquaculture in the United States: Constraints and Opportunities,* National Academy of Sciences, Washington, D.C., 1978, 123 pp.

National Research Council, *An Evaluation of Fishery and Aquaculture Programs of the Agency for International Development,* National Academy of Sciences, Washington, D.C., 1982, 161 pp.

National Research Council, *Panel on fisheries development and fisheries research: recommendations of the discussion meeting to advise the research advisory committee of the Agency for International Development,* Board on Science and Technology for International Development, Office of International Affairs, Washington, D.C., 1988, 15 pp + app.

National Research Council, *Toward Sustainability: A Plan for Collaborative Research on Agriculture and Natural Resource Management,* Board on Agriculture, Board on Science and Technology for International Development, National Academy Press, Washington, D.C., 1991, 147 pp.

Neal, R. A., US Agency for International Development strategy for aquaculture development, *Proc. Symp. Coastal Aquaculture,* 4, 1010, 1986.

Neal, R. A., A.I.D. *Strategy for Fisheries and Aquaculture,* Draft report to USAID, 1987, 23 pp.

Pond Dynamics/Aquaculture CRSP, *Fourth Annual Administrative Report,* Program Management Office, Oregon State University, Corvallis, OR, 1986, 28 pp + app.

Poponoe, H., Summary statement, in *Proceedings of the Fisheries Research Planning Workshop,* Craib, K. B. and Ketler, W. R., Eds., December 14–15, 1977, prepared for USAID under Contract No. AID/afr-C-1135-12 by Resources Development Assoc., Los Altos, CA, 1977, 89.

Research Advisory Committee, *Fisheries Background Paper,* Office of Agriculture, Bureau for Science and Technology, USAID, Washington, D.C., 1988, 4.

Skladany, M., *Research and Development in Aquaculture Biotechnology: An Approach and Interpretation from the Sociology of Scientific Knowledge,* Ph.D. Dissertation, Michigan State University, East Lansing, MI, in preparation.

Thomas, D. W., Title XII and Collaborative Research Support Programs: the BIFAD perspective, in *Proceedings of the Fisheries Research Planning Workshop,* Craib, K. B. and Ketler, W. R., Eds., December 14–15, 1977, prepared for USAID under Contract No. AID/afr-C-1135-12 by Resources Development Assoc., Los Altos, CA, 1977, 3.

Tiddens, A., *Aquaculture in America: The Role of Science, Government, and the Entrepreneur,* Westview Special Studies in Agriculture Science and Policy, Westview Press, Boulder, CO, 1990, 191 pp.

Trott, L. B., The United States Agency for International Development and mariculture, in *Coastal Aquaculture in Developing Countries,* Pollnac, R. B. and Weeks, P., Eds. ICMRD, The University of Rhode Island, Kingston, RI, 1992, 174.

United States 94th Congress, H.R. 9005, *International Development and Food Assistance Act of 1975,* Title XII: Famine Prevention and Freedom from Hunger, Public Law 94-61, an amendment to the Foreign Assistance Act of 1961, December 20, 1975, 89 Stat. 861.

United States Agency for International Development, *Managed Fish Production: An A.I.D. Strategy for Increasing Production and Improving Utilization of Food Fish,* Fisheries Division, Office of Agriculture, Technical Assistance Bureau, USAID, Washington, D.C., July 1976, 10 pp.

United States Agency for International Development, *An Inventory of US Marine Technical Assistance Programs,* Washington, D.C., draft, 1981.

United States Agency for International Development, *A.I.D. Strategy for Fisheries and Aquaculture,* Office of Agriculture, Bureau for Research and Development, USAID, Washington, D.C., draft, 1987, 23 pp.

United States Agency for International Development, *Living Aquatic Resources (Fisheries) and The Agency for International Development,* Bureau for Science and Technology, USAID, Washington, D.C., 1990b, 8 pp.

United States Agency for International Development, *Strategy for the 90's: The Fisheries Sector,* Office of Agriculture, Bureau for Research and Development, USAID, Washington, D.C., 1990a, 8 pp.

United States Agency for International Development, *The Status of Fisheries and Aquaculture Development Assistance Programs,* Office of Agriculture, Bureau for Research and Development, USAID, Washington, D.C., August 1, 1993, 49 pp.

United States Agency for International Development, *Data Base Search, AID-DISC/Development Information System,* Center for Development Information and Evaluation, USAID, Washington, D.C., 1989, 1993, 1995.

United States Department of Agriculture, *1995 Farm Bill,* Guidance of the Administration, Washington, D.C., 1995, 94 pp.

Whaley, R. S., Title XII and Collaborative Research Support Programs: The JRC perspective, in *Proceedings of the Fisheries Research Planning Workshop,* Craib, K. B. and Ketler, W. R., Eds., December 14–15, 1977, prepared for USAID under Contract No. AID/afr-C-1135-12 by Resources Development Assoc., Los Altos, CA, 1977, 12.

Yohe, J. M., Barnes-McConnell, P., Egna, H., Rowntree, J., Oxley, J., Hanson, R. G., Cummins, D., and Kirksey, A., The Collaborative Research Support Programs (CRSPs): 1978 to 1990, contributed paper for the Forum on Sustainable Agriculture and Natural Resource Management, November 13–16, 1990, National Academy of Sciences, Washington, D.C., 1990.

Yohe, J. M., Barnes-McConnell, P., Egna, H., Rowntree, J., Oxley, J., Hanson, R. G., Cummins, D., and Kirksey, A., The CRSPs: International Collaborative Research Support Programs, in Leslie, J. F. and Frederiksen, R. A., Eds., *Disease Analysis through Genetics and Biotechnology,* Iowa State University Press, Ames, IA, 1995, 321.

3   WATER QUALITY IN PONDS

James S. Diana, James P. Szyper, Ted R. Batterson, Claude E. Boyd, and Raul H. Piedrahita

INTRODUCTION

The various chemicals dissolved in the water, as well as the temperature and other physical attributes of water, all combine to form what is called water quality. For aquaculture systems, changes in water characteristics that improve the production of an aquatic crop would be considered improvements in water quality, while those changes reducing production would be considered degradation of water quality. This definition is important in aquaculture, because the utilization of water to grow aquatic crops at high densities often results in chemical attributes which, by environmental standards, may be considered reductions in water quality. Unless these changes reduce the production, safety, or value of the target organism, they would not be considered degradations of water quality for aquaculture purposes. Good water quality characteristics will be considerably different for some species than for others. Characteristics that enhance production of tilapia might be detrimental to species such as rainbow trout. Species are often chosen for aquaculture because of their tolerance to poor water quality (Chapter 8). Thus, water quality must be viewed in the context of the species cultured. This chapter reviews the quality of water in relation to production of tilapia in semi-intensive to intensive ponds.

Water quality in fish ponds is a major factor determining the production of fish. The CRSP experimental sites have been located throughout the world, and therefore they vary considerably in ambient water quality. Water quality is also dramatically influenced by pond management practices, such as stocking density of fish, fertilization strategy, and supplemental feeding. Water quality may be manipulated in attempts to overcome the limitations to pond production, by chemical or physical processes like aeration, fertilization, liming, or water exchange. Manipulation of pond water quality is a major management tool in the semi-intensive production of tilapia and may become an important limitation in intensive production.

Characteristics of water can strongly limit fish production and may be altered to some degree by aquaculture practices. Some characteristics of water, such as its dissolved mineral content, pH, alkalinity, and hardness, are strongly influenced by the source of water and the soils, as well as geological and climatic properties of the watershed. Tilapia grow best when water temperatures are relatively warm and rates of primary production are high. Other uses of the water and watershed may influence water quality. Finally, aquaculture practices influence water quality.

Water is unique in several regards. It is one of the few materials that is liquid at ambient temperatures; the only others are elemental mercury and some hydrocarbons. Water has a very high viscosity and surface tension, greater than all liquids except mercury. This surface tension strongly limits diffusion of gases into and out of the water. Water has a high specific heat, which makes it resistant to change in temperature. Our measures of energy are based on this specific heat: it takes 1 calorie of energy to increase 1 g of water by 1°C. Finally and most importantly, water has a unique relationship between temperature and density, as pure water in the liquid form reaches its maximum density at 3.94°C. Density of water decreases at lower temperatures due to a unique and weak bonding, which initiates lattice structure. This structure is completely formed in the solid state. In most liquids, density increases linearly as temperature declines and reaches a maximum in the solid state. Since water does not follow this pattern, ice floats on water and keeps the water column from completely freezing under cold ambient conditions.

The purpose of this chapter is to review the characteristics of water quality that influence tilapia production, with special emphasis on the CRSP research sites and experiments. An excellent text on water quality in aquaculture, published by Boyd (1990), evaluates water quality in general and specifically deals with chemical processes in pond water. In this chapter we will not fully evaluate all of these chemical processes, but we will review the characteristics of light intensity and penetration, nutrients and metabolites, plankton, temperature, and dissolved oxygen. After this material, a review of the effects of dissolved oxygen on tilapia in ponds and automated data collection systems will complete the chapter.

LIGHT

Depending on incidence angle, only a portion of available sunlight penetrates the water surface. At an angle of 80°, 35% of sunlight is reflected, while at 60°, only 6% is reflected. Solar radiation does not pass readily through water either but rather extinguishes rapidly with depth. In purest lake water, only 40% reaches 1 m. The penetration of light through water differs with the wavelength of light. Penetration is greatest in the middle of the spectrum and declines toward both long (red) and short (violet) wavelengths. Heat, produced by infrared radiation, is largely absorbed in the surface of a water body and rarely penetrates beyond 1 m. These characteristics result in the warming of surface waters only, while mixing of warm water to depths is influenced by wind. The absolute depth of light penetration depends largely on the abundance of dissolved and suspended materials in water. Photosynthesis requires a minimum intensity of light within the photosynthetically active radiation wavelengths (400 to 700 μ), usually about 1% of surface levels. The region in the water column where light intensity is at least 1% of the surface level is often referred to as the photic zone. The depth of light penetration can be estimated by means of a Secchi disk or a light meter; the depth of the photic zone is approximately twice the Secchi disk depth (SDD). A Secchi disk is a weighted disk, 20 cm in diameter, painted alternately in black and white quarters. It is lowered into the water until its image disappears at the SDD.

In natural lakes, SDD ranges from a few centimeters to over 40 m; in aquaculture ponds it rarely exceeds 1 m. In ponds in Thailand that received little fertilization, SDD averaged 50 cm for most of a fish grow-out period (Diana et al., 1987). However, when nutrients were added in significant amounts, SDD declined to 25 cm or less because of increased plankton abundance. Different results were found at other CRSP sites. In Panama (Hughes et al., 1991a), Indonesia (McNabb et al., 1988), and Rwanda (Hanson et al., 1989), SDD averaged 20 to 30 cm in lightly fertilized ponds, while in Honduras (Green et al., 1990a), the values were 10 to 15 cm because of high levels of clay turbidity. (These references evaluate data for Cycle 1 experiments in the CRSP and will henceforth be cited as Cycle 1 experiments). When fertilization rates were increased in these ponds, SDD remained about the same, except in Honduras (Green et al., 1990b; Diana et al., 1991a; Hanson et al., 1991; Hughes et al., 1991b; McNabb et al., 1991; henceforth cited as Cycle 2 experiments). In Honduras, use of organic fertilizer actually increased SDD by removing clay from the water and also increased primary production rates (Green et al., 1989). Use of organic material to remove colloidal clay (Avnimelech et al., 1982) is a common management practice.

SDD has been considered an indicator of relative phytoplankton abundance in ponds (Almazan and Boyd, 1978; Boyd, 1990). Since abundance of plankton decreases water clarity and SDD, this relationship may be strong within ponds in a locale or region but not necessarily between regions (for example, see Figure 1). CRSP data suggest that this relationship is weakly evident at all sites but differs between sites due to other factors like clay turbidity influencing SDD. Teichert-Coddington

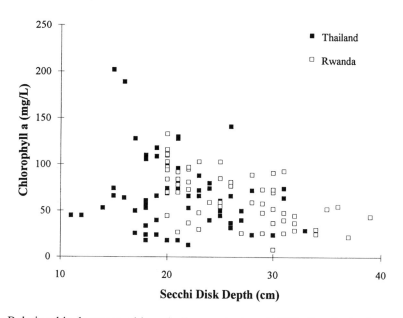

Figure 1.   Relationship between chlorophyll *a* content and SDD for Cycle 3 experiments in Thailand and Rwanda.

et al. (1992) reviewed differences in SDD between Panama and Honduras and found a typical relationship between SDD and chlorophyll *a* content in Panama but not in Honduras due to clay turbidity. Also, the relationship in Figure 1 shows considerable variability, with r^2 values for Thailand of only 0.024 ($P > 0.05$) and for Rwanda of only 0.37 ($P < 0.01$). These variations were similar on arithmetic or log scales, and were considerably higher than data from Almazan and Boyd (1978), who had r^2 values of about 0.6. Thus, the predictability of plankton standing stocks from SDD appears less precise in CRSP ponds.

NUTRIENTS AND METABOLITES

Carbon Dioxide, Alkalinity, and pH

The relationship between carbon dioxide, alkalinity, and pH is well known in limnology and in aquaculture (Wetzel, 1983; Boyd, 1990). Alkalinity is a measure of the total titratable bases in water, which include carbonate (CO_3^{-2}), bicarbonate (HCO_3^-), ammonia, hydroxide, phosphate, silicate, and some organic acids (Boyd, 1990). The source of alkalinity in water is due to the salts of weak acids becoming dissociated in water. Most of the alkalinity of freshwater is composed of carbon ions (CO_3^{-2} and HCO_3^-), which are interrelated; their proportions depend on pH. At a pH of about 4, most inorganic carbon in water is in the form of CO_2. As pH is increased, bicarbonate becomes more common, until pH 8.3, when most inorganic carbon is present as HCO_3^-. Finally, as pH continues to rise, CO_3^{-2} becomes more common. Because the noncarbonate components are relatively scarce, total alkalinity is often considered as an indicator of the inorganic carbon content of water. Since pH in aquaculture ponds is often 7 to 8, most of this inorganic carbon is in the form of bicarbonate in ponds.

Alkalinity potentially limits primary production and fish yield, since inorganic carbon is necessary for photosynthesis. Boyd (1990) showed that alkalinity below 30 mg/L as $CaCO_3$ limits primary production in well-fertilized ponds, while in unfertilized ponds alkalinities below 120 mg/L can reduce primary production (Figure 2). The latter effect is probably not a limitation by alkalinity alone but by all materials dissolved in these unfertilized ponds, since alkalinity of natural waters

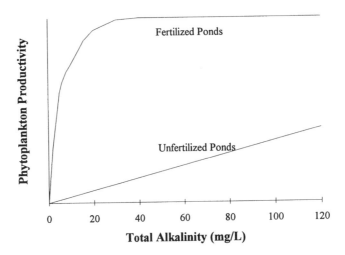

Figure 2. Relationship between phytoplankton productivity and alkalinity in ponds with differing fertilization rates. (Redrawn from Boyd, C. E., 1990.)

is related to concentration of dissolved materials in the water. The CRSP research sites encompass areas with a variety of alkalinities, from high levels in Thailand (about 120 mg/L) and Honduras (about 100 mg/L) to moderate levels in Rwanda (about 50 mg/L) and Panama (about 40 mg/L) to very low levels in Indonesia (below 30 mg/L). Thus, alkalinity values may result in differences in primary productivity among the CRSP sites.

Since pH, CO_2, and alkalinity are interrelated, diel changes in CO_2 resulting from photosynthetic and respiratory processes can result in diel changes in pH, particularly in low alkalinity ponds (Figure 3). Diel changes in pH are variable among sites in CRSP ponds. Using Cycle 2 data for Thailand and Honduras, with high alkalinity, there is a moderate change in pH within a pond over the course of the day (about 0.7 pH units). These changes also do not consistently follow the pattern

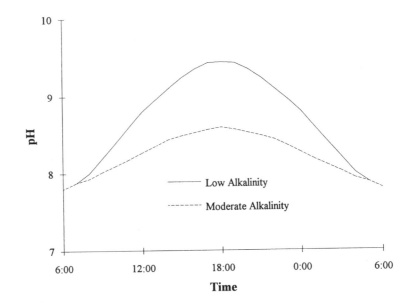

Figure 3. Changes in pH over the day in low and moderate alkalinity ponds. (Redrawn from Boyd, C. E., 1990.)

in Figure 3 but are variable over the day. In comparison, data from Indonesia show more extreme changes in pH (1.2 units) and a regular cycle, as in Figure 3, for these ponds with low alkalinity. It appears that in Indonesia and possibly Panama low alkalinity may limit primary production in well-fertilized fish ponds (McNabb et al., 1990; Teichert-Coddington et al., 1992), even though total alkalinity was greater than 30 mg/L in these ponds.

One trend in alkalinity for CRSP ponds is less well established in the literature, and that is a reduction in alkalinity over the course of a grow-out period. Diana et al. (1994a) noted that in some Thai ponds, alkalinity became low even though alkalinity had been high when the ponds were filled. This reduction in alkalinity occurred in inorganically fertilized ponds, while organically fertilized or fed ponds increased in alkalinity in Thailand, Honduras, and Panama (Teichert-Coddington et al., 1992). Carbon extraction during photosynthesis was believed to reduce CO_2 during a grow-out period, which could directly reduce total inorganic carbon in the water or could increase pH and cause $CaCO_3$ to be precipitated, reducing alkalinity. Use of organic inputs probably reversed this trend because of added CO_2 inputs and decomposition of manure in pond soils. Knud-Hansen et al. (1991) and Knud-Hansen and Batterson (1994) found similar reductions in alkalinity in Thai ponds but attributed these to carbon extraction by mollusks to deposit in shell material. The use of organic inputs may keep alkalinity at higher levels and may even result in increases in alkalinity compared to use of inorganic inputs (Knud-Hansen et al., 1991; Teichert-Coddington et al., 1992; Diana et al., 1994a).

Nitrogen

The availability of nitrogen is important to primary production in fish ponds, and several nitrogen metabolites, as well as chemical fertilizers such as NH_3, are also toxic to fish. The role of nitrogen in productivity of freshwaters is still being debated, because phosphorus originally was considered the limiting nutrient in freshwaters (Boyd, 1990; Knud-Hansen et al., 1991). Nitrogen is apparently limiting in some tropical freshwater systems (Zaret et al., 1981; Setaro and Melack, 1984), including aquaculture ponds (Diana et al., 1991b; Knud-Hansen et al., 1991; Teichert-Coddington et al., 1992).

Inorganic nitrogen in ponds exists mainly as nitrate, nitrite, ammonia, and ammonium. The sum of these products is termed dissolved inorganic nitrogen (DIN) or total inorganic nitrogen. These products are convertible through the nitrogen cycle, and the presence and abundance of different forms of nitrogen are affected by pH, oxygen concentration, and organisms that may produce or consume certain forms of N (Boyd, 1990). In addition to nitrogen inputs in these forms from source water, fertilizers, and feeds, additional nitrogen inputs may occur through nitrogen fixation of phytoplankton and bacteria (Lin et al., 1988).

Considering nitrogen as a nutrient, algal cells generally take in nitrogen as nitrate, although ammonia may also be utilized by phytoplankton (Boyd, 1990; Knud-Hansen and Pautong, 1993). Most studies in pond aquaculture measure nitrogen as DIN because of the interconversion of these forms of nitrogen. DIN values for input waters at the CRSP sites are relatively low, ranging from 0.2 to 0.5 mg/L in Thailand and Rwanda, to 0.6 to 1.0 mg/L in Indonesia, and 1.5 to 2.0 mg/L in Panama. In all cases, input waters seem higher in ammonia-N than in nitrate/nitrite.

In order to maintain high levels of primary production, it has been emphasized that nutrients must be provided in relationship to the needs of plankton and to supplies available from allochthonous sources. Several CRSP experiments have documented nitrogen limitation when ponds were fertilized with chicken manure alone at high inputs (Diana et al., 1991b; Knud-Hansen et al., 1991; Teichert-Coddington et al., 1992; Knud-Hansen and Pautong, 1993). CRSP fertilization experiments to date have demonstrated that high rates of primary production occur when N and P are both provided in inputs, usually at a rate of 4N:1P (by mass) and a total application of 28 kg N/ha/wk (Knud-Hansen and Batterson, 1994). While these inputs do provide high rates of primary production in Thailand, their global applicability has not yet been proven and is the subject of a future experiment. These inputs can be balanced by using manure, urea, and triple superphosphate as necessary, although some of these materials may degrade quality at high input levels.

Attempts in the CRSP to maintain DIN at acceptable levels have been done by altering fertilizer application frequency, amount, and quality. DIN was quickly removed from the water in controlled aquaria when inorganic materials were used, including urea (Knud-Hansen and Pautong, 1993). When manures were utilized, DIN slowly dissolved from the manure and remained at moderate levels for 2 to 6 d (Knud-Hansen et al., 1991). Knud-Hansen and Batterson (1994) varied fertilization frequency from daily to once every two weeks in Thai ponds (at the same weekly loading rate) and found no effect of fertilization frequency on primary production or fish yield. Diana and Lin (1996) attempted to alter fertilization schedules by fertilizing some ponds after DIN levels were reduced to 0.5 mg/L. Ponds in this treatment showed lower primary production and fish yields than ponds fertilized weekly, indicating that such a management system did not improve fertilization efficiency. In ponds at all CRSP sites, average DIN values remained above 0.25 mg/L when fertilized weekly with a balanced input or when urea was added into ponds coupled with chicken manure. However, data from Honduras indicates that DIN levels above 0.2 mg/L may reduce fish yields (Teichert-Coddington et al., 1992). Nitrogen fertilization remains a controversial component of CRSP research.

Nitrogen metabolites, such as ammonia, from fish and other aquatic organisms can result in reduced water quality. In semi-intensive aquaculture, deleterious ammonia levels rarely become a problem, although they can create serious toxicity problems when inorganic nitrogen or feed are added to ponds at high inputs. Toxicity of ammonia and nitrite can be exacerbated as pH rises. Both unionized ammonia and nitrite are toxic to fish at low concentrations, and unionized ammonia constitutes an increased fraction of total ammonia as pH increases above 7 (Boyd, 1990). Therefore, a synergism between forms of ammonia, nitrite toxicity, and pH may occur to cause toxicity or sublethal stress in ponds at high pH. This may be particularly problematic in lower alkalinity ponds, which have significant diel shifts in pH.

In spite of the concerns listed above, lethal or sublethal problems with nitrogen have rarely been documented in CRSP ponds. Nitrite and ammonia levels remain relatively low in fertilized ponds under most conditions. While addition of urea as a nutrient may be hypothesized to affect ammonia levels, Knud-Hansen and Pautong (1993) found few deleterious effects of using urea as a nitrogen source in aquaculture for tilapia. Nitrogen toxicity is more likely when ponds are provided with feed. However, Green (1992) and Diana et al. (1994a) used supplemental feeding for tilapia production and found no nutrient accumulation problems. In fact, Diana et al. (1994a) found limited accumulation of nutrients as long as phosphorus and nitrogen in fertilizer were added with feed to keep nutrients in balance and maintain high levels of primary production. Recent experiments (Diana et al., 1994b) indicate few toxicity problems with nitrogen even in ponds with no water exchange and supplemental feeding at densities of 9 fish/m^2.

Phosphorus

Phosphorus is most commonly considered the major limiting nutrient in freshwater, and additions of phosphorus often result in increases in primary production, whether in natural (Vallentyne, 1974) or in aquaculture systems (Boyd, 1990; Diana et al., 1991b). Unlike nitrogen, there are few problems with phosphorus toxicity, except when algal blooms occur as a result of excess phosphorus and cause depletion of oxygen. Phosphorus is mainly available to plants as orthophosphate but exists in several forms most commonly measured as total phosphorus or filterable reactive phosphorus if plant availability is being measured. Total phosphorus is relatively low in supply waters at most CRSP sites (usually 0 to 0.2 mg/L), but sometimes reaches higher levels in Indonesia and Honduras (up to 1.2 mg/L).

Early CRSP experiments indicated that fertilization with phosphorus alone at limited inputs resulted in low primary production (Cycle 1). Chicken manure was later utilized at higher rates as an input at all sites (Cycle 2), sometimes causing the ponds to become nitrogen-limited due to the surplus of phosphorus (McNabb et al., 1990; Diana et al., 1991b; Knud-Hansen et al., 1991). For most experiments, however, total phosphorus in pond waters rarely exceeded 0.5 mg/L and was often less than 0.1 mg/L. While much CRSP work has focused on nitrogen as a limit to primary production, less has been done with phosphorus. Combined inputs of both nitrogen and phosphorus are probably necessary to drive high levels of primary production.

One interesting characteristic of phosphorus is its binding with pond sediments. This characteristic was particularly noticeable in early CRSP fertilization with inorganic materials. Many of these treatments did not raise the water concentration of phosphorus by much but had a strong effect on the pond bottom. Some of these effects were noticeable for several subsequent experiments (Knud-Hansen, 1992), during which time this surplus phosphorus became redissolved in the water column. This binding and release of phosphorus from sediments has also been noted by Boyd (1990).

PLANKTON ACTIVITY

Standing Crops

Phytoplankton are probably the major instruments influencing water quality characteristics of ponds. Their utilization of nitrogen, phosphorus, and carbon was defined earlier. Photosynthesis results in production of oxygen that is critical for the life of animals in the pond. During the night, phytoplankton respiration also has a large influence on minimum oxygen levels at dawn. At times algae grow dramatically and develop a bloom, then later die and decompose, resulting in low oxygen levels, which may kill fish, or in some cases leaving potentially lethal toxins in the water. Also, the relative density of phytoplankton confines algal cells to the surface, limiting light penetration in most fertile ponds. For predominantly planktivorous fishes, like Nile tilapia, management of ponds to enhance primary productivity is critical to fish production.

Phytoplankton biomass in the water at any time results from a dynamic interaction among light, nutrient availability, growth and reproduction of plankton, and cropping by herbivores. Plankton can show dramatic changes in population density from day to day due to their short life cycle and ability to produce rapid changes in pond waters. Algal life cycles may be on the order of hours to days (Wetzel, 1983), so dramatic changes in species composition and abundance can occur very quickly. In aquaculture, most measures of phytoplankton standing crop do not attempt to measure individual species abundance but rather the entire standing crop of phytoplankton. The most common estimation method, used widely in the CRSP research, is to measure chlorophyll *a* concentration as an index of phytoplankton abundance.

Phytoplankton populations are highly responsive to nutrient concentrations and loading rates. In the Cycle 1 experiments for the CRSP, chlorophyll *a* concentrations were relatively low under low nutrient loading. Data from Thailand, Panama, and Rwanda all averaged about 10 to 20 mg/L chlorophyll *a*, while Indonesia had higher levels (40 to 50 mg/L) (Cycle 1 references). With higher fertilization rates, chlorophyll *a* contents increased to near 100 mg/L in Panama and Rwanda, exceeded 100 in Thailand (sometimes reaching 200 to 300 mg/L), remained at about 50 mg/L in Indonesia, and were relatively low in Honduras (about 20 mg/L). Ponds in Honduras were considered to be limited in primary productivity due to clay turbidity (Teichert-Coddington et al., 1992), which is reflected in the low plankton standing crops. Similarly, ponds in Indonesia were believed to be limited by low availability of inorganic carbon (McNabb et al., 1990), which was also reflected in the lack of an increase in chlorophyll *a* levels with increasing fertilization rates.

Fish farmers have often used pond color and turbidity as measures of phytoplankton standing crop (Lin, 1986; Boyd, 1990). For example, in many areas a common practice is to submerge one's hand into the water until it disappears. If the arm must be extended into the water beyond the elbow, it is time to fertilize the pond again. This practice has a basis in the relationship between density of phytoplankton and turbidity and has been more formally analyzed by Boyd (1979) and others, who examined the relationship between chlorophyll *a* content and Secchi disk depth. The latter relationship exists in CRSP ponds (see Figure 1 for an example), and while it is usually correlated with about 70% of the variation in plankton standing crop, it has less predictive value in CRSP ponds (r^2 = 0.03 to 0.4).

Photosynthesis Rate — Oxygen Production

The rate of photosynthesis is usually quite high in tropical aquaculture ponds, because of high nutrient availability, sunlight, and warm temperatures. Measurement of oxygen production is a convenient way to estimate the production of organic carbon by algae. The most common method of measurement is the light-dark bottle technique (Wetzel and Likens, 1991), where pond water is incubated in a clear (light) and an opaque (dark) bottle. Photosynthesis and respiration both occur in the light bottle, while only respiration occurs in the dark bottle. Net primary productivity (in mg O_2 per liter water per hour) equals the DO in the incubated light bottle, minus initial DO levels, divided by time. Respiration is the difference between DO of the initial sample and the sample incubated in the dark bottle, and gross primary production is the difference between samples incubated in the light and dark bottles. In unproductive natural waters, these bottles may have to be incubated for long times to get detectable differences. In ponds, the dark bottles may reach 0 DO or light bottles reach super saturation in a few hours. The resulting need for short incubation makes it difficult to extrapolate the productivity estimated to an overall daily value. For CRSP experiments, the productivity values measured in the morning are multiplied by a conversion factor based on the percent of the day's total sunlight that fell during the incubation period to calculate primary production for the daylight period (see Wetzel and Likens, 1991). This extrapolation, plus the rapid extinction of light with depth and several other factors, makes the light-dark bottle method difficult to use and interpret in aquaculture.

A second method commonly used for primary production is the free water or whole community method. In this method, changes of DO in pond water over a 24-h period, corrected for atmospheric diffusion, are used to calculate the rate of oxygen production (Hall and Moll, 1975). Usually, DO measures are taken during several time periods and at several depths and adjusted to give whole pond values each time. Measurements are necessary at least every 6 h over a 24-h period, with more frequent measurements near the daily peak time for DO. Another similar approach uses changes in pH and alkalinity, measured over 24 h, to calculate carbon uptake by chemical equilibria (Weisburd and Laws, 1990). Carbon uptake or productivity can also be estimated by converting O_2 to C using relative molecular weights and respiratory quotients.

Early CRSP experiments used the light-dark bottle method, while later ones have used the free water method. Comparisons indicate large discrepancies between the two methods over identical time periods (Figure 4), and no significant correlation exists between these two variables (in Figure 4; r^2 = 0.08, P >0.05). The free water method often produced higher values than the light-dark bottle method, both in Thailand (Figure 4) and Central America (Teichert-Coddington et al., 1992). Underestimation seems to be a consistent bias in the light-dark bottle method (Welch, 1968).

Levels of primary production in CRSP ponds are generally very high. Early experiments with low inputs gave production levels of 0.05 to 4 g $C/m^3/d$. With increasing inputs, primary production increased dramatically, reaching to 2 g $C/m^3/d$ in Rwanda, to 5 g in Honduras and Panama, and to 20 g in Thailand. Again, these were extrapolated from light-dark bottles and may be unrealistic values. However, levels of gross primary production up to 5 g $C/m^3/d$, measured by the free water method, are not uncommon in the CRSP experiments with good fertilization strategies.

Because phytoplankton are the main respirers of oxygen as well as the main producers (Boyd, 1990), their utilization of oxygen is significant in the oxygen dynamics of ponds. This factor is covered under oxygen dynamics below.

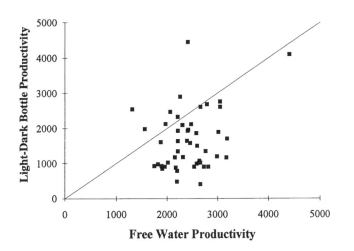

Figure 4. Relationship between light-dark bottle and free water methods for primary production (mg C/m^3/d) of ponds in Thailand during Cycle 1. Light-dark bottle measurements were made over 4 h in the morning. The line is an extrapolated equivalence line.

WATER TEMPERATURE

Tropical aquaculture is commonly done under conditions assumed to be more nearly constant in temperature than conditions in the temperate zone. However, tropical ponds warm dramatically during the day and cool at night, causing diel cycles. Also, most locations have wet and dry seasons, which may differ in temperature due to cloud cover and other factors. Daily temperature fluctuations make it difficult to compare temperature regimes, so maximum-minimum thermometers are most useful for this comparison. CRSP data indicate similar daily maximum and minimum water temperatures during wet and dry seasons at all sites or no predictable seasonal changes in temperature. Water temperatures are coolest at the high elevation site in Rwanda, with weekly maxima ranging around 22 to 30°C and minima ranging around 11 to 22°C. Panama and Honduras are intermediate, with maxima at 29 to 32°C and weekly minima at 22 to 25°C. Thailand is warmer still, with maxima at 30 to 36°C and minima at 24 to 29°C. Indonesia, the site closest to the equator, has the warmest temperatures, with maxima at 30 to 41°C and minima at 21 to 28°C. Thus, latitude and elevation have strong effects on water temperature conditions, and at all locations the water temperature changes dramatically from day to night, often as much at 10°C during one day. Temperature conditions were cool enough in Rwanda to inhibit growth and reproduction of tilapia, while conditions were best for tilapia growth in Indonesia and Thailand.

In addition to seasonal effects, water temperature can also exhibit much vertical stratification in shallow tropical ponds. In most CRSP ponds, water stratifies because of surficial warming during the day and limited winds. For example, Diana et al. (1991b) showed that pond water temperature had maximal stratification in Thailand at 1400 h, and was not stratified from about 1900 h until 0930 the next morning (Figure 5). Maximum differentials (difference between top and bottom water temperature) occurred at 1400 h, and these differentials were often 3 to 5°C between top and bottom waters. Stratification often broke down during the day because of rain or wind. During the night, stratification broke down due to cooling of surface waters. Stratification patterns in Indonesia, Honduras, and Panama appear similar to those in Thailand. In Rwanda, overall temperature is lower, but the difference in temperature with depth is even more severe, often with differentials of 5 to 7°C between top and bottom layers. At all sites, stratification occurred mainly from warming of surface waters, with little change in the temperature of the bottom water. Later CRSP experiments included more frequent measures during diel analyses, and maximum stratification occurred at either the 1600 or 1400 h sampling period.

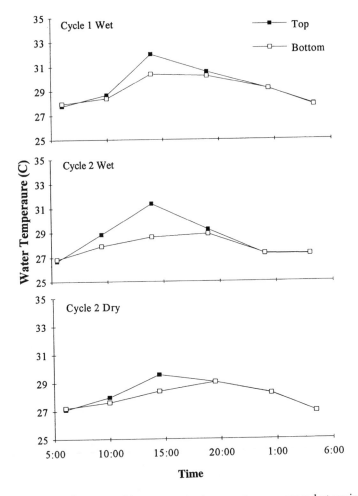

Figure 5. Average values for top and bottom water temperature measured at various times of day during Cycle 1 wet season and Cycle 2 wet and dry season experiments in Thailand.

DISSOLVED OXYGEN

Daily Fluctuations

Oxygen dynamics of semi-intensive aquaculture ponds are generally dominated by phytoplankton populations. These populations photosynthesize and respire during daylight, increasing oxygen concentrations; however, they only respire oxygen at night, reducing DO. Boyd (1990) compared values for benthic respiration, fish respiration, and biochemical oxygen demand of the water column (including plankton respiration) and found that the latter contributed most of the decline in DO at night, requiring up to 8.4 mg O_2/L water per night. Maximum mud respiration was only one-sixth of that value, and maximum fish respiration one-half. Teichert-Coddington and Green (1993c) also estimated that relative respiration of the water column (plankton) was 79% of total community respiration, benthos respiration was 17%, and fish respiration only 4% for tilapia ponds in Honduras.

Over the course of a day, oxygen levels can change even more dramatically than temperature. Diana et al. (1991b) showed oxygen variations of up to 10 mg/L from daytime lows to highs for surface waters of ponds in Thailand, and peak oxygen values over 200% saturation at about 1400 h (see Figure 6). Their data indicate increases of surface DO during daylight, with some reductions at night, but seldom did these ponds reach low DO, even in bottom waters at night. The pattern of

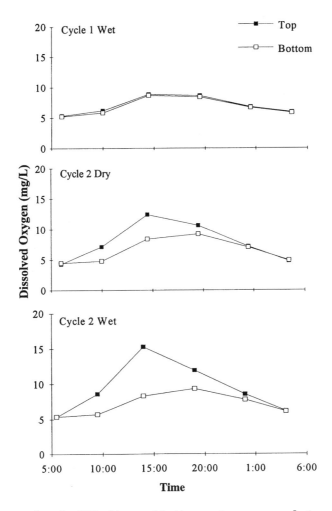

Figure 6. Average values for DO of top and bottom waters measured at various times of day during Cycle 1 wet season and Cycle 2 wet and dry season experiments in Thailand.

diel oxygen flux was similar in all other CRSP sites for Cycle 2 data, although some variations occurred. Indonesian ponds showed similar patterns, except that these ponds generally had somewhat lower surface and bottom DOs than Thai ponds. Ponds in Honduras, particularly in the wet season, had very low DOs and limited diel fluxes, probably due to the limitations on primary production caused by clay turbidity (Teichert-Coddington et al., 1992). Rwandan ponds had lower concentrations of oxygen in bottom waters, which correlated to stronger temperature stratification noted there, and lower surface DOs probably indicated reduced primary production compared to Thai and Indonesian ponds.

Depth Stratification

Diel changes in primary production and respiration also result in stratification of oxygen conditions in ponds. This has been mentioned briefly above in the section on diel oxygen changes. Diana et al. (1991b) found that oxygen differential (top to bottom) was dependent on mainly primary production ($r^2 = 0.54$), and no other measured variables correlated with degree of oxygen stratification. From this data on oxygen listed above, it appears that the degree of stratification at other CRSP sites is also dependent on primary production. However, persistent low bottom DOs at

Rwandan ponds, coupled with more stable temperature stratification mentioned earlier, indicate that these ponds may not always turn over at night. However, diurnal analyses in Cycle 3 experiments indicated that the ponds did turn over daily during that time period (Hanson et al., 1991).

Diffusion of Oxygen into Ponds

When pond water reaches conditions above or below saturation for oxygen, atmospheric diffusion can cause changes in dissolved oxygen levels. Boyd (1990) reviewed various pond conditions and predicted that diffusion of oxygen into ponds at night may reach 1.69 mg O_2/L in 12 h when ponds are only 50% saturated at dusk, while losses of oxygen can be up to 2.91 mg O_2/L in 12 h for ponds 250% saturated at dusk. Conditions in Figure 6 indicated diffusion losses would occur from 1000 h until 2000 h, while gains will occur overnight. Losses of DO are increased by oxygen stratification, which presents the most highly oxygenated water to the air–water interface where diffusion occurs. Wind also increases the rate of atmospheric diffusion with water.

Boyd and Teichert-Coddington (1992) evaluated reaeration of CRSP ponds in Honduras at different wind speeds. Ponds were deoxygenated by treatment with sodium sulfite and cobalt chloride, while formalin and copper sulfate were added to suppress biological activity in the ponds. Standard oxygen transfer coefficients were correlated to wind speed by the equation

$$K_La_{20} = 0.017x - 0.014 \tag{1}$$

where K_La_{20} is the standard oxygen transfer coefficient at 20°C (g O_2/m²/h) and x is wind speed in m/s (measured at 3 m in height above the pond). The regression was highly significant, with $r^2 = 0.882$. They also presented methods to calculate reaeration rates for ponds based on the above measures plus DO values for the pond and pond morphometry. These general formulas should allow calculation of atmospheric diffusion rates under various pond conditions at wind speeds under 5 m/s.

Mechanical Aeration

Mechanical aeration has been used by CRSP researchers in only a few cases. In ponds receiving high feed rates, lower DO levels and aeration are more common than in fertilized ponds. Boyd (1990) and others evaluated aerators and their efficiency for both continuous and emergency aeration. Most commonly these evaluations indicated that paddlewheel aerators perform best when used continuously at night.

CRSP experiments with supplemental feeds have also rarely used aeration. For example, Diana et al. (1994a) and Green (1992) found no reductions in fish growth or survival due to low DO when feed and fertilizer were added to ponds with up to three fish per m². Teichert-Coddington and Green (1993a) evaluated aeration with supplemental feeding and found significant improvements in yield and growth for tilapia when ponds were aerated after the concentrations of DO fell to below 10% saturation (emergency aeration). Many times in semi-intensive CRSP ponds, DO concentrations reached these levels; possibly the production of tilapia in these ponds could be improved by aeration.

EFFECTS OF DISSOLVED OXYGEN CONCENTRATION ON TILAPIA IN PONDS

One major characteristic of water quality that is important to the growth and survival of fish is DO. Oxygen consumption by fish and DO requirements increase with temperature and food consumption (Diana, 1995). As intensification increases, DO can often become a limiting factor.

While tilapia are generally tolerant of low DO (Chervinski, 1982; Coche, 1982), extended periods of hypoxia may cause reduced growth (Chervinski, 1982; Teichert-Coddington and Green, 1993a) and increased mortality (Coche, 1982), particularly when algal blooms occur. Many fish have shown reduced growth or survival when DO conditions fall below levels as high as 5 mg/L (Brett, 1979), or 25 to 50% saturation (Rappaport et al., 1976). However, there appear to be no guidelines as to what level of DO causes reductions in growth of tilapia (Teichert-Coddington and Green, 1993b).

One CRSP experiment addressed this effect (Teichert-Coddington and Green, 1993b). They found that aeration when DO levels were reduced to below 10% of saturation increased growth and yield of tilapia reared with supplemental feed. There were no differences between treatments where aeration was initiated at 10 or 30% saturation, indicating that aeration once DO fell to 10% of saturation was sufficient to maintain growth. Aeration also increased mud turbidity in their Honduran ponds, indicating that aeration frequency should be minimized or primary production might also suffer, further damaging oxygen production. Fish yield was about 17% lower in ponds without aerators. There were no differences in survival.

AUTOMATED DATA COLLECTION SYSTEMS

Automated monitoring of water quality parameters permits many measurements more finely resolved in space and time than manual measurements. This section provides brief background on the use of automated water quality monitoring and control systems begun in CRSP studies in 1989. Similar hardware and concepts are also used in both pond culture and smaller, more intensive systems.

Automated monitoring systems can have diverse configurations. There are two major groups of dichotomous choices for design (Ebeling and Losordo, 1989). One design choice is between deployment of sensors in the pond (Erez et al., 1990; Weisburd and Laws, 1990) or deployment of sampling pumps to deliver pond water to land-based sensors (Ebeling and Losordo, 1989). Pond deployment avoids problems with pumping and requires transmission only of electronic signals but is less convenient for calibration and maintenance (Cathcart and Wheaton, 1987; Szyper and Lin, 1990). Development of better sensors, particularly for dissolved oxygen, may well make this strategy more workable and prevalent in the future. Systems based on land-mounted sensors require care that the sample delivery system does not affect water quality. The second design choice is between transmitting signals directly to a microcomputer or to a data logger for later transfer. A logger is probably better suited to outdoor deployment. CRSP research has been conducted mainly with the land-based receivers and data loggers, based on modifications of the "Pond System" described by Ebeling and Losordo (1989).

Temperature Monitoring

In freshwater ponds, temperature can be recorded with commonly used data loggers directly from exposed thermocouple ends or simple stainless steel probes containing the wire ends. The system should contain internal calibration factors for the various bimetallic combinations. Thus, measurement is extremely inexpensive in land-based receivers or pond deployments. Exposed wire ends can remain unfouled for several weeks given daily brief cleaning, and sensors can be readily checked against standard thermometers.

Szyper and Lin (1990) evaluated natural diel cycles of thermal stratification and mixing in freshwater ponds and the response of the ponds to artificial mixing. When daytime stratification was very intense (differences of 5 to 6°C between top and bottom), natural nighttime conditions generally failed to mix water deeper than 1 m. These results were confirmed in ponds of 2.5 to 3 m depth in northeast Thailand, and show that the typical farm reservoir ponds often used for fish culture contain large hypolimnetic volumes of persistently isolated water (Figure 7).

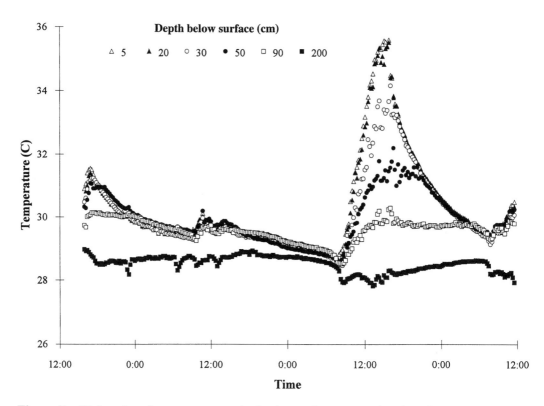

Figure 7. Diel cycles of temperature at six depths on three successive dates in a pond of 2.5 m depth at Huay Luang Fisheries Station in Thailand. The first two dates were cloudy and rainy, and showed little vertical temperature difference compared with the third (sunny) day. Note the continuous isolation of water at 200 cm depth.

Dissolved Oxygen Monitoring

Measurement of dissolved oxygen requires a more complex sensor than measurement of temperature. Most often a polarographic membrane probe is connected to a YSI oxygen meter, which sends millivolt output to the data logger. In a land-based receiver, the probe is readily maintained daily by light brushing and air calibration, which permits deployment for weeks before membrane replacement is needed. Depth resolution is limited by the number of sampling pumps that can be deployed in a pond. Green and Teichert-Coddington (1991) compared a column sampler to individual pump placement and found good correlation for pH, DO, and temperature. An efficient apparatus for in-pond sampling can be made with up to 10 pumps per meter depth. Comparisons of manual or automatic measurements were made on two dates at five depths (n = 10). Manual measurements by DO meter differed from logger system readings by 0 to 0.15 mg/L, with DO ranging from 1 to 7 mg/L. Time resolution for land receiver systems is limited by the necessity to pump water from a given pump over the sensors long enough to create a stable reading (about 60 s); thus a five-pump system can be cycled up to 12 times per hour, or such a system could sample multiple ponds each hour.

In an experiment on stratification and mixing, Szyper and Lin (1990) assessed temperature using a series of thermocouple ends attached to a plastic pipe at 8 depths (intervals of 10 to 30 cm), with temperature at each depth recorded every 15 min. Dissolved oxygen was measured by pumping pond water to a land-based receiver from two depths (10 and 130 cm). Stratification was substantially modified by pumping hypolimnetic water from about 80 cm depth to the surface with a 0.5 hp electric pump for 1 to 2 h during the day (Figure 8). Under some conditions, the deepest water was not affected by the artificial mixing. A simple temperature-exposure index (degree-hours)

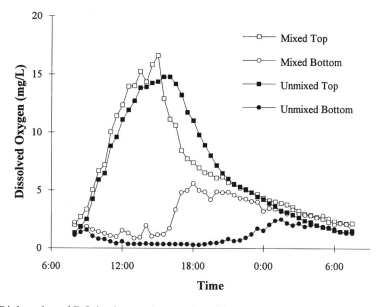

Figure 8. Diel cycles of DO in the top (squares) and bottom (circles) waters of two neighboring ponds of 1.5 m depth at the Asian Institute of Technology in Thailand. One pond (the treatment, open symbols) was mixed with a submersible pump from 1500 to 1700 h, while the other pond was unaltered (control, closed symbols). (Redrawn from Szyper, J. P. and Lin, C. K., 1990.)

was used to quantify the warming effect of mixing on the bottom layer and on the whole water column. Despite intense stratification, the warming effect of mixing in these ponds amounted only to an approximate 5% increase in degree-hours in bottom waters.

In contrast to temperature, well-timed mixing created rapid relief of DO depletion in bottom layers of a stratified pond (Figure 8). Two hours of mixing raised DO levels from about 1 mg/L to 5 mg/L at 1.3 m depth; daily DO exposure (indexed as "ppm-hours") was increased by factors of 2.5 to 8 times those of control ponds.

Short-interval DO monitoring permits detailed study of the diel cycle of DO production by photosynthesis and of respiratory consumption by all organisms. The CRSP diel sampling protocol, in which samples are taken at multiple depths 6 or 7 times daily, makes adequate assessment of net primary production (NPP), nighttime community respiration (nR), and the total daily DO balance. Gross primary production (GPP) is more difficult to assess because it includes both daytime community respiration (dR) and daytime NPP. GPP is the most appropriate quantity for assessment of photosynthetic dependence on light, knowledge of which is critical for model development (Giovannini and Piedrahita, 1990). Because calculation of GPP requires estimation of dR (which may not be equal to nR), it cannot be measured by simple water sampling.

Community respiration varies considerably during the diel cycle (Figure 9). In fertile ponds, nR often decreases with time after sunset, reaching low rates between midnight and dawn (Szyper et al., 1992), presumably because substrates such as intracellular glucose become depleted (Weger et al., 1989). Total dR is in general much higher than nR (Figure 9), with significant variation even during the day (Szyper et al., 1992; Teichert-Coddington and Green, 1993b; Giovannini, 1994).

Monitoring of Other Parameters

Automated pond monitoring has also been applied to salinity, pH, chlorophyll *a*, and total ammonia. Periodic delivery of water samples from defined depths creates the potential to study

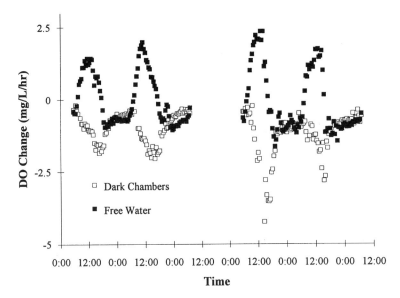

Figure 9. Rates of change in DO in a pond at the University of Hawaii during two 2-d study periods (free water rates, filled symbols), and rates of change during 20-min incubations in dark chambers (open symbols). Free water rates represent daytime net primary production during the day, and community respiration at night. Respiration chamber rates represent community respiration at all times. (Redrawn from Szyper, J. P. et al., 1992.)

diel cycles and depth distributions of sugars, amino acids, particulate carbon, nitrogen, and proximate composition of particulates. Fluorescence of phytoplankton chlorophyll *a* has been monitored by placing a flow-through fluorometer (Turner Designs model 10) in a receiver. Szyper and Ebeling (1993) used this method to establish that the large phytoplankton stock (about 400 mg chlorophyll *a* per liter) in a pond remained stable during a 2-wk study period. Vertical light extinction was sufficiently severe at 35 cm depth to prevent a diel cycle of cell fluorescence.

Automated Control

Commercial monitoring systems, as well as the data logger used in CRSP research, can transmit sensor responses to devices such as aerators and alarm systems, in response to set values of monitored parameters of water samples. Commercial systems are used in hatcheries and intensive production systems for this purpose, most often to actuate alarms in response to low water levels or DO.

Teichert-Coddington and Green (1993a) used the logger system's control capability to activate aerators when DO concentrations reached 10 or 30% saturation at the Honduras CRSP site. This experiment, described earlier, improved growth and yield of tilapia with emergency aeration. Aeration regulated by DO sensors could improve yields at lower power costs than continuous, nightly aeration. Aeration control could be combined with controlled daytime mixing to optimize a pond's DO regime. A pond could be mixed under controls using a lower power pump regulated by a monitor to prevent DO from exceeding saturation values during daylight hours, reducing diffusive loss of oxygen to the atmosphere. Szyper and Hopkins (1993) reported that ponds mixed for this purpose (though at regular hours each afternoon) had higher bottom DO levels than control ponds for at least part of each night. Such mixing would reduce the amount of time during which a DO-controlled aerator would operate.

REFERENCES

Almazan, G. and Boyd, C. E., Plankton production and tilapia yield in ponds, *Aquaculture*, 15, 75, 1978.

Avnimelech, Y., Troeger, B. W., and Reed, L. W., Mutual flocculation of algae and clay: evidence and implications, *Science*, 216, 63, 1982.

Boyd, C. E., *Water Quality in Warmwater Fish Ponds*, Agriculture Experiment Station, Auburn University, Auburn, AL, 1979.

Boyd, C. E., *Water Quality in Ponds for Aquaculture*, Alabama Agricultural Experiment Station, Auburn University, Auburn, AL, 1990.

Boyd, C. E. and Teichert-Coddington, D., Relationship between wind speed and reaeration in small aquaculture ponds, *Aquacultural Eng.,* 11, 121, 1992.

Brett, J. R., Environmental factors and growth, in Hoar, W. S., Randall, D. J., and Brett, J. R., Eds., *Fish Physiology*, Academic Press, New York, 1979, chap. 10.

Cathcart, T. P. and Wheaton, F. W., Modeling temperature distribution in freshwater ponds, *Aquacultural Eng.*, 6, 237, 1987.

Chervinski, J., Environmental physiology of tilapias, in *The Biology and Culture of Tilapias*, ICLARM Conference Proceedings 7, Pullin, R. S. V. and Lowe-McConnell, R. H., Eds., International Center for Living Aquatic Resources Management, Manila, 1982, 119.

Coche, A. G., Cage culture of tilapias, in *The Biology and Culture of Tilapias*, ICLARM Conference Proceedings 7, Pullin, R. S. V. and Lowe-McConnell, R. H., Eds., International Center for Living Aquatic Resources Management, Manila, 1982, 205.

Diana, J. S., *Biology and Ecology of Fishes*, Biological Sciences Press, Carmel, CA, 1995.

Diana, J. S. and Lin, C. K., The effects of fertilization on growth and production of Nile tilapia in rain-fed ponds, in Fourteenth Annual Technical Report, Egna, H., Burke, D., Goetze, B., and Clair, D., Eds., Pond Dynamics/CRSP Management Office, 1996, in press.

Diana, J. S., Lin, C. K., Bhukaswan, T., and Sirsuwanatach, V., Thailand Project, Cycle I of the Global Experiment, *Pond Dynamics/Aquaculture Collaborative Research Data Reports*, Volume 2, Number 1, Oregon State University, Corvallis, 1987.

Diana, J. S., Lin, C. K., Bhukaswan, T., Sirsuwanatach, V., and Buurma, B., Thailand Project, Cycle II of the Global Experiment, *Pond Dynamics/ Aquaculture Collaborative Research Data Reports,* Volume 2, Number 2, Oregon State University, Corvallis, 1991a.

Diana, J. S., Lin, C. K., and Jaiyen, K., Supplemental feeding of tilapia in fertilized ponds, *J. World Aquaculture Soc.*, 25, 497, 1994a.

Diana, J. S., Lin, C. K., and Schneeberger, P. J., Relationships among nutrient inputs, water nutrient concentrations, primary production, and yield of *Oreochromis niloticus* in ponds, *Aquaculture*, 92, 323, 1991b.

Diana, J. S., Lin, C. K., and Yi, Y., Stocking density and supplemental feeding for tilapia production, in *Twelfth Annual Technical Report,* Egna, H., Bowman, J., Goetze, B., and Weidner, N., Eds., Pond Dynamics/CRSP Management Office, Corvallis, 1994b, 153.

Ebeling, J. E. and Losordo, T. M., Continuous environmental monitoring systems for aquaculture, in *Instrumentation in Aquaculture*, Proceedings of a Special Session at the 1989 Meeting of the World Aquaculture Society, Wyban, J. A. and Antill, E., Eds., The Oceanic Institute, Honolulu, 1989, 54.

Erez, J., Krom, M. D., and Neuwirth, T., Daily oxygen variations in marine fish ponds, Elat, Israel, *Aquaculture*, 84, 289, 1990.

Giovannini, P., Water Quality Dynamics in Aquaculture Ponds: an Investigation of Photosynthetic Production and Efficiency Variations, Ph.D. Dissertation, University of California, Davis, 1994.

Giovannini, P. and Piedrahita, R. H., Measuring the Primary Production Efficiency of Aquaculture Ponds, American Society of Agricultural Engineers Paper No. 90-7034, *ASAE*, St. Joseph, MO, 1990.

Giovannini, P. and Piedrahita, R. H., Modeling net photosynthetic production optimization for aquaculture ponds, *Aquacultural Eng.*, 13, 83, 1994.

Green, B. W., Substitution of organic manure for pelleted feed in tilapia production, *Aquaculture,* 101, 213, 1992.

Green, B. W., Alvarenga, H. R., Phelps, R. P., and Espinoza, J., Honduras Project, Cycle I of the Global Experiment, *Pond Dynamics/Aquaculture Collaborative Research Data Reports*, Volume 6, Number 1, Oregon State University, Corvallis, 1990a.

Green, B. W., Alvarenga, H. R., Phelps, R. P., and Espinoza, J., Honduras Project, Cycle II of the Global Experiment, *Pond Dynamics/Aquaculture Collaborative Research Data Reports,* Volume 6, Number 2, Oregon State University, Corvallis, 1990b.

Green, B. W., Phelps, R. P., and Alvarenga, H. R., The effect of manures and chemical fertilizers on the production of *Oreochromis niloticus* in earthen ponds, *Aquaculture,* 76, 37, 1989.

Green, B. W. and Teichert-Coddington, D. R., Comparison of two samplers used with an automated data acquisition system in whole-pond, community metabolism studies, *Progr. Fish-Cult.,* 53, 236, 1991.

Hall, C. A. S. and Moll, R., Methods of assessing aquatic primary productivity, in Leith, H. and Whittaker, R. H., Eds., *Primary Productivity of the Biosphere,* Springer-Verlag, New York, 1975, 19.

Hanson, B., Ndoreyaho, V., Rwangano, F., Tubb, R., and Seim, W. K., Rwanda Project, Cycle III of the Global Experiment, *Pond Dynamics/Aquaculture Collaborative Research Data Reports,* Volume 5, Number 3, Oregon State University, Corvallis, 1991.

Hanson, B., Ndoreyaho, V., Tubb, R., Rwangano, F., and Seim, W. K., Rwanda Project, Cycle I of the Global Experiment, *Pond Dynamics/Aquaculture Collaborative Research Data Reports,* Volume 5, Number 1, Oregon State University, Corvallis, 1989.

Hughes, D. G., Diaz, A. T., Phelps, R. P., and Malca, R. P., Aguadulce, Panama Project, Cycle I of the Global Experiment, *Pond Dynamics/Aquaculture Collaborative Research Data Reports,* Volume 8, Number 1, Oregon State University, Corvallis, 1991a.

Hughes, D. G., Phelps, R. P., and Malca, R. P., Aguadulce, Panama Project, Cycle II of the Global Experiment, *Pond Dynamics/Aquaculture Collaborative Research Data Reports,* Volume 8, Number 2, Oregon State University, Corvallis, 1991b.

Knud-Hansen, C. F., Pond history as a source of error in fish culture experiments: a quantitative assessment using covariate analysis, *Aquaculture,* 105, 21, 1992.

Knud-Hansen, C. F. and Batterson, T. R., Effect of fertilization frequency on the production of Nile tilapia (*Oreochromis niloticus*), *Aquaculture,* 123, 271, 1994.

Knud-Hansen, C. F., Batterson, T. R., McNabb, C. D., Harahat, I. S., Sumantadinata, K., and Eidman, H. M., Nitrogen input, primary productivity and fish yield in fertilized freshwater ponds in Indonesia, *Aquaculture,* 94, 49, 1991.

Knud-Hansen, C. F. and Pautong, A. K., On the role of urea in pond fertilization, *Aquaculture,* 114, 273, 1993.

Lin, C. K., Acidification and reclamation of acid sulfate soil fishponds in Thailand, in *The First Asian Fisheries Forum,* MacLean, J. L., Dizon, L. B., and Hosillos, L. V., Eds., Asian Fisheries Society, Manila, 1986, 71.

Lin, C. K., Tansakul, V., and Apinhapath, C., Biological nitrogen fixation as a source of nitrogen input in fish ponds, in *The Second International Symposium on Tilapia in Aquaculture,* Pullin, R. S. V., Bhukaswan, T., Tonguthai, K., and Maclean, J. L., Eds., International Center for Living Aquatic Resources Management, Manila, 1988, 53.

McNabb, C. D., Batterson, T. R., Premo, B. J., Eidman, H. M., and Sumatadinata, K., Indonesia Project, Cycle I of the Global Experiment, Pond Dynamics/Aquaculture Collaborative Research Data Reports, Volume 3, Number 1, Oregon State University, Corvallis, 1988.

McNabb, C. D., Batterson, T. R., Premo, B. J., Eidman, H. M., and Sumatadinata, K., Indonesia Project, Cycle II of the Global Experiment, *Pond Dynamics/ Aquaculture Collaborative Research Data Reports,* Volume 3, Number 2, Oregon State University, Corvallis, 1991.

McNabb, C. D., Batterson, T. R., Premo, B. J., Knud-Hansen, C. F., Eidman, H. M., Lin, C. K., Jaiyen, K., Hanson, J. E., and Chuenpagdee, R., Managing fertilizers for fish yield in tropical ponds in Asia, in *The Second Asian Fisheries Forum,* Proceedings of the Asian Fisheries Society, Hirano, R. and Hanyu, I., Eds., Manila, 1990, 169.

Rappaport, U., Darig, S., and Marek, M., Results of tests of various aeration systems on the oxygen regime in the Genosar experimental ponds and growth of fish there in 1975, *Bamidgeh,* 28, 35, 1976.

Setaro, F. V. and Melack, J. M., Responses of phytoplankton to experimental nutrient enrichment in an Amazon floodplain lake, *Limnol. Oceanogr.,* 14, 799, 1984.

Szyper, J. P. and Ebeling, J. M., Photosynthesis and community respiration at three depths during a period of stable phytoplankton stock in a eutrophic brackish water culture pond, *Mar. Ecol. Progr. Ser.,* 94, 229, 1993.

Szyper, J. P. and Hopkins, K. D., Effects of pond depth and mechanical mixing on production of *Oreochromis niloticus* in manured earthen ponds, in *The Third International Symposium on Tilapia in Aquaculture,* Pullin, R. S. V., Lazard, J., Legendre, M., and Amon Kothias, J. B., Eds., 1993.

Szyper, J. P. and Lin, C. K., Techniques for assessment of stratification and effects of mechanical mixing in tropical fish ponds, *Aquacultural Eng.,* 9, 151, 1990.

Szyper, J. P., Rosenfeld, J. Z., Piedrahita, R. H., and Giovannini, P., Diel cycles of planktonic respiration rates in briefly-incubated water samples from a fertile earthen pond, *Limnol. Oceanogr.,* 37, 1193, 1992.

Teichert-Coddington, D. R. and Green, B. W., Tilapia yield improvement through maintenance of minimal oxygen concentrations in experimental grow-out ponds in Honduras, *Aquaculture,* 118, 63, 1993a.

Teichert-Coddington, D. R. and Green, B. W., Influence of daylight and incubation interval on water column respiration in tropical fish ponds, *Hydrobiologia,* 250, 159, 1993b.

Teichert-Coddington, D. R. and Green, B. W., Comparison of two techniques for determining community respiration in tropical fish ponds, *Aquaculture,* 114, 41, 1993c.

Teichert-Coddington, D. R., Green, B. W., and Phelps, R. P., Influence of site and season on water quality and tilapia production in Panama and Honduras, *Aquaculture,* 105, 297, 1992.

Vallentyne, J. R., Jr., The Algal Bowl, Miscellaneous Special Publication 22, Department of the Environment, Ottawa, Canada, 1974.

Weger, H. G., Herzig, P., Falkowski, P. G., and Turpin, D. H., Respiratory losses in the light in a marine diatom: measurements by short-term mass spectrometry, *Limnol. Oceanogr.,* 34, 1153, 1989.

Weisburd, R. S. J. and Laws, E. A., Free water productivity measurements in leaky mariculture ponds, *Aquacultural Eng.,* 9, 377, 1990.

Welch, H. E., Use of modified diurnal curves for the measurement of metabolism in standing water, *Limnol. Oceanogr.,* 13, 679, 1968.

Wetzel, R. G., *Limnology,* 2nd ed., Saunders College Publishing, Philadelphia, PA, 1983.

Wetzel, R. G. and Likens, G. E., *Limnological Analyses,* 2nd ed., Springer-Verlag, New York, 1991.

Zaret, T. M., Devol, A. H., and Santos, A. D., Nutrient addition experiments in Lago Jacaretinga, Central Amazon Basin, Brazil, *Proceedings of the International Association of Theoretical and Applied Limnologists,* 21, 721, 1981.

4 FERTILIZATION REGIMES

C. Kwei Lin, David R. Teichert-Coddington, Bartholomew W. Green,
and Karen L. Veverica

C. Kwei Lin, David R. Teichert-Coddington, Bartholomew W. Green,
and Karen L. Veverica

INTRODUCTION

Pond fertilization to increase fish yields has long been practiced throughout the world. It is a well-known tradition in China to utilize animal manures and human excreta as major sources of pond inputs for polyculture of the major Chinese carps (Ling, 1967; Wohlfarth and Schroeder, 1979). The uses of inorganic fertilizer were introduced more recently in temperate regions (Mortimer, 1954; Gooch, 1967). There is voluminous literature on pond fertilization, documenting many conflicting and inconsistent results based on various types of fertilizer, rates of input, and methods and frequency of application (Coleman and Edwards, 1987). Those controversial viewpoints may actually reflect the differences in the physical and chemical environments of experimental ponds as well as variations in cultured fish species and stocking densities. Some of the problems have also stemmed from the lack of proper statistical designs with sufficient replication and common protocols for pond fertilization experiments.

A major thrust of the PD/A CRSP during the past 14 years has been to develop a data base for pond dynamics and fertilization management strategies for pond culture. The common global experiments were conducted primarily on pond fertilization and its impact on water quality and fish yield. The practical goal was to provide fish growers with sound strategies and guidelines for pond fertilization. Standardized fertilizer experiments were carried out during 2 to 3 years at sites in Honduras, Indonesia, Rwanda, Panama, Philippines, and Thailand. Work plans for global experiments consisted of simple inputs of phosphorus to calibrate sites, comparison of inorganic and organic fertilizers, and comparison of various loading rates of organic fertilizers. Some site-specific experiments were also conducted, depending on the needs of the individual site and the perspective of the researchers. Standardized experimental design and work plan protocols were followed at all sites (Egna et al., 1987). Results from the various sites are comparable because Nile tilapia (*Oreochromis niloticus*) was used as the cultured species with specified 5-month grow-out cycles, during which measurements of water quality, fish sampling, and methods of fertilizer application were standardized.

PRINCIPLES OF POND FERTILIZATION

A large number of nutrients is required to stimulate phytoplankton growth. Those nutrients include major elements (C, N, K, Si, Ca, P, Mg, S, Cl) and trace elements (Fe, Mn, Zn, B, Cu, Mo, Co). Background concentrations of those nutrients in pond water are mostly derived from air, soil, source water, and rains, and these elements are normally present in limited amounts, especially the major nutrients. To maintain adequate phytoplankton production as natural food to support desirable fish yield requires fertilization by adding nutrients in either organic or inorganic forms. Theoretically, pond fertilization should be based on Liebig's law of minimum, which states that

0-56670-274-7/97/$0.00+$.50
© 1997 by CRC Press LLC

biological production is limited by the nutrient in least supply. Under limited nutrient supply, the rate of nutrient uptake by phytoplankton is concentration dependent, and total phytoplankton production is directly proportional to the initial concentration of limiting nutrient (Parsons and Takahashi, 1973). In practice, however, optimal pond fertilization can be an extremely complex matter, due to the dynamics of intrinsic and extrinsic factors in pond ecosystems. For example, the differences in nutrient requirements of individual algal species lead to a need to provide a balanced supply of major nutrients.

Nitrogen, phosphorus, and occasionally carbon are the most common limiting nutrients to phytoplankton production in natural waters and fish ponds. Under optimal growth conditions, the average nutrient compositions of phytoplankton biomass are approximately 45 to 50% C, 8 to 10% N, and 1% P, giving a typical C:N:P ratio of about 50:10:1 (Goldman, 1980). The minimal concentration of a given nutrient required to meet the optimal growth is referred to as "critical concentration." Algae are able to continue to uptake nutrients above this concentration and deposit the surplus nutrient(s) in their cells without concomitant growth, which is termed "luxury consumption." This phenomenon is particularly well known for phosphorus uptake (Kuhl, 1962). Based on a review of a wide range of published data, Coleman and Edwards (1987) concluded that a strong correlation exists between primary productivity and fish yield in ponds, and that the practical upper limit for net primary productivity is 4 g C/m^2/d or 10 to 12 g C/m^2/d for gross productivity. Under theoretically optimal conditions, to maintain this level of productivity requires 8 kg N and 0.8 kg P/ha/d. However, the nutrient inputs required to achieve such productivity in ponds is expected to be less than the theoretical nutrient quota in phytoplankton biomass for a number of reasons. First, unlike biomass production systems where phytoplankton biomass is totally harvested with all nutrients removed, the algal standing crop in fish ponds is continuously and partially consumed by filter feeding fish, such as Nile tilapia and other herbivorous organisms in the pond ecosystem. Thus, the amount of nutrients required is only to compensate for algal growth and the losses through grazing and sinking. Furthermore, only a fraction of the nutrients ingested by fish and other organisms is assimilated, and a large portion is recycled back to ponds as wastes.

The N:P ratio is an important consideration in pond fertilization. Although the typical N:P ratio in algal biomass is roughly 10:1, the fertilizer composition for pond input deviates considerably from this ratio because the natural cycles of N and P in ponds differ markedly. Nitrogen is considered to be less limited in an aquatic environment because of the input from natural nitrogen fixation through biological and atmospheric processes. Those inputs were earlier thought to be sufficient for low levels of pond production, and nitrogen fertilization was regarded as unnecessary (Hickling, 1962; Lin, 1968). In contrast, phosphorus is limited in most natural waters, as its source depends primarily on weathering of phosphorus rocks, which are not ubiquitous in the lithosphere. Furthermore, the availability of dissolved phosphorus in natural water is curtailed by its rapid reaction with cations (e.g., Fe, Ca, Mg, Al) forming relatively insoluble precipitates, which are unavailable to biota (Schlesinger, 1991). Phosphorus supply to the water column is also removed by pond muds, which strongly adsorb phosphorus (Boyd and Musig, 1981); mud adsorption capacity of P is linearly correlated to clay content of the sediments (Madhav and Lin, in press, a). Thus, the amount of phosphorus required for pond fertilization is influenced by the type of bottom soils and their %P saturation factors, which are often related to the fertilization history of the ponds (Knud-Hansen, 1992). Madhav and Lin (in press, b) demonstrated that the differential productivity of old and new ponds at AIT could be equalized by fertilizing the ponds based on their P saturation level and the percentage of clay content of bottom muds in the top 5 cm layer. In conclusion, even though a general guideline has been recommended (Knud-Hansen et al., 1993), the N:P ratio in pond fertilization is extremely variable and site-specific; therefore it can best be determined experimentally.

Surplus input of phosphorus fertilizer in ponds often results in phytoplankton "luxury consumption," forming intracellularly condensed inorganic polyphospates. In the case of deficient nutrient supply from ambient water, those reserves can serve as a potential source of nutrient for further growth through phosphate hydrolysis (Kuhl, 1974; Lin, 1978). In contrast, excessive nitrogen input in forms of ammonia or urea may lead to grave consequences caused by ammonia toxicity to fish. Upsurges in ammonia concentration are expected to occur immediately after application of those fertilizers, as they dissolve rapidly in pond water. For example, ammonia concentration in pond water may reach the stress level to tilapia of 0.5 mg/L upon a single input of ammonia fertilizer at 5 kg/ha. Many factors may cause slow and inefficient nitrogen uptake by phytoplankton, such as unfavorable weather conditions, excessive mud turbidity, and deficiencies of other nutrients. When the appropriate N:P ratio for a fertilization regime in a given pond is not known, a surplus of phosphorus is safer than a surplus of nitrogen. Nitrogen input should be lowered or postponed whenever the ammonia concentration in pond water exceeds 0.5 mg/L before fertilization. For this reason, single grade fertilizer, like urea or superphosphate, provides a greater flexibility than that of mixed grades, particularly when only nitrogen or phosphorus is needed.

Economic viability is an essential consideration for pond fertilization. Fish production can increase linearly with incremental fertilizer input in nutrient-limited ponds. However, the biomass gain per unit of fertilizer input is likely to decrease gradually with greater input, and fertilization may reach the point of diminishing return, where the economic return from fish weight gain is less than the cost of additional fertilizer input. Excessive amounts of fertilization are not only wasteful of materials but also contribute to undesirable water quality in ponds as well as in effluents discharged to natural waters.

The fertilization program for fish farmers must consider the availability and cost of source materials. For example, the single grade fertilizers (i.e., urea and TSP) may be more cost effective than mixed grades, but they are often less available in areas where the latter is commonly sold for targeted land crops. Most often, the livestock and poultry wastes are readily available on-farm fertilizers for small-scale farmers; chemical fertilizers, which require cash to purchase, may not be a viable choice for subsistence farmers. The application of animal wastes to fish ponds depends also on social and cultural acceptability. For instance, fish culture using poultry and livestock manures is commonly practiced and accepted by consumers in Thailand and many other countries. Use of human excreta, though common in China and Vietnam, is less acceptable in many other countries. Furthermore, waste products from swine are a taboo as pond inputs for certain cultures and religions.

CHEMICAL FERTILIZATION OF TILAPIA PONDS

Chemical fertilizers applied to fish ponds are analogous to those used for agriculture field crops, and their availability and variety are mostly dependent on the needs of local crop farmers. However, unlike fertilizers for field crops, which commonly include phosphorus, nitrogen, and potassium in various proportions, fertilizers used for fish ponds usually exclude potassium. The needs for potassium and other micronutrients for pond fertilization have rarely been investigated experimentally, due perhaps to earlier reports, such as Hickling (1962) that potassium fertilization was ineffective in pond production. Most pond fertilization experiments have emphasized N and P inputs. A number of chemical fertilizers with various ratios of N and P are used for land crops (Table 1). However, the availability and distribution of chemical fertilizers vary from place to place, depending mostly on the stage of agricultural development and the needs of land crops.

Traditionally, commercial fertilizers are formulated on the basis of percentages by weight of nitrogen (N), phosphorus (P_2O_5), and potassium (K_2O). For example, 20-16-8 grade fertilizer contains 20% N, 16% P_2O_5, and 8% K_2O, with an actual N, P, K content of 20, 7, and 6.6%, respectively. Those inorganic fertilizers are made in prilled or granular forms. The primary N and P nutrients contained in most compound fertilizers dissolve in water in ionic forms of nitrate (NO_3), ammonium (NH_4), and orthophosphate (PO_4). The solubility of fertilizers is an important feature in pond fertilization, as phytoplankton take up nutrients from water primarily in dissolved ionic

Table 1   Approximate Nutrient Content of
Commonly Used Chemical Fertilizers for
Agricultural Crops

Compound	Nutrient content (% weight)	
	N	P
Urea	45	0
Ammonium nitrate	35	0
Superphosphate	0	10
Triple superphosphate (TSP)	0	22
Diammonium phosphate (DAP)	18	24

forms. In general, the dissolution of phosphorus in granule fertilizers is considerably slower than that of nitrogen. Boyd (1976) determined that less than 20% of phosphorus and greater than 60% of nitrogen dissolved when various solid fertilizers sank through 2 m of water column (Table 2). Undissolved particles presumably settle and mix with muds once they sink to the pond bottom. Although not much has been documented on their pathways and fates, those residual fertilizers would eventually dissolve and react with the chemical constituents in the sediments. The resultant processes lead to either temporary or permanent loss of nutrients available in the water column. To overcome the solubility problems of conventional solid fertilizers, liquid types have recently been demonstrated to be more soluble and thus efficient for pond fertilization (Boyd, 1990). Solubility of granule fertilizers, such as monoammonium phosphate, can be enhanced by finely pulverizing the granules, making it as effective as liquid fertilizers (Rushton and Boyd, 1996). A more convenient product still is the controlled, slow-releasing fertilizers, which can be suspended in a water column over a period of months, reducing application frequency from weeks to months with similar results.

The importance of nitrogen fertilizer for fish ponds has been questioned as a result of several studies that failed to demonstrate significant benefits to fish production with applications of nitrogen (Mortimer, 1954; Hickling, 1962; Swingle et al., 1965; Boyd and Sowles, 1978; Boyd, 1981). Boyd (1976) tested the effects of nitrogen fertilizer on production of *Oreochromis aureus* in an experiment with 15 applications of 22.5 kg/ha/wk of 0-20-5, 5-20-5, and 20-20-5 compound fertilizers; the

Table 2   Percent Dissolution of Phosphorus and
Nitrogen from Selected Fertilizers After Settling
Through a 2-m Water Column in Pond at 29°C

Fertilizer	Nutrient solubility (%)	
	P	N
Superphosphate	4.6	—
Triple superphosphate (TSP)	5.1	—
Monoammonium phosphate	7.1	5.1
Diammonium phosphate (DAP)	16.8	11.7
Sodium nitrate	—	61.7
Ammonium sulfate	—	85.9
Ammonium nitrate	—	98.8
Calcium nitrate	—	98.7

After Boyd, C. E., 1981.

tests resulted in average fish production (kg/ha/30 wk) of 651, 947, and 930 kg, respectively. He concluded that although the nitrogen input at all three levels had substantially increased fish production, there was little difference between treatments of two higher inputs. The results of most of the previous studies that failed to show significant benefits of nitrogen fertilization in fish ponds were probably attributed to either relatively low stocking density (1000 to 2000 fish/ha) (Hickling, 1962; Boyd, 1976) or to stocking of nonmicrophagous filter feeding species (Mortimer, 1954; Swingle et al., 1965). Fish yields from those experiments were relatively low (<1500 kg/ha), which may not directly correlate to increase in phytoplankton production resulting from nitrogen fertilization. At low fish densities the natural food required by fish is expected to be so small that natural nitrogen fixation in ponds is probably sufficient to satisfy the nitrogen need. Phosphorus is often the first limiting nutrient to high primary productivity in freshwater and must be in sufficient supply before nitrogen supplementation is beneficial.

Application of ammonium and urea-based fertilizers can cause acidification of pond waters because of nitrification, which produces two hydrogen ions from each ammonium ion (Boyd, 1990). The potential acidity from nitrification can be neutralized by equivalent amounts of $CaCO_3$ (Table 3), as determined by Hunt and Boyd (1981). Although urea and ammonium fertilizers cost less, they can result in sufficiently high ammonia concentration to produce toxicity to aquatic animals in ponds. Residual ammonia in pond water is also subject to nitrification, which consumes oxygen and acidifies water. Alternatively, sodium nitrate is an excellent source of nitrogen for pond fertilizer without the side effects of ammonium and urea.

The optimal fertilization rate determined in CRSP experiments is much greater than that used in most previous pond experiments. Hepher (1963) observed that when large doses of inorganic fertilizers were applied to Israeli fish ponds, the concentration of nitrogen and phosphorus in pond water did not exceed 2.0 and 0.5 mg/L, respectively. Those carp culture ponds were fertilized either weekly or biweekly with 13 kg N and 5 kg P/ha, equivalent to a daily input of 1.9 kg N and 0.7 kg P/ha for weekly or 0.93 kg N and 0.36 kg P/ha for biweekly treatment. The annual fish yield over a 6-year average was 982 kg/ha for weekly and 801 kg/ha for biweekly fertilization. In comparison, the fish stocking rates in most CRSP experiments were at 10,000 to 20,000/ha, with fish yields of 3000 to 5000 kg/ha/yr. To reach such high levels of fish yield, the ponds have to be maintained at a hypereutrophic state, with phytoplankton standing crops of greater than 13 mg/L in dry biomass. Ponds were fertilized at a rate of 4 kg N and 1 kg P/ha/d, several-fold of that used by Hepher.

Despite the large number of fertilization experiments that have been conducted using chemical fertilizers as pond inputs, relatively few were conducted specifically for tilapia culture. It has been less common to use species of microphagus filter-feeding tilapia, such as Nile tilapia.

Table 3 Amount of $CaCO_3$ Required to Neutralize the Potential Acidity of Some Chemical Fertilizers

Fertilizer	Potential acidity (kg $CaCO_3$/kg fertilizer)
Urea	1.61
Ammonia sulfate	1.51
Ammonium nitrate	1.18
Diammonium phosphate	0.97
Monoammonium phosphate	0.79

Adapted from Hunt, D. and Boyd, C. E., *Trans. Am. Fish. Soc.*, 110, 81, 1981. With permission.

ANIMAL MANURES

The use of animal manures for fish culture is an extension of traditional land-crop cultivation, which uses available on-farm resources within reach of many small-scale farms in Asia (Zhu et al., 1990; Edwards, 1993). Promotion of fish culture among small-scale farmers in tropical countries requires well-established strategies, which should consider fish ponds as a component of an integrated farming system. To make those farms more efficient and sustainable, experimental data on nutrient budgets and recycling and on energy flow in various components of small-scale farms are needed.

It is well recognized that pond fertilization with animal manures stimulates production of bacteria, phytoplankton, zooplankton, and benthos. But, the nutrient availability and efficiency of animal manures to phytoplankton production have long been treated as a black box in pond dynamics. To maximize fish production with available food organisms in ponds, polyculture with a variety of fish of different feeding niches (e.g., Chinese or Indian carps) has been most commonly practiced (Ling, 1967; Prose, 1967; Olah, 1986; Wolhfarth and Hulata, 1987). Tang (1970) described multispecies polyculture as a harmonious system where the available fish foods and stocked fish species are balanced. The obvious advantages of using animal manures as a nutrient source for fish culture are that they are (1) relatively inexpensive, (2) readily available on-farm, and (3) suitable for a variety of fish in polyculture. This procedure mitigates the problem of solid waste disposal. However, there are also a number of negative aspects of using animal manures.

1. There are aesthetic objections and sanitary concerns related to the fish products from manured ponds.
2. It is time consuming to collect and apply bulk materials to ponds on a routine basis.
3. Theis procedure results in unpredictable nutrient quality and high biochemical oxygen demand, which may cause oxygen depletion of pond water when applied at high rates.
4. It is unsuitable for intensive high-yield culture systems.

Manure Composition

A large variety of animal manures have been used to fertilize fish ponds; the N and P ratio in fecal wastes of most animals remains consistently around 2, but N in urine is markedly higher than that in feces (Table 4). Human excreta and chicken feces contain relatively higher N and P than those from large ruminants, e.g., buffaloes and cattle. In general, the moisture and nutrient contents of manure may vary considerably, depending on factors such as the diet, purity and treatment of manure, and duration and conditions of storage. For example, fresh poultry droppings contain twice as much N as farmyard manure and are much richer in P; the uric acid in fresh droppings is decomposed quickly by microorganisms, releasing ammonia, which is easily lost upon exposure of manure to air (Cooke, 1982). The quality of manures depends on their composition, as they are often mixed with decomposed animal manure, plant residues, sawdust, and lime. In many instances animal manures may contain a significant amount of spilled animal feed, which contributes directly as fish diet when applied to ponds. This is particularly obvious in a poultry-fish integrated system.

Nutrient Content and Leaching

The nutrient contents of animal manures may vary over time, and nutrient availability to phytoplankton growth remains unclear (Table 5). The rate of nutrient released from animal manures over time is a key factor to decide the frequency and amount of manure required to fertilize ponds. In Indonesia, Knud-Hansen et al. (1991) showed that about two-thirds of the dissolved inorganic nitrogen (DIN) in chicken manure was leached into pond water within the first 2 d, and greater than 90% of the leachate was ammonia; 43% of total nitrogen (6 of 14 mg/g dry manure) in the

Table 4 Average Elemental Composition of Human Excreta and Common Livestock Manures

Animals	N:P ratio	N	P	K
Human				
Feces	2	3.77	1.89	1.76
Urine	11	17.14	1.57	4.86
Buffalo				
Feces	2	1.23	0.55	0.69
Urine	205	2.05	0.01	3.78
Cattle				
Feces	3	1.91	0.56	1.40
Urine	195	9.74	0.05	7.78
Goat and sheep				
Feces	2	1.50	0.72	1.38
Urine	69	9.64	0.14	—
Pig				
Feces	2	2.80	1.36	1.18
Urine	660	13.20	0.02	10.90
Duck				
Feces	2	2.15	1.13	1.15
Poultry				
Feces	2	3.77	1.89	1.76

Note: Values are expressed as % weight, fecal materials are based on dry weight; urine is liquid.
Adapted from Tacon, A. G. J., 1987.

manure was released by day 6. A similar experiment in Oregon (Nath, unpublished data) demonstrated that 51 to 57% of total nitrogen in dry chicken manure was released as DIN to water during the first 3 d, while it took 20 d to release 68 to 73% of total P as dissolved inorganic phosphorus (DIP). For duck manure, almost all soluble N and P were released to pond water within 4 d of manure application (Ullah, 1989). For buffalo manure, 90% of total Kjeldahl nitrogen (TKN) was released as DIN within 3 d, and release of DIP reached a maximum concentration on day 4, equivalent to 35% of the initial total P in manure (Shevgoor et al., 1994). Decomposition of buffalo manure resulted in dark brown water color and high concentration of suspended solids, which reduced Secchi disc transparency.

Table 5 Variability in Nutrient Content of Chicken Manure Used for Loading-Rate Experiments in Thailand in 1985

Batch	Moisture (%)	pH	N (%)	P (%)	K (%)	Available P (% citrate soluble)
1	44.1	7.0	2.8	4.0	2.6	2.8
2	30.5	8.4	2.2	3.9	2.8	3.4
3	62.7	8.8	1.7	3.6	2.4	3.5
4	49.0	6.3	2.2	4.2	0.7	3.2
5	60.8	8.9	1.8	3.1	2.2	2.6

Note: Fresh manure was obtained from Poultry Experimental Farm at Kasetsart University, and samples were analyzed by the RTG national fertilizer laboratory.

Chicken manure was chosen as a major source of animal manure for pond fertilization in CRSP experiments because it is widely available in participating countries. To assure the consistent quality of the animal manures for CRSP experiments, the materials were obtained from persistent sources and their nutrient content analyzed. The amount of inputs to fish ponds was based on standard dry weight at 65°C (Egna et al., 1987).

Fish Yields

Fish yields obtained from manure-fertilized ponds were reported to range from 7 to 36 kg ha/d (2,555 to 13,140 kg/ha/yr) (Buck et al., 1979; Wolhfarth et al., 1980; Barash et al., 1982; Wohlfarth and Hulata, 1987). Almost all the high yield ponds (>30 kg/ha/d) in those reports were polycultured with tilapia and Chinese carps, where both common and silver carp accounted for a large portion of the total yields. Most of those extremely high fish yields were obtained in Israel, (e.g., 30 kg/ha/d with cattle manure (Schroeder, 1975), 40 kg/ha/d with duck manure and waste feed (Wohlfarth, 1978), and 20 kg/ha/d with chicken manure (Milstein et al., 1991). In recent years, semi-intensive culture of tilapia has gradually gaining popularity in developing tropical countries. The yields of tilapia monoculture from manured ponds ranged from 8.6 to 19.2 kg/ha/d (Collis and Smitherman, 1978; Hopkins and Cruz, 1982; Diana et al., 1991; Green et al., 1994; Knud-Hansen and Lin, 1993). These yields are comparatively lower than those of Israeli polyculture systems. In terms of nutrient utilization efficiency, monospecies culture of tilapia, especially the planktivorous Nile tilapia, may not be the best strategy for maximum fish production in pond culture. Although Nile tilapia was shown to feed primarily on phytoplankton (Moriarty and Moriarty, 1973), its complete feeding niche and efficiency on detritus have not been critically examined.

Animal Manures as a Carbon and Detrital Source

It has been speculated that animal manures, other than being the nutrient source, may also enhance pond production through their role in detrital formation (Wolhlfarth and Shroeder, 1979; Coleman and Edwards, 1987). Many authors have demonstrated that detritus derived from animal manure benefits detritivorous species, e.g., mud carp (*Cirrhina molitorella*), mrigal (*Cirrhina mrigala*), and certain tilapia species (*Oreochromis mossambicus*). However, it is not clear to what extent Nile tilapia feed on organic detrital material, despite the fact that they are generally believed to be omnivorous feeders. To evaluate feeding choices between plankton and detritus, Xu (1989) stocked Nile tilapia at 2 fish/m^2 in open ponds, in net cages suspended in the water column and net pens set at the bottom of manure-fertilized ponds at AIT. Fish in the net cages were assumed to be deprived of detritus while fish in the other treatments would be free to feed on detritus. At the end of the 115-d experiment there were no significant differences in fish size and yield between open pond, cages, and pens. Those results indicate that Nile tilapia is primarily a column feeder and benefits little from detritus. An alternative explanation is that the pond carrying capacity exceeds fish biomass under such low stocking density. With a greater fish standing crop, the fish might have to graze or browse on other sources of food when planktonic organisms are in short supply. The questions about formation and nutritional value and about tilapia's feeding habit on detritus in manured ponds remain to be answered.

There appears to be an inverse relationship of conversion efficiency between total N and P input from chicken manure and net fish yield (Table 6). In treatments with lower chicken manure inputs at 72.5 and 145 kg N per hectare per 145 d (or 0.5 and 1 kg N/ha/d), the nitrogen gain in fish biomass was 216.7 and 116.1% of that in the manure input, respectively. This may indicate that the ponds obtained extra nitrogen from other sources (e.g., nitrogen fixation) to make up the deficiency.

Table 6   **Comparison of Conversion Efficiency of Nitrogen and Phosphorus to Fish Biomass for Various Levels of Chicken Manure Input During 145-d Grow-Out Period, Based on Average of Dry and Wet Seasons at Thailand Site**

Chicken manure (kg/ha/wk)	Net fish yield (kg/ha)	Total nutrient input (kg/ha)		Recovery in fish biomass (kg)		Conversion efficiency (%)	
		N	P	N	P	N	P
125	1619	72.5	116.5	157.1	47.9	216.7	41.1
250	1736	145.0	233.0	168.4	51.4	116.1	22.1
500	2488	290.0	446.1	241.4	73.7	83.2	15.8
1000	2454	580.0	932.1	238.0	72.6	41.0	7.8

Fertilization with Waste from Intensive Fish Culture

The waste effluents from intensive fish culture have been a major concern as a source of pollutants to natural waters (Edwards, 1993). Those wastes are potential fertilizer that can be reused in a integrated fashion to generate natural food for filter feeder species. A series of experiments were conducted in Thailand on integration of hybrid catfish (*Clarias macrocephalus x C. gariepinus*) with Nile tilapia (Figure 1). The former was cultured in cages with pelleted feed and the latter in open ponds with natural feed stimulated by the catfish wastes (Lin et al., 1989, 1995). The net yield of catfish reached 391.5 kg (144 kg/m³) and of tilapia reached 86.0 kg in a pond of 250 m² in a 4-month trial. The waste material from caged catfish contained 30.9 kg N, and 11.6 kg P, giving a fertilization rate of 5.14 kg N and 1.94 kg P/ha/d with a N:P ratio of 2.6:1. Those wastes provided nutrients in a range recommended for optimal phytoplankton growth for tilapia production (Knud-Hansen et al., 1993). The resultant tilapia production in the present experiment was equivalent to 5334 kg/ha/yr, which was comparable to ponds fertilized with conventional chicken manure (Green, 1992) or chemical fertilizers (Diana et al., 1991). A similar approach has also been attempted for integrated production of Nile tilapia, where the adults were fattened with pelleted feed in net cages and the young were nursed in open ponds fertilized with cage waste (Yang et al., in press).

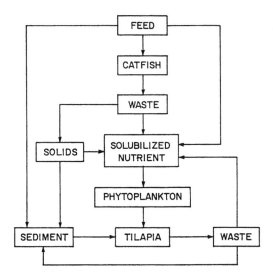

Figure 1.   A schematic nutrient pathway of an integrated catfish-tilapia culture system.

This fish-fish integrated system may be practical for small-scale fish farmers who can use small ponds for dual purposes to achieve maximum biomass production and economic return.

Nitrogen Supplement

Animal manures in general are low in N:P ratio (<3) (Table 4), and the chicken manure used for experiments in Thailand was even more skewed, with a N:P ratio of less than 1 (Table 5). Such a low N:P ratio is not optimal for phytoplankton growth, which on average requires a N:P ratio of greater than 7 (Redfield et al., 1963; Wetzel, 1983). With an estimated 10 to 13% nitrogen conversion efficiency to fish biomass in manure-fertilized ponds, 100 to 130 g of nitrogen is required to produce 1 kg of fresh fish (Edwards, 1993). In other words, a daily input of 2 to 4 kg N/ha is necessary to produce 20 to 30 kg fish/ha/d, equivalent to approximately 100 kg dry chicken manure input. In the case of manures containing less nitrogen, greater inputs are needed to meet the nitrogen requirement. Experiments conducted in Thailand showed that fish yield increased with greater chicken manure inputs, up to 500 kg/ha/wk, while 1000 kg/ha/wk did not enhance fish production (Table 6). Overfertilization may cause severe oxygen depletion and related side effects, particularly with buffalo manure (Edwards, 1993). On the other hand, nitrogen deficiency became apparent as was shown in an earlier CRSP experiment where ponds were fertilized with less than 500 kg/ha/wk of chicken manure (McNabb et al., 1990; Diana et al., 1991). Thus, ponds can be fertilized at a reduced manure rate and supplemented with inorganic N fertilizer (e.g., urea, nitrate) to provide adequate nitrogen and avoid oxygen depletion.

To test the balance of nutrients, ponds in Thailand were loaded with chicken manure at 44, 100, and 200 kg/ha/wk and supplemented with urea to make up the N:P ratio of 5:1, based on manure P concentration. The corresponding fish yield was 1172, 1449, and 2398 kg/ha per 5 months, respectively. In comparison, the fish yields in ponds receiving 500 kg of chicken manure ha/wk were similar to those receiving 44 kg manure supplemented with urea (Batterson et al., 1989). In Honduras, where organically fertilized ponds were consistently low in dissolved inorganic nitrogen, ponds receiving 750 kg/ha/wk of chicken manure were compared to those receiving the same amount of chicken manure, supplemented with 10 kg N/ha/wk (Green et al., 1994). Gross fish yield between the treatments (1401 vs. 1527 kg/ha) was not significantly different, although the dissolved nitrogen and chlorophyll *a* concentrations were greater in ponds supplemented with urea. But when the chicken manure was loaded at the same rate with various levels of nitrogen supplemented as urea, the results showed that additional urea input did not increase fish yields (Table 7). Treatments with high nitrogen input apparently triggered blooms of blue-green algae (*Anacystis* sp.) and raised ammonia concentration to a relatively high level (1.02 mg/L). Those two phenomena might have been the cause of the severe fish mortality that occurred in some of those ponds in Honduras.

GREEN FODDERS AND MANURES

Terrestrial vegetation and aquatic macrophytes are abundant and often under-utilized in many parts of the world with constantly high temperature, particularly in the tropics. These plant materials have been used in various ways as a local, low-cost means to feed herbivorous and detritivorous fish (Edwards, 1987). Herbivorous freshwater species commonly cultured in Asia are grass carp (*Ctenopharyngodon idella*), Wuchang fish (*Megalobrama amblycephala*), common carp (*Cyprinus carpio*), giant gourami (*Osphronemus gouramy*), tawes (*Puntius gonionotus*), and snake skin gourami (*Trichogaster pectoralis*) (Hora and Pillay, 1962; Edwards, 1987). Tilapia species that feed on macrophytes include *Tilapia rendalii*, *T. zillii*, and *Oreochromis mossambicus* (Bowen, 1982). But those macrophagous tilapia have not been widely cultured.

Many varieties of terrestrial vegetation commonly used for animal fodder may not be suitable for fish feed. Legume leaves, for example, though containing relative high crude protein (Table 8),

Table 7 Fish Yields in Response to Various Urea Input in Addition to Chicken Manure Over a Culture Period of 150 d in Thailand

Chicken manure N (kg/ha)	Urea N (kg/ha)	Total N (kg/ha)	Gross fish yield (kg/ha)
17.1	0.0	17.1	2825
17.1	6.3	23.3	2764
17.1	14.1	31.1	2764
17.1	29.9	46.9	2709

do not support good fish growth as feed. To determine the feed value of some common species of legumes, a series of experiments was recently done at AIT on feeding Nile tilapia with fresh leaves of pigeon pea (*Cajanus cajan*), leucaena (*Leucaena leucocephala*), gliricidia (*Gliricidia sepium*), and sesbania (*Sesbania grandiflora*). The results showed negligible fish growth when fed those green fodders over 2 months, despite the apparently sufficient level of crude protein content in the fodder (Castanares et al., 1991). Low palatability and digestibility of those legume leaves were believed to be the main reasons for poor growth of Nile tilapia.

Other than feeding vegetation directly to fish, an alternative method is to make green manure or compost, which serves as organic fertilizer to stimulate production of plankonic and benthic organisms (Edwards, 1987). The nutrient values of plants vary with growth conditions, particularly soil fertility. With the exception of the N-fixing legumes, the average content of N and P in most plant materials in the tropics is much lower than that in animal manures, particularly when measuring P (Table 9). Composting of raw plant materials can be done either on land or in ponds. Production of Nile tilapia at AIT was similar with water hyacinth loaded at a rate of 200 kg dry matter/ha/d, regardless of the forms of input, including chopped raw biomass, in-pond compost of whole plants, or on-land compost (Edwards, 1987). In addition to forming detrital matter for fish feed, the nutrient contents released from vegetation can also serve as pond fertilizers. Little et al. (1995) reported that by soaking legume leaves in water, substantial amounts of N and P can be released, and those nutrients were shown to be equally effective as inputs using inorganic fertilizers.

The CRSP site in Rwanda is an area with severe resource limitations. All agricultural byproducts have competing uses. Animal manures are especially valuable sources of organic matter for crop cultivation in land where nitrogen and phosphorus in soils are severely depleted, as in this most densely populated country of Africa.

Most Rwandan fish farmers use in-pond composting by adding grasses, kitchen wastes, and some animal manures to an enclosure in the pond to promote growth of natural food organisms. These inputs are high in carbon, and over-application can cause oxygen depletion. However, they are often the only inputs available, and they are seldom applied in excessive amounts. The most available source of vegetation is grasses collected from the field, which on average contain 67 to

Table 8 Mean Values (% Dry Weight) of N, P, and Proximate Composition of Legume Leaves

Variety	Moisture	Protein	Lipid	Fiber	N	P
Pigeon pea	70.73	27.62	3.84	18.10	4.30	0.23
Gliricidia	74.75	28.60	3.44	10.64	4.62	0.34
Leucaena	69.88	32.78	2.76	10.00	5.33	0.29
Sesbania	75.09	35.14	3.54	7.70	3.84	0.18

Adapted from Little, D., Yakupitiyage, A., Edwards, P., and Lovshin, L. L., Use of Leguminous Leaves as Fish Pond Inputs, Final Report to U.S. AID, Asian Institute of Technology, 1995.

**Table 9   Average Elemental Composition (% Dry Weight)
of Plant Residues Commonly Available in the Tropics**

Plant	N:P	N	P
Rice straw	5.7	0.58	0.10
Maize straw	1.9	0.59	0.31
Soybean straw		1.30	—
Bean straw	1.6	1.57	0.32
Weeds		2.45	—
Grass	13.6	0.41	0.03
Grass[a]		1–2	
Sugarcane trash	8.7	0.35	0.04
Ground nut shell	16.6	1.00	0.06
Coffee pulp	14.9	1.79	0.12

[a] Used in CRSP experiment in Rwanda.
From Tacon, A. G. J., The Nutrition and Feeding of Farmed Fish and
Shrimp — A Technical Manual, 1. The Essential Nutrients, FAO Rep.
GCR/RLA/ITA, Brasilia, Brazil, 1987, 117.

83% moisture, and dry matter, with 40% carbon and 1 to 2% nitrogen, giving a C:N ratio of 20 to 40:1. Phosphorus content is more variable depending on species and age of the grasses.

CRSP research was conducted to determine the most efficient way of applying the few available low-quality fertilizers/feeds as pond inputs for fish production. Fresh grass was composted in three ways: (1) direct application of organic matter to enclosures in the ponds, methods recommended by the extension service; (2) aerobic composting of mixed ingredients on land for 90 d prior to pond application; and (3) anaerobic composting (fermentation) on land for 90 d prior to application. Net production of Nile tilapia in ponds receiving 50, 80, and 100% of in-pond composted grass at 500 kg/ha/wk was 1323, 1030, and 817 kg/ha/yr, respectively. In comparison, the fish production in ponds receiving 80% grass input from aerobic and anaerobic on-land composting was 748 and 613 kg/ha/yr, respectively (Veverica et al., 1990). Fish production was significantly different between open-pond broadcasting of grasses (520 kg/ha/yr) and composting in a corner enclosure (773 kg/ha/wk) (Rurangwa et al., 1990). As the composting process results in net loss of carbon and nitrogen, in-pond composting appears to have a greater advantage in retaining those byproducts in the water column than the on-land process. Composting in enclosures also makes it easier for pond draining and fish harvest. The total enclosure area in a pond should only cover 5 to 10% of the pond surface, and the enclosures should be built with 10-cm spacing between stakes to allow fish access to feed on the compost (Veverica et al., 1990). In practice, farmers apply a large amount of compost at the beginning of the culture period so that fish can get a good start on growth and reach market size before the female fish begin to reproduce.

Fish culture using composts of plant materials in ponds is primarily dependent on production of heterotrophic detritus as well as autotrophic phytoplankton stimulated by leached nutrients. It is not clear how much each of those two pathways contributes to the growth of Nile tilapia. To increase the fish production in manure-fed ponds, polyculture of macrophagous and microphagous tilapia species should be considered.

Although those experiments demonstrated that fish could be produced with low-grade inputs (e.g., weeds and grasses) the amount of time and effort involved in gathering the raw material is staggering. Furthermore, with the apparent food conversion ratio greater than 100 (100 kg fresh material to produce 1 kg fish), the efficiency and sustainability of the terrestrial system to support pond fish culture becomes questionable.

BENEFITS FROM NITROGEN FIXATION IN TROPICAL PONDS

Nitrogen fixation by green algae has long been recognized as an important source of nitrogen input to natural waters (Dougdale and Dougdale, 1962; Horne and Viner, 1971) and to rice fields (Watanabe and Yamamoto, 1971). While the nitrogen derived from natural fixation appears to be significant in eutrophication of natural waters, the recorded amounts (<50 kg/ha/yr) are inadequate to support phytoplankton blooms for yields of filter-feeder fish in pond culture. In Hungarian carp ponds, El Samra and Olah (1979) reported that the N fixation of 5.7 kg N/ha per grow-out cycle was an insignificant input as N fertilizer. These values were relatively low due to the cool climate and short growing season that is unfavorable for blue-green algal blooms. Jiwyam (1996) showed that in laboratory culture the N fixation by *Anabaena siamensis*, a tropical nitrogen fixer, ceased at 20°C and increased with rising temperature up to 40°C.

In tropical fish ponds where blue-green algae often bloom year-round, N fixation might be a significant factor in pond fertilization. The rate of natural nitrogen fixation measured in tilapia ponds in Thailand ranged from 0.45 to 1.47 and 1.46 to 1.98 kg N/ha/d in ponds fertilized with inorganic (urea and TSP) and chicken manure, respectively (Lin et al., 1988). The lower fixation rate in ponds that received urea and TSP was probably inhibited by the high ammonia concentration. The rate of fixation also varied considerably with chicken manure loading rate (Table 10). At lower manure rates the fixed nitrogen accounted for 10 to 25% of the total nitrogen input in ponds during the 5-month grow-out cycle. This additional nitrogen income may explain the discrepancy in nitrogen content between the fertilizer input and fish biomass. Examination of CRSP experiments on chicken manure fertilization in Honduras and Thailand reveals that the nitrogen content in harvested tilapia biomass actually exceeds that in chicken manure at low input rates (17.8 and 35.7 kg/ha/d). The nitrogen content in fish biomass was calculated at 108 to 143% of that in chicken manure in Honduras, and 104 to 174% in Thailand (Table 11). At those low fertilization rates, the N fixation is perhaps a significant source of nitrogen for phytoplankton growth.

FACTORS AFFECTING RESPONSES TO FERTILIZATION

Experimental results on pond fertilization often show considerable variability among replications within treatment, so much so that aquaculturists claim each pond has its own individuality. The biological productivity of ponds in response to fertilizer inputs is governed by numerous environmental conditions; the following sections describe conditions that are most commonly encountered in tropical ponds.

Light and Temperature Regime

Although the seasonal variability of temperature and solar radiation in the tropics is minimal compared to temperate regions, considerable differences exist between low and high altitudes, as

Table 10 Comparison of Nitrogen Input (kg N/ha) from Estimated Fixation and Chicken Manure in Tilapia Ponds Fertilized with 4 Chicken-Manure Loading Rates During a 5-Month Grow-Out Period in Thailand

Nitrogen source	Chicken manure loading (kg/ha per 150/d)			
	2,500	5,000	10,000	20,000
Natural fixation	19.4–20.0	12.2–24.4	31.9–33.9	8.8–33.6
Chicken manure	75.0	150.0	300.0	600.0
Ratio	1:3.8	9.2:1	9.1:1	43:1

Note: Ratio expresses fixed N: chicken manure N input.

Table 11   Nitrogen Content in Chicken Manure Input and Harvested Fish Biomass and Calculated % N Recovery (Fish N/Manure N) During a 5-Month Period

Nitrogen input (kg/ha per 5 months)		N in Fish (kg/ha)		N recovery (%)	
Thailand	Honduras	Thailand	Honduras	Thailand	Honduras
72.5	61.9	126.7	88.8	175	143
145.0	123.8	151.2	134.0	104	108

well as rainy and dry seasons. Low temperature not only affects the productivity of food organisms but also feeding and growth of cultured fish. With the exception of Rwanda, where the temperature ranged from 19 to 23°C at 1700 m elevation, the average temperature of other CRSP sites was 27 to 30°C.

Intense solar radiation in the tropics commonly causes diurnal thermal stratification, which can potentially prevent effective water mixing and result in nutrient limitation in surface water during the daytime. This phenomenon is particularly evident in ponds where the blue-green algae are concentrated on the water surface. Such nutrient limitation, if it happens at all, is expected to be short lived because most tropical ponds with a water depth less than 1.5 m are subject to daily mixing during the nocturnal overturn (Szyper and Lin, 1990). However, little is documented on nutrient dynamics in ponds with diurnal stratification.

The effects of climatic variation on pond fertilization and fish yields are a complex phenomenon. This is demonstrated by the great inconsistency in fish yields from chicken manured ponds between dry and rainy seasons among CRSP four tropical sites (Table 12).

Acidic Soils and Water

Generally, the acid conditions of pond water originate from organic acids of peaty wetlands, acid soils in pond muds, or source water. Acid waters caused by acid sulfate soils are by far the most serious problems that prevail in large areas of the coastal environment in the tropics (Brinkman, 1982). The worldwide extent of acid sulfate soils is estimated as 13×10^6 ha. Those soils contain large concentrations of sulfate and pyrites, which upon oxidation create extremely acid top soils and surface waters. In addition, the acid waters also contain high concentrations of iron and aluminum ions which are not only toxic to fish and other aquatic organisms but also precipitate phosphate, making the fertilizer unavailable to organisms in the environment. Reclamation and utilization of acid sulfate soils in the coastal regions for aquaculture are well documented (Hickling, 1962; Brinkman and Singh, 1982; Singh, 1985).

Table 12   Comparison of Net Fish Yield Between Dry and Rainy Season Among 4 CRSP Experimental Sites Under 4 Levels of Chicken Manure Inputs in Earthen Ponds

Manure rate (kg/ha/wk)	Net fish yield (kg/ha 145–150/d)							
	Honduras		Indonesia		Panama		Thailand	
	Dry	Rainy	Dry	Rainy	Dry	Rainy	Dry	Rainy
125	781	915	391	262	852	584	1932	1306
250	1050	1381	573	433	1483	1053	1913	1558
500	1543	1643	471	711	1863	1472	3232	1845
1000	1964	2046	747	1140	2621	2495	2718	2189

Note: Values are means of triplicates in each treatment.

Table 13 Depth Profile of Chemical Characteristics of Acid Soils at Nong Sua Fisheries Station in Thailand

	Depth (cm)				
Parameter	0–20	20–40	40–60	60–80	80–100
pH	4.0	3.7	3.5	3.4	3.2
Active Fe (%)	0.7	0.7	3.0	4.4	2.9
Extractable Al (meq 100/g)	6.7	16.7	16.1	14.6	14.6
Calcium (%)	0.27	0.11	0.09	0.09	0.09
Sulfur (%)	0.49	0.17	0.43	0.66	0.78
Phosphorus (ppm)	310	155	130	110	105
Base saturation (%)	46	32	28	29	28

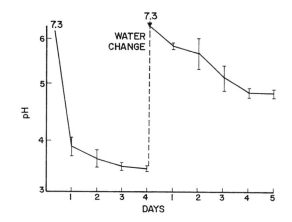

Figure 2. Improvement of pH in pond water upon repeated exchanges with water of higher pH in acid soil ponds.

One CRSP project site in Nong Sua, Thailand, was located in the alluvial plain of the Chao Phraya River delta, where 778,000 ha of acid soils exist (Pons, 1969). Analysis of soil samples reveals acidity, base saturation, and calcium concentration decreases with increasing soil depth (Table 13). Despite the potential acidity, this area has been a major center for freshwater aquaculture for many species in Thailand. A large proportion of the fish production comes from livestock-fish integrated culture systems, where organic fertilizers from animal wastes are the major source of input. The continuous deposition and active decomposition of organic matter at the pond bottom in this system may have created anaerobic conditions in surficial sediments and effectively prevented acidification of pyrite in deeper soils.

To demonstrate the impact of organic fertilizer on the reduction of acidification, an experiment was conducted at the Nong Sua station (Lin, 1986). The pH of initial pond water could be substantially increased by a repeated leaching-draining-drying treatment (Figure 2). The subsequent fertilization experiment in those acid-soil ponds with chemical fertilizer ($N_{16}P_{20}K_0$) and chicken manure showed that the latter resulted in greater stability in water pH and alkalinity (Figure 3), as well as increased phytoplankton and zooplankton production (Figure 4). The fish yield (*Oreochromis niloticus* and *Puntius gonionotus*) was 852 and 3056 kg/ha/yr in chemically and organically fertilized ponds, respectively.

To mitigate the pond acidification from acid soils, a number of measures are recommended:

- Excavate ponds above the depth of heavy acid soils.
- Remove initial acid from new ponds by repeating leaching, drying, and filling cycle before liming.
- Reduce free board area by filling ponds up to the top.
- Fertilize ponds with organic manure.
- Avoid drying the ponds between crops to oxidize acid-forming materials in bottom soils, followed by leaching and liming.

The acidity of ponds built on acid soils usually disappears after repeated pond use over a couple of years as a result of oxidation and leaching of bottom soils as well as deposition of organic matter.

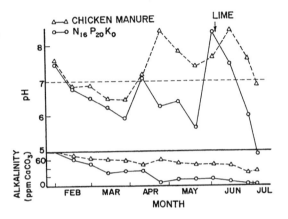

Figure 3. Comparisons of pond water pH and alkalinity in response to fertilization with inorganic and chicken manures in acid soil ponds.

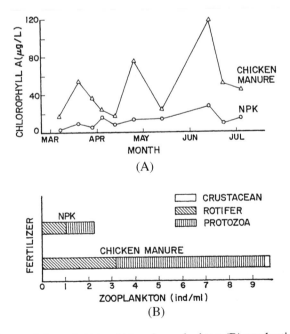

Figure 4. Comparisons of phytoplankton (A) and zooplankton (B) production in response to input of inorganic and organic fertilizers in acid soil ponds.

Clay Turbidity

Clay or mud turbidity is the most prevailing problem for fertilization management in freshwater ponds. Clay particles are microscopic colloids (<2 µm) with negatively charged ions, and Brownian motion keeps these particles in suspension. Clay turbidity in pond water originates mainly from a number of sources: (a) turbid source water, (b) rainwater runoff from pond dikes that contain dispersed clay, especially where the dike surfaces are wide and barren, (c) erosion of pond edges caused by sloshing water movement and fish grazing, and (d) resuspension of bottom muds by water and fish movements.

Clay turbidity curtails fertilizer effect as it reduces light penetration, thus limiting the photic zone in water column. It is not uncommon to find the Secchi disk visibility less than 10 cm in ponds with high clay turbidity. Furthermore, the minute colloidal particles possess relatively large surface areas for adsorbing mineral nutrients from the water (Green, 1993).

A number of applications have been suggested to mitigate clay turbidity. As the clay colloids are negatively charged, the particles can be neutralized and precipitated with inorganic compounds, such as alum (Al^{3+}), iron (Fe^{2+}), calcium (Ca^{2+}), and sodium (Na^+). They can also be coagulated with organic particles by applying animal manures, powders of plant materials, and phytoplankton. Vuthana (1995) treated turbid pond water collected from Cambodia with those materials and found that alum, quicklime, and sea salt were most effective; pulverized water hyacinth was impractical because it required a large dose that depleted dissolved oxygen in treated water, while cow manure increased turbidity (Figure 5). Chemical treatments to remove turbidity in pond water are not cost effective in most cases and can also cause undesirable side effects, such as precipitating nutrients. More effective means to alleviate turbidity are to prevent it from occurring by covering the watershed and pond dikes with vegetation, minimizing the free board area of pond edges, increasing pond water depth, and altering texture of bottom soils. As ponds age, the turbidity tends to decrease because the content of silts and organic matter in bottom soils increases. Among CRSP experimental sites, turbidity was a serious problem at Nong Sua station, Thailand, where the dikes of newly constructed ponds were barren clay soil, and at el Carao in Honduras, where wind-driven turbulence stirred up bottom sediment. The low Secchi disk visibility was mostly the result of clay turbidity.

Low Alkalinity and Liming

Besides mineral nutrients, phytoplankton production requires adequate carbon derived from biogenic and atmospheric CO_2 or lithospheric carbonate. Owing to the extremely low concentration, the supply from atmospheric CO_2 is severely limited. For example, the concentration of dissolved CO_2 in pure water under standard pressure at 25°C is 0.46 mg/L. But the actual CO_2 concentration in natural water is often an order of magnitude greater than in pure water, resulting from organismal respiration and dissolution of carbonaceous material. As alkalinity in most natural water refers to the content of free CO_2, carbonate, and bicarbonate, so its concentration indicates the reserve of inorganic carbon and the availability of CO_2 for photosynthesis. However, CO_2 may not limit phytoplankton production in most natural waters (Schindler et al., 1972), and atmospheric CO_2 supplies adequate C to support primary productivity up to 2 mg/L/d of algal biomass. However, the primary productivity for high-yield fish ponds can reach as much as 10-fold that value, and the carbon supply may well become limiting in eutrophic waters (King, 1970). For fertilized productive fish ponds the total alkalinity should be greater than 20 mg/L in temperate ponds (Boyd, 1974), and a higher concentration is probably needed for highly productive tilapia ponds in the tropics.

For pond water with low alkalinity, liming is a common practice that improves water quality in two main respects:

1. To increase pH in acid water as indicated in the following reactions

$$CaCO_3 + 2H^+ \rightarrow Ca^{2+} + H_2O + CO_2$$

2. To increase total alkalinity

$$CaCO_3 + H_2O + CO_2 \rightarrow Ca^{2+} + 2HCO_3^-$$

At most CRSP experimental sites, except the Ayutthaya station in Thailand, the alkalinity in source water was less than 30 mg/L. However, the liming requirement of a pond depends primarily on the need to neutralize the mud acidity. Several types of liming material are usually considered for fish

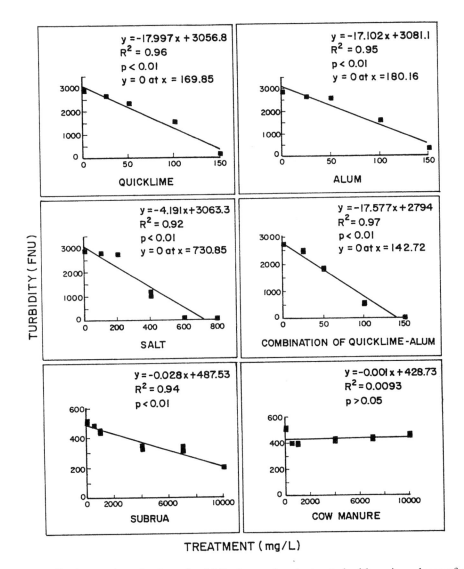

Figure 5.   Effectiveness in reduction of turbidity in pond water treated with various doses of quick-lime, alum, combined lime and alum, sea salt, subrua (water hyacinth), and cow manure.

pond use, including quicklime (CaO), hydrated lime [Ca(OH)$_2$], agricultural lime (CaCO$_3$), and dolomite [CaMg(CO$_3$)$_2$]. The efficiency of a liming material depends on its type, particle size, and neutralizing value (Boyd, 1990).

FERTILIZATION EFFECTS ON WATER QUALITY

CRSP pond fertilization experiments were designed not only to develop new management strategies to increase fish yield, but also to increase understanding of pond dynamics, specifically of the effects of the various fertilization regimes on water quality. Water quality parameters normally regarded as pertinent to fish culture in tropical ponds are nutrients and dissolved gases, especially alkalinity, pH, ammonia, and dissolved oxygen. Basic principles of water quality in relation to pond management are treated in great detail by Boyd (1990).

The goal of pond fertilization is to increase fish yield through enhancement of autotrophic production and heterotrophic production, which are widely believed to be stimulated by organic and inorganic fertilizers, respectively. Yields of the CRSP experimental fish, *Oreochromis niloticus*, are positively correlated to net primary productivity (NPP). Pond fertilization is meant to boost fertility of pond water where nutrient concentrations are often too low to support desirable fish yields. Fertilization also concomitantly affects water quality with positive or negative consequences to fish survival and growth.

Nutrients

As phosphorus was assumed to be the first limiting nutrient in tropical ponds, phosphorus was added at 8 kg P$_2$O$_5$ biweekly as TSP at all sites as in the first global experiment. Soluble reactive phosphorus (SRP) concentrations in ponds peaked within 2 d of fertilizer application and then declined to approximately prefertilization levels within 7 to 14 d (Boyd, 1990; Green et al., 1990b; Teichert-Coddington et al., 1991a; Teichert-Coddington et al., 1992a). The SRP disappeared more quickly from the water column in sites where acid sulfate soils prevail, as in Thailand and Panama (Teichert-Coddington et al., 1992b). Clay turbidity present in pond water may have been responsible for changes in SRP concentrations through adsorption/desorption of suspended colloidal clay (Green, 1993). SRP concentrations increased significantly with increased total suspended nonvolatile solids an indicator of suspended mineral particulates (Knud-Hansen and Batterson, 1994). The average SRP concentration among CRSP sites ranged from 0.01 to 0.94 mg/L PO$_4$-P.

In an experiment with elevated input of either chicken manure or chemical fertilizers containing equal amounts of N and P, the concentration of PO$_4$-P, total P and total ammonia N in pond water increased in varying degree among CRSP sites; but the NO$_2$-NO$_3$ concentration was generally low and independent of nutrient loading. In Honduras and Thailand, concentrations of NO$_2$-NO$_3$ were significantly higher in inorganically fertilized ponds than in manured ponds (Green et al., 1990a; Diana et al., 1990). These differences may have resulted from incomplete mineralization of manure nutrients as discussed in the previous section.

Total ammonia nitrogen and total nitrogen concentrations increased linearly as weekly urea fertilization increased from 0 to 28 kg/N/ha (Teichert-Coddington and Claros, 1996). At a standard weekly fertilization rate of 25 kg/N/ha and phosphorus addition to maintain a 4N to 1P input ratio, substitution of chemical fertilizer for organic fertilizer (chicken litter) resulted in significant reductions in NPP, chlorophyll *a*, total alkalinity, total P, and organic N, and significantly greater TAN (Teichert-Coddington et al., 1993).

Alkalinity

To increase total alkalinity of low-alkalinity waters, both liming and organic fertilization with chicken litter have been employed. Alkalinity is increased because bacterially generated CO$_2$ from manure decomposition dissolves calcium carbonate present in pond sediments and chicken litter (Boyd, 1990; McNabb et al., 1991; Teichert-Coddington et al., 1992a). Mean total alkalinity was greater in ponds in Honduras, Indonesia, and Panama, which were fertilized with chicken litter (Green et al., 1990a; McNabb et al., 1991; Teichert-Coddington et al., 1992a). Total alkalinity in ponds in Panama increased from 18 mg/L as $CaCO_3$ to a mean of 44 mg/L as $CaCO_3$ when 1000 kg/ha/wk of chicken litter was applied (Teichert-Coddington et al., 1990b). In Honduras, weekly additions of 1000 kg chicken litter per hectare increased mean total alkalinity from 43 to 180 mg/L as $CaCO_3$ (Green et al., 1989a). Total alkalinity in ponds in Egypt was significantly greater where organic fertilization and feed (mean of 393 mg/L as $CaCO_3$) were applied, compared to chemical fertilization alone (282 mg/L as $CaCO_3$) (Green et al., 1995). At the Ayutthaya site in Thailand, however, the effect of chicken manure on alkalinity was insignificant because the background alkalinity ranged from 50 to 100 mg/L as $CaCO_3$ (Knud-Hansen et al., 1993; Hanson et al., 1991).

Dissolved Oxygen

Dissolved oxygen (DO) concentrations in ponds are affected most by phytoplankton biomass, with greater oxygen production and consumption occurring at higher phytoplankton biomass (Boyd, 1990a). Early morning DO concentrations in phosphorus-fertilized ponds generally exceeded 60% of saturation because phytoplankton biomass, as indicated by levels of primary productivity, was low in CRSP experiments (Diana et al., 1987; McNabb et al., 1988; Hanson et al., 1989; Green et al., 1990; Teichert-Coddington et al., 1991b). Use of organic fertilizer increases the biological oxygen demand in ponds and can result in periods of low DO (Boyd, 1990). Mean early morning DO (1.2 mg/L) in organically fertilized ponds in Honduras was significantly less than in chemically fertilized ponds (2.3 mg/L) (Green et al., 1990a). While early morning DO concentrations were about 5 mg/L in ponds that received organic or inorganic fertilizer in Thailand, the oxygen differential (DO at top of water column minus DO at bottom) was significantly greater for inorganically fertilized ponds (7.6 mg/L) than for organically fertilized ponds (3.0 mg/L) (Diana et al., 1990, 1991b). Mean early morning DO was inversely related to increasing chicken litter application rate, and often was less than 10% of saturation at the highest chicken litter loading rates (Batterson et al., 1989; Green et al., 1989b; Teichert-Coddington et al., 1990b). Growth of *O. niloticus* was greater in ponds beginning aeration at 10 or 30% oxygen saturation, compared with unaerated ponds (Teichert-Coddington and Green, 1993a).

Ammonia

Ammonia, an excretory product of fish and crustaceans, also can be highly toxic to these same animals (Boyd, 1990). The 96-h LC50 (28°C) was 1.4 mg/L un-ionized ammonia for 3.4-g *O. niloticus* and 2.8 mg/L un-ionized ammonia for 45.2-g fish (Abdalla, 1990). In this same study, he also observed a linear decrease in fish growth with increasing un-ionized ammonia concentrations. Un-ionized ammonia at a concentration of 0.06 mg/L had no effect on growth at water temperatures between 28 and 33°C, but 0.8 and 0.9 mg/L caused a 50% reduction in growth at 28 and 33°C, respectively. Growth of Nile tilapia ceased at un-ionized ammonia concentrations of 1.5 and 1.7 mg/L at 28 and 33°C, respectively.

Ammonia-based chemical fertilizers, such as urea or diammonium phosphate, can be a significant source of ammonia in aquacultural ponds. Results of Cycle II research in Honduras and Thailand showed significantly greater dissolved inorganic nitrogen in chemically fertilized compared to organically fertilized ponds (Diana et al., 1990; Green et al., 1990a). In fact, NO_2-NO_3-N predominated in organically fertilized ponds in Thailand, while NH_3-N predominated in inorganically fertilized ponds (Diana et al., 1990, 1991). Ammonia concentration in ponds was independent of chicken manure loading rates from 125 to 1000 kg dry matter ha/wk, and mean NH_3-N concentrations generally were less than 0.1 mg/L (Batterson et al., 1989; Green et al., 1989b; Teichert-Coddington et al., 1990; Diana et al., 1991b; Teichert-Coddington et al., 1992). Un-ionized NH_3 concentrations greater than 0.06 mg/L under these conditions occurred rarely and only briefly occurred during afternoon periods of high photosynthetic activity and high pH. Fish mortality has been attributed to spikes in un-ionized ammonia in ponds receiving applications of 6 to 30 kgN/ha/wk as urea in addition to chicken litter (750 kg dry matter ha/wk) (Teichert-Coddington et al., 1992). Urea fertilizer hydrolyzes to ammonia, which affects pond TAN concentrations. Total ammonia concentrations increased significantly with increased applications of urea fertilizer (Dettweiler and Diana, 1991; Teichert-Coddington et al., 1992b, 1993; Newman et al., 1994; Teichert-Coddington and Claros, 1996).

COMPARATIVE EVALUATION OF CRSP FERTILIZATION EXPERIMENTS

Phytoplankton Production

The net primary productivity (NPP) in phosphorus-fertilized ponds in Thailand averaged 1.6 g C/m²/d compared to 0.9 gC/m²/d in Indonesia and 0.3 gC/m²/d in Honduras and Panama (Diana et al., 1987; McNabb et al., 1988; Green et al., 1990b; Teichert-Coddington et al., 1991, 1992a). During Cycle II NPP increased because of greater nutrient addition. Mean NPP in inorganically and organically fertilized ponds, respectively, was 1.3 and 1.7 gC/m²/d in Honduras, and 0.7 and 1.0 gC/m²/d in Indonesia (Green et al., 1989a, 1990a; McNabb et al., 1990, 1991). Net primary productivity in organically fertilized ponds in Honduras was significantly greater than in inorganically fertilized ponds (Green et al., 1989a, 1991a). Although N and P appeared to be present in adequate amounts in Indonesian ponds, primary production likely was limited by carbon (McNabb et al., 1990, 1991). In Thailand, no significant differences in gross primary productivity were detected during Cycle II between organically (11.5 gC/m²/d) and inorganically (13.0 gC/m²/d) fertilized ponds (Diana et al., 1990).

Increased loading rates of chicken manure provided increased nutrient availability, which in turn resulted in increased primary productivity. Mean NPP increased from 0.7 to 2.4 gC/m²/d in Honduras, from 0.8 to 1.9 gC/m²/d in Indonesia, and from 1.7 to 4.4 g//m²/d in Panama as chicken litter applications increased from 125 to 1000 kg dry matter ha/wk (Batterson et al., 1989; Green et al., 1989b; Teichert-Coddington et al., 1990; Knud-Hansen et al., 1991; Teichert-Coddington et al., 1992). Nitrogen may have been a limiting nutrient to primary productivity at the two highest chicken manure loading rates (Batterson et al., 1989; Diana et al., 1991b). Primary productivity in low alkalinity Indonesian ponds likely was limited by carbon at the two lowest chicken manure loading rates, but not so at the higher chicken litter application rates, because manure decomposition supplied enough dissolved inorganic carbon (Knud-Hansen et al., 1991).

Primary productivity appeared to be limited by nitrogen at high chicken manure loading rates (Batterson et al., 1989; Diana et al., 1991b). Addition of 10 kgN/ha/wk as urea to ponds fertilized weekly with chicken litter (750 kg dry matter per hectare) resulted in a significant increase in chlorophyll *a* concentration compared to ponds fertilized with chicken litter only (Teichert-Coddington et al., 1991b). Application of 28.3 kgN/ha/wk as urea to ponds receiving organic fertilization (500 kg/ha/wk) resulted in a 205% incrase in chlorophyll *a* concentration compared to organically fertilized ponds (Veverica et al., 1991). In a trial where weekly chicken litter applications (750 kg dry matter per hectare) were supplemented with 0, 6, 14, or 30 kgN/ha/wk as urea, no significant differences in NPP were observed, but occurrence of blue-green algal blooms was more frequent at the higher urea application rates (Teichert-Coddington et al., 1992a). The correlation between chlorophyll *a* and total ammonia suggested that nitrogen fertilization that results in total ammonia concentrations greater than about 0.2 mg/L was wasted (Teichert-Coddington et al., 1992a).

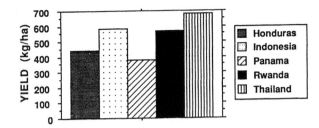

Figure 6. Mean country yields of *O. niloticus* in earthen ponds fertilized biweekly with phosphorus at 0.8 kg/ha.

Fish Yields

Experiments in Common

Many experiments were repeated during wet and dry seasons at each CRSP site. Data are generally given as treatment means from pooled dry and wet season studies. Variation from seasonal effects is therefore largely eliminated. The objectives of this comparison are to describe trends and explain some of major differences among sites.

Phosphorus

The first experiment evaluated response to applications of P as TSP (triple superphosphate) at a rate of 0.8 kg/ha every 2 weeks. Fertilizer was applied by first dissolving and splashing it over the pond surface, by placing fertilizer on a submerged platform, or by suspending the fertilizer in ponds in mesh sacks. Mean fish yields were low at all sites, ranging from 378 kg/ha in Panama to 681 kg/ha in 5 months in Thailand (Figure 6). Yields were particularly low in Panama, presumably due to acidic soils that quickly adsorbed applied P (Teichert-Coddington et al., 1992b). Honduras yields were also relatively low, most likely because of light inhibition by clay turbidity (Teichert-Coddington et al., 1992b).

Chicken Manure

Following the protocols for the global experiment, the experimental ponds were fertilized weekly with chicken manure at rates of 125, 250, 500, and 1000 kg (dry weight) per hectare. Ponds were stocked with manual sexed male tilapia fingerlings at 1 fish/m². The experimental results of fish yield obtained from various CRSP sites varied considerably (Figure 7), ranging from 262 in Indonesia to 3132 kg/ha in 5 months in Thailand. The relatively low fish yield in Indonesia was primarily caused by high rate of water loss through seepage (10% loss/d), which was further compounded with low inorganic carbon, as indicated by low alkalinity of the source water (23 mg $CaCO_3$ per liter), which was of volcanic origin (Egna et al., 1987). Data from Honduras and Panama showed the relationship between fish production (y) and manure inputs (x) could be described as $y = 797.3 + 2.945 x - 0.001 x^2$ (Green et al., 1990). The net fish yields ranged from 584 to 2495 kg/ha in 147 d in the rainy season to 781 to 2621 kg/ha in 150 d in the dry season. The maximum yield was obtained from the treatment with 1000 kg/ha/wk of chicken manure input. While fish yields were similar between dry and wet seasons in Honduras, they were significantly higher in the dry season in Panama (Table 12). In comparison, under the same fertilization regime, the range of net fish yields in Thailand was 1913 to 3132 kg/ha in 145 d in the dry season, which was significantly higher than the rainy season production of 1306 to 2189 kg/ha in 145 d. At the lowest input of 125 kg/ha/wk, the net fish yield in Thailand was 1306 and 1932 kg/ha for the rainy

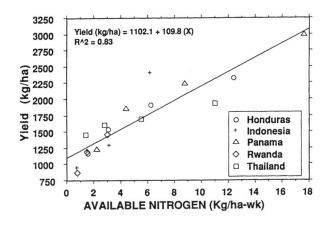

Figure 7. Mean country yields of *O. niloticus* in earthen ponds fertilized weekly with chicken manure at rates ranging from 125 to 1000 kg/ha dry matter.

and the dry season, respectively; these results were markedly higher than all other sites. Those site-specific differences were narrowed as rate of chicken manure input increased. Many factors might cause the differences between manure input and fish yield. The difference in nutrient content of chicken manure used in the various CRSP sites is likely to be a major factor. In Thailand, the manure obtained weekly from a poultry farm of a local university contained an average of 2.8% N and 4.5% P, which were much higher than the contents of manure used at other sites (e.g., 2.25 to 2.17% N and 1.21 to 1.31% P for Honduras, 2.9 to 3.5% N and 2.0 to 2.2% P for Panama, and 1.4% N for Indonesia). The P content in chicken manure used at the Thailand site (two- to three-fold greater than that of other sites), might actually meet the P requirement even at the low level of input. The manure used in Honduras was procured in one lot at the beginning of the experiment and lasted for the remainder of the season, while in Thailand it was obtained fresh weekly. The fish yields were greatly reduced in ponds with input of aged manure (Knud-Hansen et al., 1991). This perhaps resulted from nutrient losses in aged manures.

The step-wise regression of fish yield in relation to the rate of chicken manure input and available N and P revealed that available N accounted for 83% of the total variation in fish yields among treatments and sites (Figure 8). Available N and P were calculated as 45% of total N and P input in chicken manure. Nitrogen input was shown to be the key nutrient variable related to phytoplankton production, which in turn influenced fish yield (Knud-Hanson et al., 1993; Teichert-Coddington and Green, 1993b). These results do not imply that P was not necessary for algal production. Rather, P was usually applied in excess relative to N, because manure is N-poor with respect to P. Indeed, the SRP concentrations were 0.003 to 0.06 mg/L at the three lowest fertilizer treatments in Panama, which suggests that P limited productivity, probably because P was adsorbed by acidic soils (Teichert-Coddington et al., 1992b). Variable manure qualities among sites accounted for the differences in available N and P applied to ponds.

Some environmental variables not included in this data analysis would undoubtedly account for some of the variation in fish yields. Mean ambient temperatures, for example, were quite different across sites (Egna et al., 1987). Yields from the Rwanda site, located at 1700 m altitude, are expected to be lower than yields of near sea-level sites, like Panama and Thailand. The Honduras site was located at 580 m altitude, and winter temperatures became low enough to reduce fish growth compared with summer yields (Green et al., 1994). Pond soil type would also affect response to fertilization. Ponds located on acidic soils were limed to increase soil pH and total alkalinity, but there were still stark differences in water quality between acidic and nonacid sites (Teichert-Coddington et al., 1992b). Limited dissolved inorganic carbon availability was reported in Indonesia for low alkalinity ponds (McNabb et al., 1990).

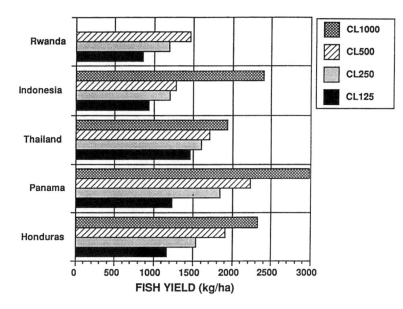

Figure 8. Regression of mean *O. niloticus* yield on available nitrogen from chicken manure applied at rates ranging from 125 to 1000 kg dry matter ha/wk; available nitrogen was calculated as 45% of total nitrogen content in chicken manure.

There were some notable differences among sites in response of fish yield to the different levels of chicken manure application. Fish yield increased linearly with chicken manure application in Panama (Green et al., 1990c), Indonesia (Batterson et al., 1989), and Rwanda (Hanson et al., 1991). The increase was curvilinear in Honduras, the rate of fish yield increase slowing with increased manure application (Green et al., 1990a). In Thailand, mean fish yields were not significantly different among the different application rates (Diana et al., 1991b).

Country-Specific Experiments

After 3 cycles of CRSP global experiments, the number of country sites was consolidated to Thailand, Honduras, and Rwanda. Thailand and Honduras sites concentrated on use of animal manures and inorganic fertilizers, while Rwanda focused on compost fertilizers. The country-specific studies are reviewed below. A comparison is then made between the Honduras and Thailand results, since their research programs were most similar.

Rwanda

Most Rwanda data was not comparable with data from other sites because compost, rather than animal manures, was employed as the principle source of nutrient input.

An experiment was designed to evaluate the effect on fish yields of increasing weekly inorganic phosphorus and nitrogen input from 7 to 36 kg/ha (Newman et al., 1994). No organic matter was applied. Mean fish yield was not significantly different among treatments. Yield tended to increase as nitrogen and phosphorus input increased to about 27 kg/ha/wk, but there was a decrease thereafter, probably because of high ammonia concentrations. High dissolved inorganic nitrogen and phosphorus in the ponds with the two highest input treatments indicated excessive fertilization. Fish yield was about as high as in ponds fertilized with chicken manure, but with a fifth or less of nitrogen and phosphorus as in chicken manure applications. The cooler growing temperature

(19 to 23°C) at high altitude (1700 m) in Rwanda was certain to moderate the inputs of nitrogen and phosphorus required for good fish growth, compared to warmer sites in Thailand and Honduras.

Thailand

The Thailand site assessed fertilization requirements by comparing the theoretical requirements of algae with nutrient inputs and dissolved inorganic C, P, and N (McNabb et al., 1990). The theoretical algal requirements were a ratio of 50:10:1 for C:N:P. Conclusions from manure fertilizer studies were that nitrogen was limiting algae and fish production in high alkalinity ponds (>100 mg/L $CaCO_3$), and that carbon was limiting production in ponds with adequate N and P concentrations when alkalinity was less than 50 mg/L $CaCO_3$ (McNabb et al., 1990). Inputs of manure alone caused nitrogen limitation because manure was poor in nitrogen relative to phosphorus. Weekly manure inputs could be decreased from 500 to 44 kg/ha as long as inorganic nitrogen was added to sufficiently replace the nitrogen in manure. It was reasoned that higher nitrogen applications were necessary, with or without manure to increase fish production.

Ponds fertilized with chicken manure at rates ranging from 44 to 200 kg/ha were supplemented with sufficient urea to decrease the N:P ratio from 7:1 to 5:1 (Batterson et al., 1990). Urea applications ranged from 5 to 22.4 kg/ha/wk, and fish production increased with increasing fertilizer input. Highest production at 2514 kg/ha per 5 months was obtained with additions of 200 kg/ha chicken manure and 22.4 kg/ha urea. These yields were greater than those obtained with chicken manure fertilization rates of 1000 kg/ha (Diana et al., 1991b).

Studies to further increase fish production were accomplished by supplementing chicken manure with both nitrogen and phosphorus (Knud-Hanson et al., 1993). Chicken manure was applied at rates ranging from 20 to 180 kg/ha/wk with urea and TSP supplement to give all treatments a nutrient application rate of 28 kg N and 7 kg P/ha/wk (4N:1P). Resultant fish yields ranged between 2186 and 3315 kg/ha per 5 months. Although yields tended to decrease with increasing manure input, differences among treatment means were insignificant. These data demonstrated that high yields could be obtained with low inputs of manure supplemented with both N and P.

A subsequent experiment evaluated chicken manure as a source of carbon in tilapia production (Knud-Hanson et al., 1993). All ponds were fertilized with urea and TSP at a rate of 28 kg N and 7 kg/P/ha/wk, respectively. Treatments consisted of control (no input), chicken manure (60 kg/ha), inorganic carbon (9.2 g $C/m^2/d$ as $NaHCO_3$), and a combined input of both. Results showed that neither the NPP nor mean fish yield was significantly different among treatments. Fish yield was correlated to NPP, however. It was concluded that chicken manure as a source of particulate carbon was of minor importance at best, and that neither manure nor bicarbonate supplementation served to increase fish production over inorganic fertilizers. With mean alkalinities in these ponds ranging from 47 to 99 mg/L/$CaCO_3$, carbon is not expected to limit algal production.

In summary, chicken manure inputs varying from 125 to 1000 kg/ha/wk resulted in tilapia yields of 1459 and 1936 kg/ha per 5 months when fish were stocked up to 1 $fish/m^2$. Total N and P inputs were 2.5 to 20 kg/ha and 3.8 to 30 kg/ha, respectively. Actual N and P available to phytoplankton after bacterial decomposition was probably 40 to 50% of total input values. Chicken manure rates were decreased to minimal levels, and inorganic sources of nitrogen and phosphorus increased to bring total N and P inputs to about 28 and 7 kg/ha/wk, respectively. Stocking rates were also increased up to about 3 $fish/m^2$. Fish yields increased to between 3500 and 4500 kg/ha in 150 d. The current recommendations for best fertilizer management given to Thai and southeast Asian farmers is to apply chicken manure weekly at 200 to 250 kg dry matter per hectare supplemented with sufficient inorganic nitrogen and phosphorus to make total N and P inputs of 28 and 7 kg/ha, respectively (James Szyper, personal communication). On more acidic soils, it is recommended that phosphorus input be increased to 10 to 14 kg/ha.

Fish yield was regressed on chicken manure input, total available N, and P input, and stocking rate in step-wise analysis. The step-one partial correlation coefficients of both stocking rate and available nitrogen input were 0.809. However, nitrogen input accounted for most (65%) of the total variation in yields. Stocking rate accounted for 16% of the residual variation. The two variables together accounted for 81% of the total variation in fish yields. Phosphorus was apparently not correlated to fish yields, probably because it was always applied in excess. However, no experiments were designed to specifically evaluate phosphorus requirements. Additional inputs of carbon as chicken manure were of apparently minor importance in the reviewed experiments, possibly because carbon was already in excess. Total alkalinities were usually greater than 100 mg/L $CaCO_3$.

Honduras

The results of fertilization experiments show that nitrogen was apparently limiting primary productivity in organically fertilized ponds, because dissolved inorganic nitrogen concentrations were persistently low (Teichert-Coddington et al., 1992a). Phosphorus effects on primary productivity could be eliminated because filterable orthophosphate concentrations were always high (>2 mg/L) (Teichert-Coddington et al., 1992a). It had been demonstrated that both fish yield and profitability increased with chicken manure input (Green et al., 1990c). A high rate of chicken manure application was therefore chosen as the best fertilization regime against which response to nitrogen supplementation with inorganic fertilizers could be evaluated. If nitrogen supplementation increased both fish yield and profitability, attempts could be made to lower organic input without sacrificing yields and profits.

An experiment was designed to evaluate supplemental nitrogen fertilization of ponds also fertilized with chicken litter at 750 kg/ha/wk (Teichert-Coddington et al., 1992a). The C:N ratio of inputs in ponds receiving only chicken litter was 11:1. Nitrogen as urea was added to reduce the C:N ratio to 8C:1N, 6C:1N, or 4C:1N. Weekly total nitrogen input ranged from 17.1 kg/ha for control ponds to 46.9 kg/ha for 4C:1N ponds. Response of fish yield to increased levels of supplemental nitrogen was discontinuous. Only the 6C:1N ratio resulted in significantly increased yield compared with the control treatment (11:1 ratio). Mean fish yields ranged between 2709 kg/ha per 5 months for 8:1 treatment to 3685 kg/ha for 6:1 treatment. Fertilizing at a 4C:1N ratio resulted in thick blue-green surface scums and reduced fish growth. Supplementing with nitrogen to achieve a C6:1N ratio helped to increase fish production to more than 50% greater than yields obtained with only chicken manure application.

It was desirable to lower organic inputs to ponds in order to decrease oxygen consumption from bacterial respiration. The next study (Teichert-Coddington et al., 1993b) evaluated effects of reducing chicken manure input while maintaining total nitrogen and phosphorus inputs at the 6C:1N ratio, as previously tested. Ponds were fertilized weekly with chicken manure at 750, 500, 250, or 0 kg dry matter per hectare. Urea and diammonium phosphate were applied to maintain total available nitrogen and phosphorus inputs at about 25 kg/ha and a N:P ratio of at least 4:1. Available nitrogen was calculated as 50% of total nitrogen in chicken manure. During both cool and hot season studies, primary productivity significantly increased with increasing chicken manure input. Mean fish yields during the cool season ranged between 1865 kg/ha at 0 manure input and 2435 kg/ha at 500 kg/ha manure input. Fish production was not correlated with primary production. Cool water temperatures are presumed to have hindered the fish growth, not allowing fish to take advantage of greater plankton biomass. During the hot season cycle, fish yield was significantly correlated with primary production, and fish yields ranged between 2079 kg/ha with no manure input to 3584 kg/ha with 500 kg/ha litter input. Chicken manure input was important for achieving high primary production and fish yields, probably by acting as a source of dissolved inorganic carbon for algae, which fish consumed (McNabb et al., 1990; Teichert-Coddington and Green, 1993b).

A step-wise regression of fish yield on chicken manure input, total available nitrogen, total available phosphorus input, and stocking density revealed that stocking density explained 49% of the total variation in yields. Chicken litter input accounted for 21% of the residual variation. The two variables together accounted for 71% of the total variation in fish yields. Total alkalinity of ponds during these experiments ranged from about 43 to 112 mg/L $CaCO_3$. Soluble reactive phosphate was always present in high concentration, so phosphorus was not responsible for the variation in fish yields observed during these trials. Nitrogen input was an important variable as described above, but chicken litter input accounted for much of the nitrogen in addition to dissolved inorganic carbon in Honduras experiments. The current recommendations for best fertilizer management given to Honduran and Central American farmers is to fertilize weekly with 500 kg/ha chicken litter (dry matter basis) and supplemental inorganic nitrogen to achieve total nitrogen inputs of 25 to 30 kg/ha. Phosphorus supplementation might be necessary on acidic soils. Fish should be stocked at least 2 fish/m^2.

Comparison of Thailand and Honduras Data

Fish yield was regressed by step-wise analysis on chicken manure input, total available nitrogen, total available phosphorus input, and stocking rate from combined Thailand and Honduras data. Results revealed that stocking rate explained the greatest amount of variation (59%) in fish yields. Available nitrogen explained 10% of the residual variation in step 2, and chicken manure application explained 4% of the remaining variation in step 3, for a total R^2 of 72%. Stocking density increased yields only if there was a sufficient nutrient supply. For example, doubling the stocking density from 1 to 2 fish/m^2 in Honduras when the only fertilizer being applied was chicken manure merely resulted in halving the mean individual fish weight without significantly increasing biomass (Green et al., 1994). However, biomass increased significantly at higher stocking density if input of total nitrogen was increased sufficiently to stimulate primary productivity. But there is a limit to increasing the stocking rate, because individual fish growth rate decreases with increasing stocking density in systems based on natural productivity. There is usually a minimum or preferential fish size for marketing. Thus, the culture period may have to be prolonged at higher stocking rates to obtain a marketable size of fish. Nutrient input, stocking rate, and culture duration must ultimately be manipulated to achieve a marketable size of fish in the most profitable manner.

The largest difference in production between sites was in response to chicken manure as a nutrient and organic carbon source. Thailand showed relatively little response to increasing weekly chicken manure input from 125 to 1000 kg/ha. Fish yields increased by 33% from 1459 to 1936 kg/ha. Honduras demonstrated a 100% gain over the same chicken manure input levels; fish yields increased from 1162 to 2329 kg/ha. Later Thailand studies indicated that chicken manure was of minor importance as a particulate or dissolved inorganic carbon source in ponds of moderate total alkalinity with high input of inorganic nitrogen and phosphorus. In comparison, Honduras demonstrated that organic matter input was important to high primary productivity and fish yields under similar conditions. One difference between the two data sets was that Thailand evaluated manure additions at low inputs (0 to 180 kg/ha) relative to Honduras (0 to 750 kg/ha). Honduras showed the greatest benefit from applications of 500 kg/ha. Thailand might have shown a positive response to organic inputs had the rates of input been higher.

Both sites demonstrated that relatively high inputs of nitrogen were necessary for high primary productivity and fish yields, although results were variable. These data are generally contrary to earlier fertilization studies that indicated little fish yield response to supplemental nitrogen inputs (Boyd, 1990). Differences in response to nitrogen fertilizers are reasonable. The current work was conducted with higher rates of nitrogen application in ponds stocked exclusively with tilapia at relatively high stocking rates for at least 150 d in tropical waters where phosphorus was in excess.

Tilapia are phytophagous and able to take advantage of primary productivity enhanced by high fertilizer inputs. Stocking rates of at least 2 fish/m^2 and sufficiently long growth periods were necessary for demonstrating increased yields relative to high fertilizer rates. Tropical water temperatures were a prerequisite to high growth rates. For example, a preliminary nitrogen supplementation experiment of 126 d with a stocking rate of 1/m^2 demonstrated significantly higher primary productivity but no difference in fish yields in Honduras (Green et al., 1994). Two similar experiments were repeated for 150 d with stocking rates of 2/m^2 (Teichert-Coddington et al., 1992a, 1993b). Results indicated that yields in addition to primary productivity could be significantly increased with nitrogen supplementation.

The fertilizer recommendations given to producers in both Thailand and Honduras are similar, despite different research approaches. These recommendations may not be optimum, however, because the optimum biological and economical nitrogen and phosphorus requirements were not established at either site. That research is underway. Nutrient requirements at acidic sites like Panama will be different because of nutrient absorption by the soils. The rate of phosphorus application relative to nitrogen application will likely be higher in these soils.

RECOMMENDATIONS FOR FERTILIZING TILAPIA PONDS

Fertilization Strategies

The PD/A CRSP experiments have used three strategies in developing fertilization guidelines for tilapia ponds. The first is fixed input strategy based on the results of a series of field trials with different quantities and types of fertilizer to establish the relationships between input levels and fish yields (i.e., production functions). The producers can then use these relationships to determine the type and level of inputs for maximal net revenue. Net revenues are calculated by subtracting the input costs from fish values produced at each level of input. As climate, soil, water, and available nutrients can differ widely among geographic regions, the production trials must be conducted in each region. The fixed input strategy can be unreliable because it does not compensate for differences between ponds in the same geographic area or for changes in nutrient limitation during the culture period (Hopkins and Knud-Hansen, unpublished data). For example, carbon limitation can occur in ponds, caused by abundant snails and clams that sequester $CaCO_3$ (Knud-Hansen et al., 1993). If N fertilization continues while C limitation is not relieved by addition of lime, NH_3 can increase to potentially lethal levels. The fixed input strategy requires a substantial up-front cost to establish the production functions. The second strategy is fixed nutrient concentrations determined by frequent water analyses for the concentration of various nutrients. Years of research in limnology and aquaculture have developed algorithms that estimate the nutrient levels required to attain maximum potential primary productivity (MPPP) at a given set of available light and temperature. The amount of fertilizer required is determined by subtracting the nutrient concentration in pond water from the estimated nutrient concentrations needed to attain MPPP. The PD/A CRSP has developed microcomputer-based expert systems, (PONDCLASS© and its successor, POND©), which can perform complex computations to implement this strategy (Lannan, 1993). The fixed nutrient concentration or variable input strategy responds to changes in nutrient limitation and, therefore, is potentially more efficient and reliable than the fixed input strategy. However, there are serious concerns about the practicality of performing frequent water quality analyses to estimate nutrient requirements for individual ponds (Boyd, 1990). The third strategy is an algal nutrient assay developed at the PD/A CRSP project in Thailand (Guttman, 1990). This method is relatively simple and can be implemented by farmers with minimal technical training. Also, it does not require access to computers or sophisticated water quality tests. To determine nutrient limitation for algal growth in a given pond, N, P, and C are added alone or combined to several pond water samples. Algal growth response to nutrient is determined after *in situ* incubation by comparing the relative algal densities. This comparison is accomplished by filtering water samples through a small disk filter using a syringe to provide the necessary pressure and visually comparing the disk filters. For example, if the filter disk from the sample supplemented with both N and P is darker than the disks from samples supplemented with N or P alone, both N and P are limiting. This strategy also responds to changes in nutrient limitation and has the potential to be more reliable and cost-effective than the fixed input strategy. The greatest drawback of nutrient bioassay is its isolation from pond muds, which often dominate the fertilization efficiency.

Fertilization Rates

The recommended rates for fertilization of tilapia ponds depend upon numerous factors, including soil and water type, relative costs of locally available fertilizers, and whether maximum or optimum production is desired. Usually, the optimum level will be less than the maximum level because of the "law of diminishing returns." Integrated livestock-fish farming systems can be an exception if the aquaculture component of the farm is used to process wastes for which a market may not exist. In that case, maximum and optimum fertilization levels will be the same. Maximum productivity using the algal assay and nutrient concentration strategies under tropical conditions in Thailand required, on average, 24 kg N/ha/wk (Hopkins and Knud-Hansen, unpublished data). The N:P ratio was approximately 7 to 12:1. In order to prevent C limitation at these high N and P levels, alkalinity needs to remain above 50 to 75 mg/L. Average extrapolated yields of male Nile tilapia stocked at 3 fish/m^2 and fertilized using these strategies ranged from 4500 to 5000 kg/ha/yr. Production functions relating nutrient inputs to tilapia yields have been developed for several different types of inputs. A parabolic relationship has been shown between the amount of chicken manure input and fish yield in Honduras (Green et al., 1994) and for pigs in the Philippines (Hopkins et al., 1982). In all of these cases, maximum yields corresponded to N loading rates of approximately 2 to 4 kg N/ha/d. A prime characteristic of these production functions is that the curves are relatively flat. Therefore, high yields are still attainable at fertilization rates considerably below the rate that produced maximum yields. The level of phosphorus fertilization is not as well defined at very high nutrient levels. Based on proximate analysis, the N:P ratio of typical algae is approximately 8 to 10:1. Thus, the fertilizers should provide that level of P. However, as pond soils can have a significant impact on the availability of P, the resultant input N:P ratio is generally much lower than that of cellular content. In ponds with acid-sulfate soils, the N:P ratio had to be decreased to levels as low as 1:1 and 2:1 when the P was sequestered in the soils. However, after these ponds had been heavily fertilized for several years, the need for P decreased to levels approximating the N:P ratio of 8:1. The relative merits of inorganic and organic fertilizers in tilapia ponds has been a point of contention within the PD/A CRSP and elsewhere. Based on experiments conducted in Thailand, Knud-Hansen et al. (1993) stated that adding chicken manure to inorganic fertilization did not increase yield of *O. niloticus*. On the other hand, the PD/A CRSP project in Honduras has found the opposite effect (Green et al., 1989). Similar contradictions can be found in the literature (Schroeder et al., 1990; Coleman and Edwards, 1987). A possible explanation for the contradictory evidence is related to the levels of fertilization intensity and the ability of manure to reduce clay turbidity. At low nutrient input levels, primary productivity is relatively low and manure-derived detritus makes a significant contribution to tilapia growth. When nutrient input levels and primary productivity increase, the relative importance of the manure-derived detritus decreases as algae and algae-based detritus become more important (Knud-Hansen et al., 1993). Also, when ponds have significant amounts of clay turbidity, adding manures will cause the clay to settle, thereby improving light penetration (Boyd, 1990). Given the shallowness and turbidity of the CRSP ponds in Honduras, the positive results from the addition of up to 500 kg chicken manure ha/wk may be a result of a reduction of that turbidity.

Fertilization Frequency

The optimization of fertilization frequencies has the potential to reduce operating costs by reducing labor requirements through less frequent fertilization, while still ensuring high algal production. An additional factor to consider when fertilizing with manures is the possibility that deoxygenation exists when large quantities of manure are added at one time. For example, Zhu et al. (1990) showed that applying pig manure daily produced higher yields of various carps than did 5- or 7-d application schedules. Fixed input fertilization schedules typically require constant fertilization (kg/ha time period) throughout the culture cycle. However, much of the early integrated animal-fish farming work was based on the number of animals instead of the quantity of manure (e.g., Hopkins and Cruz, 1982). The reasoning behind this practice was that the number of animals was much easier to manage and less noxious than measuring manure. Also, at that time, the consensus among researchers in integrated farming systems was that manure provided substantial nutritive value directly and through manure-derived detritus. Thus, small fish should have had a lesser need for manure than did large fish. Typical animal-fish farming systems started culture cycles with small animals and small fish. During the cycle, the manure levels increased as the fish grew. But we now know that, at high nutrient loads, algae and algae-derived detritus are of much more importance to fish yield than manure and manure-derived detritus. Aquaculturists who use algae to feed larval fish and shrimp are well aware of the concept "Feed the tank, not the fish!" This concept means that a fixed concentration of live food (e.g., algae) is maintained regardless of the number or size of larval organisms being fed. The reason is that feeding is the result of a chance encounter between the food and the larvae. For filter-feeding organisms, the amount of algae harvested is strongly related to the concentration of the algae in the water. Increasing manure levels throughout a culture cycle leads to lower algae concentrations at the beginning of the cycle and high algae concentrations at the end. But filter-feeding tilapia need high levels of algae throughout the culture cycle. Teichert-Coddington et al. (1990) showed that increasing inputs of pig manure led to lower primary productivity and tilapia yield than did constant or decreasing manure inputs. In this experiment, the total nutrient input was the same in all treatments. The PD/A CRSP studied the question of the most efficient frequency for fertilization with urea and triple superphosphate in Thailand (Knud-Hansen and Batterson, 1994). In that study, five fertilization frequencies were tested: daily, twice per week, weekly, twice every 3 weeks, and once every 2 weeks. The fertilization rates were 28 kg N/ha/wk and 7 kg P/ha/wk. Limitation of N or P did not occur in any of the treatments, and there was no relationship between fish yield and fertilization frequency. This indicates that fertilization with urea and TSP does not need to be more frequent than once every 2 weeks. The fertilization frequencies discussed above assume that there is a continuous need for fertilizer. However, PD/A CRSP research using nutrient concentration and algal assay fertilization strategies have shown that nutrient limitations can change throughout the culture period and between culture periods. Fertilization frequency should be based on algal need, not a set schedule.

Summary of Recommendations

1. Maximum yields of tilapia are attained with nitrogen inputs of 2 to 4 kg N/ha/d; phosphorous input levels should be sufficient to prevent P limitation. Typically, a N:P ratio of 4:1 is used. In acid-sulfate soils, the P input level might have to increase as high as an N:P ratio of 1:1.
2. Lime should be applied to maintain alkalinities above 50 to 75 mg $CaCO_3$/L when fertilizing at maximal rates.
3. The type of fertilizer, organic or inorganic, is not of particular importance. The cost of nutrients and their availability are essential factors in selecting which nutrient source to choose in a particular locale.

4. Fertilization at fixed levels of nutrient input are superior to increasing the level of nutrient input over the culture cycle.
5. If fixed levels of nutrient inputs are used, the fertilization frequency should be daily for manures and once every 1 to 2 weeks for inorganic fertilizers.
6. Quantity and frequency of fertilization is best determined by using simple techniques to detect and adjust nutrient limitation, such as nutrition concentration and algal assay methods.

REFERENCES

Abdalla, A. A. F., The effect of ammonia on *Oreochromis niloticus* (Nile tilapia) and its dynamics in fertilized tropical fish ponds (abstract), in *Seventh Annual Administrative Report, Pond Dynamics/Aquaculture CRSP 1989,* Egna, H. S., Bowman, J., and McNamara, M., Eds. Oregon State University, Corvallis, 1990, 52.

Barash, H., Plavnik, I., and Moav, R., Integration of duck and fish farming: experimental results, *Aquaculture,* 27, 129, 1982.

Batterson, T. R., McNabb, C. D., Knud-Hansen, C. F., Eidman, H. M., and Sumatadinata, K., Indonesia: Cycle III of The Global Experiment, in *Pond Dynamics/Aquaculture CRSP Data Reports,* Vol. 3, No. 3, Egna, H. S., Ed., Oregon State University, Corvallis, 1989, 135.

Batterson, T. R., McNabb, C. D., and Knud-Hansen, C. F., Yields of tilapia with nitrogen-supplemented organic fertilizers in fish ponds in Thailand, in *Seventh Annual Administrative Report,* Egna, H. S., Bowman, J., and McNamara, M., Eds., Pond Dynamics/Aquaculture CRSP 1990, Oregon State University, Corvallis, 1990, 30.

Bowen, S. H., Feeding, digestion and growth — qualitative considerations, in Pullin and Lowe-McConnel, Eds., *Proc. on the Biology and Culture of Tilapias,* ICLARM, Manila, 1982, 141.

Boyd, C. E., Lime requirements of Alabama fish ponds, Ala. Agr. Exp. Sta., Auburn University, *Ala. Bull.,* 459, 1974.

Boyd, C. E., Nitrogen fertilizer effects on production of tilapia in ponds fertilized with phosphorus and potassium, *Aquaculture,* 7, 385, 1976.

Boyd, C. E., Comparison of five fertilizer programs for fish ponds, *Trans. Am. Fish. Soc.,* 110, 541, 1981.

Boyd, C. E., *Water Quality in Ponds for Aquaculture,* Auburn University, Auburn, AL, 1990.

Boyd, C. E. and Musig, Y., Orthophosphate uptake by phytoplankton and sediment, *Aquaculture,* 22, 165, 1981.

Boyd, C. E. and Sowles, J. W., Nitrogen fertilization in ponds, *Trans. Am. Fish. Soc.,* 107, 737, 1978.

Brinckman, R., Social and economic aspects of reclamation of acid sulfate soil areas, in *Proc. Int. Symp. on Acid Sulfate Soils,* Publ. No. 32, Wageningen, The Netherlands, 1982, 21.

Brinckman, R. and Singh, V. P., Rapid reclamation of brackish water ponds in acids sulfate soils, in *Proc. of Int. Symp. on Acid Sulfate Soils,* Publ. No. 31, Wageningen, The Netherlands, 1982, 318.

Buck, D. H., Baur, R. J., and Rose, C. R., Experiments in recycling swine manure in fish ponds, in Pillay, T. V. R. and Dill, W. A., Eds., *Advances in Aquaculture,* Fishing Book News Ltd., Farnham, England, 1979.

Castanares, M. A. G., Little, D. C., Yakupitiyage, A., Edwards, P., and Lovshin, L., Feeding value of fresh perennial leguminous shrub leaves to Nile tilapia (*Oreochromis niloticus*), in *Proc. Network Meeting on Aquaculture and Schistosomiasis,* National Academy Press, Washington, D.C., 1991.

Coleman, J. A. and Edwards, P., Feeding pathways and environmental constraints in waste-fed aquaculture: balance and optimization, in Moriarty, D. J. W. and Pullin, R. S. V., Eds., *Detritus and Microbial Ecology in Aquaculture,* ICLARM, Manila, 1987, 240.

Collis, W. J. and Smitherman, R. O., Production of *Tilapia aurea* with cattle manure or a commercial diet, in *Symposium Culture of Exotic Fishes,* Shelton, W. L. and Grover, J. H., Eds., Am. Fish. Soc., 1978.

Cooke, C. W., *Fertilizing for Maximum Yield,* The English Language Book Soc., London, 1982.

Dettweiler, D. J. and Diana, J. S., The effects of stocking density and rate of fertilization on growth and reproduction of *Oreochromis niloticus* in earthen ponds, in Egna, H. S., Bowman, J., and McNamara, M., Eds., *Eighth Annual Administrative Report,* Collaborative Research Support Program 1990, Oregon State University, Corvallis, 1991, 50–54.

Diana, J. S., Lin, C. K., Bhukaswan, T., and Sirsuwanatach, V., Thailand: Cycle I of The Global Experiment, in Egna, H. S. Ed., *Pond Dynamics/Aquaculture CRSP Data Reports,* Vol. 2, Oregon State University, Corvallis, 1987, 47.

Diana, J. S., Lin, C. K., Bhukaswan, T., Sirsuwanatach, V., and Buurma, B. J., Thailand: Cycle III of The Global Experiment, in *Pond Dynamics Collaborative Research Data Reports,* Egna, H. S. and Bowman, J., Eds., Oregon State University, Corvallis, 1991, 86.

Diana, J. S., Lin, C. K., Edwards, P., and Jaiyen, K., The effect of stocking density of *Oreochromis niloticus* on the dynamics of aquaculture ponds, in *Seventh Annual Administrative Report,* Egna, H. S., Bowman, J., and McNamara, M., Eds., Pond Dynamics/Aquaculture CRSP, Oregon State University, Corvallis, 1990, 34.

Diana, J. S., Lin, C. K., and Schneeberger, P. J., Relationship among nutrient inputs, water nutrient concentrations, primary production and yield of *Oreochromis niloticus* in ponds, *Aquaculture,* 92, 323, 1991.

Dougdale, R. C. and Dougdale, V. A., Nitrogen metabolism in lakes. II. Role of nitrogen fixation in Sanctuary Lake, Pennsylvania, *Limnol. Oceanogr.,* 7, 170, 1962.

Edwards, P., Use of terrestrial vegetation and aquatic macrophytes, in Moriarty, D. J. W. and Pullin, R. S. V., Eds., *Detritus and Microbial Ecology in Aquaculture,* ICLARM, Manila, 1987, 311.

Edwards, P., Integrated fish farming, *Infofish Int.,* 5, 45, 1993.

Egna, H. S., Brown, N., and Leslie, M., Eds., General reference: site descriptions, materials and methods for the global experiment, in *Pond Dynamics/Aquaculture CRSP Data Reports,* Oregon State University, Corvallis, 1987, 84.

El Samra, M. I. and Olah, J., Significance of nitrogen fixation in fish ponds, *Aquaculture,* 18, 367, 1979.

Goldman, J. C., Physiological processes, nutrient availability and concept of relative growth rate in marine phytoplankton ecology, in Falkowski, P. G., Ed., *Primary Productivity in the Sea,* Plenum Press, New York, 1980, 179.

Gooch, B. C., Appraisal of North American fish culture fertilization studies, *FAO Fish Rep.,* 44, 13, 1967.

Green, B. W., Substitution of organic manure for pelleted feed in tilapia production, *Aquaculture,* 101, 213, 1992.

Green, B. W., Water and Chemical Budgets for Organically Fertilized Fish Ponds in the Dry Tropics, Ph.D. dissertation, Auburn University, Auburn, AL, 1993.

Green, B. W., Alvarenga, H. R., Phelps, R. P., and Espinoza, J., Honduras: Cycle III of the CRSP Global Experiment, *Pond Dynamics/Aquaculture CRSP Data Reports,* Oregon State University, Corvallis, 1989a.

Green, B. W., Alvarenga, H. R., Phelps, R. P., and Espinoza, J., Honduras: Cycle II of the CRSP Global Experiment, *Pond Dynamics/Aquaculture CRSP Data Reports,* Oregon State University, Corvallis, 1990a.

Green, B. W., Alvarenga, H. R., Phelps, R. P., and Espinoza, J., Honduras: Cycle I of the CRSP Global Experiment, *Pond Dynamics/Aquaculture CRSP Data Reports,* Oregon State University, Corvallis, 1990b.

Green, B. W., El Nagdy, Z., and El Gamal, A. R., Validation of PD/A CRSP Pond Management Strategies, *Twelfth Annual Technical Report 1994,* Egna, H. S., Bowman, J., Goetze, B., and Weidner, N., Eds., Pond Dynamics/Aquaculture Collaborative Research Support Pogram, Oregon State University, Corvallis, 1995, 12–17.

Green, B. W., Phelps, R. P., and Alvarenga, H. R., The effect of manures and chemical fertilizers on the production of *Oreochromis niloticus* in earthen ponds, *Aquaculture,* 76, 37, 1989b.

Green, B. W., Teichert-Coddington, D. R., and Hanson, T. R., Development of Semi-Intensive Aquaculture Technologies in Honduras: Summary of Freshwater Aquacultural Research Conducted from 1983 to 1992, *Research and Development Ser.* No. 39, Auburn University, Auburn, AL, 1994.

Green, B. W., Teichert-Coddington, D. R., and Phelps, R. P., Response of tilapia yield and economics to varying rates of organic fertilization and season in two Central American countries, *Aquaculture,* 90, 279, 1990.

Guttman, H., Assessment of Nutrient Limitationin Fertilized Fish Ponds by Algal Assay, M.S. thesis, Asian Institute of Technology, Bangkok, Thailand.

Hanson, B., Ndoreyaho, V., Rwangano, F., Tubb, R., and Seim, W. K., Rwanda: Cycle III of The Global Experiment, Egna, H. S. and Bowman, J., Eds., *Pond Dynamics/Aquaculture CRSP Data Reports,* Oregon State University, Corvallis, 1991, 102.

Hanson, B., Ndoreyaho, V., Tubb, R., Rwangano, F., and Seim, W. K., Rwanda: Cycle I of the CRSP Global Experiment, *Pond Dynamics/Aquaculture CRSP Data Reports,* Oregon State University, Corvallis, 1989.

Hickling, C. F., *Fish Culture,* Faber and Faber, London, 1962.

Hopkins, K. D. and Cruz, E. M., The ICLARM-CLSU Integrated Animal-Fish Farming Project: Final Report, *ICLARM Tech. Rep* 5, ICLARM, Manila, 1982.

Hora, S. L. and Pillay, T. V. R., Handbook on Fish Culture in the Indo-Pacific Region, FAO Fish. Biol. Tech. Paper No. 14, 1962.

Horne, A. J. and Viner, A. B., Nitrogen fixation and its significance in tropical Lake George, Uganda, *Nature,* 232, 417, 1971.

Hunt, D. and Boyd, C. E., Alkalinity losses from ammonium fertilizers used in fish ponds, *Trans. Am. Fish. Soc.,* 110, 81, 1981.

Jiwyam, W., Nitrogen Fixation of Blue-Green Algae and Its Nutritional Values in Fish Culture, Doctoral thesis, Asian Institute of Technology, Bangkok, Thailand, 1995.

King, D. L., The role of carbon in eutrophication, *J. Water Pollut. Contr. Fed.,* 42, 2035, 1970.

Knud-Hansen, C. F., Pond history as a source of error in fish culture experiments: a quantitative assessment using covariate analysis, *Aquaculture,* 105, 21, 1992.

Knud-Hansen, C. F. and Batterson, T. R., Effect of fertilization frequency on the production of Nile tilapia (*Oreochromis niloticus*), *Aquaculture,* 123, 271, 1994.

Knud-Hansen, C. F., Batterson, T. R., and McNabb, C. D., The role of chicken manure in the production of Nile tilapia, *Oreochromis niloticus* (L.), *Aquaculture Fish. Manage.,* 24, 483, 1993.

Knud-Hansen, C. F., Batterson, T. R., McNabb, C. D., Harahat, I. S., Sumantadinata, K., and Eidman, H. M., Nutrient input, primary productivity, and fish yield in fertilized freshwater ponds in Indonesia, *Aquaculture,* 94, 49, 1991.

Knud-Hansen, C. F. and Lin, C. K., Strategies for stocking Nile tilapia (*Oreochromis niloticus*) in fertilized ponds, in Egna, H. S., McNamara, M., Bowman, J., and Astin, N., Eds., *Tenth Annual Administrative Report,* Pond Dynamics/Aquaculture CRSP, Oregon State University, Corvallis, 1993, 275.

Kuhl, A., Inorganic phosphorus uptake and metabolism, in Lewin, R. A., Ed., *Physiology and Biochemistry of Algae,* Academic Press, New York, 1962, 211.

Kuhl, A., Phosphorus, in Stewart, W., Ed., *Algal Physiology and Biochemistry,* University of California Press, Berkeley, 1974.

Lannan, J. E., *User's Guide to PondClass© Guidelines for Fertilizing Aquaculture Ponds,* Pond Dynamics/ Aquaculture CRSP, Oregon State University, Corvallis, 1993.

Lin, C. K., Accumulation of water soluble phosphorus and hydrolysis of polyphosphates by *Cladophora glomerata, J. Phycol.,* 13, 46, 1978.

Lin, C. K., Acidification and reclamation of acid sulfate soil fishponds in Thailand, in *Proc. of the First Asian Fisheries Forum,* Mclean, J., Dizon, L. B., and Hoshilos, L. V., Eds., Asian Fisheries Society, Manila, 1986, 71.

Lin, C. K. and Diana, J. S., Co-culture of Nile tilapia (*Oreochromis niloticus*) and hybrid catfish (*Clarias mocrocephalus x C. gariepinus*) in earthen ponds, *Living Aquatic Nat. Res.,* 1995.

Lin, C. K., Jaiyen, K., and Muthuwan, V., Intergration of intensive and semi-intensive aquaculture: concept and example, *Thai Fish. Gaz.,* 425, 1989.

Lin, C. K., Tansakul, V., and Apinhapath, C., Biological nitrogen fixation as a source of nitrogen input in fishponds, in *The Second International Symposium on Tilapia in Aquaculture,* Pullin, R. S. V., Bhukaswan, T., Tonguthai, T., and Maclean, J., Eds., ICLARM Conference Proceedings, ICLARM, Manila, 1988, 53.

Lin, S. Y., *Pond Fish Culture and the Economy of Inorganic Fertilizer Application,* Chinese American Joint Commission on Rural Reconstruction, Fish. Ser. 6, 1968.

Ling, S. W., Feed and feeding of warmwater fishes in ponds in Asia and the Far East, *FAO Fish. Rep.,* 44, 291, 1967.

Little, D., Yakupitiyage, A., Edwards, P., Lovshin, L. L., Use of Leguminous Leaves as Fish Pond Inputs, Final Report to U.S. AID, Asian Institute of Technology, 1995.

Madhav, S. and Lin, C. K., Phosphorus fertilization strategy in fish pond based on sediment phosphorus saturation level, *Aquaculture,* in press, a.

Madhav, S. and Lin, C. K., Determination of phosphorus saturation level in relation to clay content in pond mud, *J. Aquaculture Eng.,* in press, b.

McNabb, C. D., Batterson, T. R., Premo, B. J., Eidman, H. M., and Sumatadinata, K., Indonesia: Cycle I of The Global Experiment, Egna, H. S., Ed., *Pond Dynamics/Aquaculture CRSP Data Reports,* Pond Dynamics CRSP, Oregon State University, Corvallis, 1988.

McNabb, C. D., Batterson, T R., Premo, B. J., Eidman, H. M., and Sumatadinata, K., Indonesia: Cycle II of The Global Experiment, in Bowman, J. and Egna, H. S., Eds., *Pond Dynamics/Aquaculture CRSP Data Reports,* Oregon State University, Corvallis, 1991, 49.

McNabb, C. D., Batterson, T. R., Premo, B. J., Knud-Hansen, C. F., Eidman, H. M., Lin, C. K., Jaiyen, K., Hanson, J. E., and Chuenpagdee, R., Managing fertilizers for fish yield in tropical ponds in Asia, in R. A. I. H., Eds., *The Proceedings of Second Asian Fisheries Forum,* The Asian Fisheries Society, Manila, 1990, 169.

Milstein, A., Alkon, A., and Karplus, I., Combined effects of fertilization rate, manuring and feed pellet application on fish performance and water quality in polyculture ponds, *Aquaculture Res.,* 26, 55, 1995.

Moriarty, D. J. W. and Moriarty, C. M., The physiology of digestion of blue-green algae in the cichlid fish *Tilapia nilotica, J. Zool.,* 171, 25, 1973.

Mortimer, C. F., *Fertilization in Fish Ponds,* Fish. Publ. No. 5, Her Majesty's Stationary Office, London, 1954.

Nerrie, B., Production of Male *Tilapia nilotica* Using Pelleted Chicken Manure, M.S. thesis, Auburn University, Auburn, AL, 1979.

Newman, J. R., Gatera, A., Seim, W. K., Popma, T. J., and Veverica, K. L., Nitrogen requirements for maximum fish production in Rwandan ponds, in Egna, H. S., Bowman, J., Goetze, B., and Weidner, N., Eds, *Eleventh Annual Administrative Report 1993,* Pond Dynamics/Aquaculture CRSP, Oregon State University, Corvallis, 1994.

Olah, J., Carp production in manured ponds, in Billard, R. and Marcel, J., Eds., *Aquaculture of Cyprinids,* IRNA, Paris, 1980, 295.

Parsons, T. R. and Takahashi, M., *Biological Oceanographic Processes,* Pergamon Press, New York, 1973.

Pons, L. J., *Acid Sulfate Soils in Thailand,* Rep. SSR-81-1969, Land Development Dept., Ministry of Agriculture and Cooperatives, Thailand, 1969.

Prose, G. A., A review of the methods of fertilizing the warmwater fish ponds in Asia and the Far East, *FAO Fish Rep.,* 44, 7, 1967.

Redfield, A. C., Ketchum, B. H., and Richards, F. A., The influence of organisms on the composition of seawater, in Hill, M. A., Ed., *The Sea,* Vol. 2, Interscience, New York, 1963, 27.

Rurangwa, E., Verheust, L., and Veverica, K., Fertilizer input method influences pond productivity, in Seventh Annual Administrative Report, Egna, H. S., Bowman, J., and McNamara, M., Eds., Pond Dynamics Aquaculture CRSP, Oregon State University, Corvallis, 1990.

Rushton, Y. and Boyd, C. E., New development in pond fertilization: liquid, controlled-release and instantly-soluble fertilizers, *WAS Abstracts, World Aquaculture 96,* 1996.

Schindler, D. W., Brunskill, G. J., Emerson, S., Broecker, W. S., and Peng, T. H., Atmospheric carbon dioxide: its role in maintaining phytoplankton standing crops, *Science,* 177, 1192, 1972.

Schlesinger, W. H., *Biogeochemistry, An Analysis of Global Change,* Academic Press, New York, 1991.

Schroeder, G. L., Cow manure in fish culture, *FAO Aquaculture Bull.,* 7, 6, 1975.

Schroeder, G. L., Wohlfarth, G. W., Alkon, A., Halevy, A., and Krueger, H., The dominance of algal-based food webs in fish ponds receiving chemical fertilizers plus organic manures, *Aquaculture,* 86, 219, 1990.

Shevgoor, L., Knud-Hansen, C. F., and Edwards, P., An assessment of the role of buffalo manure for pond culture of tilapia. III. Limiting factors, *Aquaculture,* 126, 107, 1994.

Singh, V. P., Management and Utilization of Acid Sulfate Soils for Aquaculture: a Monograph, University of the Philippines, Visaya, 1985.

Swingle, H. S., Gooch, B. C., and Rabanal, H. R., Phosphate fertilization of ponds, *Proc. Annu. Conf. Southeast Game Comm.,* 1965, 213.

Szyper, J. P. and Lin, C. K., Techniques for assessment of stratification and effects of mechanical mixing in tropical fish ponds. *Aquaculture Engineering.* 9, 151, 1990.

Tacon, A. G. J., The Nutrition and Feeding of Farmed Fish and Shrimp — A Technical Manual; 1. The Essential Nutrients, FAO Report GCR/RLA/ITA, Brasilia, Brazil, 1987, 117.

Tang, Y., Evaluation of balance between fishes and available fish food in multispecies fish culture ponds in Taiwan, *Trans. Am. Fish. Soc.,* 99, 708, 1970.

Teichert-Coddington, D., Behrends, L., and Smitherman, R., Effects of manuring regime and stocking rate on primary production and yield of tilapia using liquid swine manure, *Aquaculture,* 88, 61, 1990.

Teichert-Coddington, D. R. and Claros, N., Nitrogen fertilization in the presence of adequate phosphorus, in Thirteenth Annual Administrative Report, Egna, H. S., Goetze, B., Burke, D., McNamara, M., and Clair, D., Eds., Pond Dynamics/Aquaculture CRSP 1995, Oregon State University, Corvallis, 1996, 18–26.

Teichert-Coddington, D. R. and Green, B. W., Tilapia yield improvement through maintenance of minimal oxygen concentrations in experimental grow-out ponds in Honduras, *Aquaculture,* 118, 63, 1993a.

Teichert-Coddington, D. R. and Green, B. W., Usefulness of inorganic nitrogen in organically fertilized tilapia production ponds, *Abstract of World Aquaculture '93,* Torremolinos, Spain, May 26–28, 1993, European Aquaculture Society Special Publication No. 19, Oostende, Belgium, 1993b, 273.

Teichert-Coddington, D. R., Green, B. W., Boyd, C., Gomez, R., and Claros, N., Substitution of inorganic nitrogen and phosphorus for chicken litter in production of tilapia, in *Tenth Annual Administrative Report,* Pond Dynamics/Aquaculture CRSP, 1992, Egna, H. S., McNamara, M., Bowman, J., and Astin, N., Eds., Oregon State University, Corvallis, 1993, 19–27.

Teichert-Coddington, D. R., Green, B. W., Boyd, C. E., and Rodriguez, M. I., Supplemental nitrogen fertilization of organically fertilized ponds, in *Eighth Annual Administrative Report,* Egna, H. S., Bowman, J., and McNamara, M., Eds., Pond Dynamics Aquaculture CRSP, Oregon State University, Corvallis, 1991.

Teichert-Coddington, D. R., Green, B. W., Boyd, C. E., and Rodriguez, M. I., Supplemental nitrogen fertilization of organically fertilized ponds: variation of the C:N ratio, in *Ninth Annual Administrative Report,* Pond Dynamics/Aquaculture CRSP, Egna, H. S., McNamara, M., and Weidner, N., Eds., Oregon State University, Corvallis, 1992.

Teichert-Coddington, D. R., Green, B. W., and Phelps, R. P., Influence of site and season on water quality and tilapia production in Panama and Honduras, *Aquaculture,* 105, 297, 1992.

Teichert-Coddington, D. R., Peralta, M., Phelps, R. P., and Pretto, M. R., Panama: Cycle III of the CRSP Global Experiment 1990, *Pond Dynamics/Aquaculture CRSP Data Reports,* Oregon State University, Corvallis, 1990.

Teichert-Coddington, D. R., Peralta, M., Phelps, R. P., and Pretto, M. R., Panama: Cycle I of the CRSP Global Experiment, *Pond Dynamics/Aquaculture CRSP Data Reports,* Oregon State University, Corvallis, 1991.

Ullah, A. Md., Nutrient Release Characteristics of Duck Manure for Nile Tilapia Production, AIT thesis AE-89-43, Asian Institute of Technology, Bangkok, Thailand, 1989.

Veverica, K. L., Popma, T. J., Rwalinda, P., and Seim, W. K., Production of *Oreochromis niloticus* as a function of organic matter application rates and supplementation with inorganic nitrogen and phosphorus fertilization, in *Eighth Annual Administrative Report,* Egna, H. S., Bowman, J., and McNamara, M., Eds., Pond Dynamics/Aquaculture CRSP 1990, Oregon State University, 1991, 42–45.

Veverica, K. L., Rurangwa, E., Verteust, L., Popma, T., Seim, W., and Tubb, R., Tilapia production in Rwandan ponds is influenced by composting methods, in *Seventh Annual Administrative Report,* Egna, H. S., Bowman, J., and McNamara, M., Eds., Pond Dynamics/Aquaculture CRSP, Oregon State University, Corvallis, 1990, 24.

Vuthana, H., Fish Pond Turbidity, M.Sc. thesis (No. AE 95-28), Asian Institute of Technology, Bangkok, Thailand, 1995.

Watanabe, A. and Yamamoto, Y., Algal nitrogen fixation in the tropics, in Lei, T. A. and Mulder, E. G., Eds., *Biological Nitrogen Fixation in Natural and Agricultural Habitats, Plants and Soil Special Volume,* 403, 1971.

Wetzel, R. G., *Limnology,* 2nd ed., Saunders College Publishing, New York, 1983.

Wohlfarth, G., Utilization of manure in fish farming, *Proc. Fishfarming and Wastes Conf.,* University College, London, Janssen Services, 1978, 78.

Wohlfarth, G. and Hulata, G., Use of manures in aquaculture, in Moriarty, D. J. W. and Pullin, R. S. V., Eds., *Detritus and Microbial Ecology in Aquaculture,* ICLARM, Manila, 1987, 353.

Wohlfarth, G., Hulata, G., and Moav, R., Use of manure in aquaculture - some experimental results, *Symp. on Aquaculture in Waste Water,* Pretoria, South Africa, Paper 13, 1980.

Wohlfarth, G. W. and Schroeder, G. L., Use of manure in fish farming — a review, *Agric. Wastes,* 1, 279, 1979.

Xu, Hui, A Comparison of Tilapia (*Oreochromis niloticus*) Production in Pond, Cage and Pen Culture, AIT thesis AE-89-41, Asian Institute of Technology, Bangkok, Thailand, 1989.

Yang, Y., Lin, C. K., and Diana, J. S., Influence of Nile tilapia (*Oreochromis niloticus*) stocking density in cages on their growth and yield in cages and in ponds containing the cages, *Aquaculture,* in press.

Zhu, Y., Yang, Y., Wan, J., Hua, D., and Mathias, J. A., The effect of manure application rate and frequency upon fish yield in integrated fish farm, *Aquaculture,* 91, 233, 1990.

5   CLIMATE, SITE, AND POND DESIGN

Anita M. Kelly and Christopher C. Kohler

INTRODUCTION

The utilization of ponds for growing fish is not a novel concept, and in fact, it antedates recorded history. Early records indicate that the Chinese produced food fish in ponds some 2500 years ago, although the exact beginning of this practice is not known (Edminster, 1947; Landau, 1992). The Romans utilized fish ponds in the first century A.D., not only to raise food fish but also to raise fish for stocking in natural lakes and other ponds (Edminster, 1947). Bas-relief sculpture found in Egyptian tombs apparently depict pond culture of tilapia around 2500 B.C. (Landau, 1992).

Originally the concept of a fish pond was quite different than it is today. Early fish culturists simply attempted to make the pond resemble the natural waters in which the fish were found. Early fish culturists recognized the importance of maintaining cultural environments similar to each aquatic animal's natural environment. As a result, aquatic organisms were not cultured in the densities they are today, but in some instances prepared feeds were provided (Landau, 1992).

Early pond construction methods were essentially the same as those utilized today with respect to the use of clay, proper sloping of sides, appropriate levee width, freeboard, and methods to dispose of flood waters (Edminster, 1947). These features were recorded in the early 1700s, but the importance of a controlled water supply was not mentioned until the 1800s by Boccius (Edminster, 1947). Boccius even cautioned against utilizing trees on levees.

The prominent pond design in the early days of aquaculture is equivalent to the modern raceway. Ponds were tiered, with each pond draining into the pond below it. A slightly different method of production was utilized in the Rhine Valley in France, where farmers combined aquaculture and agriculture. After raising crops for a couple of years, farmers flooded their fields with shallow water and stocked them with fish. Once the fish were harvested, the field was drained and utilized once again for crop production (Figure 1) (Edminster, 1947).

The methods of pond construction are based on techniques that are hundreds to thousands of years old; only the refinements are new. In this chapter the relevance of climatic factors and geographical features with respect to site selection and pond design are discussed. Once the pond is designed and built, then the principles of water budgets and pond management are important and are described as they relate to the various pond types. Climatic factors, geographical features, water budgets, and pond management are important basic elements that need to be understood in order to successfully produce an aquaculture crop.

CLIMATIC FACTORS

Climate plays an important role in pond aquaculture. Climate varies from region to region and even from day to day within a given region. Climatic factors dictate the amount of rainfall, solar radiation, evaporation, and temperature ranges for a given region. The differences in climate among regions is due to (1) the inclination of the earth on its axis, which causes seasonal change; (2) the rotation of the earth, causing day and night; (3) atmospheric conditions, especially the presence of clouds, which can reflect the amount of solar radiation received; and (4) the fact that the earth is round, which decreases the amount of solar radiation that reaches the earth with increasing latitude

0-56670-274-7/97/$0.00+$.50
© 1997 by CRC Press LLC

Figure 1. A pond, near Port-au-Prince, Haiti, that is utilized for both aquaculture and agriculture. The pond is being planted with rice after the fish were harvested.

(Akin, 1991). In the following sections, various climatic factors will be considered with respect to ponds.

Solar Radiation

The earth's atmosphere is heated by three mechanisms: solar radiation; geothermal energy derived from the earth's interior as a result of radioactive decay; and gravitational fields of the earth, moon, and sun. The nonsolar energy sources provide less than 0.02% of the total atmospheric energy and have a minor effect on the earth's atmospheric systems (Akin, 1991); these will not be further considered here. Solar radiation is absorbed and reflected in similar amounts by land and water, but due to physical differences between the two surfaces, the response to absorption differs. Water and land differ with respect to specific heat, translucency, evaporation, and mobility (Akin, 1991).

The amount of energy necessary to raise the temperature of 1 g of a substance 1°C is called specific heat. Water has a high specific heat and is exceeded by only a few compounds, such as ammonia (Wheaton, 1977). Therefore, if land and water are absorbing heat at the same rate, water would heat and cool at a slower rate. Water is translucent; therefore, solar radiation will penetrate deeply in clear water, resulting in a much higher volume being heated (Akin, 1991). Water is also more mobile than land, resulting in movement of absorbed heat both vertically and horizontally within a pond.

Seasonal variations in air temperature, which are heavily influenced by seasonal variations in solar radiation, generally influence water temperature. Water temperature influences water density, which creates a natural form of circulation in aquatic systems (Wheaton, 1977). Water has a high specific heat, which buffers changes in water temperature, making the aquatic thermal environments fairly stable. Solar radiation and therefore air temperature vary with latitude and season, causing related variations in water temperature, and pond turnovers (vertical mixing). Other variations include the physical differences between freshwater and saltwater, which affect their reaction to solar radiation and air temperature, include water density, specific density, heat capacity, and vapor pressure.

Freshwater Ponds

Freshwater is most dense at 4°C; density will affect the circulation pattern of the pond. The amount of energy put into the system in the form of solar energy depends on cloud cover, latitude, altitude, season, and surface area of the pond. Losses of energy from the pond are the result of evaporation and radiation, which depend on season, altitude, latitude, water temperature, wind velocity, surface area of pond, and relative humidity. Energy exchange can also occur with the pond soil. For example, shallow ponds in the summer can have a high water temperature, resulting in heat loss to the soil.

The water flowing in and out of the pond can have an effect on the pond's temperature; water flow into a pond will add energy, whereas water flow out of the pond removes it. The location of the outflow can have a drastic effect on pond water temperature, since outflows located near the pond's surface have a tendency to remove the warmer pond water, while outlets deeper down may remove cooler water.

Circulation in freshwater ponds varies with respect to geographical location (Reid, 1961). Ponds found in temperate zones and higher altitudes in subtropical regions overturn in spring and autumn. Ponds found in warmer tropical latitudes have water temperatures, which do not fall below 4°C, resulting in only one turnover occurring in winter. Ponds typical of high mountain lakes in tropical areas exhibit continuous circulation at low temperatures (Reid, 1961).

Pond turnover is a result of the changes in water temperature throughout the year. To better illustrate this phenomenon, let us examine a temperate region pond (Figure 2). In the winter, the pond is ice covered. The water just below the ice is approximately 1 to 2°C, and the water closest to the bottom soil is approximately 4°C or slightly higher due to heat exchange between the soil and water. In spring, the increase in solar radiation results in the ice melting. The water closest to the surface will increase to 4°C. Since water is most dense at 4°C, the surface waters sink to the bottom, resulting in the spring turnover. The water is mixed throughout the pond, and the water has a constant vertical temperature distribution. The wind aids in the mixing process.

As spring turns to summer, the amount of solar radiation increases, warming the water and decreasing the surface water density. During the summer, the winds may be reduced, decreasing the amount of mixing, resulting in stratification. If the pond is deep enough, this stratification results in the formation of a thermocline, which is characterized by a rapidly decreasing temperature with increasing depth. The thermocline layer is generally thicker in early summer than in late summer, which means the temperature gradient is not as steep in late summer.

In the fall, decreases in solar radiation and air temperature, and possible increases in mean wind speed, cause surface water temperature to decrease and density to increase. Wind also directly affects water mixing through kinetic energy transfer to surface water where the resistance to mixing due to the presence of a thermocline is proportional to the magnitude of the thermocline. As surface water temperature decreases, circulation of surface waters becomes deeper until it occurs over the entire depth of the pond. This results in the fall turnover and produces uniform vertical temperatures throughout the pond.

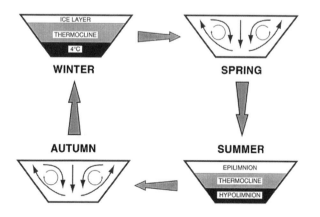

Figure 2. Diagrammatic representation of turnover in a pond or lake.

Aquaculture ponds are rarely deep enough to maintain a strong thermocline, and turnover is controlled by seasonal temperatures and winds. Aquaculture ponds are generally built to provide partial control of circulation, allowing the culturist to manipulate circulation when desired.

Saltwater Ponds

Vertical circulation in saltwater ponds is affected by both temperature and salinity gradients. Seawater is most dense at its freezing point. However, as saltwater freezes, the salts are separated from the water, resulting in near frozen saltwater having a higher salinity. This higher saline water sinks, causing mixing of the pond. Vertical salinity gradients can also occur due to salt dissolution from the soil at the bottom of the pond and from lower salinity water input to the surface of the pond from precipitation and runoff.

Rainfall

Rainfall plays an important role in ponds, particularly in some instances in which it is the only mechanism available to fill the pond (Chakroff, 1976). However, building a pond where rainfall is the only water source is not normally appropriate. During periods of drought, water losses from the pond result in higher densities of fish in the pond. This increase in density can lead to various water quality problems, resulting in the loss of all the fish. Water can be lost due to evaporation, seepage, irrigation, weed control, or fish management purposes. Therefore, most pond culturists prefer other more reliable water sources to fill their pond. Utilization of rainfall as the main water supply is generally insufficient for maintaining appropriate pond levels. However, rainfall may be adequate if a large retention pond is subsequently utilized to fill ponds.

The amount of rainfall an area receives depends on the geographic location and the climate of the area (Figure 3). In subtropical areas, rainfall is generally concentrated in one season, whereas in equatorial regions, rainfall is spread out over the year, with one or two seasons of intense rainfall (Coche and Van der Wal, 1981). Knowing the amount of rainfall an area receives prior to building a pond is important. Ponds need to be filled to meet the culturist's needs at all times; however, ponds must not be allowed to flood, which would also result in the loss of fish.

Rainfall can vary considerably over small geographic areas; therefore, if the farmer desires a more accurate indication of the amount of rainfall a pond is receiving, a rain gauge should be placed near the pond. More detailed information on the number of rain gauges and their placement as determined for agricultural studies can be obtained from Gray (1970).

Evaporation

Evaporation is the process of converting water from the liquid to the vapor state. The rate of vaporization is dependent on the water vapor to pressure gradient at the air–water interface and the interface area. Above the air–water boundary, wind and air mixing remove the water vapor, which maintains the vapor pressure gradient. The latent heat of vaporization (the amount of heat required to vaporize a given quantity of liquid water) for water is high, and the heat must come from the water (Ragotzkie, 1978). Therefore, in order to predict evaporation as accurately as possible, radiation, air temperature, vapor to pressure gradient, and temperature of the water need to be measured. However, studies utilizing these parameters are generally not employed due to their complexity; therefore, less expensive methods for determining evaporation rates have been devised.

Evaporation rates can vary from season to season and with varying geographic regions (Figure 3). For example, 16.5 cm was found to evaporate in 1 month in the southeastern U.S. when the average temperature was 32.2°C (Khosla, 1951). How much evaporation takes place depends on several factors including air and water temperature, wind, humidity, and surface area of the

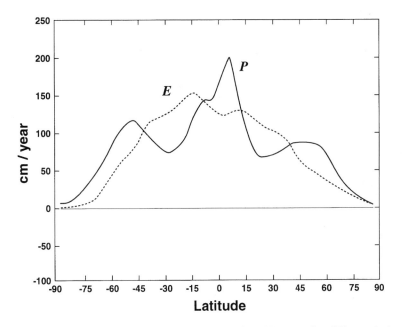

Figure 3. Average yearly precipitation (P) and evaporation (E) rates for different latitudes. (From Hartmann, D. L., *Global Physical Climatology,* Academic Press, San Diego, CA, 1994. With permission.)

pond. Elevated air temperatures, low relative humidity, gusty winds, and little cloud cover will increase evaporation rates (Coche and Van der Wal, 1981). Also the larger the pond, the more evaporation will occur. To determine the evaporation rate, many farmers rely on the rates obtained by local meteorological stations, which utilize the pan evaporation method. This method involves a Class A evaporation pan, which is 25 cm deep and 122 cm in diameter (Gray, 1970). Generally, it is not necessary to calculate evaporation rates because local meteorological stations can provide average monthly evaporation rates. However, water does evaporate faster from the Class A pan than from a pond; therefore, the evaporation rates must be multiplied by a correction factor in order to approximate water loss (Gray, 1970). Correction factors ranging from 0.60 to 0.81 have provided reliable estimates of lake evaporation (Hounam, 1973), while correction factors of 0.7 to 0.81 have been utilized to estimate water losses from small ponds in Alabama (Boyd, 1982,1984,1985a). Climatic differences between regions will influence the magnitude of the correction factor.

Evaporation estimates can be obtained utilizing the following formula modified from Khosla (1951):

$$Lm = \frac{Tm - 0.091}{2.076} \tag{1}$$

where Lm is the monthly evaporation of water (cm) and Tm is the average monthly temperature (°C). This formula only works when Tm is greater than 4.4°C. At 4.4°C, Lm = 2.13 cm, at –3.6°C, Lm = 1.78 cm, and at –21.6°C, Lm = 1.52 cm.

Boyd (1984) demonstrated that evaporation rates could be estimated based on water temperatures:

$$\text{Pond E}_{month} = -9.940 + 5.039 \; T_{month} \tag{2}$$

where E = evaporation per month and T = the average monthly water temperature. Variables can be expressed in volume or depth.

Evapotranspiration is evaporation of water from plants. Evapotranspiration rates from aquatic plants can exceed evaporation rates from a pond (Mitchell, 1974). Generally, aquaculture ponds lack aquatic plants that have leaf surfaces above the water; therefore, evapotranspiration is not considered when calculating evaporation rates.

SOIL PROPERTIES
Topography

Topography is the science of measuring the earth and its features. Topographical maps show the slope of the ground, by indicating not only the high and low areas but also how steep the land is between the high and low points. Topographical maps will provide information on the suitability of an area for pond construction. Detailed methods on measuring and constructing topographical maps for fish farming can be found in Coche (1988, 1989).

Texture and Organic Matter

Two important considerations when constructing a pond are topography and soil type. Soils utilized for ponds include clay, silty clay, and clay loams. Surfaces to avoid are sandy or gravelly soils, marl- and gypsum-containing soils, areas where limestone and shale come close to the surface and areas with anthills or holes. Marl and gypsum are highly soluble soil materials that often result in seepage and water quality problems. Limestone and shale regions are highly susceptible to fractures, which can result in leakage from the pond. Simply feeling the soil will indicate if it is suitable for a pond. Smooth, slippery soils contain significant amount of clay, whereas gritty or rough soils have sand as the main constituent. Inorganic silt is often mistaken for clay, due to the similarities in appearance between these two soils. However, appearances are deceiging in this case, as silt is very unstable when wet and should not be used to construct dams. Clay soils are very stable and are the preferred soil for dam construction. To determine if soil is silt or clay, the shaking test (Coche, 1985), as outlined below, should be performed.

1. Take sample of soil, wet it, and form it into a patty approximately 8 cm in diameter by 1.5 cm thick.
2. Place the patty in the palm of your hand, and shake it from side to side while watching the surface.
   a. If the surface is shiny, it is silt.
   b. If the surface is dull, it is clay.
3. Confirm the above result by bending the patty between your fingers; if the surface becomes dull again it is silt.
4. Finally, let the patty dry completely.
   a.If it is brittle and dust comes off when rubbing with your fingers, it is silt.
   b.If it is firm and dust does not come off when rubbing your fingers, it is clay.

A simple test to determine if the soil is adequate for ponds is to take a handful of soil, add just enough water to make it damp, and form it into a ball. Soils that form a hard ball when squeezed tightly by hand and that maintain their shape when tossed and caught are sufficient for pond construction. To determine more precisely the type of soil you have, a more definitive test called the manipulative test (Coche, 1985) can be conducted. The sequence of steps must be conducted in the following order:

1. Wet a handful of soil until it sticks together, but does not stick to your hands.
2. Roll the soil into approximately a 3-cm ball.

3. Put the ball down; if it falls apart it is sand.
4. If the ball remains intact, roll it into a 6- to 7-cm cylindrical shape; if it fails to retain the cylindrical shape, it is loamy sand.
5. If the soil retains cylindrical shape, roll it into a 15- to 16-cm long cylindrical shape; if the soil fails to retain this shape, it is sandy loam.
6. If the soil maintains this shape, attempt to bend it into a half circle; if it will not bend, then it is loam.
7. Attempt to bend the longer cylindrical mass into a full circle. If it will not bend, it is heavy loam; if it will bend but cracks slightly, it is light clay; if it bends and does not crack, it is clay.

Due to the variation between soil types, a survey of the soil should be conducted before fish ponds are constructed. This will indicate the suitability of the area for a possible pond, while saving time and money should the soils prove unsuitable for pond construction.

Soil samples for ponds are usually taken to a depth of 2 m. However, if the water table level is less than 2 m, soil samples should be taken from as deep as possible.

Soil samples are collected either by the open pit method or the auger method. The open pit method involves digging a hole, with straight sides, 0.80 m × 1.5 m × 2 m deep. If the water table is less than 2 m down, discontinue digging once the water table has been reached. Next, examine the side of the hole and sketch the soil profile. This sketch will give you the sequence and thickness of the different soil types. Each profile should then be sampled. This can be done utilizing a sample tube. Sample tubes are manufactured in lengths ranging from 30 to 60 cm, having diameters of 4 to 7 cm. Sampling tubes can be made from 16 or 18 gauge steel. These sampling tubes are thin walled, however, which makes them ineffective in loose, hard, or gravelly soils. To take the soil sample, oil the inside of the tube, and insert the sample tube horizontally into the soil as fast as possible, making sure the tube is completely inserted. Dig the tube out utilizing a knife. Seal both ends of the tube with plastic or cloth, which is tied on using string or rubber bands. The sample can then be placed in a plastic bag and labeled. In loose, hard, or gravelly soils, a garden trowel can be used to obtain the samples.

Soil samples can also be taken using an auger. Drill the auger into the soil 10 to 15 cm deep. Carefully remove the auger, and place the soil sample on a piece of paper or plastic. Continue to drill out 10- to 15-cm sections, and place them in succession on the paper or plastic. Make a sketch of the core sample collected, and write the depths of the different soil types.

When drilling with an auger, groundwater could be reached. In this case continue drilling. If the soil below the water table is sand, stop drilling. If it is clay, continue for another 30 cm.

The ability of various soils to hold water is related to soil particle size. Large particles are often difficult to compact enough to prevent large spaces, which allow water to pass through. Table 1 shows the particle size associated with the soil type. Table 2 shows the average pore space for each soil type and the seepage losses per day. Note that clay has the smallest particle size and the least amount of seepage. Ideally the soil should be 30 to 70% clay mixed with sand or silt. Avoid mixing the clay with organic material such as leaves and sticks as they may allow for water leakage.

Table 1 Particle Sizes for Various Soil Types

Soil type	Particle size (mm)
Gravel	>2.0
Coarse Sand	2.0–0.5
Medium to fine sand	0.5–0.02
Silt	0.05–0.02
Clay	<0.02

Table 2   Average Pore Space and Seepage Losses of Various Soil Types

Soil type	Average pore size (in % total vol.)	Seepage losses (mm d^{-1})
Sand	38	25–50
Sandy loam	43	13–76
Loam	47	8–20
Clay loam	49	2.5–15
Silty clay	51	1.25–10
Clay	53	0.25–5

The soil should not be 100% clay because pure clay swells when wet and contracts when dry. Once the pond is drained and the clay is allowed to dry, the resulting cracks may expose underlying permeable soils, which would facilitate seepage from the pond when it is filled again. Pure clay is also very sticky when wet and nearly rock hard when dry, making it difficult to work (Mittlemark and Landkammer, 1990).

Soil must do more than simply hold water — it should also provide nutrients and be free of toxic compounds. The French recognized this fact long ago when they flooded crop fields to provide shallow ponds to raise fish. Therefore, a good indicator of soil quality is whether it has been successfully utilized to raise crops. If the crops grow well, then the soil is probably free of harmful substances. However, soils unsuitable for agriculture due to the presence of salts can often still be used for aquaculture ponds.

Soil constituents that contribute to pond fertility include nutrients such as iron, calcium, and magnesium. Rooted aquatic plants obtain many of the essential nutrients from the soil. Conversely, phytoplankton, which does not have root systems, absorbs all of its nutrients from the water. Finfish, as well as other aquatic organisms, obtain necessary trace metals (metals that in very low concentrations are necessary for the survival of an organism) from the water. However, high concentrations of some trace metals can be harmful or toxic to fish. For example, high concentrations of iron can cause gill damage (Cruz, 1969).

Soils may also contain other chemicals that are harmful to fish. Soils are comprised of various chemicals, some of which, such as phosphorous, when in contact with water are forced down into the lower soil layers, where they accumulate. Other chemicals, such as chlorides and sulfates, are less soluble and remain in the upper soil layers. When some of the chemicals are exposed to air, they may be oxidized, forming more soluble chemicals. Hence, the chemical composition of soils is dynamic. For example, when soil containing pyrites is moved during pond construction and the pyrites are exposed to air, oxidation of the pyrites results in acidic soils having pH values less than 4.0. These types of soils, called acid sulfate soils, are found throughout Southeast Asia.

The air that is present in the pores of the soil also contains carbon dioxide, which reacts with water to form carbonic acid. This acid can react with the chemicals in the soil to form new compounds. For example, carbonic acid reacts with calcite to form calcium bicarbonate. In coal mining areas, the leftover coal mixed in with the clay will also create acidic water conditions, where solar radiation on the exposed coal releases sulfuric acid and decreases pH. The pH of the soil should be between 6.5 and 8.5. Higher or lower pH values may influence the productivity of the pond by affecting plankton growth, fish growth, and fish reproduction.

Soils can also contain excess sodium or other salts. These types of soils can be dispersed easily, are highly erosive, and are extremely susceptible to channeling. Soils containing high concentrations of salt create problems in establishing vegetation on the embankment and can result in corrosion of metal structures (Harris, 1981).

WATER SUPPLY

After the appropriate soil type has been found, the next item to consider is water supply. A pond needs a supply of water that will maintain the water level in the pond for as long as the aquaculturist needs it. The two major categories for water supply are surface water and groundwater. For the purpose of the discussion here, surface water is water that is found above ground and groundwater is water found below the surface.

Surface Water

Runoff

A common method for filling ponds is through surface water runoff, that is, water that does not seep into the soil after a rain. The amount of runoff depends on the intensity of the rainfall, the permeability of the soil, the type and amount of vegetation present, and the topography of the area. More runoff will occur in heavy rains because the water cannot be absorbed by the soil quickly enough. In areas with heavy vegetation, the vegetation slows down the water, allowing more to be absorbed into the soil. In areas with a sloping topography, rain that is not absorbed immediately will run off following the slope. The advantage to utilizing water runoff is that no pumping costs are incurred. Monthly runoff (Rm) in cm can be calculated utilizing the formula:

$$Rm = Pm - Lm \qquad (3)$$

where Pm is the average monthly rainfall (cm) and Lm is the monthly evaporation (cm) (Khosla, 1951). Annual runoff (Ra) can also be calculated by utilizing the formula

$$Ra = Pa - XTa \qquad (4)$$

where Pa is the annual rainfall (cm), Ta is the average annual temperature (°C), and X is a constant for a catchment basin (cm/°C) (Khosla, 1951). For example, Auburn, AL, has an average yearly temperature of 20.8°C , an average of 137.2 cm of rain, and a catchment constant of 4.15 cm/°C; the annual runoff would be 50.8 cm.

There are two disadvantages to utilizing runoff to fill a pond. The first is that it is dependent on rainfall. Rainfall varies with climate and geographic locations, as previously described. The second disadvantage is that the water may contain pollutants. Water running into ponds from nearby crop fields can absorb and carry pesticides into the pond. These are extremely difficult to remove from the water column, and many are highly toxic to fish.

Rivers and Streams

Rivers and streams are important water sources for some ponds. There are four factors to consider when utilizing rivers and streams as water supplies for ponds. First, the flow must be sufficient to fill the pond and subsequently maintain water level. Obviously, streams that may dry up during seasonal droughts will not adequately provide enough water to compensate for evaporation and seepage losses. Second, the river or stream must not be subject to excess flooding. Excess flooding could result in the overfilling of the pond, resulting in loss of organisms over the spillways and levees. In cases of extreme flooding, where the river or stream completely covers the pond, the entire culture crop could escape. Third, the river or stream should be well vegetated to prevent excess soil being washed into the stream, thereby increasing turbidity, and fourth, the river or stream should have only a light silt load, especially during periods of flooding.

The advantages to utilizing water from rivers and streams is that the water usually has high oxygen concentrations and, if the topography is right, pumping water into the ponds may be unnecessary. The disadvantages to rivers and streams as supply water are many. First, they are subject to runoff from a wide variety of sources, increasing the probability of pollutants being present. Second, the quantity of water available can fluctuate throughout the year. The amount of rainfall and the other uses of the stream, such as irrigation or livestock watering, may limit the amount of water available to fill the pond. Third, temperature of the river or stream can also fluctuate throughout the year. If the water supply is shallow, water temperature can increase rapidly. Fourth, other organisms can be introduced into the pond, which may be harmful to the species being cultured. Fifth, the water diverted for the pond could affect the amount of water available or the water quality downstream. And finally, rivers and streams are often considered public property and are often heavily regulated.

Lakes

The advantage to utilizing lakes as a source of water is that they are generally high in oxygen. Unfortunately, the disadvantages may outweigh the advantages. Generally, pumping the water is required, which increases the cost of production. As with rivers and streams, other organisms can also be introduced into the pond, and the temperature may vary. Depending on the size and depth of the lake, the water may be cooler than that found in the pond.

Springs

Springs are free-flowing waters that are a result of aquifers intersecting the surface of the ground. Because the water flows freely, no pumping is required. The water temperature is generally relatively constant, and the water is usually free of pollutants. However, water flow can vary with the seasons. The water obtained from springs is often high in carbon dioxide and low in oxygen.

Groundwater

Artesian Well

An artesian well is a well that has been dug into an aquifer. The pressure of the water in the aquifer forces the water up and out, eliminating pumping. This water source, like springs, has a fairly constant temperature and is usually free of pollutants. The water is also generally high in carbon dioxide and low in oxygen. Unfortunately, this good source of water is not generally found in areas suitable for aquaculture.

Water Table

Below the ground's surface, in the permeable layers, water may be held. This area is called the water table. If the table is high enough, the pond may extend into it. The water from the water table in this case would not have to be pumped and is generally free of pollutants. However, the aquaculturist has no control over the water level in the pond and is not capable of draining the pond. Again, as with other underground water sources, the water is generally high in carbon dioxide and low in oxygen.

Well Water

Generally, the water table is deep; therefore, a well must be dug in order to pump the water into a pond. This water is usually pollutant free and has a relatively constant temperature, and the water supply will not vary. Conversely, the water is high in carbon dioxide, low in oxygen, and may contain high concentrations of other dissolved gases. Naturally, the water must be pumped, and it could be very expensive to dig the well.

Water Supply Quality

Once an adequate water supply has been located, it must be checked to ensure it is suitable for pond culture. A detailed discussion on the parameters and methods of measuring water quality can be found in Chapter 3.

GUIDELINES FOR SITE SELECTION AND POND DESIGN

Pond construction techniques are given in detail elsewhere (Chakroff, 1976; Coche, 1992; Wheaton, 1977); therefore, only the basic construction methods as well as the suitability and basic pond design for a given area will be discussed.

Most ponds utilized for aquaculture are manmade. The two main types of ponds are embankment and excavated ponds. Embankment ponds are constructed by building a dam or dike to impound water (Figure 4). They are the most frequent type of pond used in aquaculture because they can be built in a wide range of topographic conditions. Excavated ponds are constructed by removing soil to form a hole (Figure 5). If the soil is acceptable for construction, the soil is compacted with either heavy equipment or, in areas where heavy equipment is not feasible, by using livestock (Hickling, 1971). Excavated ponds are utilized only in areas of flat topography. The disadvantages to excavated ponds are that it is costly to remove the soil necessary to build the pond and the ponds often must be pumped in order to drain. However, drainage ditches can be dug, allowing for emptying of ponds by gravity, or pipes can be buried for the same purpose.

The shape of the pond is a function of the species being cultured and local topography. However, there are other factors (characteristics) that must be considered when constructing a pond.

There are several reasons to build a pond with sloped sides. First, water pressure increases with depth; therefore, the bottom of the dike must be wider to compensate for the increased pressure. Secondly, sloping sides prevent erosion and limit aquatic vegetative growth. Third, slopes provide access for harvesting, aeration, etc., and fourth, some species require sloped sides in order to reproduce.

Figure 4.   Embankment pond at Carbondale, Illinois.

Figure 5.   Excavated pond at Iquitos, Peru.

Aquaculture ponds should be constructed so that rooted aquatic vegetation is limited; therefore, the pond should have a minimum depth of not less than 1 m. In balance with advantages of shallower slopes, pond dikes should be steep to limit the number of shallow areas at the pond margins to minimize aquatic vegetative growth and maximize usable pond volume. Commonly, pond banks are built with an inside slope of 2:1 to 4:1 slope (horizontal to vertical) and outside slopes of 2:1 to 3:1. More stable soils can withstand a 2:1 slope, whereas unstable soils require a 4:1 slope. Vertical sides can be constructed from concrete (Figure 6), masonry, or other similar manmade materials, but these increase the cost of the pond.

In order to drain the pond completely, the bottom must slope toward the drain. The pond bottom slope generally ranges between 1000:3 to 1000:6. Stable bottom soils will accommodate higher slopes (Wheaton, 1977). A harvest or catchment basin near the drain outlet is often advisable (Figure 7). The size of the harvest basin is generally 1 to 10% of the total pond area. The harvest basin should be 45 to 60 cm deeper than the pond bottom. The depth of the pond depends on the

Figure 6.   Culture pond with concrete sides, Port-au-Prince, Haiti.

Figure 7. Harvest basin in a pond that is lined with a polyethylene liner to prevent seepage.

climate. Table 3 provides the recommended minimum depths for watershed ponds in different climates.

Barrage Ponds

Barrage ponds are filled by rainfall or by spring water. The pond is built by constructing a dam to enclose the water in the low places. The dam prevents water from entering or leaving except when needed.

Barrage ponds should not be built where the flow from a spring is too great, as the water flow may break down the dam. Small streams that do not flow too strongly are good sources for barrage ponds. Because barrage ponds are built in low areas, they fill up readily in heavy rains. As a consequence, overflow channels are used to regulate the water level.

Embankment Ponds

Embankment ponds are constructed by building a dam across a narrow valley with a stream. Constructing ponds in this way allows for a large volume of water to be retained with very little work. The main problem with embankment ponds is that water is constantly flowing through; as a

Table 3 Minimum Pond Depths for Various Climates[a]

Climate	Minimum pond depth (m)
Wet	1.5
Humid	1.8–2.1
Moist humid	2.1–2.4
Dry sub-humid	2.4–3.0
Semiarid	3.0–3.7
Arid	3.7–4.3

[a] Data from USDA, 1973.

result, undesirable fish and contaminants from upstream can be introduced into the pond. Outgoing water also takes away nutrients, sometimes including feeds, that may be essential to the successful culture of a species.

Watershed Ponds

Watershed ponds are built in areas where local topography allows runoff to collect (Figure 8). Dikes are built across small valleys, and the pond is filled with runoff. Like embankment ponds, watershed ponds are relatively inexpensive to build. However, these ponds rely on rain to be filled. Unless alternate methods to fill these ponds are available, this could be the major disadvantage. If water quality problems arise, there is essentially no way to flush the pond. Also, during periods of drought, the pond level may drop, which could prove detrimental to the species being cultured. Another major drawback to watershed ponds is that chemicals applied to surrounding grounds may wash into the pond.

Diversion Ponds

Diversion ponds are built usually by excavating earth and building dikes. Water for these ponds is diverted from another water source, such as a stream or lake. These types of ponds allow control of the water within the pond.

POND MANAGEMENT

Erosion

Once the pond has been constructed, the embankment must be protected from erosion. Erosion can be caused by heavy rains, wind, wave action, or by livestock wading into the pond. Fencing is often necessary to prevent erosion due to livestock. To prevent erosion from rain, wind, or wave action, the pond banks should be planted with grass, not only to keep the soil out of the pond but also the nutrients that can stimulate further aquatic weed growth.

High concentrations of soils washing into a pond result in high levels of suspended particles. Suspended solids limit light penetration within the water column, which reduces the oxygen production by plants. Suspended particles can also cause increased water temperature, which may

Figure 8.   A watershed pond in southern Illinois.

exceed the upper lethal temperatures of the fish being cultured. Suspended solids will eventually settle, causing such detrimental effects as smothering of eggs, suffocation of bottom-dwelling fish, and reduction in fish growth. Fish growth can be hindered in two different ways. First, the amount of natural food available decreases due to lack of light penetration, and second, the turbidity of the water interferes with the fish's ability to see and catch the available food.

Turbidity in water can be corrected by adding chemicals that bind to the suspended particles, thereby clearing up the water (coagulation and flocculation). Chemicals and organic material that can be utilized to reduce turbidity are given in Table 4, with suggested application rates. However, chemicals react differently under different conditions; therefore, concentrations for individual ponds should be predetermined by treating a small sample (jar) of the water. The concentration of chemical to use should be the minimum amount of the chemical that would cause the suspended particles to precipitate out in 1 h. Some chemicals react in a way that is harmful to fish if they are added in large quantities. For example, alum ($Al_2(SO_4)_3 \cdot 14\ H_2O$) lowers the pH in soft water ponds (Boyd, 1990). Remember that chemical treatments only provide a temporary solution, and the source of the turbidity needs to be identified and eliminated.

Table 4 Chemicals and Organic Material Used to Reduce Turbidity in Ponds

Chemical	Application rate
Hydrated aluminum sulfate	5.6 kg/ha
Agricultural limestone	560 kg/ha
Agricultural gypsum	560 kg/ha
Barnyard manure	2440 kg/ha
Cottonseed meal	85 kg/ha
Superphosphate	28 kg/ha

Erosion due to wind forces alone can be corrected by planting vegetation along the embankment. In Israel, trees are planted on the sides of the pond toward the prevailing winds in order to reduce the wind force (Hickling, 1971). Caution should be taken not to plant trees in the levee or embankment because the roots could destabilize the soil. Ponds should not be too protected from the wind, as the wind provides a method of circulation.

In large ponds, very large swells can be produced by the wind. These swells can undercut the bank around the water line. Grass cover in this situation will not help much, as it will be drowned at water level or just above it. Stones, cement plates, and polyethylene sheeting have been utilized along the banks to deter erosion (Figures 6, 7, and 9). Sticks and bamboo can be driven into the pond bottom along the lee embankments to create a wind break and thereby reduce erosion as well (Hickling, 1971).

Aquatic Weeds

Aquatic vegetation may require control in fish culture ponds. Excessive growth of aquatic vegetation interferes with natural food production, fish growth, and fish harvesting, and may cause oxygen depletion in the pond.

Aquatic weeds can be controlled in a variety of ways. Probably the least detrimental to the system is hand harvesting and removing the weeds from the pond. Hand pulling the weeds does not require addition of chemicals, which can be harmful to cultured species if added in improper quantities. Removal of the pulled weeds from the pond also removes nutrients that could be utilized to stimulate further weed growth within the pond and prevents possible oxygen depletion in the pond as a result of decomposition.

Shading is another method of controlling weeds in the pond. Limiting the amount of sunlight retards aquatic plant growth. Chemical dyes such as Aquashade, Aquashadow, and Blue Vail can be utilized to color the water and reduce light penetration. However, it is important to read the label carefully, as not all dyes will work in all water. For example, Aquashade will not work in water that has traces of chlorine, as chlorine causes the color to be lost.

Fertilizing the pond is a method to control the growth of pond bottom weeds. Fertilization increases the amount of phytoplankton, thereby shading rooted vegetation. This must be done prior to the growth of rooted vegetation; otherwise, the fertilization regime will stimulate the growth of aquatic macrophytes rather than the microphyte or phytoplanton.

Biological controls such as grass carp can be utilized to control aquatic weed growth as well. However, many governments regulate the introduction of exotic species, and local regulations concerning non-native animals should be investigated prior to stocking.

Ducks and geese are effective at removing water weeds and keeping the grass on pond banks cut short (Hickling, 1971). Use of fowl as aquatic weed controllers has two other benefits; they fertilize the water by adding manure, and they are an additional source of meat (Hickling, 1971).

There are a variety of chemicals on the market for treating aquatic weeds. Caution should be taken when applying these chemicals to culture ponds. Too much of the chemical could be harmful or even fatal to the cultured species. Again, chemicals react in different ways under different conditions. For example, copper sulfate is more toxic to fish in low alkalinity water than in water having high alkalinity (Inglis and Davis, 1972).

WATER BUDGETS

In order to maintain a relatively constant water level in a pond, the amount of water entering the pond must be equivalent to the amount of water lost from the pond. This is called the water budget. The hydrologic equation as described by Winter (1981) and Boyd (1985b) is:

$$\text{gains} = \text{losses} \pm \Delta \text{ storage} \qquad (5)$$

Aquaculture ponds usually have a small watershed and are generally constructed in areas where there is no groundwater inflow (Boyd, 1982). Therefore, gains in culture ponds are the result of rainfall, runoff, and water supplied from an outside source such as well, stream, or reservoir. Water losses in a pond are incurred through evaporation, seepage, and overflow or regulated discharge (Green and Boyd, 1995). Therefore, the above hydrologic equation can be rewritten for levee and watershed ponds. The watershed of a levee pond is simply the inside slope of the levee, and as a result runoff is a minor factor (Yoo and Boyd, 1994). Therefore, the water budget for a levee pond is

$$P + I + RO = (S + E) \pm \Delta V \qquad (6)$$

where P = precipitation; I = inflow from well, stream, or reservoir; RO = runoff; S = seepage; E = pond evaporation; and V = change in storage volume. Variables are expressed in volume or depth of water. The water budget equation for a watershed pond is

$$P + RO = (S + E + OF) \pm \Delta V \qquad (7)$$

where OF = spillway overflow.

Measurements of the above variables, excluding seepage, have been previously given and are explained in greater detail elsewhere (see Yoo and Boyd, 1994). Seepage is the loss of water from a pond through the soil. Seepage can result in ponds that have soils that are too permeable to hold water. Although analysis of soil samples taken prior to pond construction can prevent usage of permeable soils, occasionally the site may be less than satisfactory, and permeable soils must be used. In these cases, methods of reducing seepage should be considered. Measurement of seepage from a pond is very difficult; therefore, it is generally estimated utilizing a variation of the water budget formula:

$$P + RO = E + S + \Delta H \tag{8}$$

where H = stage change, i.e., change in water level depth (Boyd, 1985b). Due to the difficulty in estimating runoff, it is best to determine seepage during dry periods, when inflow and outflow from the pond are essentially zero (Boyd, 1985b). The estimation of seepage can then be simplified to the following formula:

$$S = \Delta H - E \tag{9}$$

where the variables are generally expressed in depth. Stage changes in the pond can be made with a staff gauge that has been mounted in the pond. Reported seepage rates from ponds range from 1 mm to more than 25 mm per day (Parsons, 1949; Allred et al., 1971; Boyd, 1982; Boyd and Shelton, 1984).

Seepage Control

Seepage can be reduced in a variety of ways. The following methods come from those presented by Holtan (1950a, 1950b) and Renfro (1952). Sealing a pond can be accomplished by compaction alone if the site has a wide range of particle sizes and approximately 10% clay to effect a seal. Utilizing this method requires that the pond be drained and dried and all vegetation and large rocks removed. The bottom is then pulverized to a depth of 20 to 25 cm with a disk. Under optimal moisture conditions, the pulverized soil is compacted. A detailed discussion on the effects of moisture and compaction can be found in McCarty (1981). In ponds that are less than 3 m deep, the thickness of the compacted soil should be no less than 20 cm. In ponds greater than 3 m in depth, two or more 20 to 25 cm layers of soil should be compacted. This is accomplished by removing the top layer and stockpiling it while the bottom layer is being compacted.

If the soil of the pond bottom consists of mainly coarse particles, compaction of the soils will not seal the pond. A layer of well-graded soils consisting of at least 20% clay can be installed over the pond bottom. The well-graded soils should be compacted to a depth of 0.3 m for water depths up to 3 m. For ponds greater than 3 m, an additional 5.0 cm of compacted soil should be added for every 0.3 m increase in water depth. The compacted area in near inflow and outflow areas should be protected by riprap.

Bentonite can be utilized to reduce seepage in ponds that do not have a widely fluctuating water level. Bentonite works well when it is incorporated into well-graded coarse soils. Bentonite is a montmorillonitic clay that can swell 8 to 20 times its original volume. Therefore, when bentonite is mixed with coarse, highly permeable soil and wetted, it will swell and fill the void spaces in the soil. Bentonite, like clay, shrinks and cracks when dried; therefore, it should only be used in ponds where the water level does not fluctuate enough to expose the bottom allowing it to dry.

Rates of application of bentonite vary from 4.8 to 34.4 kg/m^2. The best estimates for treatment rates are based on laboratory analysis of permeability. Generally, the more permeable the soil, the more bentonite must be utilized in order to adequately control seepage. Once the rate has been determined, bentonite should be spread uniformly over the entire pond bottom. It should be mixed with pond bottom soil to a depth of 15 cm. The soil mixture should then be compacted. The bentonite layer should then be covered with 0.3 m of soil, and the pond should be filled immediately to prevent drying of the bentonite.

Excessive seepage can occur in clay soils because sometimes the particles of clay associate in such a manner as to form open, porous, honeycomb shapes. Applying chemicals known as dispersing agents causes the honeycomb shapes to collapse and thereby reduces seepage.

Treatment with dispersing agents can be costly and time consuming. Laboratory permeability tests can be utilized to determine the best dispersing agent, as well as the optimal treatment rate of the dispersant. Common dispersing agents include salt (sodium chloride), soda ash (sodium carbonate), tetrasodium pyrophosphate (TSPP), and sodium tripolyphosphate (STPP). Salt and soda ash can only be utilized with soils having a high cation exchange. Salt should be added at a rate of 1.6 kg/m^2, and soda ash should be applied at the rate of 1.0 kg/m^2. TSPP and STPP are good general soil dispersing agents and should be applied at a rate of 0.5 kg/m^2. Since dispersing agents are only effective in fine-grained soils, soils for treatment with dispersing agents should contain 50% or more of particles with diameters of less than 8.0 mm and at least 15% of particles with diameters less than 0.03 mm.

In order to apply dispersing agents, the pond must be drained and dried. The dispersing agent is applied uniformly over the pond bottom. The rates given previously are for 15 cm of soil. The dispersing agent is mixed with the upper 15 cm of soil. With the soil at optimum soil moisture, the soil should be compacted. Again, if the water depth exceeds 3 m, two or more 15-cm layers must be treated with the dispersing agent and compacted. The high water line and the inflow and outflow areas should be protected by riprap (Figure 9). Ponds should be filled slowly to allow the dispersing agent time to work with the soils. Ponds treated with dispersing agents are often turbid when filled. The turbidity will subside, and ponds should not be treated with chemicals to reduce turbidity.

Flexible membranes of polyethylene, vinyl, and butyl rubber have all been utilized to seal ponds (Figure 7). Polyethylene liners are fairly resistant to weathering but are very susceptible to mechanical damage. Polyethylene liners can only be joined together or patched utilizing heat-sealing methods. Vinyl plastic and butyl rubber liners are fairly resistant to mechanical damage and can be joined with a special cement.

To install a liner, the pond must be drained, dried, and cleared of all vegetation, rocks, or other sharp objects. If the pond bottom is covered with coarse soil, a layer of fine-grained material should be deposited to provide a cushion for the membrane. The membrane should be flexible enough to

Figure 9. Pond with riprap along the edge to prevent undercutting of the embankment by waves.

adjust to small settlements in the pond. The banks that will be lined with an exposed membrane should not be sloped more than 1:1. Banks that will be lined with covered membranes should be sloped no more than 3:1 because steeper slopes may result in the soil covering the membrane to slide off.

To apply the membrane, a trench measuring approximately 0.3 m by 0.3 m is dug around the pond just above the water line and the membrane edges are buried. The liners are laid in strips, with an approximately 15-cm overlap for seaming. Allow at least 10% slack in polyethylene liners. The membrane should then be covered with a 10- to 15-cm layer of soil that is free of any objects that could puncture the liner.

Polyethylene and vinyl pond liners do not prevent vegetative growth in a pond. Pond banks may be treated with a herbicide to reduce the establishment of vegetation prior to installing the liner.

Organic matter such as manure, grass, paper, or cardboard can also be used to seal a pond; however, this method is less reliable than the other methods. Organic matter can be spread over the pond bottom at a rate of 2.4 to 4.9 kg/m^2 and mixed with the soil with a disk. The soil should be allowed to stand for a few weeks and then compacted before filling the pond.

POND DESIGN

Commercial aquaculture ponds are usually rectangular in shape, have flat bottoms free of aquatic vegetation, and have a slope of 1 or 2% to the drain. Water depth ranges from 1 to 1.2 m at the shallow end to 1.5 to 1.8 m at the drain. Water levels will vary as pond depths vary with climate (Table 3). The banks of the ponds generally have a slope of 2:1 to 4:1, depending on the stability of the soil. If the pond sides are constructed utilizing rock or any other stable substance, the wall of the pond can be vertical.

The dams or dikes of a pond are usually less than 8 m in height. The soil utilized in dams must be able to support the weight of the dam and should have low permeability to prevent seepage. Soils that are permeable to water can be utilized in the foundations of dams, provided that there is some method to control water loss.

In ponds where permeable soils cover fairly impermeable soils, a foundation cutoff, or key, is utilized to minimize seepage losses (Wheaton, 1977). The cutoff is formed by digging a trench beneath the area where the dam will be located and deep enough to penetrate well into the impermeable layer (Figure 10). The trench bottom should have a minimum width of 1.25 cm and a side slope of no more than 1:1. Once the trench is dug, a 15- to 20-cm layer of impermeable soil is placed and compacted in the trench. This process is repeated until the trench is filled, which now forms the key. Since the key is sealed to the impermeable soil layer by compaction, seepage through the dam through the permeable layer is prevented. Once the key has been constructed, the dam can be built.

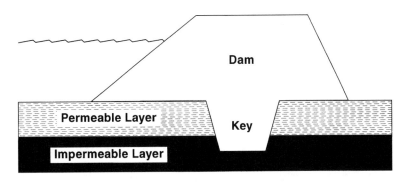

Figure 10.   A key for a dam built on permeable soil and extending into the underlying impermeable soil layer to prevent seepage through the dam.

In areas where impermeable soil is in short supply, the key can be extended to the top of the dam. In this situation, the key can be utilized with or without the foundation cutoff, depending on the type of bottom soils (Wheaton, 1977). Since the key is constructed of impermeable material, the rest of the dam can be constructed of permeable soils. The key should have a minimum side slope of 1.5:1 and should be wide enough at the top to allow for compaction (Wheaton, 1977).

The minimum top width of a dam up to 3 m in height is 2.5 m or greater. However, if the dam is to be utilized as a roadway, the top width should be 4 to 4.5 m. A more in-depth discussion of the calculations utilized to determine the top width of the dam can be found in Wheaton (1977).

The slopes of the dam sides are a function of the soil utilized to construct them. Generally, a slope of 3:1 is utilized on the upstream side of the dam and a 2:1 slope is utilized on the downstream side (Wheaton, 1977). As the instability of the soil increases, the slope of the dam will decrease.

Added height to a dam to prevent water from topping over the dam is known as freeboard. Freeboard is defined as the vertical distance from the pond surface to the top of the dam when the spillway is discharging at the designed depth. The recommended amount of freeboard is provided in Table 5. However, local regulations may dictate the amount of freeboard a dam must have and should be checked prior to construction. The dam should also be constructed high enough to prevent waves from going over the top. Dam height is discussed in detail by Wheaton (1977).

Table 5   The Recommended Amount of Freeboard Based on Pond Depth[a]

Pond length (m)	Freeboard (m)
200	0.3
200–400	0.5
400–800	0.6

[a] USDA, 1971.

Even though the soil in the dam has been compacted, soil settlement will occur. The amount of settlement depends on soil material utilized in construction, moisture content of the soil, amount of compaction, speed of construction, and properties of the foundation materials (Wheaton, 1977). The settlement allowance should not be less than 5% of the dam height, and when utilizing poor materials or methods, 10% of the dam height should be used for settlement allowance estimates (Wheaton, 1977).

Well-designed aquaculture ponds have two outlets, a mechanical spillway for normal outflow and an emergency spillway for handling water during peak storm periods. These two structures can be combined into one, but generally are not.

Ideally, it should take only 24 to 28 h to drain a pond. For ponds less than 0.4 ha in surface area, a 15-cm PVC pipe with a drain control is suitable (Figure 11). For ponds up to 0.8 ha in surface area, a 20-cm drain line is required to achieve adequate drainage. Ponds that are greater than 0.8 ha in surface area require drain pipes of 25 to 30 cm. For larger sized pipe, using reclaimed steel or cast iron pipe is more economical than PVC. For drain pipes 20 cm and above, an alfalfa valve set in concrete is an inexpensive type of drain control (Figures 12 and 13). Caution should be taken when installing drain pipes 30 cm and above, as flooding can result below the discharge point.

In existing ponds that do not have a drain or lack a drain of adequate size, a portable siphon can be used to drain the pond (Figure 14). If it is necessary to move the siphon, it can be cut into sections and reassembled at the new location. Generally, portable siphons are replaced by permanent siphons, as shown in Figure 15. This type of drainage system not only allows the water level to be lowered but also acts as a spillway. This spillway prevents damage to the dam and reduces fish escapes over the emergency spillway.

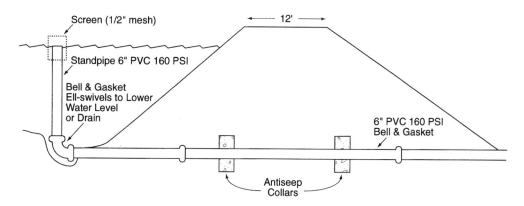

Figure 11. Drain system satisfactory for ponds up to 0.4 ha in size and not exceeding 1.8 to 2.1 m in depth. Overflow pipe swivels to reduce water level or to drain. (From Lewis, W. L., Use of farm ponds for the production of food fish for home use and specialized marketing, *Southern Ill. Fish. Bull.*, No. 6, 1981. With permission.)

The function of a permanent siphon drain as a mechanical spillway results from rising water in the pond, which seals the vent pipe and primes the siphon. The siphon loses its prime and flow stops when the water has been siphoned down to the level of the open end of the vent pipe. By placing a plate over the discharge of the tube and closing off the spillway valve, the siphon can be primed with a self-priming centrifugal pump connected to the priming port on the siphon, and the pond can be drained to the desirable level.

A monk is a common outlet structure used for ponds. It is a three-sided structure generally constructed of wood or concrete (Figure 16). At the bottom of the back side is an outlet pipe, which traverses the embankment, allowing for drainage. The fourth or open side generally has two rows of grooves into which wooden slats can be inserted, and the top edge of the board determines the water level. The wood used for these slats must be soft wood, which swells when wet, forming a tighter seal. It is advisable to have two rows of wooden slats with clay packed between them. This configuration makes a watertight dam yet allows excess water to spill over the top.

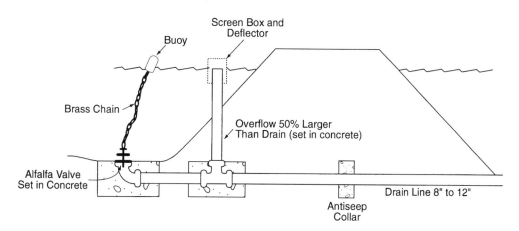

Figure 12. Drain suitable for ponds greater than 0.4 ha in size. This system can also be used for partial drainage of the pond. (From Lewis, W. L., Use of farm ponds for the production of food fish for home use and specialized marketing, *Southern Ill. Fish. Bull.*, No. 6, 1981. With permission.)

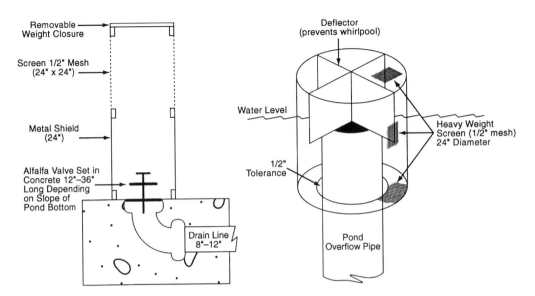

Figure 13.   Shield and screen used for draining the pond. Shield prevents complete draining of the pond and avoids the necessity of constant tending. Unit is turned upside down for final draining. Pond overflow and deflector screen to prevent escapement of fish. (From Lewis, W. L., Use of farm ponds for the production of food fish for home use and specialized marketing, *Southern Ill. Fish. Bull.*, No. 6, 1981. With permission.)

The two rows of slats allow the monk to be set up as a bottom draw. In the front slot, a screen is placed in the bottom to prevent fish from escaping, and boards are then set above the screen, extending above the water surface. The rear set of boards extend from the bottom of the pond to the desired water level. When the water level rises above the desired depth of the pond, then it flows under the front boards and up and over the rear boards.

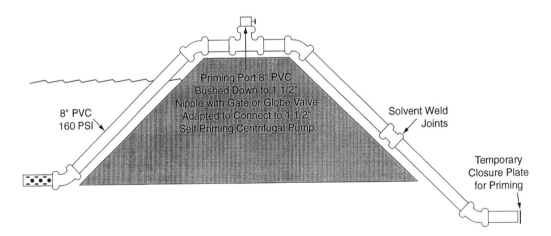

Figure 14.   Portable siphon system that can be installed in existing ponds to permit lowering of the water level. (From Lewis, W. L., Use of farm ponds for the production of food fish for home use and specialized marketing, *Southern Ill. Fish. Bull.*, No. 6, 1981. With permission.)

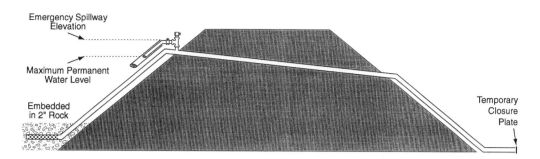

Figure 15.   A combination siphon/trickle tube that can be installed in an existing pond to protect the vegetal spillway, reduce loss of fish over the spillway, and permit lowering of the water level. (From Lewis, W. L., Use of farm ponds for the production of food fish for home use and specialized marketing, *Southern Ill. Fish. Bull.,* No. 6, 1981. With permission.)

Wooden monks are cheaper and lighter than concrete or brick monks, and if made of suitable wood, will last as long as 20 years (Hickling, 1971). An alternative to the monk in small ponds is to use a pipe, sometimes a hollow tree trunk or bamboo, buried in the bank and stuffed with grass and clay. The stuffing is removed when it is time to drain the pond (Hickling, 1971).

CONCLUSION

The importance of climatic factors, geographical features, soil, and water as they relate to aquaculture ponds varies throughout the world. These variabilities play an important role in pond design, as they are all intricately linked. Ponds that are built range from designs that are simple and fairly inexpensive to those that are complex and expensive. This flexibility in design and construction is what makes aquaculture available and feasible to all economic sectors of the world.

Figure 16.   A concrete monk for draining a pond. Note the two grooves for wooden slats and the clay that is packed between them.

REFERENCES

Akin, W. E., *Global Patterns: Climate, Vegetation, and Soils,* University of Oklahoma Press, Norman, 1991.

Allred, E. R., Manson, P. W., Schwartz, G. M., Golany, P., and Reinke, J. W., Continuation of Studies on the Hydrology of Ponds and Small Lakes, Minnesota Agricultural Experimental Station Technical Bulletin 274, University of Minnesota, Minneapolis, 1971.

Boyd, C. E., Hydrology of small experimental fish ponds at Auburn, Alabama, *Trans. Am. Fish. Soc.,* 111, 638, 1982.

Boyd, C. E., Hydrology of channel catfish ponds, *Miss. Agric. For. Exper. Sta. Bull.,* 290, 44, 1984.

Boyd, C. E., Pond evaporation, *Trans. Am. Fish. Soc.,* 114, 299, 1985a.

Boyd, C. E., Hydrology and pond construction, in Tucker, C. S., Ed., *Channel Catfish Culture,* Elsevier, New York, 1985b, chap. 4.

Boyd, C. E., *Water Quality in Ponds for Aquaculture,* Birmingham Publishing Co., Birmingham, AL, 1990.

Boyd, C. E. and Shelton, J. R., Jr., Observations on the Hydrology and Morphometry of Ponds on the Auburn University Fisheries Research Unit, Alabama Agricultural Experimental Station Bulletin 558, Auburn University, Auburn, AL, 1984.

Chakroff, M., *Freshwater Fish Pond Culture and Management,* Volunteers in Technical Assistance, Mt. Ranier, MD, 1976.

Coche, A. G., *Soil and Freshwater Fish Culture,* Food and Agriculture Organization of the United Nations, Volume 6, Rome, 1985.

Coche, A. G., *Topography for Freshwater Fish Culture,* Topography Tools, Food and Agriculture Organization of the United Nations, Volume 16/1, Rome, 1988.

Coche, A. G., *Topography for Freshwater Fish Culture,* Topographical Surveys, Food and Agriculture Organization of the United Nations, Volume 16/2, Rome, 1989.

Coche, A. G., *Pond Construction for Freshwater Fish Culture,* Food and Agriculture Organization of the United Nations, Volume 20/2, Rome, 1992.

Coche, A. G. and Van der Wal, H., *Water for Freshwater Fish Culture,* Food and Agriculture Organization of the United Nations, Vol. 4, Rome, 1981.

Cruz, J. A., About the possibility of iron absorption by fishes and its pathological action, *Ann. Limnol.,* 5, 187, 1969.

Edminster, F. C., *Fish Ponds for the Farm,* Charles Scribner's Sons, New York, 1947.

Gray, D. M., *Handbook on the Principles of Hydrology, Water Information Center,* Port Washington, NY, 1970.

Green, B. W. and Boyd, C. E., Water budgets for fish ponds in the dry tropics, *J. Aquaculture Eng.,* 14, 347, 1995.

Harris, J., Federal regulation of fish hatchery effluent quality, *Am. Fish. Soc. Bio-Eng. Symp. Fish Culture,* 1, 157, 1981.

Hartmann, D. L., *Global Physical Climatology,* Academic Press, San Diego, CA, 1994.

Hickling, C. F., *Fish Culture,* Faber and Faber, London, 1971.

Holtan, H. N., Sealing farm ponds, *Agr. Eng.,* 31, 125, 1950a.

Holtan, H. N., Holding Water in Farm Ponds, Soil Conservation Service Tech. Publ. 93, U.S. Department of Agriculture, Washington, D.C., 1950b.

Hounam, C. E., Comparisons Between Pan and Lake Evaporation, World Meteorological Organization Technical Note 126, Geneva, 1973.

Inglis, A. and Davis, E. L., Effects of Water Hardness on the Toxicity of Several Organic and Inorganic Herbicides to Fish, U.S. Fish and Wildlife Service, Technical Paper 67, Washington, D.C., 1972.

Khosla, A. N., Appraisal of Water Resources: Analysis and Utilization of Data, U.N. Sci. Conf. Conserv. Utilization of Resources, IV, 1951, 64.

Landau, M., *Introduction to Aquaculture,* John Wiley and Sons, New York, 1992.

Lewis, W. L., Use of farm ponds for the production of food fish for home use and specialized marketing, *Southern Ill. Fish. Bull.,* No. 6, 1981.

McCarty, D. F., *Essentials of Soil Mechanics and Foundations,* Prentice Hall, Englewood Cliffs, NJ, 1982.

Mitchell, D. S., The effects of excessive aquatic plant populations, in *Aquatic Vegetation and Its Use and Control,* Mitchell, D. S., Ed., United Nations Educational, Scientific, and Cultural Organization, Paris, 1974.

Mittlemark, J. and Landkammer, D., Design and Construction of Diversion Ponds for Aquaculture, Minnesota Sea Grant Publication AQUA17.90, University of Minnesota, St. Paul, 1990.

Parsons, D. A., The Hydrology of a Small Area Near Auburn, Alabama, Technical Publication 85, U.S. Soil Conservation Service, Washington, D.C., 1949.

Ragotzkie, R. A., Heat budgets of lakes, in *Lakes Chemistry, Geology, and Physics*, Lerman, A., Ed., Springer-Verlag, New York, 1978.

Reid, G. K., *Ecology of Inland Waters and Estuaries*, Reinhold, New York, 1961.

Renfro, G., Sealing Leaky Ponds and Reservoirs, Soil Conservation Service Technical Publication 150, U.S. Department of Agriculture, Washington, D.C., 1952.

U.S. Department of Agriculture, Ponds for Water Supply and Recreation, Agriculture Handbook 387, Soil Conservation Service, U.S. Department of Agriculture, Washington, D.C., 1971.

U.S. Department of Agriculture, Building a Pond, Farmers Bulleton No. 2256, Soil Conservation Service, U.S. Department of Agriculture, Washington, D.C., 1973.

Wheaton, F. W., *Aquaculture Engineering*, Krieger Publishing, Malabar, FL, 1977.

Winter, T. C., Uncertainties in estimating water balance in lakes, *Water Res. Bull.*, 17, 82, 1981.

Yoo, K. H. and Boyd, C. E., *Hydrology and Water Supply for Pond Aquaculture*, Chapman and Hall, New York, 1994.

6   POND BOTTOM SOILS

Claude E. Boyd and James R. Bowman

INTRODUCTION

The importance of soil characteristics in pond construction was discussed in Chapter 5. Characteristics and condition of bottom soil are also important in pond management. The exchange of substances between soil and water affects water quality, which in turn influences fish production. Although it is generally recognized that there are strong effects and interactions among soil characteristics, water quality, and fish production in ponds, much more attention has been given to water quality than to soil condition as a factor limiting fish production in ponds. The PD/A CRSP has placed emphasis on measuring the effects of pond water quality variables on fish production and developing management procedures for improving water quality. However, some research has been conducted on pond soils, and future work on pond dynamics will no doubt have a greater focus on pond soil condition.

The purpose of this brief chapter is to provide the reader with an overview of the role of bottom soil in pond aquaculture and to provide a summary of PD/A CRSP research on pond soils.

POND SOILS AND FISH PRODUCTION

There are many soil properties that can be measured, and each property can exhibit a wide range of values. At present, only a few properties are known to play a major role in aquaculture; the following discussion will be limited to those properties.

Soil Texture

Soil texture refers to the size distribution of mineral particles that comprise the soil. The most reactive fraction of the soil is comprised of the smallest particles, the clay. The clay particles offer large surfaces for adsorption and ion exchange. The best agricultural soils normally are loams, comprised of a mixture of particles of different size classes. A sandy soil will not retain much water and has a small capacity to adsorb and hold nutrients. A soil with a majority of clay-sized particles binds water and nutrients tightly, and is often sticky and difficult to till. A loamy soil is intermediate between the two extremes. Loam has a greater capacity to adsorb water and nutrients than a sandy soil, but the water and nutrients are not held as tightly as by clay soils. Loam soils are not as sticky as clay soils and are easier to till. Pond soils must contain 20 to 30% clay-sized particles in order to provide a barrier against seepage, but clayey-loam soils are generally better for fish production than heavy clay soils. As a general rule, soils that are considered desirable for terrestrial agriculture, that occur in suitable locations for pond sites, and that contain enough clay to provide a barrier to seepage can be used for aquaculture ponds.

0-56670-274-7/97/$0.00+$.50
© 1997 by CRC Press LLC

Other than general statements about problems with too much sand or too much clay in pond bottom soils (Jhingran, 1975; Boyd, 1990), little has been done to relate soil texture to fish production.

Organic Matter

Soils with high concentrations of organic matter (e.g., >18%) are not ideal materials with which to build ponds. Organic soils are normally poorly drained, low in nitrogen, and acidic, so the organic matter decomposes slowly and accumulates. In addition, excessive settling occurs in ponds that are constructed with organic soils. However, when ponds are constructed on soils with high concentrations of native organic matter, better aeration, higher pH, and greater nitrogen availability resulting from management inputs usually accelerate organic matter decomposition. Soil organic matter concentration will also decrease until a new equilibrium concentration is reached. In ponds constructed on mineral soils, bottom soils usually have low initial concentrations of organic matter. Organic matter concentrations tend to increase over time until equilibrium between inputs and outputs of organic matter in the pond bottom is reached. The equilibrium concentration of organic matter in bottom soil of aquaculture ponds will depend upon the amount and type of organic matter reaching the pond bottom. Ponds that are infested with aquatic vascular plants or receive high inputs of manure will tend to have higher concentrations of soil organic matter than ponds that receive inorganic fertilizers or feeds. This is because manure and higher order plants are less suitable as substrates for microorganisms than are phytoplankton, fish feces, and uneaten feed (Boyd, 1995). Large inputs of nutrients in intensive aquaculture can cause such high rates of primary productivity that organic matter accumulates in bottom soils, even though phytoplankton is easily biodegradable.

A small amount of organic matter in pond soils is beneficial. It contributes to the cation exchange capacity of bottom soil, chelates trace metals, provides food for benthic organisms, and releases inorganic nutrients upon decomposition. However, too much organic matter in pond soils can be detrimental because microbial decomposition can lead to the development of anaerobic conditions at the soil–water interface. When the surface layer of the bottom soil is anaerobic, reduced substances (NO_2^-, NH_3, FE^{2+}, Mn^{2+}, H_2S, and CH_4) may be released into the overlaying water. Some reduced substances such as NO_2^--N, NH_3, and H_2S are toxic to fish at relatively low concentrations.

Banerjea (1967) evaluated the potential of a large number of ponds in India for fish production. Ponds with the greatest fish production had soil organic carbon concentrations of 1.5 to 2.5%. At lower concentrations of organic carbon, there was little organic matter to support adequate production of benthic fish food organisms. Higher concentrations of soil organic carbon were associated with oxygen depletion in the bottom soil. In most ponds, soil organic carbon concentrations are less than 2%, but much higher concentrations are sometimes found in intensively managed ponds. It is important to manage ponds so that excessive amounts of organic matter do not accumulate in bottom soils and cause anaerobic conditions at the soil–water interface.

Soil Acidity

Total alkalinity is an important variable regulating fish production, and ponds with acidic bottom soils typically exhibit low total alkalinity levels. Boyd (1974) and Murad and Boyd (1991) stated that ponds should have at least 20 mg/L total alkalinity for good fish production. There is evidence that total alkalinity should be 50 mg/L or more for intensive production of tilapia in fertilized or manured ponds. Boyd (1974) showed that the total alkalinity and total hardness of pond water were related to the base unsaturation of pond soils. The base unsaturation is the percentage of the cation exchange capacity of a soil that is occupied by acidic ions. Clay particles and organic matter have a negative charge that is satisfied by attraction of swarms of cations from the soil solution. These ions can be exchanged depending upon concentrations on the soil and in the water. This ability of the soils to exchange ions is called the cation exchange capacity (CEC). The CEC is measured in meq/100 g soil, and observed values typically range from less than 1 to more than 100 meq/100 g, depending upon the amounts and types of clay and organic matter in the soil. Aluminum ions are a major source of exchangeable acidity in soil, but some ferric iron and hydrogen ions may occur on cation exchange sites. Among pond soils studied by Boyd (1974), total alkalinity of pond waters was less than 20 mg/L when base unsaturation was greater than 20%, but total alkalinity was usually above 50 mg/L when base unsaturation of bottom soils was 10% or less.

The relationship between soil pH and degree of base unsaturation differs with soil properties, and the total alkalinity of the water is a better index of the potential of a pond for fish production than is soil pH. Nevertheless, Banerjea (1967) found that the optimum soil pH for fish production in ponds in India was 6.5 to 7.5. When pond soils are too acidic and total alkalinity of water is too low for good fish production, agricultural limestone or other liming materials may be used to neutralize soil acidity and enhance fish production.

Soil Nutrients

When ponds are filled with water from surface runoff, streams, or well water, the inflowing water will contain plant nutrients. In the pond, there will be an exchange of nutrients between soil and water until equilibrium is established. In the acidic soils typical of humid regions, concentrations of inorganic carbon (carbon dioxide and bicarbonate) have often been reduced to very low levels by leaching, and concentrations in surface runoff are correspondingly low (Boyd, 1995). As a result, equilibrium concentrations of inorganic carbon in ponds in these areas are often too low to permit high rates of primary productivity. In semiarid areas, on the other hand, many soils contain calcium carbonate and are alkaline, and surface waters typically contain more bicarbonate, so equilibrium concentrations of inorganic carbon in ponds in these regions are often higher than those in humid areas. Most pond waters, regardless of pH and alkalinity, will not contain enough dissolved inorganic nitrogen (nitrate and total ammonia nitrogen) or phosphorus to support adequate primary productivity for high levels of fish production. Acidic ponds should be limed to elevate pH and total alkalinity and provide more inorganic carbon for phytoplankton. Chemical fertilizers and manures can be applied to ponds to enhance the availability of nitrogen and phosphorus for phytoplankton. Nitrogen fixation is an abundant source of nitrogen in many ponds, and nitrogen fertilization is often not necessary. Bottom soils strongly adsorb phosphorus, and most ponds require phosphorus fertilization. Soils with a low pH and high concentrations of iron and aluminum oxides are especially adsorptive of phosphorus. Phosphorus also is quickly lost from solution in ponds where waters have a high pH and a large calcium concentration (Boyd, 1990).

Banerjea (1967) studied the relationships between pond soil characteristics and fish production in many ponds in India. He stated that available nitrogen concentrations should exceed 250 ppm and available phosphorus concentrations should be above 60 ppm to support good fish production. Similar estimates have not been made for other regions, but almost all studies have shown that phosphorus applications will enhance phytoplankton productivity and increase fish production, regardless of soil phosphorus concentrations. Nitrogen fertilization alone usually will not enhance fish production, even in ponds with low soil nitrogen concentrations. However, in some cases, nitrogen plus phosphorus fertilization will provide greater fish production than phosphorus fertilization alone (Boyd, 1990).

Soil Classification

There has been little effort to classify pond soils. Sometimes the procedures of *Soil Taxonomy* (Soil Survey Staff, 1990) are used to classify particular pond soils according to the nomenclature used in soil science. Computerized decision support systems under development by the CRSP use an adaptation of the *Soil Taxonomy* system proposed by Bowman (1992), which is discussed later in this chapter. Hajek and Boyd (1994) developed a system for rating the limiting features of pond soil for use in pond construction and management, but this system must be considered a preliminary effort.

POND SOILS AT PD/A CRSP SITES

The soils of the CRSP ponds are as diverse as the geographic areas in which they are found; these represent a wide range of climatic, rainfall, elevational, and temperature differences while still remaining within the zone of the tropics (Table 1). The resulting diversity of soils is apparent in their physical composition and chemical characteristics, as well as in the nature of the supply waters and the soil–water interactions at each site.

Table 1   Geographic Locations and Site Characteristics of the PD/A CRSP Research Sites

Site	Latitude	Longitude	Elevation (m)	Mean annual rainfall (m)	Mean annual air temp. (°C)
El Carao, Honduras	14°26′N	87°41′W	583	765	19.6–31.0
Bogor, Indonesia	6°6′S	106°7′E	220	350	23–33
Aguadulce, Panama	8°15′N	80°29′W	0	1453	23.8–33.5
Gualaca, Panama	8°31′N	82°19′W	100	4320	17–34
Iloilo, Philippines	10°45′N	122°30′E	4	2100	27
FAC[a], Philippines	15°43′N	120°54′E	30	n.a.	n.a.
Butare, Rwanda	2°40′S	29°45′E	1625	1200	14–28
Ayutthaya, Thailand	14°11′N	100°30′E	5	1372	28
AIT[b], Thailand	14°20′N	100°35′E	0	1174	28
Abbassa, Egypt	30°32′N	31°44′E	6	51.5	20.5

[a]   Freshwater Aquaculture Center, Central Luzon State University.

[b]   Asian Institute of Technology, Bangkok.

Physical Composition of Soils

Clay content is the physical feature of pond soils most frequently of concern to aquaculturists. The soil's clay content is critical, not only from the standpoint of construction (ease of construction as well as stability of the embankments) but also with respect to hydrology (seepage and its effects of the water budget of a pond). The CRSP ponds display a wide range of values in this characteristic, with clay contents ranging from a low of 24% in one pond at Gualaca, Panama (average 29.6%) to a high of 70.7% in one pond in Bogor, Indonesia, where the average clay content was 60.7% at the initiation of CRSP experimental work (Table 2). All of these clay percentages fall within the range considered acceptable for pond construction by most aquaculturists (>20% clay).

As noted frequently in the aquaculture literature, a high clay content alone does not guarantee low permeability and seepage in a pond. The type of clay present is also of importance (Coche, 1985). The CRSP experience has again made this fact clear. For example, whereas pond soils at Bogor, Indonesia, are composed predominantly of clays, they are notably nonswelling and quite permeable. These characteristics are typical of clays of volcanic origin, which are composed predominantly of halloysite (McNabb et al., 1988). During the first set of experiments conducted at the Bogor site, from late 1983 to early 1984, the average daily water loss from the ponds was 19% (McNabb et al., 1988). Similarly, the ponds at El Carao, Honduras, contain from 50 to 57%

Table 2 Initial Physical Composition of Pond Soils at the PD/A CRSP Research Sites

Site	Pond	% Organic matter	% Sand	% Silt	% Clay
El Carao, Honduras	B–1	1.27	19.6	28.4	52
	B–2	0.51	21.6	24.4	54
	B–3	0.70	25.6	24.4	50
	B–4	0.71	13.6	36.4	50
	B–5	0.92	19.6	30.4	50
	B–6	0.89	13.6	30.4	56
	B–7	1.15	17.6	30.4	52
	B–8	0.80	18.4	25.2	56.4
	B–9	1.20	20.8	24.4	54.8
	B–10	0.96	16.8	26.8	56.4
	B–11	1.12	16.8	26.8	56.4
	B–12	1.12	14.8	28.4	56.8
Bogor, Indonesia	C–1	3.7	9.0	29.3	61.7
	C–2	3.1	11.3	26.8	61.9
	C–3	3.0	13.4	28.7	57.9
	C–4	2.9	9.9	27.7	62.4
	C–5	3.1	9.9	31.8	58.3
	C–6	3.0	11.8	29.3	58.9
	C–7	2.8	16.0	32.0	52.0
	C–8	2.5	21.7	27.9	50.4
	C–9	0.8	6.4	29.6	64.0
	D–5	2.3	8.8	24.9	66.3
	D–6	2.6	10.1	26.4	63.5
	D–7	2.1	8.9	20.4	70.0
Aguadulce, Panama	4	1.2	58	10	32
	7	1.2	42	10	48
	13	1.2	44	12	44
	14	0.9	50	8	42
	16	1.1	44	10	46
	21	1.5	48	18	34
	25	1.0	50	14	36
	28	1.3	56	10	34
	34	1.3	58	14	28
	35	0.69	60	6	34
	37	0.74	58	6	36
	42	0.85	60	10	30
Gualaca, Panama	1	2.6	42	24	34
	2	2.7	48	24	28
	3	2.3	44	26	30
	4	3.1	44	26	30
	5	2.1	42	26	32
	6	3.9	54	22	24
	7	2.3	40	24	36
	8	2.0	52	22	26

continues

Table 2   (continued)

Site	Pond	% Organic matter	% Sand	% Silt	% Clay
	9	3.0	50	24	26
	10	3.2	46	24	30
	17	2.01	—	—	—
	18	2.01	28	22	50
Iloilo, Philippines	B01	1.64			
	B02	1.34			
	B03	3.13			
	B04	3.43			
	B05	2.69			
	B06	2.21	Particle size data		
	B07	3.03	not available		
	B08	2.25			
	B09	1.98			
	B10	2.51			
	B11	1.61			
	B14	2.76			
	B15	2.75			
	B16	4.52			
	B17	3.00			
	B18	4.48			
	B19	3.04			
	B20	3.40			
Butare, Rwanda	B1	2.9	37.1	16.1	46.8
	C2	7.6	45.3	16.9	37.8
	C3	0.5	61.3	9.5	29.2
	C5	5.9	33.5	23.3	43.2
	C7	2.4	40.5	10.3	49.2
	C8	1.4	28.2	16.0	55.8
	D1	1.0	51.2	10.6	38.2
	D3	3.6	43.3	17.5	39.2
	D5	3.1	40.0	16.2	43.8
	D6	1.0	52.6	16.5	30.9
	D9	1.9	46.2	14.6	39.2
	D10	1.6	47.2	17.1	35.7
Nong Sua, Thailand	1	1.85	12	18	70
	2	1.51	10	24	66
	3	1.61	10	24	66
	4	1.38	10	24	66
	5	1.24	12	23	65
	6	1.48	8	26	66
	7	1.28	10	24	66
	8	1.28	10	24	66
	9	1.31	12	24	64
	10	1.38	8	25	67
	11	1.34	9	23	68

continues

Table 2   (continued)

Site	Pond	% Organic matter	% Sand	% Silt	% Clay
	12	1.28	7	24	69
Ayutthaya, Thailand	1	0.77	10	23	67
	2	1.04	14	24	62
	3	0.91	10	24	66
	4	0.64	10	22	68
	5	0.70	13	23	64
	6	0.64	11	20	69
	7	0.50	12	22	66
	8	0.91	10	24	66
	9	0.50	12	20	68
	10	0.64	10	21	69
	11	0.77	9	21	70
	12	1.04	14	23	63
El Carao, Honduras	B–1	1.27	19.6	28.4	52
	B–2	0.51	21.6	24.4	54
	B–3	0.70	25.6	24.4	50
	B–4	0.71	13.6	36.4	50

From CRSP Data Reports series, earliest available data; except as noted, all data are from Volume I (General Reference) or Experimental Cycle I.

clay but have displayed seepage rates that are uncharacteristically high for ponds with high clay percentages. In this case, the high permeability is believed to be due to the predominance of kaolinitic clay (Teichert-Coddington et al., 1988). Like the clays of volcanic origin, soils dominated by kaolinite and similar (highly weathered, 1:1 types) clays are also known to be more permeable than, for example, soils dominated by 2:1 type clays.

A compositional feature sometimes overlooked but of importance to aquaculturists in certain areas is the organic matter content of the soil. From the standpoint of physical characteristics, high levels of organic matter would be of concern when ponds are built in areas of peat deposits, because such areas are often characterized by standing water and poor drainage. A high water table is often to blame for this, rather than soil impermeability, because peats are in fact highly permeable. None of the pond soils at the CRSP sites have sufficient quantities of organic matter to be classified as peats (Table 2).

Chemical Characteristics of Soils

There is also a great deal of site-to-site variation in the chemical characteristics of CRSP pond soils. Soil acidity is perhaps the most significant chemical characteristic with respect to the productivity of aquaculture ponds. The acidity of the soils ranges from pH 3.9 to 4.4 in Butare (Rwanda) and Nong Sua (Thailand) to pH 8.0 to 8.8 at El Carao (Honduras). The lowest cation exchange capacities (CECs) were those found in Bogor (Indonesia), Nong Sua and Ayutthaya (Thailand), and Gualaca (Panama), which all had values <24 meq per 100 g. Values of CEC reported for El Carao (Honduras) were mostly in the range of 21 to 30 meq per 100 g, whereas CEC values for ponds at the Butare (Rwanda) site were variable (15 to 39 meq per 100 g). CEC values for several of the sites were not reported.

The ponds at both Nong Sua (Thailand) and Iloilo (Philippines) were built on acid sulfate soils. The pH values and levels of aluminum (Al) and iron (Fe) at Nong Sua reflect the highly acidic nature of the soil there, but several years of intensive liming at the Iloilo site resulted in the reduction of aluminum to trace levels and pH values mostly in the range of 6.2 to 7.0 by the beginning of CRSP research there. The Butare (Rwanda) ponds also had very low pH values and high levels of Al and Fe at the initiation of CRSP research (Table 3), but the soils of that site were not reported to be acid sulfate soils.

In Thailand, research activities were started at Nong Sua but were transferred to the site at Ayutthaya soon after the CRSP began Cycle I of its research. Marked differences between the soil chemistry of the two sites were apparent. Ponds at Nong Sua were characterized by acid sulfate soils, with pH values of 3.9 to 4.4, whereas those at Ayutthaya are alkaline, clayey soils, with pH values of 7.0 to 7.6. These extreme differences in soil type were reflected in the levels of aluminum and iron in the soil. Concentrations of these substances were 180 to 2560 ppm and 95.2 to 165.6 ppm, respectively, at Nong Sua, whereas 8 to 38 ppm and 22 to 96 ppm were the corresponding values at the Ayutthaya site (Table 3). The low input levels of the CRSP Cycle I experiments (designated baseline experiments) limited productivity in both of these systems and precluded reaching conclusions about the effects of pH or Fe and Al concentrations on phosphorus availability or pond productivity.

The volcanic-origin pond soils at the Bogor (Indonesia) site had initial pH values in the range of 5.5 to 6.3 and CECs of 18 to 20 meq per 100 g. The production of algae and fish in these ponds, built on leached, highly-permeable soils and supplied with water of low alkalinity, was significantly limited until measures were undertaken to raise pond water alkalinity to 55 mg/L (McNabb et al., 1988, 1991).

In addition to the physical considerations related to soil organic matter contents, organic matter content also affects pond productivity. Earlier in this chapter it was noted that the ponds with the greatest fish production are usually those with soil organic carbon concentrations of 1.5 to 2.5% (Banerjea, 1967). These values are roughly equivalent to organic matter contents of 2.5 to 4.2%. It was also noted that the organic carbon contents of most aquaculture ponds is less than 2%, or less than about 3.3% organic matter. Most of the experimental ponds at the CRSP sites indeed fall within these typical ranges. Two of the ponds at the Butare site (Rwanda) were reported to have organic matter contents of 7.6% (Pond C2) and 5.9% (Pond C5) at the beginning of CRSP experimental work (Table 2), but the levels reported for the same ponds 2 years after CRSP research began were considerably lower. By that time the value for Pond C2 was reported as 1.2% and that for Pond C5 was 1.7% (Hanson et al., 1989).

Effects of Soil on Supply Water

It was mentioned above that the soils at the Bogor (Indonesia) site were highly leached and of volcanic origin. This type of soil is common throughout the surrounding area, and its highly leached nature leads to surface and groundwaters that are low in carbonate-bicarbonate alkalinity and in minerals (McNabb et al., 1988, 1991). Thus, the water available for supplying ponds at Bogor had a total hardness of just 21.2 mg CaCO$_3$/L and an alkalinity of 30.89 mg CaCO$_3$/L (Egna et al., 1987). In contrast, other areas in Indonesia with uplifted limestone ridges had surface and groundwaters characterized by considerably higher levels of minerals (McNabb et al., 1988).

In general, the other freshwater CRSP research sites also have source water supplies with alkalinity and hardness values in the ranges that might be expected given the local types of soils. For example, soil pH values at Gualaca (Panama) were in the range of 4.7 to 5.2 (Table 3) and the supply water at that site had total hardness and alkalinity levels of just 14.6 and 18.4 mg/L, respectively. At Butare (Rwanda), where soil pH values ranged from 3.9 to 4.4 (Table 3), the total hardness and alkalinity of the supply water were 43.3 and 17.0 mg/L, respectively, and at Ayutthaya

Table 3   Initial Chemical Characteristics of Pond Soils at the PD/A CRSP Research Sites

Site	Pond	pH	CEC (meq per 100 g)	CaCO$_3$ (%)	Al (ppm)	Fe (ppm)	L.R.[a]
El Carao, Honduras	B–1	8.0	22.3	5.1	n.a.[b]	9.3	7.5
	B–2	8.6	21.0	4.7	n.a.	5.3	7.5
	B–3	8.8	24.7	5.1	n.a.	7.9	7.5
	B–4	8.0	30.0	9.5	n.a.	6.1	n.a.
	B–5	8.0	17.8	9.6	n.a.	6.2	n.a.
	B–6	8.3	29.4	5.5	n.a.	6.6	7.5
	B–7	8.9	24.9	3.8	n.a.	10.2	7.5
	B–8	8.4	26.7	4.3	n.a.	10.0	7.5
	B–9	8.3	30.0	5.5	n.a.	10.9	7.5
	B–10	8.6	33.6	6.4	n.a.	8.8	7.5
	B–11	8.8	26.3	6.8	n.a.	8.5	7.5
	B–12	8.7	23.1	6.4	n.a.	10.2	7.5
Bogor, Indonesia	C–1	5.8	18.0–20.4[c]	<0.2	n.a.	66–74[c]	6.5
	C–2	6.1		0.2	n.a.		6.7
	C–3	6.2		<0.2	n.a.		6.8
	C–4	6.0		<0.2	n.a.		6.8
	C–5	6.0		<0.2	n.a.		2.5
	C–6	5.8		<0.2	n.a.		6.5
	C–7	5.5		<0.2	n.a.		6.5
	C–8	6.1		0.3	n.a.		6.7
	C–9	6.2		<0.2	n.a.		6.7
	D–5	6.2		<0.2	n.a.		6.6
	D–6	6.2		<0.2	n.a.		6.5
	D–7	6.3		<0.2	n.a.		6.5
Aguadulce, Panama	4	6.5	n.a.	0.2	tr.[d]	408	n.a.
	7	7.2	n.a.	0.2	tr.	529	n.a.
	13	6.4	n.a.	0.2	tr.	276	n.a.
	14	6.5	n.a.	0.2	tr.	356	n.a.
	16	6.9	n.a.	0.2	tr.	392	n.a.
	21	7.6	n.a.	0.2	tr.	658	n.a.
	25	6.9	n.a.	0.2	tr.	223	n.a.
	28	6.8	n.a.	0.2	tr.	222	n.a.
	34	6.9	n.a.	0.2	tr.	n.a.	n.a.
	35	7.3	n.a.	0.2	tr.	n.a.	n.a.
	37	7.3	n.a.	0.2	tr.	n.a.	n.a.
	42	7.6	n.a.	0.2	tr.	n.a.	n.a.
Gualaca, Panama (after one application of limestone)	1	5.1	<24	tr.[d]	2.8	48.2	n.a.[b]
	2	5.1	<24	tr.	4.5	67.8	n.a.
	3	4.9	<24	tr.	7.4	42.8	n.a.
	4	4.7	<24	tr.	7.8	32.8	n.a.
	5	4.9	<24	tr.	7.8	11.2	n.a.
	6	5.0	<24	tr.	4.5	39.6	n.a.
	7	4.9	<24	tr.	5.0	36.7	n.a.
	8	5.0	<24	tr.	6.9	48.2	n.a.
	9	5.0	<24	tr.	6.2	15.8	n.a.
	10	5.2	<24	tr.	3.6	41.7	n.a.

continues

Table 3   (continued)

Site	Pond	pH	CEC (meq per 100 g)	CaCO$_3$ (%)	Al (ppm)	Fe (ppm)	L.R.[a]
	17	5.3	<24	n.a.	n.a.	48.2	n.a.
	18	5.8	<24	n.a.	n.a.	35.6	n.a.
Iloilo, Philippines	B01	6.61	n.a.	n.a.	<0.1	201.6	n.a.
	B02	6.74	n.a.	n.a.	<0.1	68.7	n.a.
	B03	6.67	n.a.	n.a.	<0.1	77.9	n.a.
	B04	6.70	n.a.	n.a.	<0.1	242.9	n.a.
	B05	6.25	n.a.	n.a.	<0.1	64.0	n.a.
	B06	5.52	n.a.	n.a.	<0.1	36.5	n.a.
	B07	6.76	n.a.	n.a.	<0.1	57.3	n.a.
	B08	6.54	n.a.	n.a.	<0.1	41.1	n.a.
	B09	6.44	n.a.	n.a.	<0.1	40.9	n.a.
	B10	5.21	n.a.	n.a.	<0.1	41.2	n.a.
	B11	6.52	n.a.	n.a.	<0.1	54.9	n.a.
	B14	6.85	n.a.	n.a.	<0.1	454.2	n.a.
	B15	6.81	n.a.	n.a.	<0.1	582.8	n.a.
	B16	6.92	n.a.	n.a.	<0.1	618.8	n.a.
	B17	7.21	n.a.	n.a.	<0.1	2469	n.a.
	B18	7.05	n.a.	n.a.	<0.1	1054	n.a.
	B19	6.51	n.a.	n.a.	<0.1	1510	n.a.
	B20	6.79	n.a.	n.a.	<0.1	1142	n.a.
Butare, Rwanda	B1	4.4	30	n.a.	108	8000	6.3
	C2	4.2	24	n.a.	238	5000	5.8
	C3	4.3	14	n.a.	114	1900	6.8
	C5	4.0	39	n.a.	193	3500	5.9
	C7	4.1	29	n.a.	263	400	5.8
	C8	4.1	21	n.a.	182	300	6.4
	D1	4.3	18	n.a.	137	300	6.2
	D3	3.9	27	n.a.	249	400	n.a.
	D5	4.1	23	n.a.	177	9000	n.a.
	D6	4.3	15	n.a.	150	800	n.a.
	D9	4.2	22	n.a.	270	300	n.a.
	D10	4.1	25	n.a.	252	1000	n.a.
Nong Sua, Thailand	1	4.4	20.8	n.a.[b]	1160	127.2	n.a.
	2	4.3	21.1	n.a.	680	115.2	n.a.
	3	4.0	19.9	n.a.	840	165.6	n.a.
	4	4.3	20.0	n.a.	2560	106.4	n.a.
	5	4.2	20.3	n.a.	140	95.2	n.a.
	6	4.0	19.8	n.a.	180	127.2	n.a.
	7	3.9	20.1	n.a.	1150	142.4	n.a.
	8	4.0	16.6	n.a.	960	137.6	n.a.
	9	4.1	19.6	n.a.	620	131.2	n.a.
	10	4.1	17.6	n.a.	2000	131.2	n.a.
	11	4.1	20.1	n.a.	1280	104.8	n.a.
	12	4.1	20.4	n.a.	2000	154.8	n.a.
Ayutthaya, Thailand	1	7.0	20.8	n.a.	25	20	—
	2	7.6	21.1	n.a.	36	28	—

continues

Table 3   (continued)

Site	Pond	pH	CEC (meq per 100 g)	CaCO₃ (%)	Al (ppm)	Fe (ppm)	L.R.ᵃ
	3	7.5	19.9	n.a.	44	8	—
	4	7.5	20.0	n.a.	22	20	—
	5	7.0	20.3	n.a.	96	35	—
	6	7.5	19.8	n.a.	36	8	—
	7	7.4	20.1	n.a.	52	35	—
	8	7.4	16.6	n.a.	48	38	—
	9	7.4	19.6	n.a.	48	20	—
	10	7.5	17.6	n.a.	30	20	—
	11	7.6	20.1	n.a.	22	16	—
	12	7.3	20.4	n.a.	28	20	—

[a] Lime Requirement: soil pH in SMP buffer.

[b] n.a. = data not available.

[c] Pooled data, all ponds, Cycle II experiments.

[d] tr. = trace.

From CRSP Data Reports series, earliest available data; except as noted, all data are from Volume I (General Reference) or Experimental Cycle I.

(Thailand), soil pH values were from 7.0 to 7.6 while source water pH, alkalinity, and total hardness values were 8.6, 92 mg/L, and 184 mg/L, respectively (Egna et al., 1987).

Soil–Water Interactions

In laboratory studies, soil samples from three of the current CRSP research sites were treated with deionized water and/or suspensions of $CaCO_3$ in deionized water to evaluate soil–water interactions with respect to alkalinity in the water column. Samples were obtained from Butare (Rwanda), the Freshwater Aquaculture Center of the Philippines (FAC), and El Carao (Honduras). All samples from the Butare station had very low initial pH values (4.5 to 4.9) and base saturation percentages (18 to 48), whereas samples from El Carao and FAC were alkaline (pH 7.9 and 7.7, respectively, and initial base saturation 100%). Samples from El Carao and the FAC were therefore treated only with deionized water. After 24 h of treatment (with agitation to promote complete reaction), the mean total alkalinity values (three replications) of the supernatant solutions for these soils were 62 and 32 mg $CaCO_3$/L, respectively. The levels of calcium and sodium in these soil samples were high relative to those of other soils tested, suggesting that in addition to calcium carbonate, sodium carbonate may have been present in quantities sufficient to release considerable alkalinity into solution. In addition, the exchangeable magnesium present in the FAC sample was higher than most other samples, whereas the amount of exchangeable potassium present in the sample from El Carao was unusually high. In contrast, soil samples from the Butare site treated only with deionized water had insignificant amounts of alkalinity (all less than 2.5 mg $CaCO_3$/L); these samples all required treatment with concentrations of at least 0.75 millimolar $CaCO_3$ to reach alkalinities of 20 mg $CaCO_3$/L or greater.

BOTTOM SOIL RESPIRATION

One of the most important factors in intensive pond aquaculture is the condition of the surface layer of bottom soil. This layer should remain oxygenated to encourage decomposition of organic matter by aerobic bacteria, prevent the release of toxic, inorganic metabolites into pond water, and provide a good habitat for benthic organisms. The rate of organic matter input to the pond bottom is determined by quantities of manure or feed applied, amount of feces produced by fish, and amount of dead plankton that settles to the bottom. The organic matter input to the pond bottom increases as the intensity of aquaculture increases. The amount of aerobic respiration will increase with the amount of organic matter in the soil, up to the point that surface soil becomes anaerobic.

A study was conducted in Honduras to determine aerobic respiration during and between fish crops in heavily fertilized tilapia ponds. During the fish grow-out period, rates of soil respiration were estimated for 24-h periods from the oxygen loss from 7.6-cm diameter, opaque tubes of oxygenated well water inserted into the bottom soil. Oxygen loss rates were converted to carbon dioxide evolution rates by multiplying by 44/32, the molecular weight ratio of CO_2/O_2 (Teichert-Coddington and Green, 1993). Soil respiration during the fallow period was measured by an *in situ* technique for estimating the rate of carbon dioxide evolution from undisturbed soil (Anderson, 1982).

Before ponds were drained, bottom soil respiration rates were 0.56 g $CO_2/m^2/d$ in Pond B9 and 1.21 g $CO_2/m^2/d$ in Pond B10 (Figure 1). Ponds were drained on 19 August 1991, and soil respiration in both ponds was above 2 g $CO_2/m^2/d$ on 20 August. As the bottom soils dried, soil respiration rates increased to a maximum 7.14 g $CO_2/m^2/d$ and 10.16 g $CO_2/m^2/d$ in Ponds B9 and B10, respectively, with soil moisture contents of 45.8% in Pond B9 and 48.0% in Pond B10. As soil moisture content continued to decline, respiration also declined. On 28 August, respiration rates were 3.35 and 3.97 g $CO_2/m^2/d$ in the two ponds, and soil moisture contents were 29.1 and 26.5%. Rainfall on 29 August wetted the pond bottoms and respiration increased. Ponds were refilled on 3 September, and soil respiration immediately dropped to less than 1 mg $CO_2/m^2/d$ (Figure 1). During the fish culture period, the highest soil respiration rate in either pond was 1.62 g $CO_2/m^2/d$. When ponds were drained on 3 February 1992, soil respiration rates again increased as the soils dried. Peak soil respiration during the second fallow period was 4.75 g $CO_2/m^2/d$ in Pond B9 and 7.21 g $CO_2/m^2/d$ in Pond B10. These values occurred at soil moisture contents of 35% and 47%, respectively.

Based on these findings, it can be concluded that soil respiration rates were greater during fallow periods than during culture periods. Pond water contains no more than 10 to 20 mg/L (0.001 to 0.002%) dissolved oxygen (DO), and DO concentrations at the pond bottom may be much lower. Air is 20.95% oxygen by volume, so draining water from ponds and drying soil so that water content is below saturation permits air containing a high concentration of oxygen to contact the soil surface and to enter the pore spaces among the soil particles. Enhanced oxygen availability permits greater rates of microbial decomposition of organic matter.

The average rates of soil respiration (Table 4) were higher during the first fallow period than during the second ($P < 0.05$). Air temperature was normally higher in August than in February at El Carao, but mean soil temperatures for the two fallow periods did not differ (Table 4). Average soil moisture content did not differ ($P < 0.05$) for the two fallow periods (Table 4). Therefore, the greater mean rate of soil respiration during the first fallow period apparently was related to a greater input of chicken litter in the culture period preceding the first fallow period than in the culture period before the second fallow period (Table 4).

Even though rates of soil respiration were much greater during the fallow period than during the culture period, the culture period was much longer, and more organic matter was oxidized during the culture period than during either of the fallow periods (Table 4). These observations suggest that methods for enhancing organic matter decomposition during the culture period could have considerable benefits.

BOTTOM SOIL TREATMENTS

Studies were conducted in Honduras to develop techniques for reducing the rate of accumulation of soil organic matter in ponds. Application of nitrogen fertilizer and aeration were considered

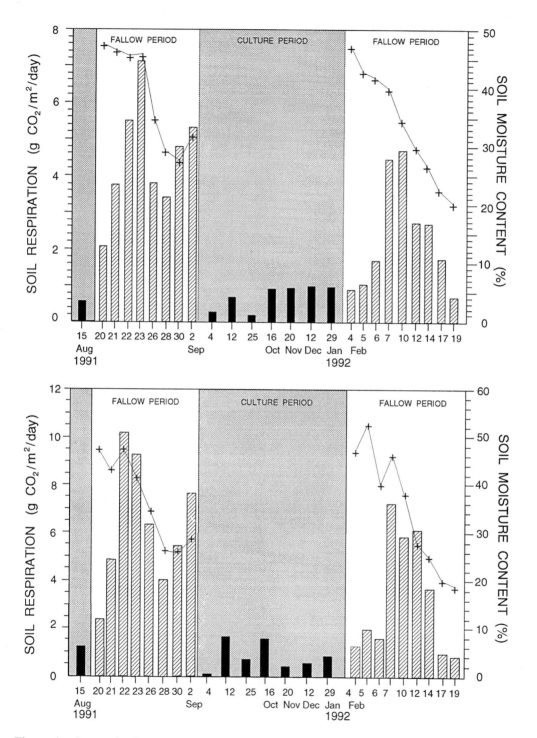

Figure 1. Rates of soil respiration during culture and fallow periods in two fish ponds (B-9 and B-10). Soil moisture concentrations (dry weight basis) are provided for the fallow period. (From Boyd, C. E. and Teichert-Coddington, D. J., *J. World Aquaculture Soc.,* 25, 417, 1994.)

Table 4   **Average Daily Rates of Soil Respiration, Total Soil Respiration, Soil Moisture Content, and Soil Temperature (2-cm depth) in Fish Ponds During Culture and Fallow Periods**

		Period		
		Fallow	**Culture**	**Fallow**
		Aug. 20 to	**Sept. 4, 1991**	**Feb. 4 to 19,**
Variable	**Pond**	**Sept. 2, 1991**	**to Feb. 3, 1992**	**1992**
Soil respiration rate	B–9	5.0	0.7	2.9
(g CO_2/m^2/d)	B–10	6.9	0.8	4.1
Total soil respiration	B–9	65	106	44
(g CO_2/m^2)	B–10	89	121	62
Soil moisture (%)	B–9	39.6	Flooded	35.0
	B–10	37.8	Flooded	36.6
Soil temperature (°C)	Both	27.7	25.7	25.8

during the grow-out period, and the reduction in soil organic matter through processes occurring during pond draining and drying also was determined (Ayub et al., 1993).

Nitrogen Application

Ponds were treated with chicken litter at 750 kg dry weight per hectare each week, so urea was added to determine if additional nitrogen would enhance the degradation of the chicken litter and decrease the rate of organic matter accumulation in soils. As shown in Figure 2, there was an increase in soil carbon in the 0- to 5-cm soil layer in ponds treated with chicken litter, and in ponds treated with chicken litter and 22 kg urea (45% N) per hectare each week. Soil carbon concentrations did not change in the 5- to 15-cm soil layer. Thus, it appears that nitrogen fertilization did not stimulate microbial activity and increase the decomposition of organic matter in pond bottom soils.

Pond soil for the nitrogen application trial described above had a C:N ratio of about 10. This suggests that the soil had adequate nitrogen to support rapid microbial growth even before the nitrogen fertilizer was applied. Experience with organic soils, which have a much larger C:N ratio, suggests that nitrogen application will stimulate decomposition (Boyd, 1990).

Aeration

An aeration experiment was conducted in which all ponds were treated weekly with chicken litter at 1000 kg/ha (dry weight) for 2 months, followed by application of a pelleted, 20% protein feed at 3% of fish body weight per day for 3 months. Three of the ponds were aerated, and the other three were unaerated controls. Both aerated and control ponds had increases in soil carbon in the 0.5-cm soil layer (Figure 2), but the increase was larger in the control ponds than in the aerated ponds. In the 5- to 15-cm soil layer, there was an increase in the soil carbon of control ponds, but no significant increase was measured in the aerated ponds. This difference suggests that aeration to supply dissolved oxygen at the pond bottom can accelerate organic matter decomposition in bottom soils.

Draining

When ponds are drained, water currents created by the outflowing water suspend soil particles and sweep them out of ponds. Schwartz and Boyd (1994) showed that the weight of the solids that

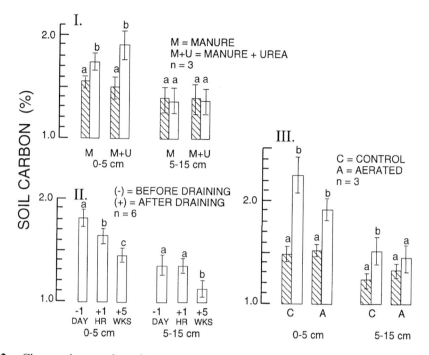

Figure 2. Changes in organic carbon concentrations in (I) manured ponds and ponds treated with manure plus urea; (II) ponds with bottoms dried between crops; and (III) aerated and control ponds. Nine samples were taken from 0- to 5-cm and 5- to 15-cm depths on each sampling date. Number of ponds per treatment (n) is given in the figure. In I and III, shaded bars represent the initial concentrations and open bars give concentrations at the end of the experiments. Vertical lines in ends of bars represent 95% confidence intervals for the means. Two bars indicated by the same lower case letter do not differ significantly at the 5% level of probability (comparison valid only for bars within a set). (From Ayub, M., Boyd, C. E., and Teichert-Coddington, D., *Prog. Fish-Cult.*, 55, 210, 1993.)

might settle that were lost during a single draining of channel catfish ponds averaged 9400 kg/ha. Small particles swept from the pond bottom come from the surface soil layer and are enriched in organic matter. When ponds in Honduras were drained, soil carbon concentration in the 0- to 5-cm layer declined from 1.82 to 1.65%, but draining had no influence on soil carbon concentration in the 5- to 15-cm layer. The loss of soil carbon during draining may be considered a good effect on the pond environment, but the suspended soil particles in pond effluent represent a sediment load to receiving bodies of water.

Drying

In the Honduras experiment, after ponds were drained, the soil was allowed to dry for 5 weeks. When the soil dried, deep cracks developed. Drying the soil removed free water from pore spaces and allowed air to enter the soil. The cracks also enhanced aeration. Soil carbon in the 0- to 5-cm layer decreased from 1.65 to 1.44% during 5 weeks of drying; the corresponding decrease of soil carbon in the 5-to 15-cm layer was from 1.34 to 1.12% (Figure 2). Soil carbon in the 0- to 15-cm layer declined from 1.44 to 1.23%; this decline represented a 14.5% decrease in the concentration present immediately after draining. During the grow-out period in the aeration experiment, the concentration of carbon in the 0- to 15- cm layer of aerated pond soil increased from 1.39 to 1.62%

(16.5% increase over initial concentration). Soil carbon in unaerated ponds increased from 1.33 to 1.76% (32.3% increase over initial concentration). Comparing gains in carbon during the grow-out period with losses of carbon during the fallow period suggests that a fallow period between grow-out periods can greatly retard the rate of organic matter accumulation in pond bottom soils.

Although the results shown in Figure 2 suggest that a drying period permits considerable decomposition of organic matter, it may not be necessary to dry pond bottoms for 5 weeks to obtain benefits. More information on factors affecting soil organic matter decomposition during the fallow period were obtained from laboratory studies described below.

Moisture Concentration

Eight soil samples from Alabama were dried to different moisture concentrations and soil respiration determined (Boyd and Pippopinyo, 1994). The optimum soil moisture concentration for microbial respiration had a range of 12 to 20%. This range suggests that although drying the soil accelerates decomposition, if soils are dried too much, there will be insufficient moisture for bacterial activity, and soil respiration will decline. The optimum moisture concentration for soil respiration in ponds in Honduras was around 35 to 45%, which was much higher than optimum concentrations found for pond soils in Alabama (Boyd and Teichert-Coddington, 1994). The difference in optimum moisture concentration between the two sites was related to the nature of the clay minerals in soils at the two sites. Soils near Auburn, AL, have a high clay content, but the clay particles are fairly large and consist of kaolinite, mica, and hydrous oxides (McNutt, 1981). The specific surface area of clays in soils at Auburn is typically around 100 m^2/g clay. The clay fraction of soils from bottoms of fish ponds at the El Carao research station in Honduras consisted of 41.8% smectite (a montmorillonite), and the specific surface area was 439 m^2/g clay (Green, 1993). Because of the higher surface area, soils from ponds at El Carao hold water more tightly and have a larger amount of biologically unavailable water than soils of ponds at Auburn. The optimum moisture content for soil respiration will vary from site to site, depending upon the type and amount of clay and other colloidal particles present in the soil.

pH

The optimum pH for organic matter decomposition in Auburn's ponds was 7.5 to 8.0, but pH values between 7 and 8.5 were all associated with much greater rates of soil respiration than lower or higher soil pH (Figure 3). Liming acidic soil to raise the pH above 7 appreciably increased soil respiration (Boyd and Pippopinyo, 1994); a treatment rate of 1000 to 2000 kg/ha is probably sufficient for most soils. Agricultural limestone is the best liming material to use in most cases, because even when applied at a rate of 2000 kg/ha, its use does not increase soil pH above 7.5. When calcium hydroxide was applied at 2000 kg/ha, there was an initial increase in soil pH above 10. High pH caused an initial inhibition of soil respiration, but the inhibition was temporary. Calcium hydroxide or calcium oxide are useful for killing fish pathogens, which may lie dormant in pond soils between crops.

Duration of Fallow Period

When the soils from Alabama were incubated at optimum moisture concentration and pH, the respiration rate decreased over time. About 70% of the total respiration for a 32-d period occurred during the first 16 d (Boyd and Pippopinyo, 1994). The decline in respiration over time resulted from a decline in labile organic substrate in response to microbial activity. A fallow period of 2 to 4 weeks is probably adequate for most soils. Those soils that dry quickly would need a shorter fallow period.

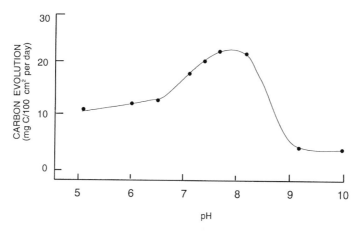

Figure 3. Effect of pH on soil respiration.(From Boyd, C. E. and Pippopinyo, S., *Aquaculture*, 120, 1994.

Tilling

A trial also was conducted in the laboratory to stimulate the influence of tilling on soil respiration (Boyd and Pippopinyo, 1994). When soil was crumbled in respiration chambers, carbon dioxide evolution increased about 35% above that of soil which was allowed to develop a dry crust on its surface. Crumbling the soil provided better contact with air and encouraged aerobic microbial activity. Tilling of bottom soils should have a similar influence in fallow ponds.

POND SOIL CLASSIFICATION

Classification is one of the tools that scientists use to help them understand natural systems, to provide a framework to guide research efforts, and to enable them to make predictions about the behavior of soils of particular classes. On a practical level, classification systems can be used to organize management practices that are suitable for particular classes of pond systems. Classification systems can have particular (narrow) purposes or multiple (broad) purposes. A multiple-purpose classification of soils for aquaculture might be one that helps determine soil suitability for pond construction while simultaneously providing information about pond fertility (through soil–water interactions), potential productivity, probable management problems (physical problems, such as seepage, but also chemical problems, such as phosphorus absorption/fixation or potassium limitations), and appropriate management practices (treatments for acidity or alkalinity, seepage, or nutrient limitations). In contrast, a single-purpose classification would deal with only one of these problem areas.

Textural Classification

Perhaps the most commonly used and best understood classification of soils is the textural classification system for mineral soils used by the U.S. Department of Agriculture. In that system, a sample of soil is classified according to the proportions of the three main particle size fractions (sand, silt, and clay) that it contains. Samples in which the proportions of these three components are similar are grouped together in textural classes with specific names; a sample with a very high percentage of one of the three components is called a sand, a silt, or a clay, whereas samples with more even mixtures of the components are given compound names that reflect the relative proportions of the components, for example, sandy loam or silty clay. The relationships between the various classes are shown in Figure 4.

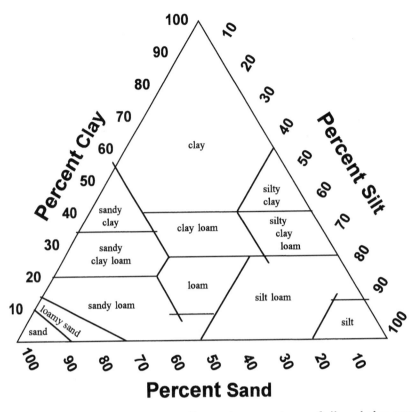

Figure 4. A soil triangle. Points corresponding to the percentages of silt and clay present in the soil are located on the silt and clay lines. Lines are then projected inward, parallel in the case of silt to the clay side of the triangle and in the case of clay parallel to the sand side of the triangle. The compartment in which the two inwardly projected lines intersect names the soil.

One advantage of this system is that it is familiar to many aquaculturists. The textural classes are most useful as indicators of the physical properties of a soil and are therefore widely used in the evaluation of the suitability of soils for pond construction. For example, it is known that sandy soils are highly permeable and that they do not make good embankment construction material unless mixed with a suitable amount of clay. On the other hand, soils with clay contents that are too high are usually very difficult to work, and unless they are provided with a "skeleton" of coarser soil material (e.g., sand), embankments constructed from these soils tend to slump after some time. It is thus clear that some combination of the three fractions is the best for pond construction purposes. Coche (1985) stated that the textural classes most suitable for pond construction included sandy clay, sandy clay loam, clay loam, silty clay, and silty clay loam. It is easy to remember the appropriate groups because their class names all contain the term "clay" in combination with some other textural term.

Textural classification does not consider some of the other soil properties that are of concern to aquaculturists, however. Classification according to texture is based only on the single physical property of particle size, and can thus only indirectly address the chemical characteristics of relevance to pond dynamics. It is commonly assumed, for example, that virtually all of a mineral soil's CEC comes from the clay fraction, and that the sand and silt fractions do not make any appreciable contribution to it. Because textural classification does not consider the mineralogy of the clay fraction, knowledge of the textural class of a given soil does not tell one much about its CEC. Mineralogy is also important because 1:1 type clay minerals or iron and aluminum oxides can tend to form sand-like granules, making soils dominated by them much more permeable than soils high in 2:1 type clay minerals. Further, because textural classification deals only with mineral soils, it is of little use where organic soils such as peats or clayey peats are of concern. Finally, textural classification does not consider the reaction (acidity or alkalinity) of a soil, the chemical factor of perhaps the greatest importance to aquaculturists.

The CRSP Soil Classification System

In an effort to classify soils in a way that alleviates some of the shortcomings of the basic textural classification without becoming too complex, Bowman (1992) proposed adapting portions of the *Soil Taxonomy* (Soil Survey Staff, 1975, 1990) classification for use in aquaculture. In this system, soil materials are classified at the family level according to criteria that are directly relevant to the concerns of aquaculturists. In fact, the intent of classification at the family level in *Soil Taxonomy* was "to group the soils . . . having similar physical and chemical properties that affect their responses to management and manipulation for use" (Soil Survey Staff, 1975, p. 80). The criteria used to distinguish the groups included particle size, mineralogy, reaction, soil temperature, soil depth, soil slope, soil consistency, soil coatings, and cracks in the soil (Soil Survey Staff, 1975). Of greatest importance among these criteria, from an aquacultural standpoint, are soil reaction (pH), particle size distribution, and mineralogy.

Bowman (1992) proposed using the above criteria, together with organic matter content and the source of acidity or alkalinity in the soil, to classify pond soils (or the soils of proposed pond sites) in two categories, or levels, as shown in Figure 5a. The first level is concerned with soil reaction (pH) and divides soils into three classes, acid, neutral, and alkaline. The second level is concerned with soil composition and is actually condensed from categories dealing with the source of soil acidity or alkalinity (exchangeable cations vs. free minerals), organic matter contents (organic vs. mineral soils), particle-size distribution (sandy, coarse-loamy, fine-loamy, and clayey soils), and mineralogy of the clay fraction (1:1, 2:1, and mixed clay minerals). (Part b of Figure 5 shows these categories for acid soils.) The definitions of the proposed soil classes are given in Table 5, and the approximate correspondence of the particle-size classes of this system with the textural classes used by the USDA are shown in Figure 6. Criteria used to distinguish soils at the family level in *Soil Taxonomy* (Soil Survey Staff, 1975) that were not considered essential for classifying aquaculture pond soils included soil temperature, soil depth, soil slope, soil consistency, soil coatings, and cracks in the soil.

This classification is a multiple-purpose classification; it was intended to provide information relevant not only to the physical properties of soil classes and their suitability as construction materials but also about important chemical properties of the classes, such as soil pH, CEC, pH-base saturation relationships, and nutrient limitations. For example, knowing that a pond soil (or the soil at a potential pond site) belongs to the class of soils that are acidic, clayey, and dominated by 1:1 (kaolinitic) clays reveals not only that its permeability will probably be greater than that of other clayey soils but also that its natural pH, CEC, and percent base saturation will probably all be low, that its anion exchange capacity (AEC) will be relatively high, and that it will have a relatively high potential for absorbing and fixing phosphates (Bowman, 1992). In addition, the lime requirement of a pond built in this type of soil will probably be much lower than that of a pond built on an acidic peaty-clay or a 2:1-dominated clay soil with the same initial pH, primarily because the CEC of this soil type is relatively low. Such information is useful to workers evaluating potential pond sites as well as to aquaculturists responsible for the management of ponds already in production. Once the soil class of a new pond has been identified, the management practices known to be appropriate for that class can be applied.

Identification of the class to which a specific soil belongs is straightforward. At the first level a simple soil pH determination establishes the reaction class of a soil. Extremely low pH readings further suggest that a soil may belong to either the organic or the acid sulfate class. Organic soils

Level 1:   SOIL REACTION

Level 2:   SOIL COMPOSITION

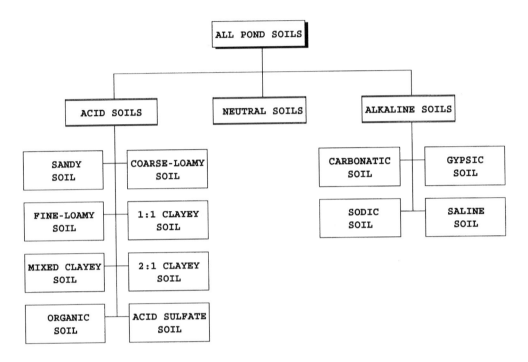

Figure 5a.

Figure 5.   The CRSP classification of pond soils, as proposed by Bowman (1992) and implemented in the CRSP PONDCLASS (Lannan et al., 1991, 1993) and POND (Bolte et
al., 1994) decision support systems. Soil classification is based on soil reaction, source
of acidity or alkalinity, organic matter content, particle-size distribution, and clay
mineralogy. Part A: Soil classes as defined in Table 5. Part B: The full hierarchy of
acid soils, showing soils classified according to source of acidity, organic matter content,
particle-size distribution, and clay mineralogy. (From Bowman, J. R. and Lannan, J. E.,
J. World Aquaculture Soc., 26(2), 176, 1995. With permission.)

are usually identified by their dark (often almost black) color and the high amount of fibrous,
partially decomposed plant material present, whereas acid sulfate soils are typically mineral soils
(low in organic materials) that contain yellowish mottles. Typical pH values for organic soils are
in the range of 4.0 to 5.6, whereas those for acid sulfate soils generally fall below 3.5, and sometimes
lower than 2.0. Standard "texturing" by feel determines the particle-size class of a soil sample, and
the mineralogy of the clay fraction is inferred from soil color, local soil names, climate, topography,
and vegetation. Knowledge from local farmers and agricultural workers as well as soil survey data,
where available, provide valuable supplements to the clues listed above for class identification. If
necessary, laboratory analyses can also be used to assist in this process.

The classification described above was used as the basis for the lime requirement estimation
routines in the PD/A CRSP's computerized PONDCLASS© (Lannan et al., 1991, 1993) and POND©
(Bolte et al., 1994) decision support systems. Users of these systems can specify the soil type of
a specific pond, together with its initial pH, and the system returns an estimated lime requirement

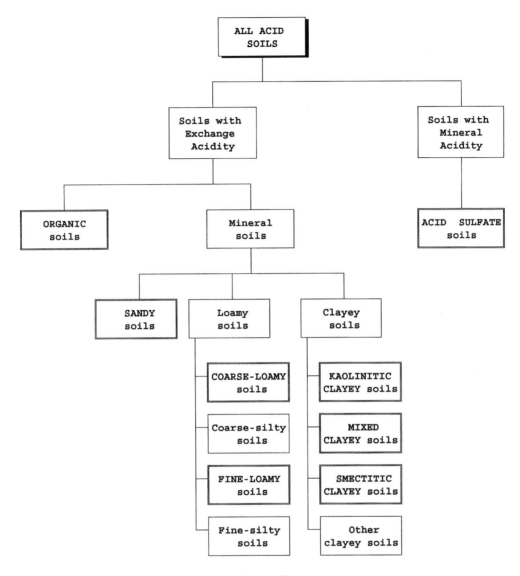

Figure 5b.

for that pond. The details of this method are described in the next part of this chapter. This classification of soils, with information about the chemical characteristics typical of each class, can also provide the basis for the development of buffer-type lime requirement techniques similar to those developed by Boyd (1974) for Alabama ponds.

LIME REQUIREMENT METHODS

Pond liming is commonly practiced in areas with acid soils and soft waters but is not necessary in areas where water supplies or soils have high concentrations of carbonates ("hard" waters and "alkaline" soils). The immediate goal of liming is to neutralize soil acidity, that is, to decrease base unsaturation to nearly 0% and increase soil pH to the near neutral range. Benefits that may result from applications of lime to the soil include increases in the availability of nutrients from the soil (Truog, 1948), increased benthic production (Bowling, 1962), and increased microbial activity in the muds (Pamatmat, 1960).

Table 5   A Classification of Soils According to Reaction (pH) and Composition[a]

Level 1: Soil Reaction	
Acid soil:	Soil with a pH of less than 7.0 (practically, pH <6.6), or soil in which percent base saturation is <100.
Neutral soil:	Soil with a pH close to 7.0 (practically, pH between 6.6 and 7.3) and a base saturation percentage close to 100.
Alkaline soil:	Soil with a pH of greater than 7.0 (practically, pH >7.3), a base saturation percentage of 100, AND containing appreciable quantities of salts, usually carbonates of calcium, magnesium, or sodium.
Level 2: Soil Composition	
Sandy soil:	Mineral soil material with not less than 70-85% sand, but less than 50% of the sand is fine or very fine sand, and the quantity [% silt + (2 × % clay)] is less than 30.
Coarse-loamy soil:	Mineral soil material with less than 18% clay that is not sandy soil.
Fine-loamy soil:	Mineral soil material with 18–34% (<35%) clay.
1:1 Clayey soil:	Mineral soil material that contains 35% or more clay, and the clay fraction is dominated by (contains 50% or more) kaolinite, other 1:1 layer minerals, oxides of iron or aluminum (e.g., gibbsite, goethite), nonexpanding 2:1 layer minerals, or amorphous substances (e.g., allophane); less than 10% of the clay fraction is composed of montmorillonite or other expanding 2:1 layer minerals.
Mixed clayey soil:	Mineral soil material that is 35% or more clay, but in which the clay fraction is not dominated by (contains less than 50% of) any one clay mineral.
2:1 Clayey soil:	Mineral soil material that is 35% or more clay, and that is dominated by (contains 50% or more) smectite, montmorillonite, or nontonite.
Organic soil:	Soil that contains at least 20% organic matter by weight, and at least 30% organic matter when clay content is as high as 60% (Brady, 1984)
Acid sulfate soil:	Wet clay soil high in reduced forms of sulfur. Includes potential acid sulfate soils.
Carbonatic soil:	Mineral soil material that contains more than 40% carbonates (as $CaCO_3$) plus gypsum, and the carbonates are >65% of the sum of carbonates and gypsum.
Gypsic soil:	Mineral soil materials containing more than 40% carbonates (as $CaCO_3$) plus gypsum, and the gypsum is >35% of the sum of carbonates plus gypsum.
Saline soil:	Non-sodic soil material that contains sufficient soluble salts to impair its productivity.
Sodic soil:	Soil material that contains sufficient sodium to interfere with most crops, and in which the exchangeable sodium percentage is 15% or greater.

[a]   After Bowman (1992).

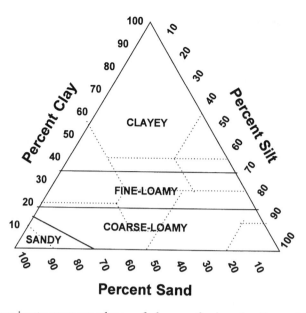

Figure 6. The approximate correspondence of classes of mineral soils used in the CRSP classification with the USDA soil texture classes. (From Bowman, J. R. and Lannan, J. E., *J. World Aquaculture Soc.*, 26(2), 175, 1995.)

Beneficial changes in water quality occur as a result of liming as well. Foremost among them are increases in alkalinity and hardness, pH buffering, and the clearing of humic stains, which may inhibit photosynthesis (Boyd, 1979). Indeed, it is often stated that the main purpose of lime applications is to raise the alkalinity of pond water, thereby removing carbon deficiencies, which limit phytoplankton growth (Huet, 1972; Boyd, 1979). Moyle's (1945, 1946) investigations suggest that better productivity can be expected from ponds in which alkalinities are greater than about 40 mg/L, and PD/A CRSP research in Indonesia and Thailand has shown carbon limitation in ponds when alkalinity was lower than about 33 mg/L (McNabb et al., 1990).

Low pH values (<6.0) for pond soils and low total alkalinity values (<20 mg/L as $CaCO_3$) in pond waters are indications that ponds need lime (Boyd, 1979). Unfortunately, the amount of lime required for the neutralization of acidity or for specific changes in water column alkalinity are not indicated by pH and in fact may vary considerably from site to site, depending on soil and source water characteristics. Because of the relationship between water alkalinity and the base unsaturation of the soil, alkalinity control is often a soil management problem rather than a water management problem.

Historically, pond liming rates have often been estimated using agricultural liming tables or agricultural laboratory lime requirement (LR) techniques such as the SMP (Shoemaker et al., 1961) and Adams-Evans (Adams and Evans, 1962) buffer methods. One potential problem associated with the use of agricultural methods is that, although a particular method may have been well suited to the geographical area for which it was developed, it may under- or overestimate LRs in another area. For example, the "SMP" buffer method was developed for agricultural soils in the midwestern U.S., but has been found to overestimate LRs in some other areas (Tisdale et al., 1985). Another potential problem is that a method that is suited to the agricultural soils of a particular area may actually overestimate the LRs of fish ponds in that area. For example, Boyd (1974) found that the Adams-Evans buffer method (Adams and Evans, 1962) gave much higher estimates of the LRs of Alabama fish ponds than were actually needed in those ponds. These and other types of problems have prompted several workers to develop lime requirement estimation techniques adapted specifically for use in fish ponds.

Buffer Method for Ponds

Based on information from ponds in Alabama, Boyd (1974) modified the Adams-Evans (1962) agricultural method for use in ponds in Alabama and other areas with similar soil–water relationships. The new procedure used the same laboratory procedures as the Adams-Evans method but was modified to provide LR estimates for raising pond water total hardness (and alkalinity) to 20 mg/L, rather than to raise soil (mud) pH to a particular level. Some refinements to the method were made by Boyd and Cuenco (1980). This approach has been widely and successfully used for the determination of pond LRs both in the southeastern U.S. and in other areas. Details of the laboratory procedure and the lime requirement table are provided by Boyd (1979).

This method is applicable only in cases where soil/mud acidity is due to the presence of acidic cations on the exchange complex. In soils with acidity due to the oxidation of iron pyrite or sulfide deposits, other methods for determining LRs are necessary (Boyd, 1979).

Modified Buffer Method for Ponds

Pillay and Boyd (1985) proposed a modification of Boyd's original procedure to broaden the applicability of the approach to a wider range of soil types. In the modified procedure, the total exchange acidity of a pond mud is estimated using a buffer, and the amount of limestone required to completely neutralize that acidity (decrease base unsaturation to 0%) is calculated. The complete procedure is described in Pillay and Boyd (1985). This approach resulted in higher LR estimates for ponds in Alabama than did the use of the original method, even though the 1.5x liming factor was not included, and it always resulted in a higher requirement than necessary (Pillay and Boyd, 1985). The advantages of the modified method were that it was rapid and simple, and it did not require knowledge about the pH-base unsaturation relationship of the specific pond soil (mud) in question.

Direct Calculation of Lime Requirements

Boyd (1979) pointed out that the LR of a pond could be calculated directly if sufficient information about the soil and water properties of the pond were known. Unfortunately, the needed information for a wide variety of ponds was not known. Of particular importance were information about the pH-base saturation and base saturation–alkalinity relationships of ponds on diverse soil types. In a CRSP-supported study, Bowman (1992) undertook an evaluation of the pH–base saturation relationships of subsurface soil horizons and used them to directly calculate theoretical LRs for ponds with different bottom soil types.

Bowman (1992) compiled a soils data base from data published in Soil Survey Investigations Reports (SCS/USDA, 1967, 1970, 1973, 1974, 1976, 1978–1981). Soil horizons in the data base were classified according to organic matter content, particle-size distribution, and mineralogy (Table 5), and a theoretical pH-base saturation curve, based on work by Peech and Bradfield (1948) and Sposito (1989), was fit to the data for each of seven classes of acid soils. The parameters of the fitted equations were then used to solve a lime requirement equation based on the work of Peech (1965) and others, in which

$$LR = CEC \cdot \left(BS_{des} - BS_{in}\right) \cdot CF \cdot D \cdot D_b \qquad (1)$$

where LR = lime requirement, in kg/ha; CEC = cation exchange capacity, in meq/100 g soil; BS_{des} = desired (target) base saturation, as a percentage of CEC; BS_{in} = initial base saturation, as a percentage of CEC; CF = correction factor; D = depth (of soil reaction), in cm; and D_b = bulk density of the soil, g/cm^3. Values for CEC and D_b were derived from the soils data base developed for the evaluation of pH–base saturation relationships, initial levels of base saturation were calculated from the same pH–base saturation relationships using initial values of pH, CF was the equivalent weight of $CaCO_3$ (50), used to convert units of exchange acidity (meq/100 g soil or cmol/kg soil) to lime application rates (kg/ha-cm), and D was set at 15 cm (based on the work of Cuenco, 1977, reported in Boyd, 1979).

Tables of theoretical lime requirements for seven of the soil composition classes described in Table 5 were developed by solving Equation 1 for the normal ranges of pH values found in each type of soil and using target base saturations equivalent to a pH of 6.5 in mineral soils and 5.5 in organic (peat) soils. The base saturation/pH targets were chosen based on nutrient availability information for mineral soils (Truog, 1948) and organic soils (Lucas and Davis, 1961). A summary of the results for the seven types of acid soils is given in Table 6. These tables are not meant to replace the use of more accurate LR methods, such as the buffer methods developed by Boyd (1974) and Pillai and Boyd (1985); rather they are intended to be available to field workers who do not have access to laboratory facilities but who can make reasonable estimates of soil type and

Table 6　Summary of Provisional Lime Requirements for Seven Classes of Acid Soils at the Common Target pH of 6.5 (5.5 for organic soils)

Soil pH	Soil Class[a]						
	SAND	COCO	FIFI	KKCL	MXCL	MTCL	HIST
6.5	0	0	0	0	0	0	0
6.4	131	249	202	494	306	481	0
6.3	302	498	605	1006	611	1005	0
6.2	522	871	807	1530	917	1576	0
6.1	796	1120	1211	2061	1528	2194	0
6.0	1124	1493	1413	2595	1834	2862	0
5.9	1501	1742	1816	3126	2139	3581	0
5.8	1911	2239	2220	3647	2751	4353	0
5.7	2335	2613	2624	4155	3056	5177	0
5.6	2748	3110	3027	4645	3668	6055	0
5.5	3129	3732	3431	5113	4279	6984	0
5.4	3462	4355	4037	5557	4890	7964	1412
5.3	3741	5101	4642	5973	5502	8992	2842
5.2	3967	5972	5247	6360	6418	10064	4272
5.1	4143	6843	6257	6719	7335	11176	5682
5.0	4277	7589	7468	7047	8864	12323	7054
4.9	4378	8336	9284	7347	11003	13499	8374
4.8	4452	8834	11908	7619	18644	14698	9627
4.7	4507	9456	13926	7865	21395	15913	10803
4.6	4547	9829	15137	8085	22923	17135	11896
4.5	4575	—	16146	8282	24146	18357	12900
4.4	4596	—	16751	8458	25062	19572	13815
4.3	4610	—	—	8614	25674	20772	14642
4.2	4621	—	—	8752	—	21950	15382
4.1	4629	—	—	8873	—	23099	16041
4.0	4634	—	—	8981	—	24214	16625

Note: All lime requirements are given as kilograms per hectare of surface area (kg $CaCO_3$/ha).

[a] Class name abbreviations are as follows: SAND: sandy; COCO: coarse-loamy; FIFI: fine-loamy; KKCL: 1:1 clayey; MXCL: mixed clayey; MTCL: 2:1 clayey; and HIST: organic.

soil pH in the field. They can and should be modified if workers have direct knowledge of soil characteristics such as CEC, depth of reaction (D), or soil density (D_b).

The general approach to estimating lime requirements described here has been incorporated into the PD/A CRSP's computer program PONDCLASS©: *Guidelines for Fertilizing Aquaculture Ponds* (Lannan et al., 1991, 1993) and the POND© decision support system (Bolte et al., 1994). In these programs the user need not refer to tables and make manual adjustments to the tabular values. Rather, PONDCLASS© and POND© are interactive, allowing the user to make changes in the program's assumptions (such as depth of reaction) prior to having the lime requirement estimates calculated and displayed.

Current CRSP research topics focus on the relationships between soil alkalinity and water alkalinity for a variety of different types of acid soils, on comparisons of several methods for estimating pond soil lime requirements, and on validation of the lime requirement estimates produced by the POND© and PONDCLASS© programs. The results of this research will enable the refinement of procedures for estimating the lime requirements of these different types of soils.

REFERENCES

Adams, F. and Evans, C. E., A rapid method for measuring lime requirement of red-yellow podzolic soils, *Soil Sci. Soc. Am. Proc.,* 26, 355–357, 1962.

Anderson, J. P. E., Soil respiration, in Page, A. L., Miller, R. H., and Keeney, D. R., Eds., *Methods of Soil Analysis, Part 2, Chemical and Microbiological Properties,* American Society of Agronomy, Madison, WI, 1982, 831–871.

Ayub, M., Boyd, C. E., and Teichert-Coddington, D., Effects of urea application, aeration, and drying on total carbon concentrations in pond bottom soils, *Prog. Fish-Cult.,* 55, 210–213, 1993.

Banerjea, S. M., Water quality and soil condition of fish ponds in some states of India in relation to fish production, *Ind. J. Fish.,* 14, 113–144, 1967.

Bolte, J., Nath, S., Ernst, D., and Tice, M., *POND© Version 2.0. Biosystems Analysis Group,* Department of Bioresource Engineering, Oregon State University, Corvallis, 1994.

Bowling, M. L., The Effects of Lime Treatment on Benthos Production in Georgia Farm Ponds, in *Proc. Annu. Conf. Southeastern Game and Fish Commissioners,* 16, 418–424, 1962.

Bowman, J. R., *Classification and Management of Earthen Aquaculture Ponds, with Emphasis on the Role of the Soil,* Ph.D. dissertation, Oregon State University, Corvallis, 1992, 211 pp.

Bowman, J. R. and Lannan, J. E., Evaluation of soil pH-percent base saturation relationships for use in estimating the lime requirements of earthen aquaculture ponds, *J. World Aquaculture Soc.,* 26(21), 172–182, 1995.

Boyd, C. E., *Bottom Soils, Sediment, and Pond Aquaculture,* Chapman & Hall, New York, 1995, 348 pp.

Boyd, C. E., Lime requirements of Alabama fish ponds, *Alabama Agricultural Experiment Station Bulletin* 459, Auburn University, Auburn, AL, 1974, 20 pp.

Boyd, C. E, *Water Quality in Warmwater Fish Ponds, Alabama Agricultural Experiment Station,* Auburn University, Auburn, AL, 1979, 359 pp.

Boyd, C. E., *Water Quality in Ponds for Aquaculture, Alabama Agricultural Experiment Station,* Auburn University, Auburn, AL, 1990, 482 pp.

Boyd, C. E. and Cuenco, M. L., Refinements of the lime requirement procedure for fish ponds, *Aquaculture,* 21, 293–299, 1980.

Boyd, C. E. and Pippopinyo, S., Factors affecting respiration in dry pond bottom soils, *Aquaculture,* 120, 283–294, 1994.

Boyd, C. E. and Teichert-Coddington, D., Pond bottom soil respiration during fallow and culture periods in heavily-fertilized tropical fish ponds, *J. World Aquaculture Soc.,* 25, 417–423, 1994

Brady, N. C., The Nature and Properties of Soils, 9th ed., Macmillan, New York, 1984, 750 pp.

Buol, S. W., Hole, F. D., and McCracken, R. J., *Soil Genesis and Classification,* 2nd ed., The Iowa State University Press, Ames, IA, 1980, 406 pp.

Coche, A. G., Simple Methods for Aquaculture, Soil and Freshwater Fish Culture, *FAO Training Ser.* 6, Food and Agriculture Organization of the United Nations, Rome, 1985, 174 pp.

Cuenco, M. L., *Application of Lime in Ponds, its Penetration in Muds, and Effect on Water Hardness,* Masters thesis, Auburn University, Auburn, AL, 1977.

Egna, H. S., Brown, N., and Leslie, M., General Reference: Site Descriptions, Materials and Methods for the Global Experiment, Pond Dynamics/Aquaculture CRSP Data Reports, Vol. 1, Office of International Research and Development, Oregon State University, Corvallis, 1987, 84 pp.

Green, B. W., Water and Chemical Budgets for Organically Fertilized Fish Ponds in the Dry Tropics, Ph.D. dissertation, Auburn University, Auburn, AL, 1993.

Hajek, B. F. and Boyd, C. E., Rating soil and water information for aquaculture, *Aquacultural Eng.,* 13, 115–128, 1994.

Hanson, B., Ndoreyaho, V., Tubb, R., Rwangano, F., and Seim, W. K., Rwanda: Cycle I of the Global Experiment, Pond Dynamics/Aquaculture Collaborative Research Support Program, Office of International Research and Development, Oregon State University, Corvallis, 1989, 62 pp.

Huet, M., Breeding and cultivation of fish, in *Textbook of Fish Culture,* Fishing News (Books) Ltd., West Byfleet, Surrey, England, 1972, 436 pp.

Jhingran, V. G., *Fish and Fisheries of India,* Hindustan Publ., New Delhi, 1975, 954 pp.

Lannan, J. E., Nath, S., Manickam, F., Bowman, J., and Snow, A., *PONDCLASS© Version 1.1 PC,* Pond Dynamics/Aquaculture CRSP, Office of International Research and Development, Oregon State University, Corvallis, 1991.

Lannan, J. E., Nath, S., Manickam, F., Bowman, J., and Snow, A., *PONDCLASS© Version 1.2 PC,* Pond Dynamics/Aquaculture CRSP, Office of International Research and Development, Oregon State University, Corvallis, 1993.

Lucas, R. E. and Davis, J. F., Relationships between pH values of organic soils and availabilities of 12 plant nutrients, *Soil Sci.,* 92, 177–182, 1961.

McNabb, C. D., Batterson, T. R., Premo, B. J., Eidman, H. M., and Sumatadinata, K., Indonesia Project: Cycle I of the Global Experiment, *Pond Dynamics/Aquaculture CRSP Research Data Reports,* Vol. 3, Office of International Research and Development, Oregon State University, Corvallis, 1988, 67 pp.

McNabb, C. D., Batterson, T. R., Premo, B. J., Eidman, H. M., and Sumatadinata, K., Indonesia Project: Cycle II of the Global Experiment, *Pond Dynamics/Aquaculture CRSP Research Data Reports,* Vol. 3, No. 2, Office of International Research and Development, Oregon State University, Corvallis, 1991, 49 pp.

McNabb, C. D., Batterson, T. R., Premo, B. J., Knud-Hansen, C. F., Eidman, H. M., Lin, C. K., Jaiyen, K., Hanson, J. E., and Chuenpagdee, R., Managing fertilizers for fish yield in tropical ponds in Asia, in The Second Asian Fisheries Forum, Hirano, R. and Hanyu, I., Eds., *Proc. Second Asian Fisheries Forum,* Tokyo, Japan, 17–22 April 1989, The Asian Fisheries Society, Manila, 1990, 169–172.

McNutt, R. B., Soil Survey of Lee County, Alabama, U.S. Department of Agriculture, Soil Conservation Service, Washington, D.C., 1981.

Moyle, J. B., Some chemical factors influencing the distribution of aquatic plants in Minnesota, *Am. Midland Nat.,* 34, 402–420, 1945.

Moyle, J. B., Some indices of lake productivity, *Trans. Am. Fish. Soc.,* 76, 322–334, 1946.

Murad, H. and Boyd, C. E., Production of sunfish (*Lepomis* spp.) and channel catfish (*Ictalurus punctatus*) in acidified ponds, *Aquaculture,* 94, 381–388, 1991.

Pamatmat, M. M., *The Effects of Basic Slag and Agricultural Limestone on the Chemistry and Productivity of Fertilized Ponds,* Masters thesis, Auburn University, Auburn, AL, 1960.

Peech, M., Lime requirement, in Black, C. A., Ed., *Methods of Soil Analysis, Part 2,* American Society of Agronomy, Madison, WI, 1965, 927–932.

Peech, M. and Bradfield, R., Chemical methods for estimating lime needs of soils, *Soil Sci.,* 65, 35–55, 1948.

Pillay, V. K. and Boyd, C. E., A simple method for calculating liming rates for fish ponds, *Aquaculture,* 46, 157–162, 1985.

Schaeperclaus, W., *Textbook of Pond Culture,* Rearing and Keeping of Carp, Trout and Allied Fishes, Book Publishing House Paul Parey, Berlin, 1933, 261 pp. (English translation by Hund, Frederick, Stanford University, Fish and Wildlife Service Fishery Leaflet 311, U.S. Department of the Interior.)

Schwartz, M. and Boyd, C. E., Effluent quality during harvest of channel catfish from watershed ponds, *Prog. Fish-Cult.,* 56, 25–32, 1994.

SCS/USDA (Soil Conservation Service, U.S. Department of Agriculture), *Soil Survey Investigations Reports* Numbers 12 and 23–37, Soil Conservation Service, U.S. Department of Agriculture, Washington, D.C., 1967, 1970, 1973, 1974, 1976, 1978–1981.

Shoemaker, H. E., McLean, E. O., and Pratt, P. F., Buffer methods for determining lime requirement of soils with appreciable amounts of extractable aluminum, *Soil Sci. Soc. Am. Proc.,* 25, 274–277, 1961.

Soil Survey Staff, *Soil Taxonomy, A Basic System of Soil Classification for Making and Interpreting Soil Surveys,* Agriculture Handbook 436, Soil Conservation Service, U.S. Department of Agriculture, Washington, D.C., 1975, 743 pp.

Soil Survey Staff, *Keys to Soil Taxonomy,* Soil Mgt. Support Serv., Tech. Monog. No. 19, 4th ed., Virginia Polytechnical Institute and State University, Blacksburg, VA, 1990.

Sposito, G., *The Chemistry of Soils,* Oxford University Press, New York, 1989, 277 pp.

Teichert-Coddington, D. and Green, B., Comparison of two techniques for determining community respiration in tropical fish ponds, *Aquaculture,* 1993.

Teichert-Coddington, D. R., Stone, N., and Phelps, R. P., Hydrology of fish culture ponds in Gualaca, Panama, *Aquacultural Eng.,* 7, 309–320, 1988.

Tisdale, S. L., Nelson, W. L., and Beaton, J. D., *Soil Fertility and Fertilizers,* 4th ed., Macmillan, New York, 1985, 754 pp.

Truog, E., Soil reaction influence on availability of plant nutrients, *Soil Sci. Soc. Am. Proc.,* 11, 305–308, 1948.

7 ENVIRONMENTAL CONSIDERATIONS

Wayne K. Seim, Claude E. Boyd, and James S. Diana

INTRODUCTION

Water, air, soil, and aquatic organisms can be viewed as overlapping compartments in the pond environment. Substances exchange between these compartments through their closely associated interfaces. Aquaculture, with its intimacy of contact with receiving waters, should include evaluation and mitigation of potential environmental risks. Such action will reduce constraints on development by this emerging industry and protect it from ineffective or unnecessary restrictions. Other water users, such as intensive agriculturists and wastewater managers, have been forced to consider the environmental concerns of society and the restrictions on both the characteristics and the amount of effluent released and the type of treatment required. The aquaculture industry may be in the process of following a similar pathway as it responds to environmental legislation while incorporating its own sense of environmental responsibility and awareness. Aquacultural scientists from industrialized nations, where environmental legislation may be well established, are helping to shape aquacultural practices in countries where legislation and environmental concern may be just emerging. Although regulation to limit the environmental impacts of aquaculture may be defensible, such restrictions have often been unnecessarily burdensome, complex, and purposeless. Surveys of U.S. aquaculturists indicate that aspects of the permitting process and operational regulations are often inappropriate extensions of regulations designed for other forms of agriculture. The aquaculture community may find it necessary to lobby and build influence within the political community, while conducting convincing environmental research, to effect changes in these regulations.

Aquaculture has expanded rapidly throughout the world, doubling production to about 14 million metric tons between 1984 and 1992 (FAO, 1994). As production has increased, aquacultural practices have begun to resemble traditional animal agriculture in terms of increased intensity. Aquaculturists have been able to operate aquatic systems near the water quality limitations of the species cultured. Careful monitoring of dissolved oxygen and ammonia, sometimes coupled with mechanical systems to prevent lethal conditions, have allowed production at high levels within the limits of real but allowable risk of crop failure. Traditional agriculture has increased production in much the same way, increasing inputs and stock density to maximize profits per unit area. However, just as traditional agriculture has encountered problems, such as point and nonpoint discharge of wastes, fertilizers, drugs, and other potentially toxic substances, aquaculture may now face such problems.

Environmental problems associated with aquaculture may lead to legislative constraints beyond those encountered by traditional animal culture. Brackishwater shrimp culture, for instance, has attracted the attention of environmentalists and others concerned about the loss of wetlands, particularly valuable mangrove ecosystems, by improper culture practices. Such concerns, if ignored, may result in aquaculture facing confining legislation or public rejection of the product, as sometimes occurs in other areas of agriculture. Many examples already exist: consumers rejected apples from the Pacific Northwest because of concern over the use of the additive Alar; consumers have demanded "dolphin safe" tuna. Reduced red meat consumption may derive not only from

0-56670-274-7/97/$0.00+$.50
© 1997 by CRC Press LLC

health concerns but also from the perception that beef production is associated with poor rangeland practices and high resource commitments that impact other species.

Recent appreciation for the value of preserving indigenous species must be addressed by aquaculturists. Cultured exotic species, such as tilapia, introduced into receiving waters have endangered native species through competition, disease, or predation. Even where cultured fish are indigenous, loss of genetic diversity in native fish has occurred from hybridization with cultured stocks. When choosing which species to culture, aquaculturists must now take potential impacts on indigenous species into account, along with performance and economical considerations.

Aquaculture need not be environmentally damaging; instead, it can play a prominent role in conserving natural resources while contributing much needed social benefits. Attaining a more sustainable, productive, and environmentally sound food production base throughout the world may not be achievable, in fact, without the incorporation of an expanded aquaculture industry into the agricultural process. By avoiding negative impacts on the environment, aquaculture can solve rather than create environmental problems on a global scale.

POND EFFLUENT

Volume

When ponds are drained for harvest, effluent volume is the same as pond volume. Ponds seldom have gauged overflow structures for estimating overflows at other times, but it is possible to make rough estimates of total effluent volume (overflow volume + volume drained for harvest) with the basic hydrologic equation

$$\text{Inflows} = \text{Outflows} \pm \Delta S \tag{1}$$

where ΔS = change in storage, or the initial water volume minus the final water volume.

Important inflows for aquaculture are precipitation (P), runoff (RO), seepage (S_{in}), and regulated inflow (RI), while evaporation (E), seepage (S_{out}), overflow (OF), and regulated discharge (RD) are the main outflows (Boyd, 1982).

Aquaculture ponds usually are either watershed ponds formed by building dams across watersheds to impound runoff or levee ponds formed by building levees around the areas in which water is impounded (Boyd, 1985a). Watershed ponds seldom have a source of regulated inflow and depend upon runoff from watersheds for water supply. Levee ponds have no watershed other than the insides and tops of levees; they must be filled and maintained by water from wells, streams, lakes, or estuaries.

For watershed ponds, the hydrologic equation may be written as

$$P + RO + S_{in} = \left(E + S_{out} + OF + RD\right) \pm \Delta S \tag{2}$$

Watershed ponds often have water seeping in and out at the same time. During parts of the year, net seepage may be into the pond (positive), and at other times it may be out of the pond (negative), but on an annual basis net seepage is usually an outflow (Boyd and Shelton, 1984; Shelton and Boyd, 1993). Seepage is difficult to measure, but its volume is usually small enough that it can be neglected in effluent estimates. Total discharge or effluent volume (Q) is OF + RD. The period that ponds are empty between crops is both the end of one crop and the beginning of the next, and ΔS may be considered 0 between crops. Neglecting seepage, setting $\Delta S = 0$, and substituting Q for OF + RD gives

$$Q = (P - E) + RO \tag{3}$$

Boyd (1985b) showed that pond evaporation for use in Equation 3 can be estimated as $0.81 \times$ pan evaporation. Precipitation can be measured with a rain gauge, and runoff can be estimated by the curve number technique, water accounting method, or other procedures (Yoo and Boyd, 1993). Values for P, E, and RO normally are measured in water depth, and they must be converted to volume based on pond and watershed areas.

The total discharge from watershed ponds will be large in high rainfall climates, for ponds on watersheds with a high runoff potential, and for ponds with a large watershed to pond area. Notice that when pond evaporation equals precipitation, effluent volume will equal runoff volume (discounting seepage, of course).

Suppose that a 1-ha pond of 1 m average depth has a 20-ha watershed; annual rainfall is 150 cm, annual pan evaporation is 140 cm, and annual runoff is 25% of precipitation. Runoff will be 1.50 m $\times$ 0.25 $\times$ 20 ha watershed $\times$ 10,000 m^2/ha = 75,000 m^3. Pond evaporation is 1.40 m $\times$ 0.81 = 1.13 m, and precipitation is 0.37 m greater than pond evaporation. The precipitation excess (P – E) is 0.37 m $\times$ 1 ha $\times$ 10,000 m^2/ha = 3700 m^3. Total discharge will be 75,000 m^3 + 3700 m^3 = 78,700 m^3.

The approach for estimating effluent volume for watershed ponds can be applied to levee ponds. The hydrologic equation for levee ponds is

$$P + RI + S_{in} = (E + S_{out} + OF + RD) \pm \Delta S \tag{4}$$

Neglecting seepage, setting $\Delta S = 0$, and substituting Q for OF + RD, as was done above for watershed ponds, gives total discharge from levee ponds as

$$Q = (P - E) + RI \tag{5}$$

In levee ponds where water exchange is not used as a management technique, RI is kept to a minimum, and it will be approximately equal to the volume of water needed to fill ponds and replace evaporation losses. In climates where pond evaporation exceeds precipitation, water must be added to replace evaporation, and the only effluent will be that drained from ponds at harvest. In climates where precipitation exceeds pond evaporation, water will overflow from ponds during rainy periods, and total effluent volume will be greater than the final volume drained from ponds for harvest. For example, consider a 1-ha levee pond, which is initially filled to an average depth of 1.5 m and to which no more water is added during the fish grow-out period because precipitation exceeded pond evaporation by 20 cm. Ignoring seepage, effluent will be (P – E) + RI = 1.70 m, and the effluent volume will be 1.70 m $\times$ 1 ha $\times$ 10,000 m^2/ha = 17,000 m^3.

If levee ponds are not drained for fish harvest, overflow will equal (P – E). Levee ponds should not be filled by regulated inflow until the water level is well below the top of the water outflow control structure, because if it is, all precipitation falling on the pond surface will overflow and not be stored (Boyd, 1982). In most areas, normal daily precipitation in excess of 5 to 10 cm is rare, so providing 10 to 15 cm storage below the overflow structures can eliminate considerable overflow after rains.

Water exchange is routinely used in marine shrimp farming, and it is occasionally used in other types of aquaculture. Water exchange greatly increases effluent volume, but it does not substantially increase the total amount of chemical substances discharged as compared to ponds without water exchange that are drained for harvest.

Chemical Characteristics

Water that overflows from ponds following rains, during water exchange, and during the initial phase of draining for harvest is normal aquaculture pond surface water. The potential pollutional strength of aquaculture pond water tends to increase as the level of fish or shrimp production increases (Boyd, 1990; Boyd and Musig, 1992). Although aquaculture pond waters are suitable for raising high densities of certain aquatic animals under specific management regimes, pond water may be much more concentrated in nutrients, organic matter, ammonia, and suspended solids and less concentrated in dissolved oxygen than receiving water bodies. Therefore, aquaculture ponds are potential sources of pollution.

A robust body of literature exists discussing concentrations of water quality variables in waters of aquaculture ponds (Hinshaw, 1973; Barker et al., 1974; Schneider, 1974; Tackett, 1974; Boyd, 1978; Ellis et al., 1978; Tucker and Lloyd, 1985; Schwartz and Boyd, 1994a, 1994b). These reports consistently show that aquaculture pond water contains high levels of particulate organic matter in the form of plankton (primarily phytoplankton), considerable nitrogen and phosphorus bound in plankton cells, elevated concentrations of total ammonia nitrogen, and appreciable chemical and biochemical oxygen demand. Concentrations of water quality variables increase in response to increasing levels of fish or shrimp production, so that the pollution potential of intensive aquaculture is much greater than that of extensive or semi-intensive aquaculture. Typical concentration ranges for water quality variables in pond waters are provided in Table 1.

Table 1   Typical Ranges for Selected Water Quality Variables in Surface Waters of Aquaculture Ponds at Three Levels of Production

	Level of aquacultural production		
Water quality variable	Extensive <1000 kg/ha	Semi-intensive 1000–5000 kg/ha	Intensive >5000 kg/ha
Chlorophyll *a* (µg/L)	10–50	50–150	150–500
Biochemical oxygen demand (mg/L)	2–5	5–20	20–40
Volatile solids (mg/L)	5–10	10–20	20–50
Turbidity (NTU)	5–10	10–25	25–50
Nitrate-nitrogen (mg/L)	0.01–0.1	0.1–0.2	0.2–0.3
Total ammonia nitrogen (mg/L)	0.1–0.5	0.5–2	2–5
Total Kjeldahl nitrogen (mg/L)	0.5–2.0	2–4	4–10
Total phosphorus (mg/L)	0.05–0.1	0.1–0.3	0.3–0.7
Settleable solids (mL/L)	0.0–0.05	0.05–0.1	0.1–0.5

When a pond is drained for fish or shrimp harvest, the first 75 to 85% of the water released will have a composition similar to that depicted in Table 1. However, the final water released is much more concentrated, because of the activity of fish crowded in a small water volume, erosion of the pond bottom by outflowing water, and disturbance of bottom soil by seining operations. For example, Schwartz and Boyd (1994b) reported the following maximum concentration of water quality variables in the final 5 to 10% of effluent from semi-intensive chemical catfish ponds: settleable solids, 10 mL/L; Kjeldahl nitrogen, 14.5 mg/L; total ammonia nitrogen, 3.7 mg/L; total phosphorus, 3.3 mg/L; biochemical oxygen demand, 55.4 mg/L; total dissolved solids, 1948 mg/L. The three channel catfish ponds studied by Schwartz and Boyd (1994b) averaged 1.5 m deep. Complete draining released the following quantities of substances per hectare: total suspended solids, 5400 kg; settleable solids, 39 m^3; Kjeldahl nitrogen, 78.7 kg; total ammonia nitrogen, 17.7 kg; nitrate-nitrogen, 0.8 kg; nitrite-nitrogen, 0.5 kg; total phosphorus, 12.1 kg; biochemical oxygen demand, 448 kg.

In terms of wastewater, pond effluent primarily contains suspended solids, nutrients, and organic matter. The organic matter is primarily particulate and in the form of plankton cells. Therefore, the biochemical oxygen demand exerted by pond water is primarily plankton respiration instead of microbial decomposition of dissolved organic matter. Much of the suspended inorganic matter in pond water is clay-sized particles, which settle slowly. Although aquaculture pond effluent usually has higher concentrations of dissolved and suspended substances than receiving waters and may exceed effluent limitation guidelines recommended by pollution control agencies, they are much less concentrated than pollutants in industrial and municipal wastewaters.

Treatment

Waste substances produced by shellfish and cage culture in open systems are discharged so diffusely as to preclude practical methods of treatment. Settled organic wastes may be pumped or dredged for removal or dispersal, although the outcome of such actions may result in toxic conditions of low dissolved oxygen and elevated hydrogen sulfide concentrations at the treated site. Continual resiting of the cage culture system may be necessary to prevent build-up of organic deposits of fecal wastes, uneaten feed, and dead organisms. Where necessary or required, effluent from ponds or raceways is amenable to well-known treatment techniques utilized for organic wastes for many years. The underlying physical and biological processes of waste treatment are processes already occurring within a pond production system, although a unique ecology of chemical and biological succession is incorporated in these systems (Hawkes, 1963).

The three major stages of organic waste treatment include removal of settleable solids; decomposition of dissolved or colloidal materials by organisms utilizing hydrolysis, oxidation, and reduction processes; and removal of nutrients. Removal of settleable solids is the most practical and easily accomplished of the three. Merely allowing sediments to settle within a production pond for several days after harvest activities and before final draining may remove sufficient BOD, solids, and phosphorus to meet effluent standards. Sediments accumulating on pond bottoms can then be removed for use as fertilizer on other crops. Exposed soils are subject to oxidation and mineralization of accumulated materials. Colt (1983) points out that periodic draining of ponds and subsequent drying of the pond soils can significantly contribute to in-pond water quality management. If these approaches are insufficient, or where more intensive culture systems utilizing flow-through of supply water are used, a treatment system external to the pond may be necessary. The choices for treatment are can broadly be categorized as either filtration or sedimentation by gravity.

Filtration by sand or other means is practical only where space or other limitations prevent application of the more cost-efficient sedimentation methods (Pillay, 1992). Combining sand filtration with nutrient processing and removal by emergent vegetation is a proven and effective combination. Reeds and bulrushes rooted in sand/gravel filtering substrate supply oxygen via passages within the plant structure for micro-organisms decomposing waste organic matter. The roots also grow into any accumulating material and assimilate nutrients (National Academy of Science, 1976). Solids can also effectively be removed by sedimentation. The nature and composition of the waste material should drive the design of treatment systems. Fish culture solids from intensive systems are denser and faster settling than domestic wastes, for instance, and tend to form a viscous sludge (Mudrak, 1979). Sedimentation rate is determined by size and density of particles and the temperature and velocity of the water. Settling times from a few hours to several days are necessary to remove significant amounts of settleable solids. A rectangular-shaped sedimentation pond with a length-width ratio greater than four enhances settling efficiency (Arceivale, 1983).

After removal of settleable organic materials, no further need for decomposition of dissolved organics may be necessary. Nutrient removal, either for the purpose of meeting effluent standards or to enable recycling of otherwise lost nitrogen and phosphorus, is more applicable to recirculating or highly intensive flow-through systems. Pond systems with relatively low nutrient levels and often intermittent discharge are less amenable to this approach.

Retaining nutrients within the pond by efficient fertilization methods and by culturing several species with differing feeding behavior to partition the environment may be a more sustainable strategy than external treatments designed specifically for nutrient removal. Utilizing herbivorous filter-feeders, zooplankton consumers, and piscivorous and detritivorous species may more efficiently recycle nutrients and limit losses to discharge waters. In Chinese and Indian polyculture of carps, scavenger species are utilized to recycle feces and detritus (Pillay, 1992). Such strategies to recycle or retain in[pond solids also result in nutrient loss reduction.

CONTAMINATION OF POND SYSTEMS

Heavy Metals

Metals are common contaminants in the water supplies of ponds and other aquatic systems. Metal sources include weathering of natural deposits, volcanic eruptions, and a variety of human activities such as mining operations, mine tailings, point discharges of industrial processes and sewage, runoff from solid wastes (garbage), erosion of metal pipes, and atmospheric sources. Atmospheric sources include combustion of fossil fuels and smelting processes (Laws, 1981; Edwards and Densem, 1980).

Metals, such as copper, chromium, iron, nickel, tin, and zinc, and metalloids such as selenium are essential trace substances important in plant and animal nutrition. However, optimal concentrations span narrow ranges, and too little induces nutrient deficiency or too much induces toxicity. Nonessential heavy metals, such as lead and mercury, are also toxic and occur in aquatic systems from natural or anthropogenic sources.

An aquaculture operation may also be the source of metal contamination. Copper may occur from $CuSO_3$ used as an algicide, other copper compounds used as wood or net preservatives, or erosion from copper pipe, brass valves, or pump parts. Organotin used as a netdip or antifouling paint may also originate from the culture process.

Acute toxicity expressed as the LC50 (the concentration resulting in 50% mortality of test organisms typically exposed for 48 or 96 h) for plants, invertebrates, and fish varies from less than 1 µg/L (ppb) for mercury to over 500 mg/L for lead (Hodson et al., 1979). Reported LC50 values for heavy metals may vary several orders of magnitude because of differences in susceptibility of species and because of the strong influence of water chemistry on toxicity. Heavy metals such as copper are less toxic at high alkalinity or hardness. Complexation with carbonate ion is apparently the major cause, as the free ion form (Cu^{+2}) contributes most of the toxicity (Chapman and McCrady, 1977). Thus, at alkalinity of 30 mg/L (as $CaCO_3$) the LC50 for salmonid fish is about 20 µg/L total copper, but if alkalinity is 200 mg/L, the LC50 rises to about 100 µg/L, a fourfold decrease in toxicity (Chapman and McCrady, 1977). Benthic filter feeders, often exposed to the highest metal concentrations, bioaccumulate metals from sediments but pass little of these residues on to higher trophic levels. Therefore, heavy metals tend not to biomagnify up the food chain at the high biomagnification factors of some very lipid-soluble pesticides or organometals (Rand and Petrocelli, 1985; Leland and McNurney, 1974); however, heavy metals may occur in the environment at concentrations sufficient to accumulate harmful residues in fish consuming contaminated prey organisms, even though bioaccumulation factors are small (Woodward et al., 1994). Low growth and survival resulted for young fish consuming invertebrates that had accumulated arsenic, cadmium, copper, lead, and zinc in a study by Woodward et al. (1994). Losses of metal-sensitive food organisms and excessive metal residues occurring in tolerant fish food organisms has been documented (Dallinger et al., 1987).

Inorganic mercury, as metallic mercury or mercuric salts, can be methylated in aquatic sediments by methogenic bacteria to methyl mercury, an extremely dangerous neurotoxicant with solubility properties much like DDT. Methyl mercury biomagnifies up food chains, with long half-lives in fish of 200 d or more (Massaro and Giblin, 1972). Elevated methyl mercury in fish is common, and the FDA has set levels from 0.5 to 1 ppm for human consumption (EPA, 1972). As humans only excrete about 1%/d of their methyl mercury body burden, only about 0.03 mg/d of methyl mercury can safely be consumed on a regular basis (Grant, 1971).

Pesticides

Although pesticide contamination has been a major problem in natural waters, there have been comparatively fewer problems with pesticides in aquaculture ponds. However, some problems have occurred, and unless care is taken when pesticides are used on agricultural land surrounding ponds or in application of pesticides in aquaculture management, fish and shrimp mortality or contamination can result. Pesticides include herbicides, insecticides, fungicides (and many other specific agents such as piscicides and molluscicides), and disinfectants.

Sources

Agricultural use on the watersheds of aquaculture ponds constitutes the major external sources of pesticides. Pesticides may reach ponds by aerial drift, runoff directly into ponds, runoff into lakes or tributaries of estuaries that serve as water supplies, and by improper disposal of pesticides in water that ultimately enters ponds. The use of pesticides with long residual lives is declining in most nations, and most pesticides that are used today degrade to nontoxic forms within a few days. However, some compounds are extremely toxic, and use of pesticides in aquacultural areas should be avoided if possible. Key factors for protecting ponds from pesticides are as follows: build ponds a considerable distance from pesticide-treated fields; plant trees or other tall vegetative cover between ponds and pesticide-treated areas to intercept aerial drift of pesticides; construct topographic barriers (ditches or terraces) to prevent pesticide-contaminated runoff from entering ponds; where feasible, use groundwater to fill and maintain water levels in ponds; use proper methods of pesticide application on crops; dispose of pesticides and pesticide containers in a safe manner.

When ponds are constructed on land that has been used for crop production, pesticide levels in the soil should be tested. If significant amounts of pesticide residues are detected, the surface soil should be excavated and buried in the centers of dams or levees to prevent it from contacting the pond water.

It also is possible for manures used as fertilizers or for feed ingredients to be contaminated with pesticides. Again, pesticide residue testing may be used to alert aquaculturists of a potential pesticide problem.

Of the pesticides, insecticides pose the greater danger to fish and shrimp production, as some of these materials are toxic at minute concentrations. However, herbicides and fungicides can also be toxic to aquatic animals. Herbicides and other pesticides may also inhibit photosynthetic production of oxygen in ponds (Vickers and Boyd, 1971; Tucker, 1987).

If proper precautions are exercised, pond aquaculture can be conducted in areas with intensive agriculture. For example, the Yazoo Basin of Mississippi (often called the Mississippi Delta) is the location of 70 to 80% of the channel catfish production in the U.S. This area also has intensive cotton, rice, and soybean production with heavy pesticide use. Pesticide contamination of some ponds has occurred, but as a management problem, pesticide contamination has not been a major concern. Coexistence of catfish production and traditional agriculture has been possible through careful location of ponds, removal of pesticide-contaminated soil during pond construction, use of groundwater to fill ponds, and reasonable cooperation from cotton, rice, and soybean farmers in proper use and disposal of pesticides.

Chemicals are used in ponds to kill unwanted fish, treat fish disease, and control unwanted aquatic plants. Substances such as rotenone, antimycin, ammonia, chlorine, teaseed cake, and oil cake have long been used to destroy wild fish before ponds are stocked (Boyd, 1990). These substances normally have a short residual life, and aquaculturists are generally familiar with the use of these materials and have encountered few problems. Insecticides have sometimes been used to eradicate wild fish from ponds, but this practice should be discouraged because of high toxicity and potential for residues in pond soils and aquatic animals.

Chemicals used for treating fish diseases should be used carefully, because concentrations slightly higher than those needed to combat the disease may be toxic to fish. For example, potassium permanganate is useful for treating bacterial diseases at 2 to 4 mg/L, but concentrations of 6 to 8 mg/L can be toxic to fish (Tucker and Boyd, 1977). Some compounds may have adverse effects on water quality. Formalin application at 10 to 15 mg/L is a common pond treatment for external parasites of fish, but such formalin concentrations can kill phytoplankton and cause dissolved oxygen depletion (Allison, 1962).

Herbicides have been applied to aquaculture ponds to control aquatic weeds with some success but at high risk. Ponds become infested with a particular kind of vegetation because the pond environment is suitable for that particular type of vegetation. For example, ponds with clear water often are infested with underwater weeds, and ponds with high nutrient concentrations can have dense phytoplankton blooms. Application of herbicides and algicides can cause a temporary improvement, but when the toxicity of the herbicide or algicides has dissipated, environment conditions will remain the same, and the unwanted species will quickly become re-established. Attempts to use algicides such as copper sulfate and simazine as continually applied chemical controls (chemostats) on phytoplankton growth in aquaculture ponds have been largely unsuccessful (Tucker and Boyd, 1978, 1979; Masuda and Boyd, 1993). Limited success has been achieved in the use of copper sulfate to reduce the abundance of species of blue-green algae associated with off-flavor in channel catfish ponds (van der Ploeg, 1989; van der Ploeg and Boyd, 1991; Boyd, 1990) (other organisms such as actinomycetes may also contribute to off-flavor). Plants killed by herbicide and algicide treatments may deplete oxygen as they decompose. Some compounds (e.g., copper algicides) have a high toxicity to fish and other aquatic animals, and severe fish mortality has occasionally resulted from their use. Some herbicides and algicides have a long residual life and may be a pollution hazard in pond effluent.

Toxicity

The toxicity of different pesticides to fish and other aquatic organisms varies greatly. Some compounds have 96-h LC50 concentrations as low as 0.1 to 0.2 µg/L, while the LC50 concentrations of other compounds may be several orders of magnitude higher. (Even lower concentrations may have harmful sublethal effects.) It is beyond the scope of this chapter to provide toxicity data. However, in order to obtain the necessary permit to use pesticides in the U.S. and in many other countries, a large body of toxicity data must be generated. The U.S. Fish and Wildlife Service evaluated the toxicity of more than 400 chemicals to fish and aquatic invertebrates and published the data in a report (Johnson and Finley, 1980). *The Pesticide Manual* (1991) contains a wealth of information on pesticides. Data on pesticide toxicity provided to the U.S. Environmental Protection Agency (U.S. EPA) by companies seeking registration of their chemicals for use in the U.S. is available to anyone through the Freedom of Information Office of the U.S. EPA.

Sufficient evidence that an observed mortality or reduction in growth resulted from pesticide contamination can be elusive. Data must show that there is a source of contamination with the suspected compound, that the concentrations of the suspected compound are high enough to cause the problem, and that all other possible causes of the problem have been discounted. To illustrate the difficulty that may be encountered in proving that a pesticide caused a problem in aquaculture, consider the severe and widespread mortality of shrimp on farms in the Gulf of Guayaquil, Ecuador, which began about mid-1992 and continued into 1994. The problem started in an area known as Taura, and shrimp from ponds experiencing the problem exhibited a consistent group of symptoms which became known as the Taura Syndrome. The Taura Syndrome began at roughly the same time that some fungicides, previously unused in Ecuador, were introduced to combat a new fungal disease of bananas. There is much banana production in the Taura area, and some shrimp farms are close to the banana plantations. During 18 to 20 months, the Taura Syndrome had spread out from the Taura area to affect shrimp farms throughout the Gulf of Guayaquil. Intensive studies of shrimp larvae used for stocking, shrimp from ponds, water quality in the estuary, water quality and bottom soils in ponds, management practices, feed quality, etc., failed to provide conclusive evidence of the cause of the Taura Syndrome. Minute concentrations of suspected fungicides were found in water, soil, and shrimp, but the concentrations were much lower than concentrations necessary to cause toxicity in laboratory toxicity tests. Some laboratories claimed to be able to reproduce the symptoms of Taura Syndrome with comparatively high concentrations of the suspected chemicals. Nevertheless, after two years of concentrated effort, the cause of Taura Syndrome has been identified as a virus; some Ecuadorian shrimp farmers still claim that one or more fungicides are a major factor in the Taura Syndrome.

Tissue Residues

Many pesticides that enter aquaculture ponds from external sources or are applied for management purposes can be absorbed, concentrated, and stored in fish and shrimp. The U.S. and most other developed nations have programs for sampling food products and analyzing them for pesticides. If concentrations are considered too high for human safety, the products may be seized to prevent their consumption. If pesticide residues are found in a food product for import into the U.S. or other countries, it may cast doubt on the safety of the particular product from anywhere within the country of origin and cause serious problems in importation of the product. Therefore, it is important to guard against the contamination of aquaculture ponds with pesticides and to limit the use of pesticides for management purposes to only essential applications.

Pesticides available in some countries may no longer be registered for use in the U.S. because of some environmental risk associated with their use. Pesticides belonging to the group known as organochlorines should especially be avoided because of their tendency to bioaccumulate in organisms. These have characteristics of high lipid solubility and low water solubility and can accumulate in tissues over 100,000 times the water concentration (Jarvinen et al., 1977). The major route of pesticide uptake is directly from the water by passive absorption across the gill and body surface, but fish can also absorb toxicants by consuming contaminated food (Rand and Petrocelli, 1985). Only some highly lipophilic substances, such as DDT and methyl mercury, readily cross the gut wall and biomagnify to this extent. Obtaining uncontaminated feeds is a necessary step to prevent dangerous or illegal residues from occurring.

DRUGS AND OTHER CHEMICALS USED IN AQUACULTURE

The process of producing food by aquaculture receives equal or greater scrutiny by regulatory agencies than other agricultural processes. Public safety considerations are a legal requirement, an ethical standard, and a self-interest objective. Any food producer values consumer trust; therefore, the use of the various substances under government regulation to maintain healthy, productive aquatic animals requires considerable care, knowledge, and sense of responsibility.

Chemicals used in aquaculture include drugs, vaccines, and pesticides. The registration, distribution, and use of new animal drugs is regulated by the U.S. Food and Drug Administration (FDA) under the Federal Food, Drug, and Cosmetic Act. This act defines a drug as an "article" used for "diagnosis, cure, mitigation, treatment, or prevention of disease" or an "article" other than food intended to affect some function or structure in man or animals, or a substance listed in "official drug compendia." There are no drugs for aquacultural use that FDA has determined to be generally recognized as safe and effective by qualified investigators. Drugs without this status are referred to as new animal drugs (Federal Joint Subcommittee on Aquaculture, 1994). Some (five) new animal drugs have been approved by FDA for use in aquaculture of animal food organisms.

Unapproved drugs are to be classified as high or low regulatory priority. Use of drugs with high regulatory priority is carefully regulated, and unapproved uses may subject the user to strict enforcement or to increased risk of liability.

A list follows of FDA-regulated drugs (Table 2) taken from the current Guide to Drug, Vaccine, and Piscicide Use in Aquaculture (Federal Joint Subcommittee on Aquaculture, 1994). Readers are encouraged to read that document (or the most current source) in full for a more exhaustive treatment of this issue, and a more complete listing of drugs. Consultation with knowledgeable veterinarians is also recommended before using any drug or other substance with aquatic animals to be used as human food or as food or food supplements (e.g., fish meal) for other animals, such as chickens, which would in turn be used for human food.

Table 2   New Animal Drugs Approved for Use in Aquaculture by FDA

New animal drugs approved by FDA

1. Tricaine methanesulfonate (MS 222)
    Use: anesthetic for fish.
2. Formalin (under several trade names)
    Use: control of external parasites on fish, control of fungi on eggs.
3. Romet 30 (Sulfadimethoxine and ormetroprim)
    Use: control of enteric septicemia in catfish, control of furunculosis in salmonids.
4. Sulfamerazine
    Use: control of furunculosis in trout.
5. Terramycin (Oxytetracycline)
    Use: control of a variety of diseases.

Note: Formalin (formaldehyde, alcohol mixture) is a human carcinogen, and inhalation of vapors is dangerous to users. The U.S. Occupational Safety and Health Administration (OSHA) regulates storage and use of formalin and requires eyewashes and showers where it is being handled, and ventilation where it is being stored to protect employees.

Consult label instructions or other sources before use, as FDA approval is not blanket but may apply only to specific uses, species, or life stages. Labels are more than suggestions for use, they define the only legal applications (including handling, dosages, withdrawal times, and other aspects) of regulated drugs and chemicals. Using regulated materials in ways or on species or life stages not in conformance to labeling instructions could represent an illegal act or an ineffective or dangerous practice.

Drugs of high regulatory priority that have not been approved for use, such as methyltestosterone (MT), may be used under a closely regulated exemption. These drugs are classified as investigational new animal drugs (INAD) and require a permit from FDA's Center for Veterinary Medicine (CVM) for purchase, transport, or use. According to the Guide to Drug, Vaccine, and Pesticide Use in Aquaculture (Federal Joint Subcommittee on Aquaculture, 1994), INAD exemptions are of two types, standard and compassionate, and these can be either routine or emergency. Compassionate INAD exemptions are issued in response to concern for the health of aquatic animals. INAD exemptions in general are mechanisms to allow the development of data to document approval of a new animal drug classification. The collection of data in support of an INAD must be done using methods approved by the FDA. Therefore, consultations with the FDA and/or the National INAD Coordinator is essential before studies to support a INAD exemption are undertaken.

The CRSP has contributed to the accumulating data on the use of MT in aquaculture, examining such issues as immersion applications (Gale et al., 1995) and tissue retention time (Curtis et al., 1991).

Some drugs do not have FDA approval but are of low regulatory priority. These drugs have not been shown to be effective and safe by well-documented studies. However, FDA has no convincing evidence of safety problems with these drugs. If these are used with good management practices and no local environmental problems result, regulatory action by FDA is unlikely. The Federal Joint Subcommittee on Aquaculture (1994) outlines five conditions that should be met to avoid regulatory action and to ensure safety of application:

1. Apply only for the prescribed indications on the specified life stages and species.
2. Use in accordance with good management practices.
3. Use products of appropriate grade for use with food animals.
4. Use at the prescribed dosages.
5. Adverse environmental responses are unlikely.

A partial listing of these drugs is shown in Table 3.

LANDSCAPE ALTERATION

Wetland Destruction

In much of the world landscape, alteration has accelerated to the point of alarming conservationists (Cunningham and Saigo, 1990). For example the removal of massive tracts of tropical rain

Table 3 Drugs without FDA Approval but with Low Regulatory Priority

Drug	Permitted use
1. Acetic acid	Parasiticide used as a dip at specified dosage.
2. Calcium chloride	Increase water hardness for proper egg hardening or improving osmotic balance in fish. Concentrations used are prescribed.
3. Calcium oxide	External protozoacide for fingerling to adult fish at a prescribed concentration.
4. Carbon dioxide gas	Anesthetic for fish.
5. Fuller's earth	Reduce adhesiveness of fish eggs.
6. Whole garlic	Control of helminth and copepod parasitation in marine salmonids.
7. Hydrogen peroxide	Control of fungi in all fish life stages at specified dosage.
8. Ice	Reduce fish metabolic rate.
9. Magnesium sulfate	Treat external monogenetic trematode parasitism and external (Epsom salts) crustacean parasitism in fish at all life stages.
10. Onion	Treat external crustacean infestation and prevent external parasitism by copepods in fish at all life stages.
11. Paprain	Removal of the gelatinous matrix coating of fish eggs to improve hatchability and resistance to disease.
12. Potassium chloride	Relieve stress and reduce osmoregulatory shock.
13. Providone iodine compounds	Disinfection for fish eggs at specified dose.
14. Sodium bicarbonate (baking soda)	Anesthetic use by increasing carbon dioxide concentration.
15. Sodium chloride (salt)	Parasiticide and osmoregulatory aid.
16. Sodium sulfite	Increase hatchability of fish eggs as a 15% solution exposing eggs for 5–8 min.
17. Urea and tannic acid	Denature the adhesive substances on fish eggs at recommended application rates.

Note: Use of recommended concentrations and time of exposure may increase effectiveness and safety and ensure legal compliance.

Adapted from Federal Joint Subcommittee on Aquaculture, Guide to Drug, Vaccine, and Pesticide Use in Aquaculture, Texas Agricultural Extension Service Pub. No. B-5085, 1994.

forests has many scientists concerned that this amazing diversity of flora and fauna will be lost, which would ultimately influence the world's climate. In the U.S., land reclamation and agricultural development have damaged wetlands, such as in the Everglades National Park and other habitats (Morehead, 1982). In developed cropland areas such as west-central Illinois, millions of tons of soil erode from the land each year (Johnson, 1978). Although ponds have contributed to problems associated with landscape alteration, ponds can provide effective solutions to some of these problems. Destruction of wetlands has been particularly damaging to natural resources. Wetlands and other transitional areas between upper watersheds and natural streams have areas with distinct species compositions; these play a key role in the hydraulics and nutrient dynamics of the system (Holland, 1991). These areas are also likely sites for ponds. Extreme low and high flow patterns with alternating drought/flood cycles can result as watershed storage capacity is lost from converted wetlands. Ponds can enhance water storage and flow moderation of such transitional zones. Landscape considerations should be part of the process of selecting sites for pond development. Locating ponds in accordance with regional planning and existing land use regulations may be the only coherent approach to protect or enhance sustainability in terms of landform function and diversity.

For instance, mangrove swamps are generally not appropriate sites for aquaculture farms because of the loss of these naturally productive marine systems and because of the occurrence of acid sulfate soils, high organic content, and sulfur-reducing bacteria associated with this environment (Pillay, 1992). Soil pyrites accumulated by bacterial reduction of sulfate from seawater, when exposed to atmospheric oxidation, release ferrous iron, which oxidizes to ferric iron at low pH. It can then oxidize pyrite, elemental sulfur, and sulfide, to produce sulfuric acid. Bacteria also contribute to oxidation of elemental sulfur in these soils, which produces extreme acid conditions following drainage for pond construction. Selecting swamp areas at the margins of mangrove ecosystems may be a more sustainable practice for pond development and protect primary mangrove environments. Acid sulfate soils may be utilized in aquaculture with proper management and reclamation. Flushing, liming, and fertilization have been used to improve such soils for use in aquaculture (Lin, 1986; Brinkman and Singh, 1982).

Land Reclamation

Ponds can be a part of rational agricultural development through water and soil conservation. In west-central Illinois, a reclamation program demonstrated the role of ponds in reducing erosion and improving watersheds (Johnson, 1978). The damaging cycle of deforestation followed by short-term use and abandonment can be reversed through pond development to provide alternative and sustainable means of subsistence. Most of the world's aquaculture is in low intensity production of indigenous fishes and aquatic plants (Nash and Kensler, 1990). Such aquaculture can use aquatic communities as agroecosystems with limited land use and efficient conversion of inputs to useable products. Unproductive acid sulfate soils in both freshwater and marine environments have been reclaimed for use in aquaculture by tidal or freshwater flushing, and by liming and enriching ponds with organic and inorganic fertilizers (Lin, 1986; Brinkman and Singh, 1982). Therefore, aquaculture may play a central role in appropriate development and reclamation of damaged tropical environments.

SUSTAINABILITY IN AQUACULTURE

The complex concept of sustainability may elicit some common sharing of meaning; however, agreement between users of the term may be illusory. For instance, ecological sustainability defined as concern for maintaining ecosystem composition, structure, and function (including landform diversity), may conflict with agricultural sustainability, defined as increasing production while maintaining the production base of soil fertility. "Sustainable agriculture should involve the successful management of resources for agriculture to satisfy changing human needs while maintaining or enhancing the quality of the environment and conserving natural resources" (TAC/CGIAR, 1988). This may be a suitable conceptual framework for pond aquaculture (see also FAO, 1990). Still, it remains a complex concept with different interpretations of such concepts as conserving natural resources. Here, "sustainability" is used in its most inclusive form, which is maintaining system productivity and contributing to aspects of ecological conservation, such as landform and species genetic diversity and conservation of water and water quality. Including the human population as part of the ecosystem provides a more balanced view of sustainability (Barica, 1992). Shell (1993) promotes incorporating an ecosystem perspective in aquaculture because the complex interactions between aquatic organisms and the physical environment of the production system are embedded within the surrounding human cultural and sociological system. A systematic appreciation of the many linkages and knowledge of the flow of resources within this many-layered network will serve as a conceptual framework for sustainability in the conduct of aquaculture.

The pond aquaculturist applies sustainability as a point of view rather than as a list of necessary steps to include in the production process. Water and water quality conservation, biological conservation, landscape considerations, and the whole spectrum of production activities impinge not only on the production site but on the environment of the pond. Responsibility for all of nature and society seems an extravagant burden to place on the pond aquaculturist, who, like other business people, works within economic constraints where selection of the fittest is as real as in ecological systems. But reasonable application of sustainability concepts should in the long run contribute to successful fish farming. For instance, recycling of nutrients within a farming system is both economically efficient and sustainable. Ponds typically have effluent with excessive nutrient and solids content. Traditional crop production needs both water and the substances considered pollutants in water discharged from ponds. Nutrients needed by the pond systems can be replaced by waste products of animal and plant production, increasing the efficiency of utilization of purchased fertilizers and feeds. Integrated farm systems conserve water, soils, and nutrients, protect the receiving aquatic community, and exemplify sustainability.

Aquaculture and Human Diseases

Human diseases associated with water and aquatic organisms pose too large and diverse a topic to treat adequately is this short section. Use of human sewage as fertilizer poses additional complexity to risks of contamination by human pathogens. Food- or waterborne infection by trematodes (flatworms or flukes) is perhaps the most important human health concern associated with aquaculture; a problem of such magnitude that every aquaculturist in broad global areas should be aware of the potential risks.

Two general groups of trematode-caused diseases will be considered here, (1) opisthorchiasis (sometimes clonorchiasis) from liver and intestinal flukes (fishborne trematodes, FBTs) that are transmitted to humans via consumption of raw or improperly prepared fish or shellfish and (2) schistosomiasis, transmitted to humans via water contact (waterborne trematodes, WBTs). Of the FBTs, liver fluke species such as *Clonorchis sinensis, Opisthorchis felineus*, and *O. viverrini* (Asiatic liver fluke, endemic in Southeast Asia, China, Korea, Japan, and countries formerly in the USSR) are of more human health importance than the intestinal FBTs such as *Fasciolopsis* (giant intestinal fluke of Asia) (Larsson, 1994; Santos, 1994). Intestinal trematodes, though of less importance, still represent over 70 species of flukes infecting over 1 million people. Liver flukes infect over 50 million people worldwide, especially in eastern and southern Asia (Santos, 1994). Symptoms of chronic human infection includes gastrointestinal dysfunction, jaundice, fatigue, fever, and respiratory problems. Intermediate hosts are snails and freshwater fish; final hosts are mammals, including humans, cats, dogs, pigs, and wild animals, where consumed cercariae or metacercaria develop to adults in the liver bile ducts. Adults produce eggs that enter aquatic systems via defecated wastes, continuing the cycle where suitable intermediate hosts exist.

Costs of treatment, loss of labor in production, disability, and training and costs of control are high. Thailand's health care costs for opisthorchiasis for 3 million people was about $8,300,000 (Santos, 1994). Control and prevention measures vary with country but include promoting adequate food inspection and processing, detection and treatment of patients, snail control, and improving disposal of human wastes. The U.S. FDA approach is visual examination by candling and physical removal of parasites, including cestodes and nematodes (FDA, 1994). The World Health Organization recommends the application of Hazard Analysis and Critical Control Point (HACCP) concepts to FBT problems (FAO/WHO, 1993). This strategy is an attempt to identify critical points in the fish production to consumption process amenable to control, providing a more analytical and systematic approach.

Schistosomiasis (or Bilharzia) results from infection by WBTs of the genus *Schistosoma*. Approximately 200 million people in 76 countries are infected, with an annual mortality of about 200,000 (CTD/TDR, 1990). Mammals in nearly all world areas are subject to infection by *Schistosoma*, a genus of digenetic trematodes that includes the group known as blood flukes, due to their unusual location in the smaller blood vessels of their host. At least four species are known to infect humans via direct contact with water containing trematode cercaria released from the intermediate snail host, snails of the subfamily Bulininae (*Bulinus, Physopsis, Pyrgophysa*) or planorbid snails of the genus *Biomphalaria* and the amphibious *Oncomelania*, of the family Amnicolidae (Anon., 1982). Miracidium hatches after eggs reach water bodies from human urine or feces, depending on the trematode species. The ciliated miracidia enter aquatic snails and reproduce, producing many cercaria, which are released back into the water. Cercaria enter the human host through the skin after direct contact with infected water. Cercaria enter the blood stream, moving from the lungs to small abdominal blood vessels, where they develop to adults, mate, and produce eggs, which leave the host via urine or feces. Control is by removal of snails by chemical or biological means, minimizing water contact by humans, treating patients, reducing contamination of water by human wastes, and maintaining fish in ponds to consume some cercaria. A variety of biological controls on snails are possible, including use of fish that prey on snails, screening water sources, vegetation removal, control of water level fluctuations, and pond drying, for instance (Larsson, 1994). Aquaculture can contribute to greater infection by extending the distribution of suitable snail habitat and the opportunity for water contact by humans. However, proper management, education, and greater intensification of aquaculture can contribute to reduced risk of infection by this parasite.

Regulations

Society increasingly attempts to accomplish social goals through regulation. In the U.S., the pond aquaculturist faces a formidable array of regulations on the variety of activities that fish culture involves. Many of these regulations differ from state to state and sometimes within a state. Others, such as those administered directly by the federal government, are uniform. In some cases, a variety of state and federal agencies may regulate the same activity. For example, aquacultural development on land possibly classified as wetlands may require approval of the U.S. Army Corps of Engineers, a state land development commission, and perhaps the county government, in addition to meeting siting regulations imposed by state agencies. On marginal wetlands, a good chance exists that the various agencies will not be in agreement on whether a specific piece of land qualifies as protected wetlands or can be developed for pond production. In some countries, environmental impact statements must also be completed and approved before development may occur.

If the land is approved for pond construction, other permits and regulations must be faced. The state department of natural resources or fisheries and wildlife must approve a permit to engage in the culture of each species. The state department regulating water use may have to issue a permit approving water rights before any diversion of the state's water may occur. The state department regulating discharges to the state waters must then be solicited for a waste discharge permit. These permits specify the amount of each potential pollutant allowed and often require periodic sampling and reporting. Waste discharge permits are the basic tool regulating environmental discharges in the U.S. The U.S. EPA, under the 1972 Federal Clean Water Act and amendments, requires states to regulate point and nonpoint source emissions to air, land, and water. The states administer NPDES (National Pollution Discharge Elimination System) permits in accordance with federal guidelines issued under the Clean Water Act or contained in other EPA documents and under EPA oversight. According to Pillay (1992), the rules published in the Federal Register (EPA, 1974) for permit applications to discharge from aquaculture production systems require

1. Identification of the kind and quantity of pollutant(s) to be used in the aquaculture project
2. Available information on
   a. The conversion efficiency of the pollutant to harvestable product
   b. The potential increased yield of the species being cultured
   c. Any identifiable new product to be produced, including anticipated quantity of harvestable product
3. Identification of the species of organisms to be cultured
4. Identification of the water quality parameters required for the growth and propagation of the cultured species, including, but not limited to, dissolved oxygen, salinity, temperature, and nutrients such as nitrogen (nitrates, nitrites, and ammonia), total phosphorus and total organic carbon
5. Identification of possible health effects of the proposed aquaculture project, including
   a. Diseases or parasites associated with the crop, which could affect aquatic life outside the designated project area and which could become established in the designated project area and/or in the species under cultivation
   b. The potential effect on human health
   c. Bioconcentrations in the crop, including but not limited to radionuclides, heavy metals, and pathogenic organisms associated with the pollutant used
   d. Potential for escape of nonindigenous species from the designated project area
6. Identification of pollutants produced by the species under culture, especially those which may be channeled into waste effluent such as ammonia, hydrogen sulfide, organic residues, phosphates and nitrates
7. Identification of the disposal method to be used, should there be a necessity for intentional destruction or a massive natural death of the organisms under culture

However, EPA is requiring only tests for settleable solids, and Pillay (1992) indicates the EPA does not appear to be applying any rigid water quality standards except in the case of marine siting regulations. Nevertheless, states may impose more stringent requirements before granting discharge permits.

The presence of specific substances regulated by EPA may also restrict discharges. National Water Quality Criteria documents are issued by EPA for individual substances, such as dissolved oxygen, ammonia, temperature, specific metals, etc. (EPA, 1986a, 1986b). The state must respond by following the directives in these documents, which list short-term and long-term "safe" levels for individual pollutants that must be met in receiving waters. The limits may vary in relation to water quality of the receiving water. For instance, ammonia no-effect levels vary with stream pH. Waters that have unique species or especially high values in water quality are given more stringent regulatory attention. The state issues permits if these criteria can be achieved. If unsafe or illegal residues might occur in plant or animal products produced within the culture process or in organisms in receiving waters and consumed by humans, then Federal Food, Drug, and Cosmetic Act regulations may restrict such residues. Those restrictions extend to residues in the fish cultured that otherwise meet legal requirements but might result in secondary contamination of some other agricultural product via contribution as a feed or feed supplement. In general, drugs are regulated by the U.S. FDA under the Federal Food, Drug, and Cosmetic Act. FDAs Center for Veterinary Medicine (CVM) is specifically responsible for manufacture, distribution, and use of animal drugs (Federal Joint Subcommittee on Aquaculture, 1994). Pesticides, including disinfectants, are regulated by U.S. EPA under the Federal Fungicide, Insecticide, and Rodenticide Act (FIFRA), which requires pesticide registration and enforces labeling and application methods. Users may also have to be certified to apply specific pesticides.

The U.S. Department of Agriculture regulates veterinary biologics, including vaccines and other therapeutants and disease diagnostics, through its Animal and Plant Health Inspection Service (APHIS). Only veterinary biologics licensed by this agency can be used in aquaculture. State departments of agriculture may impose additional restrictions or may also register federally approved pesticides to permit their use in that state. Such agencies may also regulate vaccines or other chemicals or require administration by licensed veterinarians.

The Occupational Health and Safety Administration could conceivably also impose regulatory restrictions. The use of formalin (formaldehyde in alcohol solution) is restricted by OSHA, where employees might be exposed to the carcinogenic vapors. Ventilation, eyewash stations, and showers might be required to ensure occupational safety.

Thus, both federal and state regulatory requirements must be met before chemicals may be legally used in food production. This maze of regulatory layers is designed to ensure a safe food supply and prevent environmental contamination or degradation. In this regard, fish farming has some advantage over poultry, pork, and beef production, where medicated feeds are common. Aquacultural products that are not exposed to such medications may continue further market inroads on traditional meat products through greater consumer acceptance.

The regulatory role of government in the aquacultural industry, although sometimes viewed as a constraint to the industry, is appropriate and necessary. Unnecessary constraints develop when government, in the conduct of its reasonable and necessary roles, errors in specific applications of its responsibility, particularly in its regulatory role. In the National Aquaculture Development, Research and Promotion Act, the U.S. government describes its role in promotion, establishing policy, research, data collection, and communication (Subcommittee on Agriculture, 1993).

BIOLOGICAL CONSERVATION

Exotic Species or Strains

Improper aquacultural practices can work against efforts to protect species diversity present in natural waters by introducing new species or genetically altered strains of indigenous species. Native populations have characteristics and capacities for each life history stage that are linked to historic habitats and stored within the population gene pool. For instance, salmon that ascend long, powerful streams to spawn are larger than salmon populations that ascend shorter, lower gradient streams, because the former require greater energy stores to reach distant spawning grounds (Thorpe and Mitchell, 1981).

Selective breeding to improve production or marketability of cultured species results in genetic alteration of manipulated populations. In addition, individual characteristics that allow for survival in the culture environment may be different than those needed for survival in the natural environment, which suggests that cultured populations eventually differ from their wild counterparts after generations of exposure to the culture environment, even without systematic efforts to selectively breed desired characteristics. Now transgenesis, or genetic engineering, has provided the capability to instantaneously introduce new phenotypes (Devlin and Donaldson, 1992). Genetically altered stocks released into the environment could alter existing genetic resources of native species, even when cultured species are indigenous at that site. For example, Atlantic salmon (*Salmo salar*) farmed in sea cages in Northern Ireland demonstrate genetic differences when compared to wild salmon. Crozier (1993) evaluated potential genetic impacts of escaped cage-reared salmon on wild stocks in the Glenarm River. He found that at four of the seven loci examined in wild salmon, allele frequencies had shifted in the direction of those in the farmed strain. This reduced the genetic heterogeneity between wild and farmed strains and suggested that escaped salmon were interbreeding with wild stock. The emerging concept of biodiversity encompasses species diversity, genetic diversity, and habitat diversity. Alteration of any component contributes to altered and perhaps less fit communities.

Protection of local gene pools can be accomplished by physical containment and by biological controls such as sterilization. Physical containment is notoriously unreliable, as weather or human error or equipment failure overcome planning and diligence. Avoiding culture of particular species in sensitive areas may be necessary in some cases.

Conserving diversity in subject aquatic systems is a precursor for sustainable development in aquaculture. Accidental or purposeful introductions of exotic species around the world have rarely been viewed as positive events. Introduction of the common carp in North America is an example. Lake Victoria in east Africa is often used as a warning of the potential for damaging impacts from introduced species. There, following introductions in the 1950s of the Nile perch (*Lates niloticus*) and tilapia, including *Oreochromis niloticus*, *Tilapia melanopleura*, and *Tilapia zilli*, hundreds of Victoria's endemic fishes are now listed as endangered (Kaufman, 1992; Ogutu-Ohwayo and Hecky, 1990). Nile perch and introduced tilapia dominate the fisheries, and trophic diversity declined along with species diversity. Subsequent eutrophication including fish kills may be related to changes in the lake fish community, although other factors are also suspect. Damage to endemic species by introductions of exotics is well established, and numerous other examples are well documented. Damaging results from species introductions are not universal. The planktivorous fish, *Limnothrissa miodon*, was introduced into Lake Kivu in Rwanda and the Kariba reservoir and established a valuable fishery with no apparent effect on native fishes (Ogutu-Ohwayo and Hecky, 1990). Other examples exist here also of positive results of accidental or purposeful introductions. The outcome is difficult to predict, however, suggesting environments are at risk when exotic species are introduced. Practices such as utilizing native species in culture provide a sustainable approach to protecting existing aquatic communities. Aquaculturists, through proper management and an appreciation of the value of natural aquatic systems, can be important players in the protection of native species.

REFERENCES

Allison, R., The effects of formalin and other parasiticides upon oxygen concentrations in ponds, *Proc. Annu. Conf. S.E. Assoc. Game Fish Comm.*, 16, 446–449, 1962.
Anonymous, *Stedman's Medical Dictionary,* 24th ed., Williams and Wilkins, Baltimore, MD, 1982, 1260.
Arceivale, S. J., Hydraulic modelling for waste stabilization ponds, *J. Environ. Eng.*, 109(5), 265–268, 1983.
Barica, J., Sustainable management of urban lakes: a new environmental challenge, *Central Canadian Symp. on Water Pollution Research,* Burlington, Ontario, 1992, 211.
Barker, J. E., Chesness, J. L., and Smith, R. E., Pollution Aspects of Catfish Production — Review and Projections, U.S. Environmental Protection Agency, Technical Series, EPA-660/2-74-064, 1974, 212 pp.
Boyd, C. E., Effluent from catfish ponds during fish harvest, *J. Environ. Qual.*, 7, 59–62, 1978.
Boyd, C. E., Hydrology of small experimental fish ponds at Auburn, Alabama, *Trans. Am. Fish. Soc.*, 111, 638–644, 1982.
Boyd, C. E., Hydrology and pond construction, in *Channel Catfish Culture,* Tucker, C. S., Ed., Elsevier/North Holland, Amsterdam, 1985a, 107–133.
Boyd, C. E., Pond evaporation, *Trans. Am. Fish. Soc.*, 114, 299–303, 1985b.
Boyd, C. E., Water Quality in Ponds for Aquaculture, Alabama Agricultural Experiment Station, Auburn University, Auburn, AL, 1990, 482 pp.
Boyd, C. E. and Musig, Y., Shrimp pond effluent, in *Proc. Special Session on Shrimp Farming,* Wyban, J.A., Ed., World Aquaculture Society, 1992, 195–197.

Boyd, C. E. and Shelton, J. L., Jr., Observations on the hydrology and morphometry of ponds on the Auburn University Fisheries Research Unit, Alabama Agricultural Experiment Station, Auburn University, Auburn, AL, Bulletin 588, 1984, 64 pp.

Brinkman, R. and Singh, V. P., Rapid reclamation of brackish water ponds in acid sulfate soils, in *Proc. Int. Symp. Acid Sulfate Soils,* Publ. No. 31, The International Institute for Land Reclamation and Improvement, Wageningen, The Netherlands, 1982, 318–330.

CTD/TDR, *Tropical Diseases 1990,* Center for Tropical Disease CTD/TDR, 1990.

Colt, J., Pond culture practices, in *Principles and Practices of Pond Aquaculture: a State of the Art Review,* Lannan, J. I., Smitherman, R. O., and Tchobanoglous, G., Eds., Pond Dynamics/Aquaculture CRSP, Oregon State University, Corvallis, 1983, 187–197.

Crozier, W. W., Evidence of genetic interaction between escaped farmed salmon and wild Atlantic salmon (*Salmo salar* L.) in a Northern Irish river, *Aquaculture,* 113, 19–29, 1993.

Cunningham, W. P. and Saigo, B. W., *Environmental Science,* William C. Brown, Dubuque, IA, 1990.

Curtis, L. R., Diren, F. T., Hurley, M. D., Seim, W. K., and Tubb, R. A., Disposition and elimination of 17a-methyltestosterone in Nile tilapia (*Oreochromis niloticus*), *Aquaculture,* 99, 192–201, 1991.

Dallinger, R., Prosi, F., Segner, J., and Back, H., Contaminated food and uptake of heavy metals by fish: a review and a proposal for further research, *Oecologia,* 73(1), 91–98, 1987.

Devlin, R. H. and Donaldson, E. M., Containment of genetically altered fish with emphasis on salmonids, in Hew, C. L. and Fletcher, G. L., Eds., *Transgenic Fish,* Singapore World Scientific, Singapore, 1992, 229–266 (abstract only).

Edwards, R. W. and Densem, J. W., Fish from sewage, *Appl. Biol.,* 5, 221–270, 1980.

Ellis, J. E., Tackett, D. L., and Carter, R. R., Discharge of solids from fish ponds, *Prog. Fish-Cult.,* 40, 165–166, 1978.

EPA (Environmental Protection Agency), Water Quality Criteria, EPA-R3-73-033, Washington, D.C., 1972, 594 pp.

EPA, Aquaculture projects — requirements for approval of discharges, *Fed. Reg.,* 39(115), Part II, 20770–20775, 1974.

EPA, Ambient Water Quality Criteria for Dissolved Oxygen, EPA 44/5-86-003, Office of Water Regulations and Standards, Criteria and Standards Division, Environmental Protection Agency, Washington, D.C., 1986a.

EPA, Water Quality Standards Criteria Summaries: A Compilation of State/Federal Criteria, U.S. Environmental Protection Agency, Washington, D.C., 1986b.

FAO (Food and Agriculture Organization of the United Nations), FAO Activities Related to Environment and Sustainable Development, FAO Council Document CL98/6, 1990.

FAO/WHO (Food and Agriculture Organization of the United Nations/World Health Organization), Codex Guidelines for the Application of the Hazard Analysis Critical Control Point (HACCP) System, 20th Session of the Joint FAO/WHO Codex Alimentarius Commission, WHO/FNU/FOS/93.3, U.N. Food and Agriculture Organization, Rome, 1993.

FAO (Food and Agriculture Organization of the United Nations), Aquaculture Production, 1986–1994, FAO Fisheries Circular No. 815, Rev. 6, XI, 1994.

FDA (Food and Drug Administration), Fish and Fishery Products Hazards and Controls Guide, Food and Drug Administration, Washington, D.C., 1994, 228 pp.

Federal Joint Subcommittee on Aquaculture, Working Group on Quality Assurance in Aquaculture Production and the Extension Service, U.S. Department of Agriculture, Guide to Drug, Vaccine, and Pesticide Use in Aquaculture, Texas Agricultural Extension Service Pub. No. B-5085, Texas A&M University, 1994.

Gale, W. L., Fitzpatrick, M. S., and Schreck, C. B., in Proc. Fifth International Symposium on Reproductive Physiology of Fish, Goetz, F. W., Ed., in press.

Grant, N., Mercury and man, *Environment,* 13(4), 3–15 1971.

Hawkes, H. A., *The Ecology of Waste Water Treatment,* Pergamon Press, Oxford, 1963, 203 pp.

Hinshaw, R. N., Pollution as a Result of Fish Cultural Activities, U.S. Environmental Protection Agency, EPA-R3-73-009, 1973, 209 pp.

Hodson, P. V., Borgmann, U., and Shear, H., Toxicity of copper to aquatic biota, in Nriagu, J. O., Ed., *Copper in the Environment,* Part II, *Health Effects,* Wiley-Interscience, 1979, 307–372.

Holland, M. M., Management of land/inland water ecotones: needs for regional approaches to achieve sustainable systems, Conf. Phosphorus and Nitrogen Dynamics and Retention in Ecotones of Lowland Temperate Lakes and Rivers, Masurian Lakeland (Poland), 1991.

Jarvinen, A. W., Joffman, M. J., and Thorslund, T. W., Long-term toxic effects of DDT food and water exposure on fathead minnows (*Pimephales promelas*), *J. Fish. Res. Bd. Can.,* 34, 2089–2103, 1977.

Johnson, V., Watershed ponds, *Farm Pond Harvest,* 12(4), 6, 1978.

Johnson, W. and Finley, M., Handbook of Acute Toxicity of Chemicals to Fish and Aquatic Invertebrates, U.S. Department of Interior, Fish Wildlife Service, Resource Publ. No. 137, Washington, D.C., 1980, 98 pp.

Kane, H., Growing fish in fields, *World Watch,* Sept./Oct, 20–27, 1993.

Kaufman, L., Catastrophic change in species-rich ecosystems: the lessons of Lake Victoria, *Bioscience,* 42(11), 846–858, 1992.

Larsson, B., Three Overviews on Environment and Aquaculture in the Tropics and Subtropics, ALCOM Field Document No. 27, Food and Agriculture Organizatiion of the United Nations, 1994, 46 pp.

Laws, E. A., *Aquatic Pollution,* John Wiley and Sons, New York, 1981, 301–335, 482.

Lin, C. K., Acidification and reclamation of acid sulfate soil fish ponds in Thailand, in *The First Asian Fisheries Forum,* Maclean, J. L., Dizon, L. B., and Hosillos, L. V., Eds., Asian Fisheries Society, Manila, 1986, 71–74.

Massaro, E. J. and Giblin, F. J., Uptake, distribution and concentration of methylmercury by rainbow trout (*Salmo gairdneri*) tissues, in Hemphill, D. D., Ed., *Trace Substances in Environmental Health,* University of Missouri Press, Columbia, MO, 1972, 107–112.

Masuda, K. and Boyd, C. E., Comparative evaluation of the solubility and algal toxicity of copper sulfate and chelated copper, *Aquaculture,* 117, 287–302, 1993.

Morehead, J. M., National parks, conservation and development, in *The Role of Protected Areas in Sustaining Society,* McNeely, J. A. and Miller, K. R., Eds., Proc. of the World Congress on National Parks, Bali, Indonesia, 1984, 496–502.

Mudrak, V. A., Guidelines for economical commercial fish hatchery wastewater treatment systems, in *Proc. Bio-Engineering Symp. for Fish Culture,* Allen, L J. and Kinney, E. C., Eds., Fish Culture Section Publ. 1, American Fisheries Soc. and the Northeast Soc. of Conservation Engineers, 1979, 174–189.

Nash, C. E. and Kensler, C. B., A global overview of aquaculture production in 1987, *World Aquaculture,* 21(2), 104–112, 1990.

National Academy of Science, Advisory Committee on Technology Innovation of the Board on Science and Technology for International Development, *Making Aquatic Weeds Useful: Some Perspectives for Developing Countries,* National Academy of Science, Library of Congress Cat. No. 76-53285, 1976, 175 pp.

Ogutu-Ohwayo, R. and Hecky, R. E., Fish introductions in African and some of their implications, in *Int. Symp. on the Ecological and Genetic Implications of Fish Introductions (FIN),* Billington, N. and Herbert, P. D. N., Eds., Bol 48, No. Suppl. 1, 1990, 8–12.

Pillay, T. V. R., *Aquaculture and the Environment,* Fishing News Books, Oxford, 1992, 189 pp.

Plucknett, D. L., International agricultural research for the next century, *Bioscience,* 43(7), 432, 1993.

Rand, G. M. and Petrocelli, S. R., *Fundamentals of Aquatic Toxicology,* Hemisphere Publishing, 1985, 666 pp.

Santos, C. A. L., Prevention and control of food borne trematode infections in cultured fish, *FAO Aquaculture Newslett.,* 8, 11–15, Dec. 1994.

Schneider, R. F., Development Document for Proposed Effluent Limitations Guidelines and Standards of Performance for the Fish Hatcheries and Farms Point Source Category, Draft Report, EPA National Field Investigations Center, Denver CO, 1974, 237 pp.

Schwartz, M. F. and Boyd, C. E., Channel catfish pond effluent, *Prog. Fish-Cult.,* 1994a, in press.

Schwartz, M. F. and Boyd, C. E., Effluent quality during harvest of channel catfish from watershed ponds, *Prog. Fish-Cult.,* 56, 25–32, 1994b.

Shell, E. W., The development of aquaculture: an ecosystem perspective, Alabama Agricultural Experiment Station, Auburn University, Auburn, AL, 1993, 265 pp.

Shelton, J. L., Jr. and Boyd, C. E., Water budgets for aquaculture ponds supplied by runoff with reference to effluent volume, *J. Appl. Aquaculture,* 2, 1–27, 1993.

Subcommittee on Agricultural Research, Conservation, Forestry, and General Legislation, United States Senate, 103 Congress, Hearing on Senate Bill S.1288, 1993.

Tackett, D. I., Yield of channel catfish and composition of effluent from shallow-water raceways, *Prog. Fish-Cult.,* 36, 46–48, 1974.

Technical Advisory Committee/CGIAR (TAC/CGIAR), Priorities for International Support to Agricultural Research in Developing Countries, TAC Secretariat, FAO, Rome, 1988.

Thorpe, J. E. and Mitchell, K. A., Stocks of Atlantic salmon (*Salmo salar*) in Britain and Ireland: discreteness, and current management, *Can. J. Fish. Aquat. Sci.,* 38, 1576–1590, 1981.

Tucker, C. S., Short-term effects of propanil on oxygen production by plankton communities from catfish ponds, *Bull. Environ. Contam. Toxicol.,* 39, 245–250, 1987.

Tucker, C. S. and Boyd, C. E., Relationships between potassium permanganate treatment and water quality, *Trans. Am. Fish. Soc.,* 106, 481–488, 1977.

Tucker, C. S. and Boyd, C. E., Consequences of periodic applications of copper sulfate and simazine for phytoplankton control in catfish ponds, *Trans. Am. Fish. Soc.,* 107, 316–320, 1978.

Tucker, C. S. and Boyd, C. E., Effects of simazine treatment on channel catfish and bluegill production in ponds, *Aquaculture,* 15, 345–352, 1979.

Tucker, C. S. and Lloyd, S. W., Water Quality in Streams and Channel Catfish (*Ictalurus punctatus*) Ponds in West-Central Mississippi, Mississippi Agriculture and Forestry Experimental Station, Mississippi State University, Research Report, 10, 1–4, 1985.

Van der Ploeg, M., Studies of Cause and Control of Off-Flavor in Water and Pond-Raised Fish, Ph.D. dissertation, Auburn University, Auburn, AL, 1989, 100 pp.

Van der Ploeg, M. and Boyd, C. E., Effects of copper sulfate treatments on off-flavor and levels of dissolved copper in channel catfish ponds, in Proc. Louisiana Aquaculture Conference, 1991, 17–18.

Vickers, D. H. and Boyd, C. E., Effects of organic insecticides upon carbon 14 uptake by freshwater phytoplankton, in *Radionuclides in Ecosystems, Vol. 1,* Nelson, D. J., Ed., Proc. 3rd Nat. Symposium on Radioecology, Oak Ridge Nat. Lab., Oak Ridge, TN, 1971, 492–496.

Woodward, D. F., DeLonay, A. J., Little, E. E., and Smith, C. E., Effects on rainbow trout fry of a metal-contaminated diet of benthic invertebrates from the ClarkFork River, Montana, *Trans. Am. Fish. Soc.,* 123, 51–62, 1994.

Worthing, C. and Hance, R., Eds., The Pesticide Manual, A World Compendium, 9th ed., British Crop Protection Council, Surrey, Great Britain, 1991, 1141 pp.

Yoo, K. H. and Boyd, C. E., *Hydrology and Water Supply for Aquaculture,* Chapman and Hall, New York, 1993, 483 pp.

8

ATTRIBUTES OF TROPICAL POND-CULTURED FISH

David R. Teichert-Coddington, Thomas J. Popma, and Leonard L. Lovshin

INTRODUCTION

Food fish culture may be described as environmental management for growth of an edible product. The environment is managed contingent on a complex interaction of social, biological, and physical factors (Shell, 1993). The pond environment is demanding, and the marketing requirements for fish are even more restrictive, so only a few species of fish have been found suitable for culture. There are 20,000 to 40,000 species of fish (Lagler et al., 1977), but Hepher and Pruginin (1981) estimated that the number of documented species being cultured was only about 70. Fewer species are commercially cultivated.

Fish must meet certain requirements for culture. The requirements may differ depending on the culture system and aquacultural experience of the fish culturist. Bardach et al. (1972) listed four areas of consideration when selecting a fish for culture: reproductive habits, requirements of the eggs and larvae, feeding habits, and adaptability to crowding. Huet (1972) listed seven requisite conditions for cultured fish. Fish must adapt to the climate where cultured, have fast growth rate (reach a large size), reproduce under culture conditions, accept feed, be satisfactory to the consumer, support high population densities, and resist disease.

This chapter describes seven attributes that a fish should have for semi-intensive culture in tropical or subtropical ponds:

- Handling tolerance
- Crowding tolerance
- Low susceptibility to disease
- Toleration of poor water quality
- Efficient conversion of natural foods and feed
- Controllable reproduction
- Marketability

This list does not include some of the requirements listed by Bardach et al. (1972) and Huet (1972), because CRSP experience and culture systems differ from theirs. For example, the requirements of eggs and larvae are important considerations for marine species that produce very small, delicate eggs and larvae that are difficult to propagate. Marine mollusks and invertebrates have various larval developmental stages as a further complication to seed production. However, larvae of most tropical freshwater fish species are comparatively large and uncomplicated to raise, so larval culture is not a chief concern. Huet (1972) emphasized climate adaptability. Indeed, adaptation to a particular temperature regime is an assumption underlying all other attributes. A tropical fish should be raised at tropical temperatures for the best expression of growth and other attributes. Failure to

0-56670-274-7/97/$0.00+$.50
© 1997 by CRC Press LLC

possess all attributes does not imply that the fish cannot be cultured, but success may be marginal, dependent on local conditions, and not easily transferable to other sites. This chapter is organized to (1) discuss attributes of culture fish and suggest best management practices to enhance the attributes and (2) describe the attributes and biology of tilapia, a model culture fish.

ATTRIBUTES OF POND FISH

Based on the experiences of the authors and a survey of the literature, seven attributes were identified for successful fish culture. These attributes are not inflexible but serve to categorize key characteristics.

Handling Tolerance

Handling stress can lead to major fish mortality by causing physiological dysfunction or reducing resistance to disease. Fish are usually handled several times during growth from fry to a marketable size. Fry must be collected and redistributed to nursery ponds and then harvested and redistributed for one or more phases of grow-out. During grow-out, fish are sampled for growth. After harvest fish sometimes must withstand a day or two in holding facilities before being processed for market. All of these procedures require that fish be seined, concentrated, and usually removed from the water for short periods of time. During seining and transfer, fish may sustain epidermal injury and probably a certain degree of injury to internal organs from weight of other fish in the nets. Handling may induce periods of hypoxia when fish are concentrated in small volumes of water, transferred out of water, and transported in small tanks. Workers may not be well trained or motivated to treat fish well, and proper facilities for rapid transport are unavailable in many countries. The ideal culture fish must therefore tolerate varying degrees of physical handling and oxygen deprivation.

Tolerance to handling is appreciated by the field worker. Those accustomed to handling tilapia are often dismayed by lower handling tolerance exhibited by fish such as tambaquí (*Colossoma macropomum*) or Chinese carps. Even within the same species, fry and fingerlings are often able to tolerate more handling than adult fish. Small fish, because of higher gill surface to body weight ratios, are less susceptible to low dissolved oxygen, and do not appear to sustain as much injury during transport. Differences among species of the same genus and even among strains of the same species are sometimes apparent. Moehl (1989) concluded that *Oreochromios macrochir* were less tolerant of environmental stress than *O. niloticus* because of higher mortality during pond culture in Rwanda. Experience with several strains of red tilapia in Honduras indicates that all are less tolerant of handling than pure *O. niloticus*, even when reds are crossed with the pure line. Lower handling tolerance results in lower survival during pond culture (Green et al., 1994) and higher mortality following handling.

Handling stress can often be minimized by modifying handling technique. Stress from hypoxia can be avoided with aeration and by minimizing the time fish are concentrated. Shading fish during transport and working during the cool of the day take advantage of lower metabolic rates of fish at cooler temperatures. An exception to this rule is encountered in high altitude culture of tilapia, such as in Rwanda, where fish handling is timed to take advantage of higher water temperatures (Veverica, personal communication). Early morning handling of tilapia was avoided because water temperatures were typically 19 to 20°C and transport water might be 17°C.

Stress can be ameliorated by adding salt to transport water. Freshwater fish blood, with a salinity of about 8 ppt, is hypertonic in freshwater. Stress induces a loss of ions from fish and the disruption of osmoregulation (Heath, 1987). Additions of sodium chloride to transport water helps to minimize the disruption. Piper et al. (1986) and Tucker and Robinson (1990) recommend 1 to 3 ppt, but Jensen and Durborow (1984) prescribed 0.2 to 0.5 ppt for indefinite treatment. Tilapia could easily tolerate treatments up to 3 ppt, but lower concentrations are probably more appropriate.

Food grade sodium bicarbonate, calcium sulfate, and potassium chloride may also be added at 0.2 ppt as complements to sodium chloride.

Tranquilizers may be used to calm fish, thereby reducing oxygen consumption and possible injury to the fish. A common and safe tranquilizer is MS-222 at 50 to 100 mg/L (Stickney, 1979; Woynarovich and Horváth, 1980). Tranquilizers are most commonly used with the larger carps and colossomas, but not recommended for routine use with tilapias because they are expensive and may not be available.

Crowding Tolerance

Costs related to pond construction and farm management are high. Fish must be stocked at high enough rates to generate acceptable returns on investment. Stocking rates are higher where land and water are expensive because greater quantities of fish must be produced per area of pond to overcome higher costs.

A candidate pond fish must therefore prosper under population densities orders of magnitude higher than those found in nature. Most pond fish are herbivorous or omnivorous, and gregarious. Highly aggressive, territorial, or predacious fish are less appropriate for monoculture but may be included at low densities to control reproduction of fish like tilapia. Clariid and pangasiid catfish of Asia and Africa are exceptions. Although primarily carnivorous, they grow well at high stocking densities (Csavas, 1994).

Low Susceptibility to Disease

Disease usually occurs following stressful events such as stocking, sampling, hypoxia, or subacute concentrations of chemicals like ammonia. Disease may be manifested as an acute bacterial or viral infection, or more likely as a slower acting parasitic or fungal infection. Tilapia can usually be handled without mortality from disease if temperatures are between 25 and 32°C. *Saprolegnia*, a fungal infection, is more common when tilapia are handled at suboptimum water temperatures. The ideal culture fish must have low susceptibility to disease, especially if cultured in regions where knowledge of pathogens is limited, or where therapeutics are unavailable or prohibitively expensive.

Disease outbreaks are more likely at high tilapia stocking rates (Roberts and Sommerville, 1982; Hubbert, 1989), probably because of increased frequency of contact with infected animals, greater concentrations of ambient organic matter where pathogens may incubate, and increased probability of stress from poor water quality. For example, diseases at semi-intensive levels ($\leq 3/m^2$) in the tropics is infrequent, although pathogenic organisms might be isolated from ponds (Roberts and Sommerville, 1982). However, *Streptococcus* outbreaks are becoming a serious threat at high stocking densities used in intensive aquaculture (Hubbert, 1989; Chang, 1994).

Toleration of Poor Water Quality

Fish acquire oxygen and nutrients from water and excrete wastes into the same medium, so fish culture becomes an environmental control problem. Water is fertilized to increase natural pond productivity. Water is also the repository of cellular metabolites from fish and bacterial decomposition of wasted feed. Some fertilizers and metabolites, like ammonia, can slow fish growth and induce mortality if present in excess (Boyd, 1990; Abdalla et al., 1992). Hydrogen sulfide (H_2S), which forms under anoxic conditions in pond sediments, particularly in brackish water, is toxic to fish at low concentrations.

Oxygen is consumed during phytoplankton and bacterial respiration, as well as by fish. Hypoxia or anoxia results when respiration by the aquatic community exceeds oxygen production by phytoplankton and diffusion from the atmosphere. Hypoxia reduces fish growth by decreasing feed consumption (Hepher, 1988; Lai-fa and Boyd, 1988; Teichert-Coddington and Green, 1993b) and/or by decreasing feed utilization efficiency (Hepher, 1988; Lai-fa and Boyd, 1988). Anoxia results in fish death, of course, since oxygen is essential to cellular metabolism.

It seems paradoxical that while proper pond management involves maintaining good water quality, a candidate culture fish should tolerate poor water quality. However, the ideal pond environment is rarely achieved for long in most systems. The natural variability of plankton communities alone can result in low dissolved oxygen events that fish must survive (Boyd, 1990). Overfertilization with urea or ammonia-based fertilizers, or high feeding rates may result in periods of high ammonia concentrations (Teichert-Coddington et al., 1992; Abdalla et al., 1992). Intensification of fish stocking rates dramatically increases the difficulty of good water quality management, because nutrient inputs, metabolites, and community oxygen consumption increase concomitantly.

Low dissolved oxygen concentrations in ponds can be corrected by mechanical aeration or by flushing with oxygen-rich water. Intensively managed fish ponds require regular aeration and/or water exchange. However, mechanical aeration and pumping are expensive and beyond the means of many small farmers in less developed countries. Water is often a scarce resource, so a culture system cannot usually be based upon water exchange. Many small producers are insufficiently trained to understand fish pond ecology and the interrelationships between nutrient input and water quality. A candidate culture fish should therefore tolerate limited periods of inferior water quality.

Efficient Conversion of Natural Foods and Feed

Fish ponds must be fertilized to enhance growth of natural food organisms, or fish must be offered a prepared feed. Culture systems in the Philippines reviewed by Schmittou et al. (1985) were categorized by level of nutrient input. Systems ranged from extensive without fertilizers or feeds, to intensive feeding where fish nutrition was gained entirely from feeds. Lovell (1989) categorized fish culture by three levels: production from natural foods, production from natural foods supplemented with feed, and intensive culture under artificial conditions. Both classification systems centered on nutrient input because nutrient management is usually the predominant concern of aquaculturists. Nutrients are usually the largest variable cost of a culture system, particularly where feeds are employed.

Regardless of nutrient input, candidate culture fish should efficiently convert ingesta to fish flesh and grow rapidly to a marketable size. Commonly cultured fish like tilapias and carps are equipped with mouth parts and pharyngeal plates that grind vegetation to rupture cell walls, and possess long intestines to efficiently digest vegetable matter. The pH of stomach fluids in tilapias may be as low as 1.25, sufficient to chemically lyse cell walls of blue-green phytoplankton (Bowen, 1982).

Tambaquí (*Colossoma macropomum*) is an example of a fish that grows well in its natural habitat on a diet comprised largely of fruits, nuts (Goulding, 1980), and cladoceran zooplankton (Goulding and Carvalho, 1982) but lack the capacity to graze on phytoplankton. They therefore grow poorly in fertilized ponds (Green et al., 1994) but grow as well as or better than tilapia when offered a supplemental feed (Peralta and Teichert-Coddington, 1989; Teichert-Coddington, 1996).

Carnivores are not usually good primary culture fish because feeds high in animal protein are expensive, requiring that fish be marketed at high prices. For example, hybrid clariid catfish are a preferred fish in southeast Asia, but market prices of catfish are high because of their carnivorous diet relative to tilapia and carps. Production of clariids is actually limited in some areas by lack of fisheries by-catch, a relatively cheap protein source (Csavas, 1994).

Higher quantities of prepared feeds are used as stocking densities increase, because natural pond productivity is insufficient to support high fish growth. Prepared diets may be used to supplement fish in organically fertilized systems (Green, 1992) especially in the final stages of a culture cycle initiated with fertilization (Green et al., 1994; Diana and Lin, 1996), or offered as a primary nutrient exclusive of natural foods. Aquaculture today often involves supplemental feeding unless the fish are for family consumption or cultivated under resource-poor conditions. Readers looking for comprehensive treatments of feeding and fish nutrition are referred to Hepher (1988) and Lovell (1989), and Chapter 11 of this book.

Controllable Reproduction

A culturist should be able to reproduce fish to provide adequate numbers of fingerlings for grow-out at desired, scheduled times. Otherwise, the culturist must depend on the supply of wild-caught offspring, which may only be seasonal, unpredictable, or simply insufficient for aquaculture. Having control over fish reproduction also allows the culturist to genetically improve fish stocks by selective breeding, hybridization, or some other form of genetic manipulation. Increased aquacultural yields from genetic improvement are sometimes large. For example, hybridization of prized indigenous clariid catfish with the fast growing, hardy *Clarius gariepinus* from Africa revitalized clariid culture in Southeast Asia (Csavas, 1994).

Some fish like cichlids and common carp can be reproduced at a young age by people with little training, so they are good for use in areas where technical expertise is unavailable or where there is poor access to hatcheries. Other fish, like colossomas and Chinese carps, reach sexual maturity at an older age and must be hormonally induced to spawn. More expertise and pond facilities are therefore required, and the fingerlings may only be available from hatcheries.

Fish reproductive efficiency is an important culture attribute related to controlled reproduction (see Chapter 10). Reproductive efficiency is the product of fecundity, spawning frequency, hatchability of eggs, and fry survival (Siraj et al., 1983). Usually, egg size, larvae size, and fry survival decrease as fish fecundity increases. Fecundity varies widely among fish. For example, colossomas (Woynarovich, 1986) and Chinese carps (Woynarovich and Horváth, 1980; Shireman and Smith, 1983) produce 100,000 to 300,000 eggs/kg, while *Oreochromis niloticus* produces between 2200 (Mires, 1982) and 13,200 eggs/kg (Siraj et al., 1983) per clutch. Tilapia fecundity per gram of fish is size dependent, decreasing as fish weight increases (Siraj et al., 1983). The reproductive efficiency of a given biomass of small tilapia may actually approach that of a similar biomass of carp, because tilapia spawn frequently year round, while carp spawn once in the spring. For example, the annual fecundity of 5 kg of 50-g tilapia, each of which produces 13,000 eggs/kg ten times-a-year, is similar to that of one 5-kg Chinese carp spawning once.

Practical fry production per kilogram of female is actually much less than egg production, especially when weight of nonspawning females is included in calculations. For example, field data in Honduras (Green et al., 1994) and Ecuador (Popma, 1987) indicated that actual fry production from pond-spawned tilapia averaged 1700 fry per kilogram stocked female in ponds drain harvested every 2.5 to 3.5 weeks.

Egg incubation and fry nursing are other important reproductive characteristics. Eggs from fish induced to spawn require incubation, usually in running, oxygenated water. Incubation facilities are an added capital expense and require added labor and expertise to manage. Small size fry like those of colossoma and carps require more care in nursery ponds to eliminate crustacean and insect predators and to produce small size zooplankton as food (Woynarovich and Horváth, 1980).

Marketable and Profitable

Economics of production including marketing is a nonbiological attribute often ignored by the culturist. A fish could arguably meet all biological criteria for culture but be unprofitable because the consumer does not accept it. Or, a fish may grow well and be highly marketable, but the inputs are too difficult or expensive to obtain for producer adoption. Successful, sustainable culture is predicated on producing fish profitably, whether at the subsistence or commercial level.

Although early aquacultural development projects were often conceived for nutritional improvement (Lovshin, 1975; Crance and Leary, 1979; Lovshin et al., 1986) and not necessarily for income generation, fish are a high-value, marketable product used to generate income even at the subsistence level. Aquaculture development for laudable objectives like nutritional improvement will probably stagnate without attention to marketing and economics (Parkman and McCoy, 1977; Street, 1978).

Edwards et al. (1996) discovered that fish culture in resource-poor, northeast Thailand was unpopular and faltering because fish yields were too low to justify effort. Nutrient input levels had to be increased to high enough levels to make yields from fish ponds economically attractive. On the other hand, integrated culture of ducks and fish was highly productive but eventually abandoned because of difficulty with obtaining feed ingredients for ducks and marketing eggs. Green et al. (1994) provided enterprise budgets for 41 different semi-intensive production systems with tilapia in Honduras. Income above variable costs for a 5-month cycle ranged from $–680/ha for low rates of inorganic fertilization to $3170/ha for moderately high applications of chicken litter and urea. These data illustrate that aquacultural development must incorporate production cost analyses and product marketing to improve chances of success.

TILAPIA: A MODEL CULTURE FISH

Tilapia possess many attributes desired in a culture fish. This explains why they are cultured so widely and why they were chosen for simultaneous testing at all CRSP sites. The biology and culture of tilapia will be reviewed in this section as a case study of pond fish attributes. Note that attributes related to handling and crowding tolerance were presented in the first part of this chapter; hence, the focus of this case study will be on disease tolerance, water quality, feed conversion, controlled reproduction, and marketability.

Nomenclature

Tilapia is a colloquial name for a large group of fish within the family Cichlidae. Tilapias are endemic to Africa and the Middle East, where more than 70 species have been identified. The scientific classification for tilapias has not been standardized, so a summary of the nomenclature is provided.

The American Fisheries Society includes all tilapias under the single genus *Tilapia* (Robbins et al., 1991). Most of the aquaculture scientific community, including the authors of this book, follow an alternative classification scheme proposed by Trewavas (1982). This scheme divides the tilapias into three major taxonomic groups based primarily on reproductive behavior:

- Substrate incubators (*Tilapia* spp.),
- Maternal mouthbrooders (*Oreochromis* spp.)
- Paternal and biparental mouthbrooders (*Sarotherodon* spp.)

The common and scientific names of the few commercially important species follow:

Nile tilapia	*O. niloticus*
Silver perch (Jamaica)	*O. niloticus*
Mojarra plateada (Colombia)	*O. niloticus*
Blue tilapia	*O. aureus*
Java tilapia	*O. mossambicus*
Mossambique tilapia	*O. mossambicus*
Zanzibar tilapia	*O. hornorum*

"Red" tilapia	*O. hybrids*[a]
Galilee tilapia	*S. galilaeus*
Black-chinned tilapia	*S. melanotheron*
—	*O. esculenta*
—	*O. macrochir*
Congo tilapia	*T. rendalli*
—	*T. zillii*
—	*O. andersonii*
—	*O. spilurus*

[a] Usually developed from *O. mossambicus* and/or *O. hornorum* stocks (for the red color), and often crossed with *O. niloticus* and/or *O. aureus* to improve growth and other aquacultural characteristics.

Large-scale commercial culture of tilapia worldwide is limited almost exclusively to the first four species listed above and the "red" tilapia. The remaining tilapia species listed above are utilized in small-scale commercial or subsistence aquaculture, primarily in Africa.

Disease Tolerance

Tilapias are apparently more resistant to viral, bacterial, and parasitic diseases than other commonly cultured fish (see Chapter 12). Tilapia rarely show signs of disease at temperatures above 22°C. However, viral, bacterial, and parasitic problems have all been reported, especially after stress from low temperature, handling, severe crowding, or poor water quality. For example, *Streptococcus* outbreaks are a serious threat at the high stocking densities used in intensive tilapia aquaculture (Hubbert, 1989; Chang, 1994). Fungal infections, especially from *Saprolegnia*, are particularly common after handling when water temperature is below 20°C.

No viral disease other than *Lymphocystis* has been reported for tilapia. Under high temperature and ammonia stress, myxobacterial infections, especially *Columnaris*, may be problematic. Myxobacterial gill infections may also cause heavy losses among fry, especially at low temperatures. The most common bacterial diseases are hemorrhagic septicemias, especially from *Aeromonas hydrophila* and, under hyperintensive culture, *Edwardsiella tarda*.

"Ich" or "White spot," caused by the protozoan parasite *Ichthyopthirius multifiliis*, can be serious among fry and juveniles in intensive recirculating systems. Ich epizootics are much less likely in tropical areas where the greatest commercial production of tilapia occurs because water temperatures are warmer than 20 to 24°C, which is the optimal temperature range for this disease organism. The protozoan *Trichodina* may also reach debilitating densities on stressed fish at low water temperatures.

Monogenean and digenean helminthic parasites are not uncommon on tilapia but are normally of low pathogenicity with little effect on fish growth.

Parasitic crustaceans, such as *Argulus*, *Ergasilus*, and *Lernaea*, have caused some serious losses, but most reports are from Africa where tilapias are endemic and from Israel where tilapias are associated with common carp.

Water Quality

Tilapia are more tolerant to poor water quality than most commonly cultured fish. This is an important reason for the successful culture of these fish in many different kinds of systems. Impacts of selected water quality variables on tilapia are reviewed in this section.

Dissolved Oxygen

Low dissolved oxygen is usually the first encountered water quality constraint to pond fish growth. Tilapia are commonly cultured in heavily fertilized ponds where community respiration is similar to or exceeds oxygen production by phytoplankton. Fish are often seen at the water surface in the early mornings gulping oxygen-rich water, a response to low dissolved oxygen (Kramer and McClure, 1982). Tilapia routinely survive dawn dissolved oxygen (DO) concentrations of less than 0.5 mg/L, levels considerably below the tolerance levels for most other cultured fish. Survival for up to 6 h at 0 mg/L has been reported by Teichert-Coddington and Green (1993a). In spite of tilapias' ability to survive hypoxia, ponds should be managed to maintain DO above 1 to 2 mg/L, because growth becomes depressed when concentrations fall chronically below this level (Teichert-Coddington and Green, 1993b).

pH

Tilapia seem to grow best in water that is near neutral or slightly alkaline. Growth is reduced in acidic waters (Teichert-Coddington and Phelps, 1989; Boyd, 1990), probably in part because production of natural food organisms is reduced. Acidity should be neutralized by liming ponds (Boyd, 1990). In the absence of soil testing or other technical assistance, lime application rates used on nearby agricultural fields can be used on ponds.

The upper limit on pH is not well known, but afternoon increases of pH to 10 apparently do not seriously affect tilapia production. The lethal alkaline limit is probably pH 11 or greater.

Ammonia

Massive mortality of tilapia occurs within a couple of days when fish are suddenly transferred to water with un-ionized ammonia levels greater than 2 mg/L. However, when acclimated to sublethal levels, tilapia can survive a couple of days at un-ionized ammonia concentrations of as high as 3 mg/L (Redner and Stickney, 1979)

Prolonged exposure (several weeks) to un-ionized ammonia levels greater than 1 mg/L causes heavy losses, especially among fry and juveniles in water with low DO. The first mortalities from prolonged exposure begin at un-ionized ammonia concentrations as low as 0.1 or 0.2 mg/L. Appetite and growth of tilapia fingerlings is depressed at un-ionized ammonia concentrations as low as 0.08 mg/L (Abdalla et al., 1992).

Salinity

Tilapias are freshwater species, but all can tolerate brackish water. *O. niloticus* is the least saline tolerant of the commercially important species, but has been reported to grow well at salinities of 18 to 25 ppt (Suresh and Lin, 1992). The "red" tilapia, which are composed partly of salinity tolerant *O. mossambicus*, have grown well in sea strength water (Watanabe et al., 1990). A consequence of differences in salinity tolerance is that *O. niloticus* may be the tilapia species of choice in brackish water penaeid shrimp ponds, where the release of tilapia to estuaries is undesirable. *O. niloticus* are unlikely to reproduce at salinities higher than 25 ppt.

Temperature

Inability to tolerate low temperatures is a serious constraint for commercial culture of tilapia in temperate regions. The lethal low temperature for most species is 10 or 11°C. Feeding generally ceases when temperature falls below 16 or 17°C. Low reproduction and disease-induced mortalities after handling may result at temperatures below 21°C. Preferred water temperatures for tilapia growth are approximately 28 to 32°C, but the range varies depending on what species of tilapia is cultured. When fish are fed to satiation, growth at the preferred temperature is typically three times greater than at 20 to 22°C. Maximum feed consumption at 22°C is only 50 to 60% as great as at 26°C (Newman et al., 1996). Mean yield of tilapia stocked at 2/m^2 and grown for 150 d with organic fertilization in central Honduras at 580 m altitude was 2800 kg/ha during the hot, wet season, compared with 2200 kg/ha during the cool, dry season (Green et al., 1994). Tilapias reportedly tolerate temperatures up to 40°C, but stress-induced disease and mortality are problematic when temperatures exceed 37 or 38°C.

Conversion of Natural Foods and Feed

Tilapia are excellent culture species, partly because they grow well on a variety of natural food organisms, including plankton, green leaves, benthic organisms, aquatic invertebrates, larval fish, detritus, and decomposing organic matter. In ponds with heavy supplemental feeding, natural food organisms typically account for 30 to 50% of tilapia growth (Schroeder, 1978), whereas in full-fed channel catfish ponds only 5 to 10% of the fish growth is traced to ingestion of natural food organisms (Chuapoehuk, 1977).

Tilapias are often considered filter-feeders because they can efficiently harvest planktonic organisms from the water column. However, "filter-feeder" is somewhat of a misnomer because tilapias do not physically filter the water through gill rakers, as do silver and bighead carps. The gills of tilapia secrete a mucous that entraps planktonic cells; the plankton-rich bolus is then ingested (Fryer and Iles, 1972). This mechanism allows tilapia to harvest micro-phytoplankton as small as 5 μm in diameter (Popma, 1982). Digestion and assimilation of plant material occurs along the length of a long intestine, usually at least six times the total length of the fish. Macrophytes are not considered a preferred food of *O. niloticus*, but they are sufficiently herbivorous to prevent establishment of most emergent plants in aquaculture ponds.

Ingestion of plant tissue does not automatically imply digestion and assimilation. Most fish species derive little nutrition from plant tissue ingested accidentally in pursuit of other food. Tilapia, however, obtain substantial nutritional benefit from plant material. Digestion of filamentous and planktonic algae and higher plants is aided by two mechanisms: physical grinding of plant tissues between two pharyngeal plates of fine teeth (Bowen, 1982), and a stomach pH below 1.5, which lyses the cell walls of blue-green algae and bacteria (Moriarty, 1973; Bowen, 1982). Tilapia digest 30 to 60% of the protein in algae, with blue-green algae being digested more efficiently than green algae (Popma, 1982).

Tilapia utilize natural food organisms so efficiently that standing crops of fish exceeding 3500 kg/ha in 150 d can be obtained in well-fertilized unfed ponds in Honduras (Teichert-Coddington and Green, 1993c) and Thailand (Knud-Hansen and Lin, 1993). Tilapia effectively utilize natural food organisms not ingested by many other fish. Some people, therefore, erroneously conclude that tilapia must have simple nutritional requirements. The nutritional requirements of tilapia, however, are very similar to other warmwater fish (Lim, 1989; Luquet, 1991).

Protein and energy contents of diets are primary nutritional considerations for commercial production of tilapia. Adequate digestible energy (DE) spares maximum dietary protein for growth. Additional dietary protein, without sufficient additional energy, actually depresses growth. Optimum DE levels in tilapia rations with reasonable protein quality are 8.3 to 9.3 kcal DE per gram of crude protein (Kubaryk, 1980; Winfree and Stickney, 1981). The high end of the energy range would be recommended for diets with protein of high quality. Nonprotein dietary energy can be obtained from carbohydrates and lipids.

Tilapias are similar to channel catfish in assimilating starch carbohydrate in cereal grains (e.g., corn, whole wheat) but are significantly more efficient than channel catfish in the digestion of the more complex carbohydrates in highly fibrous foodstuffs. Tilapias obtain little or no digestible energy from cellulose. Lipids, highly concentrated sources of dietary energy, are highly digestible by tilapia, especially lipids in the form of animal and vegetable oils that remain liquid at water temperature (Popma, 1982).

Maximum growth of tilapia is achieved at crude protein dietary levels of 35 to 50% (Kubaryk, 1980; Lim, 1989; Winfree and Stickney, 1981), but economically optimum levels in commercial diets for juveniles and adults are usually 25 to 35%. The low end of the protein range is most appropriate at suboptimal DE levels (diets with little lipid and/or a high percentage of the more complex carbohydrates).

Protein quality of a tilapia diet is a function of the combination of amino acids, the building blocks of all proteins. Tilapias, like other fish, shrimp, and terrestrial animals, require ten essential amino acids (Lim, 1989; Santiago and Lovell, 1988). Plant protein, in general, is deficient in two of the essential amino acids, methionine and lysine. For optimum growth of tilapia, approximately 5% of the dietary protein should be lysine and about 3% should be methionine+cystine. To supplement these deficiencies, feedstuffs of animal origin (fish meal, meat, bone meals, etc.) usually constitute 7 to 15% of economically optimum supplemental rations for commercial production of tilapia. Soybean meal is probably the most complete of the oilseed meals for meeting the essential amino acid needs of tilapia (although it is slightly deficient in methionine+cystine), but it seldom replaces all animal protein in economically optimum rations. Cottonseed meal can be added to tilapia diets to a level of at least 15% without causing gossypol toxicity. Sunflower seed meal, peanut meal, and rapeseed are reasonably good protein sources for tilapia diets, but have one or more drawbacks, including antinutritional factors and toxic substances; these drawbacks, however, are no more serious for tilapia than for channel catfish or carp.

Mineral requirements for tilapia are not completely understood. Recommended available phosphorus in rations is 0.5% (NRC, 1993). Only about one-third of the phosphorus from grains and plant material is nutritionally available to tilapia, but inorganic phosphorus is readily absorbed. Most required calcium is obtained from the water by absorption across the gill membranes (NRC, 1993), but additional dietary calcium is usually added because dicalcium phosphate, added at 15 g/kg to provide adequate phosphorus in intensive culture, is a relatively inexpensive mineral for animal rations. Magnesium, sodium, potassium, iron zinc, copper, iodine, selenium, and other trace minerals are generally derived from the water to satisfy most nutritional requirements. However, because of the information gaps about tilapia nutritional requirements and the relatively low cost of mineral supplements, these supplements are usually added to tilapia rations.

Requirements for a few vitamins are known for tilapia (NRC, 1993). Because of the possible consequences of vitamin deficiency and the relatively low cost of vitamins, vitamin premixes are usually added to tilapia rations.

Recommended feeding rates for tilapia are a function of fish size, water temperature, fish biomass density, and abundance of natural food organisms. As with other fish species, the optimum feeding rate is inversely related to fish weight. At 27 to 29°C, common feeding rates (Hepher and Pruginin, 1981; Lovell, 1989) for high quality feeds are

Weight (g)	Feeding rate (% body weight/d)
1–5	10–6
5–20	6–4
20–100	4–2.5
100–200	2.5–2
200–400	2–1.5

Appetite decreases rapidly at lower temperatures. Maximum feed consumption of tilapia at 22°C is only 50 to 60% of maximum feed consumption at 26°C (Newman et al., 1996).

When agricultural byproducts are fed, ponds are usually fertilized and feeding rates are often reduced by nearly 50% so that natural food organisms adequately supplement the low quality feedstuff. Feed conversion efficiency with agricultural byproducts may become uneconomical when the quantity of the feedstuff "overpowers" the ability of natural food organisms to supplement the amino acid deficiencies of the nutritionally incomplete ration.

Manuring has long been practiced in small-scale and subsistence culture of tilapia, but this practice is also often a cost effective means of increasing the food supply in large-scale commercial tilapia ponds during the early part of a production cycle when fish biomass is still low. Where logistically convenient, ponds are usually manured until the combined input of manure plus feed begins to cause water quality problems.

Tilapias are continuous feeders during daylight hours (Moriarty and Moriarty, 1973), in contrast with other carnivorous species that are opportunistic and capable of taking a single, large meal. This consideration and field observations have led researchers to conclude that a daily ration should optimally be divided into three to four meals (Jauncey and Ross, 1982; Lim, 1989). Logistical considerations on commercial farms, however, commonly limit feeding frequency to twice daily.

Reproduction Characteristics of Tilapia

The four principal cultured tilapias, all of the genus *Oreochromis*, are maternal mouthbrooders. In all cases, the polygamous male excavates a nest in the pond bottom, generally in water less than 1 m deep. After a short mating ritual, the female spawns in the nest and incubates the externally fertilized eggs in her buccal cavity until they hatch. Fry remain in the female's mouth through yolk sac absorption, and often seek refuge in her mouth for several days after swim bladder inflation.

Sexual maturity in tilapia species is a function of age and size. Tilapia populations in large lakes mature at a later age and larger size than the same populations raised in culture ponds. For example, *O. niloticus* matures at about 10 to 12 months and 350 to 500 g in several East African lakes. The same population under conditions of near maximum growth will reach sexual maturity in culture ponds at an age of 4 to 6 months and about 150 g. Under food-limited conditions *O. niloticus* and *O. aureus* can reach sexual maturity in small ponds at weights as low as 20 g.

The ease with which tilapia reproduce is reason for both its popularity as a culture species and a hindrance to its commercialization. Reproduction in culture ponds can occur within 3 months of fingerling stocking. Ponds can rapidly become overpopulated with young fish that compete with adults for a limited food supply. This is not a major problem where small fish are marketable; otherwise, it is a great impediment to profit making.

Male adult tilapia grow faster than females. Therefore, many culture regimes seek to stock only males to eliminate reproduction and improve growth rates. Techniques to produce all-male populations are discussed in Chapter 10 of this book. They include masculinization by androgen treatment, segregation of males by visual examination of external genitalia, and hybridization. An alternative to preventing reproduction is to stock predator fish that prey on reproduction. Guapote tigre (*Cichlasoma managuense*) have been used for this purpose in Central America (Dunseth and Bayne, 1978). Stocking 500 fingerlings of 0.5 to 5 g/ha has resulted in good control of tilapia reproduction in ponds stocked with populations of hormonally masculinized male tilapia at 2 to 3/m^2 with about 2% females (Green et al., 1994). McGinty (1983) stocked 75 to 125 peacock bass (*Cichla ocellaris*) per hectare to control tilapia recruitment in ponds stocked with up to 2800 mixed sex tilapia per hectare. Peacock bass are native to the Amazon basin of South America but have been transplanted to some areas of Central America and the Caribbean. Snakehead (*Ophicephalus* spp.) are commonly stocked at 380 to 500 fingerlings per hectare in southeast Asia to control tilapia reproduction (Ling, 1977).

Marketability of Tilapia

Marketing of tilapia at the subsistence level usually occurs at the pond bank. There is normally little price premium for larger fish. Fish sizes greater than 150 g are marketable in rural areas of Central America, and even smaller fish are marketable in some African countries.

Commercial culture requires more sophisticated marketing. Size, fish color, presentation, flavor, and distribution are some considerations. The North American market serves as an example. Buyers of whole and processed tilapia are very conscious of flesh and gustatory quality. Consumers prefer white-fleshed fish. Tilapia flesh is light gray to white. Some believe that the "red" tilapia and the less common "white" tilapia have a whiter flesh than the pure tilapia species, such as *O. niloticus*. However, once the fillet is frozen, differences are less discernible. Tilapia have some dark or red muscle under the skin that accumulates along the lateral line and disperses over the surface of the light colored muscle. The red muscle can give skinned tilapia a darker color that is unappealing to consumers looking for white-fleshed fish. The peritoneal lining of pure tilapia species (*O. niloticus*, *O. aureus*, and especially *O. mossambicus*) is black and objectionable to many buyers of whole or headed, gutted, and skinned tilapia. The black peritoneal lining is easily removed with a scrub brush but requires extra labor or specialized equipment. Red and white tilapias have a clear peritoneal lining that does not need removal. Removal of the rib cage during filleting also removes the peritoneal lining so peritoneal color is not a problem when tilapia are filleted. Thus, red and white tilapia may have color and flesh qualities that enhance its marketability. However, buyers should remember that red tilapia are more difficult to raise than normal colored tilapia and the cost per kilogram live weight to raise the red tilapia is commonly higher than normal colored tilapia.

Gustatory quality is also important. Taste tests have demonstrated that tilapia do not absorb off-flavor from animal manures, but they do absorb off-flavors produced by certain blue-green algae and other microorganisms. Muddy flavored flesh is more prevalent in tilapia cultured in freshwater than in saline water, because blue-green algae are less prevalent above 5 ppt salinity. Muddy flavor can be found in any cultured fish, but tilapia appear to have a higher incidence of off-flavor than most freshwater fish because they are reared in highly fertile waters with higher incidence of blue-green algae blooms. Processors and buyers must be vigilant to assure that off-flavored fish are not processed and sold. Processors can generally assure on-flavor fish by placing the live tilapia in clean water for 3 to 5 d to purge them of off-flavors before processing. Bleeding the live fish before processing will also reduce off-flavors.

Red muscle found under the skin accumulates lipids that turn rancid when fish are stored for 1 to 2 months. Tilapia larger than 600 g accumulate more red muscle than smaller fish. Rancidity of the red muscle is not a problem when whole tilapia or tilapia fillets are iced and sold fresh. Frozen tilapia products have more problems with a fishy flavor, and the red muscle may have to be removed before freezing to maintain gustatory quality for periods longer than a month.

A major disadvantage of tilapia is low dress-out and fillet yield compared with some other cultured fish. In general, the dress-out yield of tilapia increases slightly with fish size and if the tilapia is well fed and robust. Interspecific differences in dress-out and fillet yield are minimal. After removal of scales and viscera, heads-on dress-out weight is 76 to 80%. Average yield of headed, gutted and skinless tilapia is about 51% (Clement and Lovell, 1994).

Fillet yield depends on the method employed to fillet the fish. Small pin bones located on the median line between the tenderloin and the rib cage can be troublesome to the consumer. Fillets that have the ribs removed but retain some flesh from the rib cage and the pin bones yield about 36 to 38% of total live weight. The same fillet with a triangular notch on the median line to remove the pin bones yields about 32 to 35% of live weight. Fillets with pin bones and rib cage flesh removed so that only the tenderloin and caudal area remain yield about 25% of live weight (Clement and Lovell, 1994). Some large tilapia are "deep skinned" to remove some of the red muscle found under the skin. Deep skinning reduces the filet yield to 22 to 25% of the live weight. Assuming a

35% fillet yield, whole tilapia must weigh 480 to 800 g to provide two 3- to 5-oz fillets (1 oz = 28 g) and 800 to 1120 g to provide two 5- to 7-oz fillets. Fish yielding the larger fillet convert feed less efficiently and require more sophisticated technology, thus increasing production costs per unit weight.

REFERENCES

Abdalla, A. A. F., McNabb, C., Knud-Hansen, C., and Batterson, T., Growth of *Oreochromis niloticus* in the presence of un-ionized ammonia, in *Ninth Annual Administrative Report,* Pond Dynamics/Aquaculture CRSP, 1991; Egna, H. S., McNamara, M., and Weidner, N., Eds., Office of International Research & Development, Oregon State University, Corvallis, 1992.

Bardach, J. E., Ryther, J. H., and McLarney, W. O., *Aquaculture: The Farming and Husbandry of Freshwater and Marine Organisms,* John Wiley & Sons, New York, 1972, 868.

Bowen, S. H., Feeding, digestion and growth: quantitative considerations, in *The Biology and Culture of Tilapias;* Pullin, R. S. V. and Lowe-McConnell, R. H., Eds., ICLARM Conference Proceedings 7, International Center for Living Aquatic Resources Management, Manila, Philippines, 1982.

Boyd, C. E., *Water Quality in Ponds for Aquaculture,* Alabama Agricultural Experiment Station, Auburn University, Auburn, AL, 1990, 482.

Chang, P. H., *Streptococcus* Infection in Fish, Ph.D. dissertation, Auburn University, Auburn, AL, 1994.

Chuapoehuk, W., Nutritional contribution of natural pond organisms to channel catfish growth in intensively fed ponds, Ph.D. dissertation, Department of Fisheries and Allied Aquacultures, Auburn University, Auburn, AL, 1977.

Clement, S. and Lovell, R. T., Comparison of processing yield and nutrient composition of cultured Nile tilapia (*Oreochromis niloticus*) and channel catfish (*Ictalurus punctatus*), *Aquaculture,* 119, 299, 1994.

Crance, J. H. and Leary, D. R., *The Philippine Inland Fisheries Project and Aquaculture Production Project: Completion Report,* Research and Development Series No. 25, International Center for Aquaculture, Agricultural Experiment Station, Auburn University, Auburn, AL, 1979, 23.

Csavas, I., Status and perspectives of culturing catfishes in East and Southeast Asia, in *Biological Bases for Aquaculture of Siluriformes,* 23–27 May, Montpellier, France, 1994.

Diana, J. S. and Lin, C. K., Timing of supplemental feeding for tilapia production, *J. World Aquaculture Soc.,* in press, 1996.

Dunseth, D. R. and Bayne, D. R., Recruitment control and production of *Tilapia aurea* (Steindachner) with the predator, *Cichlasoma managuense* (Gunther), *Aquaculture,* 14, 383, 1978.

Edwards, P., Demaine, H., Innes-Taylor, N., and Turongruang, D., Sustainable Aquaculture for Small-Scale Farmers: Need for a Balanced Paradigm. *Outlook on Agriculture,* 25 (1), 19, 1996.

Fryer, G. and Iles, T. D., *The Cichlid Fishes of the Great Lakes of Africa: Their Biology and Evolution,* T. F. H. Publications, Hong Kong, 1972, 641.

Goulding, M., *The Fishes and the Forest,* University of California Press, Los Angeles, CA, 1980.

Goulding, M. and Carvalho, M. L., Life history and management of the tambaqui (*Colossoma macropomum,* Characidae): an important Amazonian food fish, *Revista Brasileira Zool.,* 1, 107, 1982.

Green, B. W., Substitution of organic manure for pelleted feed in tilapia production, *Aquaculture,* 101, 213, 1992.

Green, B. W., Teichert-Coddington, D. R., and Hanson, T. R., *Development of Semi-Intensive Aquaculture Technologies in Honduras: Summary of Freshwater Aquacultural Research Conducted from 1983 to 1992,* Research and Development Series Number 39, International Center for Aquaculture and Aquatic Environments, Auburn University, Auburn, AL, 1994, 48.

Heath, A. G., *Water Pollution and Fish Physiology,* CRC Press, Boca Raton, FL, 1987, 245.

Hepher, B., *Nutrition of Pond Fishes,* Cambridge University Press, New York, 1988, 388.

Hepher, B. and Pruginin, Y., *Commercial Fish Farming, with a Special Reference to Fish Culture in Israel,* Wiley Interscience, New York, 1981, 261.

Hubbert, R. M., Bacterial diseases in warwater aquaculture, in Shilo, M. and Sarig, S., Eds., *Fish Culture in Warm Water Systems: Problems and Trends,* CRC Press, Boca Raton, FL, 1989.

Huet, M., *Textbook of Fish Culture: Breeding and Cultivation of Fish,* Fishing News Books Ltd., Farnham, Surrey, 1972, 436.

Jauncey, K. and Ross, B., *A Guide to Tilapia Feed and Feeding,* Institute of Aquaculture, University of Sterling, Scotland, 1982.

Jensen, J. and Durborow, R., *Tables for Applying Common Fishpond Chemicals,* Circular ANR-414, Alabama Cooperative Extension Service, Auburn University, Auburn, AL, 1984, 11.

Knud-Hansen, C. F. and Lin, C. K., Strategies for stocking Nile tilapia (*Oreochromis niloticus*) in fertilized ponds, in *Tenth Annual Administrative Report,* Pond Dynamics/Aquaculture CRSP, 1992; Egna, H. S., McNamara, M., Bowman, J., and Astin, N., Eds., International Research & Development, Oregon State University, Corvallis, 1993.

Kramer, D. L. and McClure, M., Aquatic surface respiration, a wide spread adaptation to hypoxia in tropical freshwater fish, *Environ. Biol. Fish,* 7, 47, 1982.

Kubaryk, J. M., Effect of Diet, Feeding Schedule and Sex on Food Consumption, Growth, and Retention of Protein and Energy by Tilapia, Ph.D. dissertation, Department of Fisheries and Allied Aquacultures, Auburn University, Auburn, AL, 1980.

Lagler, K. F., Bardach, J. E., Miller, R. R., and Passino, D. R. M., *Ichthyology,* John Wiley & Sons, New York, 1977, 506 pp.

Lai-fa, Z. and Boyd, C. E., Nightly aeration to increase the efficiency of channel catfish production, *Prog. Fish-Cult.,* 50, 237, 1988.

Lim, C., Practical feeding — tilapias, in Lovell, T., Ed., *Nutrition and Feeding of Fish,* Von Nostrand Reinhold, New York, 1989.

Ling, S.-W., *Aquaculture in Southeast Asia: A Historical Overview,* University of Washington Press, Seattle, 1977, 108.

Lovell, T., *Nutrition and Feeding of Fish,* Van Nostrand Reinhold, New York, 1989, 260.

Lovshin, L., *Progress Report on Fisheries Development in Northeast Brazil,* Research and Development Series No. 9, International Center for Aquaculture, Agricultural Experiment Station, Auburn University, Auburn, AL, 1975, 11.

Lovshin, L. L., Schwartz, N. B., Castillo, V. G., Engle, C. R., and Hatch, U. L., Cooperatively Managed Rural Panamanian Fish Ponds: the Integrated Approach, Research and Development Series No. 33, International Center for Aquaculture, Agricultural Experiment Station, Auburn University, Auburn, AL, 1986, 47.

Luquet, P., Tilapia, *Oreochromis* spp., in Wilson, R. P., Ed., *Nutrient Requirements of Finfish,* CRC Press, Boca Raton, FL, 1991.

McGinty, A. S., Population dynamics of peacock bass, *Cichla ocellaris* and *Tilapia nilotica* in fertilized ponds, in *International Symposium on Tilapia in Aquaculture. Nazareth, Israel 8–13 May 1983;* Fishelson, L. and Yaron, Z., Eds., Tel Aviv University, Tel Aviv, Israel, 1983.

Mires, D., A study of the problems of the mass production of hybrid tilapia fry, in *The Biology and Culture of Tilapias;* Pullin, R. S. V. and Lowe-McConnell, R. H., Eds., ICLARM Conference Proceedings 7, International Center for Living Aquatic Resources Management, Manila, Philippines, 1982.

Moehl, J. F., Jr., Evaluation of *Tilapia macrochir* and *Tilapia nilotica* for pond culture in Rwanda, Masters thesis, Auburn University, Auburn, AL, 1989.

Moriarty, C. M. and Moriarty, D. J. W., Quantitative estimation of the daily ingestion of phytoplankton by *Tilapia nilotica* and *Haplochromis nigripinnis* in Lake George, Uganda, *J. Zool. London,* 171, 15, 1973.

Moriarty, D. J. W., The physiology of digestion of blue-green algae in the cichlid fish *Tilapia nilotica, J. Zool.,* 171, 25, 1973.

Newman, J. R., Popma, T. J., and Seim, W. K., Effects of temperature on maximum feed consumption and growth of juvenile Nile tilapia, in *World Aquaculture Society, Book of Abstracts, World Aquaculture '96, Bangkok, Thailand,* Creswell, R. L., Ed., World Aquaculture Society, Louisiana State University, Baton Rouge, 1996.

NRC (National Research Council), *Nutrient Requirements of Fish,* National Academy Press, Washington, D.C., 1993, 114.

Parkman, R. W. and McCoy, E. W., *Fish Marketing in El Salvador,* Research and Development Series No. 12, International Center for Aquaculture, Agricultural Experiment Station, Auburn University, Auburn, AL, 1977, 19.

Peralta, M. and Teichert-Coddington, D. R., Comparative production of *Colossoma macropomum* and *Tilapia nilotica* in Panama, *J. World Aquaculture Soc.,* 20, 236, 1989.

Piper, R. G., McElwain, I. B., Orme, L. E., McCraren, J. P., Fowler, L. G., and Leonard, J. R., *Fish Hatchery Management,* United States Fish and Wildlife Service, Washington, D.C., 1986, 517.

Popma, T., *Freshwater Fish Culture Development Project, Final Technical Report,* Auburn University/ University of Florida/USAID Technical Assistance Contract, Auburn University, Auburn, AL, 1987.

Popma, T. J., Digestibility of selected feedstuffs and naturally occurring algae by tilapia, Ph.D. dissertation, Department of Fisheries and Allied Aquacultures, Auburn University, Auburn, AL, 1982.

Redner, B. D. and Stickney, R. R., Acclimation to ammonia by *Tilapia aurea, Trans. Am. Fish. Soc.,* 108, 383, 1979.

Robbins, C. R., Bailey, R. M., Bond, C. E., Brooker, J. R., Lachner, E. A., Lea, R. N., and Scott, W. B., *Common and Scientific Names of Fishes from the United States and Canada,* 5th ed., Special Publication 20, American Fisheries Society, Bethesda, MD, 1991, 183.

Roberts, R. J. and Sommerville, C., Diseases of tilapias, in *The Biology and Culture of Tilapias,* Pullin, R. S. V. and Lowe-McConnell, R. H., Eds., ICLARM Conference Proceedings 7, International Center for Living Aquatic Resources Management, Manila, Philippines, 1982.

Santiago, C. B. and Lovell, T., Amino acid requirements for growth of Nile tilapia, *J. Nutr.,* 118, 1540, 1988.

Schmittou, H. R., Grover, J. H., Peterson, S. B., Librero, A. R., Rabanal, H. B., Portugal, A. A., and Adriano, M., *Development of Aquaculture in the Philippines,* Research and Development Series No. 35, International Center for Aquaculture, Agricultural Experiment Station, Auburn University, Auburn, AL, 1985, 31.

Schroeder, G. L., Autotrophic and heterotrophic production of micro-organisms in intensely-manured fish ponds, and related fish yields, *Aquaculture,* 14, 303, 1978.

Shell, E. W., *The Development of Aquaculture: an Ecosystems Perspective,* Alabama Agricultural Experiment Station, Department of Fisheries and Allied Aquacultures, Auburn University, AL, 1993, 265.

Shireman, J. V. and Smith, C. R., *Synopsis of Biological Data on the Grass Carp, Ctenopharyngodon idella (Cuvier and Valenciennes, 1844),* FAO Fisheries Synopsis No. 135, Food and Agriculture Organization of the United Nations, Rome, 1983, 86.

Siraj, S. S., Smitherman, R. O., Castillo-Galluser, S., and Dunham, R. A., Reproductive traits for three year classes of *Tilapia nilotica* and maternal effects of their progeny, in *International Symposium on Tilapia in Aquaculture,* Nazareth, Israel, Fishelson, L. and Yaron, Z., Eds., Tel Aviv University, Tel Aviv, Israel, 1983.

Stickney, R. R., *Principles of Warmwater Aquaculture,* John Wiley & Sons, New York, 1979, 375.

Street, D. R., *An Economic Assessment of Fisheries Development in Colombia,* Research and Development Series No. 20, International Center for Aquaculture, Agricultural Experiment Station, Auburn University, Auburn, AL, 1978, 10.

Suresh, A. V. and Lin, C. K., Tilapia culture in saline waters: a review, *Aquaculture,* 106, 201, 1992.

Teichert-Coddington, D. R., Effect of stocking ratio on semi-intensive polyculture of *Colossoma macropomum* and *Oreochromis niloticus* in Honduras, Central America, *Aquaculture,* 143. 291, 1996.

Teichert-Coddington, D. R. and Green, B. W., Influence of daylight and incubation interval on water column respiration in tropical fish ponds, *Hydrobiologia,* 250, 159, 1993a.

Teichert-Coddington, D. R. and Green, B. W., Tilapia yield improvement through maintenance of minimal oxygen concentrations in experimental grow-out ponds in Honduras, *Aquaculture,* 118, 63, 1993b.

Teichert-Coddington, D. R. and Green, B. W., Usefulness of inorganic nitrogen in organically fertilized tilapia production ponds, in *Abstracts of World Aquaculture '93,* European Aquaculture Society Special Publication No. 19, Oostende, Belgium, Torremolinos, Spain, May 26–28, 1993, 1993c.

Teichert-Coddington, D. R., Green, B., Boyd, C., and Rodriguez, M. I., Supplemental nitrogen fertilization of organically fertilized ponds: variation of the C:N ratio, in *Ninth Annual Administrative Report,* Pond Dynamics/Aquaculture CRSP, 1991, Egna, H. S., McNamara, M., and Weidner, N., Eds., International Research & Development, Oregon State University, Corvallis, 1992.

Teichert-Coddington, D. R. and Phelps, R. P., Effects of seepage on water quality and productivity of inorganically fertilized tropical ponds, *J. Aquaculture Tropics,* 4, 85, 1989.

Trewavas, E., Tilapias: taxonomy and speciation, in *The Biology and Culture of Tilapias,* Pullin, R. S. V. and Lowe-McConnell, R. H., Eds., ICLARM Conference Proceedings 7, International Center for Living Aquatic Resources Management, Manila, Philippines, 1982.

Tucker, C. S. and Robinson, E. H., *Channel Catfish Farming Handbook,* Van Nostrand Reinhold, New York, 1990, 454.

Watanabe, W. O., Clark, J. H., Dunham, J. B., Wicklund, R. I., and Olla, B. L., Culture of Florida red tilapia in marine cages: the effect of stocking density and dietary protein on growth, *Aquaculture*, 90, 123, 1990.

Winfree, R. A. and Stickney, R. R., Effects of dietary protein and energy on growth, feed conversion efficiency and body composition of *Tilapia aurea*, *J. Nutr.*, 111, 1001, 1981.

Woynarovich, E., *Tambaqui e pirapitinga: Propagação de alevinos*, Publicação 2, Assessoria de Comunicação Social da Companhia de Desenvolvimento do Vale do São Francisco (CODEVASF), SGAN Quadra 601, Bloco I, Sala 204, Brasília, DF, 1986.

Woynarovich, E. and Horváth, L., *The Artificial Propagation of Warmwater Finfishes — A Manual for Extension*, FAO Fisheries Technical Paper No. 201, Food and Agriculture Organization of the United Nations, Rome, 1980, 183.

9   FACTORS AFFECTING FISH GROWTH AND PRODUCTION

Richard W. Soderberg

INTRODUCTION

Growth in fishes in aquaculture is a complex process by which ingested energy is converted to biomass. The efficiency of this conversion is regulated by the growth potential of the organism, its trophic status, and various abiotic factors such as food supply, temperature, and adverse environmental factors brought about by the conditions in which the fish are cultured.

DESCRIPTION OF GROWTH

The growth of fishes follows an asymmetric sigmoid curve described by von Bertalanffy (1938; Figure 1). This stylized description of growth, when considered over the life span of the fish, ignores the seasonal effects of cold or cloudy weather and spawning periods in which growth may slow or cease during portions of each year. The resulting oscillations in the growth curve are generally not of importance in aquaculture because growth is usually considered over a relatively brief culture period. Seasonal fluctuations are usually nonexistent in salmonid culture due to constant temperatures of the groundwater supplies used and the complete reliance on artificial diets. In tropical static water aquacultures, which is the emphasis of the present discussion, seasonal fluctuations in growth may be caused by rainy periods which reduce the photosynthetic production of natural foods. However, by choosing species, such as tilapia, for culture that complete growth to a marketable size within a single rainy or dry season, separate crops of fish can be reared in the two seasons. The resulting growth curves then both follow the von Bertalanffy pattern in form if not magnitude.

Fish growth, thus considered, is characterized by an exponential growth surge in the juvenile stage, followed by a period during which growth is relatively linear, which gradually leads to a dampened phase where growth is slowed as the fish approaches a theoretical asymptotic maximum size (Figure 1). Production is growth multiplied by fish number and follows a similar pattern as described in Chapters 11 and 13 of this book.

When trout growth, measured in units of length/time is regressed against temperature within a sufficiently restricted optimum range, a linear relationship results (Haskell, 1959; Soderberg, 1995). Soderberg (1990) showed that this linear relationship also applied to tilapia when grown in flowing water on complete diets at densities which preclude spawning behavior. Soderberg (1992) presented linear equations for several species of cultured fish from which growth rates in mm/day can be accurately computed from daily temperature. The equation for blue tilapia (*Oreochromis aureus*), for example, is

$$\Delta L = -0.853 + 0.048T \tag{1}$$

0-56670-274-7/97/$0.00+$.50
© 1997 by CRC Press LLC

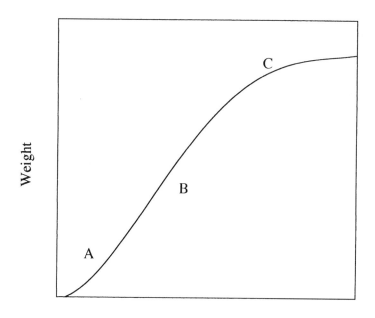

Figure 1. Idealized fish growth curve. A = exponential phase, B = linear phase, C = asymptotic phase.

where ΔL = growth rate in mm/day and T = temperature in °C (Soderberg, 1990). Projected length can be converted to weight using the approximately cubic relationship between the two.

According to this model, the growth rates of fish at constant temperatures, in length units, are linear over time regardless of fish size. When lengths are converted to weights, it is clear that linear growth in length units is exponential in weight units and corresponds to the exponential phase of the von Bertalanffy growth function (Figure 2). Most descriptions of fish growth (in weight) report that the exponential growth stanza is of short duration while low biomasses of larval fish utilize vast food resources. Thereafter, fish growth (in weight) becomes linear. This is typically the case in wild and some extensively managed pond fisheries. Exponential growth may be prolonged for the entire culture period in aquacultures where fish are not raised to sizes where gonadal development competes with somatic growth. Rainbow trout (*Onchorynchus mykiss*), for example, are rarely cultured to greater than 30 cm (313 g), a size to which growth is exponential in flowing water culture conditions, although wild specimens in excess of 17 kg have been reported for this species. In contrast, tilapias are generally cultured to sizes at which maturation can occur. Tilapias vary considerably in their growth potentials and only those species which grow to the larger sizes have been widely cultured. Fish which have the potential to attain larger maximum sizes are larger at the end of their exponential phase of growth and therefore grow faster to desired harvest sizes. Legendre and Albaret (1991) found that the growth rates of fishes were positively correlated with their maximum observed lengths and suggested that this relationship be used to evaluate fish species for aquaculture. While growth patterns of juvenile tilapias in ponds are lacking in the literature, and it is not possible to isolate endogenous from exogenous influences on the characteristics of fish growth patterns, the foregoing discussion indicates that the shift from exponential to linear growth is primarily an endogenous phenomenon.

Most reports on the pond culture of tilapias depict a linear pattern of weight increase over time up to a size of about 250 g (Anderson and Smitherman 1978; Collis and Smitherman 1978; Fram

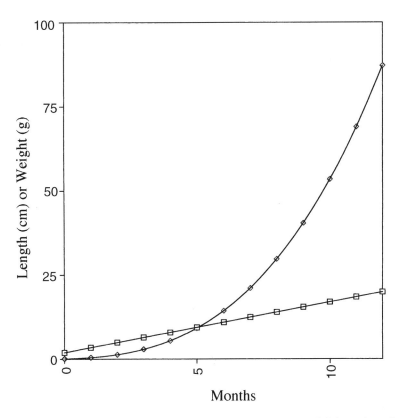

Figure 2. Fish growth, when linear in units of length, is exponential in units of weight and indicates maximum growth rate, unhindered by biotic or abiotic factors. □ = Length (cm); ◇ = Weight (g).

and Pagan-Font 1978; Teichert-Coddington and Green, 1993; Green and Teichert-Coddington, 1994). In most tropical locations, daily and seasonal temperature variations are slight. Selection of slow-maturing species or culture of a single sex of a rapidly maturing species may prevent some food resources from being partitioned into gonadal growth and the linear phase of growth may be prolonged until growth is slowed due to food limitations or degraded water quality. Thus the point at which relatively linear growth is dampened to an asymptotic pattern depends largely upon exogenous factors. Collis and Smitherman (1978) showed that tilapia grew in a linear fashion to a size of about 275 g on a commercial diet, but growth became asymptotic at about 200 g when natural food was supplied through pond fertilization with cattle manure. The growth of tilapia in aerated ponds appeared to be linear for fish up to nearly 250 g, but in unaerated ponds, growth became asymptotic at about 175 g (Teichert-Coddington and Green, 1993). In both of these studies, linear growth became asymptotic due to the dampening effects of endogenous factors, food supply in the first case and dissolved oxygen in the latter.

FACTORS AFFECTING GROWTH

The duration of the exponential growth stanza, the deviations of growth from exponential to linear to asymptotic phases and the magnitude of the asymptotic maximum size vary widely among species and strains of fish and among similar fish in different environmental conditions. Understanding and controlling these factors is the underlying principal of aquaculture.

Endogenous Factors

It is obvious that some fishes grow larger than others. The biological growth potential for any organism is programmed into its genome by its evolutionary history. This simple observation has a profound influence in aquaculture. Balarin (1979) reported a maximum size of 2.5 kg (50 cm) for *O. niloticus*. The maximum size reported for *O. rendalli* is 1.3 kg (40 cm) and for *O. macrochir* is 1.2 kg (40 cm) (Balarin, 1979). This difference in growth potential is one of the reasons that *O. niloticus* is one of the most popular tilapias for aquaculture.

As fish mature, endogenous changes in growth are caused by the diversion of energy from growth to gonadal development, spawning and other activities related to reproduction. As fish approach their asymptotic maximum, most ingested energy is used for maintenance and very little is used for growth.

Fish typically have sexual differences in growth. The males of tilapias grow to larger sizes than the females and are often grown alone in mono-sex cultures.

Exogenous Factors

Food Supply

Growth of fish having identical endogenous growth characteristics varies greatly among different environmental conditions. This is due, most importantly, to differences in the quantity and quality of the food resources provided by or to these environments. The influence of pond management practices on the production, and hence growth, of fishes in aquaculture is described in Chapters 4 and 11 of this book.

Temperature

Food consumption and growth in fishes increase with temperature to a maximum, then fall abruptly because the energy required for maintenance increases rapidly, thus decreasing the energy available for growth. The temperature at which growth is maximum is called the optimum temperature for growth and is determined in laboratory conditions with fish fed to excess. Optimum temperatures for growth range from 13°–17°C for salmonids and 25°–30°C for centrarchids (Brett, 1979). Suffern et al. (1978) reported that the optimum temperature for growth of the hybrid tilapia, *O. mossambicus* (F) X *O. hornorum* (M) was around 30°C. Maximum growth for *Tilapia zillii* fed lettuce was 31.4°C (Platt and Houser, 1978).

The y-intercept of a temperature (y) vs. growth (x) plot shows theoretical zero growth temperatures of 3.7°C for brook trout (*Salvelinus fontinalis*; Haskell, 1959) and 17.8°C for *O. aureus* (Soderberg, 1990). At temperatures between the intercept of the temperature vs. growth regression and the optimum temperature for growth, growth is proportional to temperature when fish are fed maximum rations and can be accurately predicted from a linear equation (Soderberg, 1992).

Dissolved Oxygen

The oxygen consumption rates of fishes respond to decreases in environmental dissolved oxygen tensions in the manner explained by Fry (1957; Figure 3). Oxygen consumption is at a maximum at tensions between saturation and an incipient limiting level. Mortality results from prolonged exposures to a low level of environmental oxygen, defined as the incipient lethal level. At dissolved oxygen tensions between the incipient limiting and lethal levels, the oxygen consumption rate is dependent upon the environmental oxygen tension.

Reported incipient limiting oxygen tensions range from 60 mm Hg for brown bullhead (*Ictalurus nebulosis*; Grigg, 1969) to 100 mm Hg for brook trout (Graham, 1949) and rainbow trout

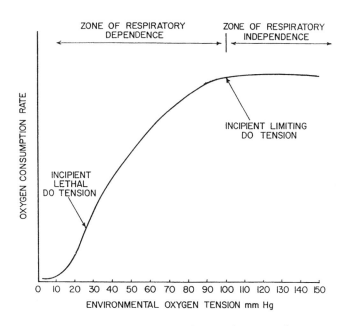

Figure 3.   Oxygen consumption by fish as related to environmental oxygen tension. After Fry
(1957). (From Soderberg, R. W., *Flowing Water Fish Culture*, CRC Press, Boca Raton,
FL, 1995. With permission.)

(Itazawa, 1970). Incipient limiting levels of oxygen generally average 73 mm Hg for warmwater
fish and 90 mm Hg for salmonids (Davis, 1975). An oxygen tension of 73 mm Hg at 1 atm of
pressure is equal to 4.1 mg/L at 20°C, 3.7 mg/L at 25°C and 3.5 mg/L at 30°C. Instructions for
converting dissolved oxygen tensions to concentrations are provided by Soderberg (1995).

Because exposure to dissolved oxygen tensions below the incipient limiting level results in
reduced oxygen consumption and therefore reduced delivery of oxygen to the tissues, this value
should correspond to the oxygen tension below which growth is reduced. This assertion is
generally supported by the literature on the effects of constant low levels of dissolved oxygen
on fish growth (Andrews et al., 1973; Larmoyeaux and Piper, 1973; Carlson et al., 1980; Downey
and Klontz, 1981), but fish in static water ponds are exposed to fluctuating dissolved oxygen
levels and therefore, the impact on growth may be more complex. The diurnal cycle of photosynthesis and respiration in ponds results in low oxygen concentrations occurring briefly in the
early morning hours before photosynthesis begins to return oxygen to the water (Figure 4). The
degree of hypoxia to which the fish are exposed is determined by the density of the plankton
bloom in the pond.

Studies on the effects of periodic hypoxia are severely lacking in the literature, but early
morning oxygen concentrations below 3 mg/L are generally considered to be undesirable
(Boyd, 1990). Tucker et al. (1979) reported that channel catfish reduced their food intake following
mornings when dissolved oxygen concentrations fell below 1–2 mg/L. Rappaport et al. (1976)
found that carp growth decreased when morning dissolved oxygen levels were regularly below
25% of saturation (2 mg/L at 26°C).

The tolerance of tilapia to hypoxic conditions (Fernandes and Rantin, 1994) have made them
popular culture species. Channel catfish ponds are aerated when dissolved oxygen levels drop to
24 mg/L (40–80 mm Hg at 26°C; Tucker and Robinson, 1990). Teichert-Coddington and Green
(1993) found that tilapia grew as well when aeration began at 10% of saturation (0.8 mg/L at 26°C)
as they did when aeration was started when pond dissolved oxygen levels reached 30% of saturation
(2.4 mg/L at 26°C).

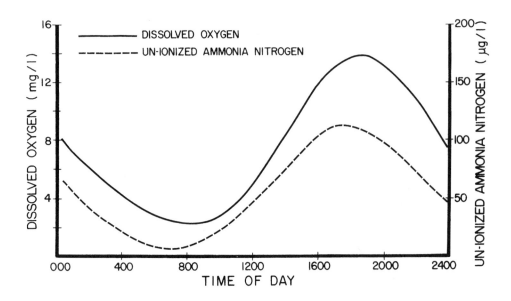

Figure 4. Typical diet fluctuations in dissolved oxygen (——) and un-ionized ammonia (– – – –) in intensive pond aquacultures. (From Soderberg, R. W., *Fish Biol.*, 24, 683, 1984b. With permission.)

Ammonia

Most fish excrete their nitrogenous waste in the form of ammonia which can accumulate in intensively managed ponds to levels as high as 1–2 mg/L (Boyd, 1982). Ammonia ionizes in water and the proportion of the un-ionized form (NH_3) in an ammonia solution increases with pH. It is well known that long term exposure to un-ionized ammonia reduces fish growth. Fractions of NH_3 in an ammonia solution at various pH and temperature values are shown in Table 1. A concentration of 0.016 mg/L NH_3-N is generally considered to be the maximum safe level for prolonged exposure by salmonids (Soderberg, 1995). Robinette (1976) found that exposure to 0.06 mg/L NH_3 did not affect the growth of channel catfish, but that growth was reduced at exposures of 0.12 mg/L. Colt and Tchobanoglous (1978) reported that channel catfish growth was reduced by 50% when exposed to 0.52 mg/L NH_3. Abdalla et al. (1992) gives 0.08 mg/L as the maximum level of NH_3 that *O. niloticus* could tolerate without showing reduced growth.

Table 1.   Fraction of NH_3 in a Total Ammonia Solution as a Function of pH and Temperature

	Temperature °C		
pH	**20**	**25**	**30**
6.5	0.00125	0.00179	0.00253
7.0	0.00394	0.00564	0.00796
7.5	0.01236	0.01762	0.02476
8.0	0.03807	0.05366	0.07432
8.5	0.11123	0.15205	0.20248
9.0	0.28355	0.36186	0.44532

Note: Calculated using the procedures summarized by Soderberg (1995).

In fish ponds, pH and thus NH_3, fluctuate with the diel cycle of CO_2 caused by photosynthesis and respiration by pond organisms. Fish in fertile ponds with high levels of ammonia may be exposed to toxic levels of NH_3 for short periods in the afternoon of each day when CO_2 concentrations are lowest (Figure 4). Fish in intensive static water aquacultures exhibit gill lesions characteristic of ammonia exposure that undoubtedly retard respiration and therefore adversely affect growth (Soderberg et al., 1984a). Thus ammonia is probably one of the exogenous factors that dampens fish growth in the later stages of some pond aquacultures. Quantification of the degree to which ammonia reduces growth is difficult to document due to the diel patterns of exposure.

When Soderberg (1985) intensively reared rainbow trout in static water ponds, total ammonia concentrations ranged from 0.3–1.3 mg/L. Average daily exposures to NH_3 were below 0.0125 mg/L, but the average daily maxima ranged from 0.0083 to 0.0487 mg/L. Fish growth and survival were not correlated to average daily exposure to NH_3, but were related ($P < 0.05$) to average daily maxima ($r = -0.62$ for growth and $r = -0.75$ for survival; Soderberg et al., 1983).

Soderberg et al. (1984b) monitored ammonia dynamics in ponds where channel catfish were intensively reared. The total ammonia concentrations averaged less than 1 mg/L and the average daily minimum, median and maximum NH_3 concentrations were 0.004, 0.042, and 0.110 mg/L, respectively. Fish growth was not correlated with NH_3 exposure even though afternoon levels in most ponds were commonly higher than those reported to reduce fish growth.

Other exogenous factors, such as salinity and competition, have also been shown to affect fish growth (Moyle and Cech, 1996), but food supplies, temperature, dissolved oxygen and ammonia are the most important abiotic factors affecting fish growth in tropical aquaculture.

Bioenergetic Budgets for Aquaculture

Bioenergetics is the study of the energy budget by which ingested energy is partitioned into energy lost as feces or excretory products and energy used for maintenance, activity and growth or the elaboration of new body tissues. Fish growth can be expressed in terms of energy partitioning as follows: $I = M + G + E$ where I = ingested food energy, M = energy expended for metabolism, G = energy expended for growth and E = energy excreted (Jobling, 1994). Growth may be further partitioned into somatic and gonadal components. Thus, bioenergetics is the physiological framework for the relationship between feeding and growth.

Complete combustion of dietary fat, carbohydrate and protein yields approximately 9.3, 4.1 and 5.7 C/g, respectively (Jobling, 1994). The energy available to fish is species specific and depends upon trophic status. Herbivorous species are able to extract more energy from carbohydrates than carnivorous species which depend more heavily upon protein for their energy requirements. Trout are able to utilize 8 C/g from fat, 1.6 C/g from carbohydrates and 3.9 C/g from protein; whereas channel catfish (*Ictalurus punctatus*) can use 8.5 C/g from fat, 2.9 C/g from carbohydrates and 4.5 C/g from protein (Piper et al., 1982). The difference between total combustible energy and available energy in these food components represents that portion which is either indigestible, and is passed as feces, or non-metabolizable nitrogenous compounds which are excreted from the gills or kidneys. From the above examples, it is clear that the digestibility (expressed as a percentage of total combustible energy) varies among species. The available energy is then partitioned among maintenance, activity, and growth.

Bioenergetics may be applied to aquaculture to evaluate the relative importance of factors which influence growth. It has been determined through bioenergetic budgeting that the young of carnivorous species deposit 20–35% of ingested energy as growth in laboratory conditions (Jobling, 1994).

Liu and Chang (1992) developed a bioenergetic model that accounts for the integrated effects of fertilization rate, stocking density and spawning on the growth of *O. niloticus*. The relationships between growth and anabolism, catabolism, food assimilation, food availability and food consumption were synthesized from existing expressions for fish growth models. Since mixed-sex populations of tilapia spawn in production ponds before reaching harvestable size, an expression was added to the model to account for resources used by pond-spawned juveniles. Required parameters and coefficients were fit to the model using data collected from CRSP experiments conducted in Thailand.

The resulting bioenergetic model closely fit the observed data and showed that fish growth is most sensitive to the anabolism and catabolism exponents. Food assimilation efficiency and food consumption had lesser effects on growth. The model indicated that *O. niloticus* ponds should be fertilized with 500 kg wet chicken manure/ha/week and stocked at 10,000/ha. Elimination of spawning substantially increases fish growth. Bioenergetic models for aquaculture can also incorporate economic components as shown by Springborn et al. (1992) who developed a model for determining the optimum time to harvest fish.

Jobling (1994) reported that the difficulty in quantifying energy expenditure of fish outside of laboratory conditions has resulted in most of the problems with energy partitioning in wild or pond-reared fish being unsolved. The work of Liu and Chang (1992) and others shows, however, that the principles of bioenergetics, theoretically applied to pond aquaculture, can form a framework for the quantitative study of pond management practices.

Bioenergetic modeling of pond fish populations is described in depth in Chapter 13 of this book.

FISH GROWTH IN TROPICAL AQUACULTURE

Reporting Fish Growth

Fish growth in nearly all tropical aquaculture studies is reported as absolute rates in g/day (Hopkins, 1992). That is, the difference between the mean harvest weight and the mean stocking weight is divided by the number of days in the culture period. This measurement is useful for comparing treatment means within an experiment, but has little utility for comparing growth among different studies. This is due to the non-linear relationship between length and weight. As fish increase in size, their incremental weight gain increases so that comparisons among studies are only useful if the initial weights are similar.

Relative growth is sometimes used to report fish growth rates in aquaculture studies. In this case, the absolute growth in weight is divided by the initial weight and expressed as a percentage. Using relative growth rate to compare fish growth among different studies would be difficult because the measure is only pertinent to the time interval over which it was made.

Growth may also be expressed exponentially as:

$$Wt = Wi \ e^G \tag{2}$$

where Wt is final weight and Wi is initial weight. In this case the growth rate is reported as G, the instantaneous rate of growth, or G is expressed as a percentage and reported as specific growth rate in percent per day. Hopkins (1992) discouraged the use of specific growth rate to compare fish growth in aquaculture because only the growth of smaller fish is exponential. The instantaneous growth rate, G, because it varies with fish size, is most useful when used to express annual growth among cohorts in a population such as is usually the case when studying natural fish populations.

In studies where intermediate data are collected, growth among treatments are usefully compared by weight vs. time plots. The data presented by Green and Teichert-Coddington (1994) for male and mixed sex *O. niloticus* grown at three stocking rates clearly show the treatment effects and that growth was linear over time (Figure 5). A graphic representation of exponential growth is shown by De Silva et al. (1991) for *O. mossambicus* X *O. niloticus* hybrids fed a pelleted diet in aquaria (Figure 6). Pouomogne and Mbongblang (1993) showed graphically that growth was asymptotic when *O. niloticus* were fed a 27% plant protein diet at three rates (Figure 7). Often treatment effects show different patterns of growth such as in the case of a study reported by Teichert-Coddington and Green (1993) in which *O. niloticus* were grown in aerated and non-aerated

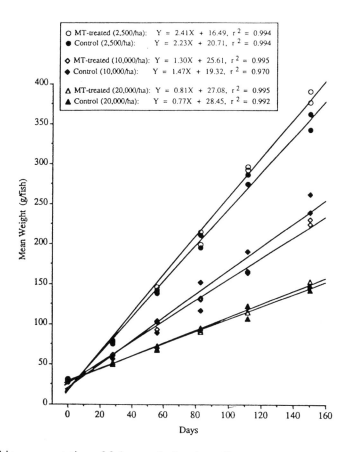

Figure 5. Graphic representation of fish growth showing a linear pattern of weight gain over time. Data are for normal and methyltestosterone-treated *Oreochromis niloticus* grown at stocking rates of 2,500, 10,000, and 20,000 per ha. (From Green, B. W. and Teichert-Coddington, D., *Aquaculture Fish Manage.*, 25, 613, 1994. With permission.)

ponds. Growth was linear in the aerated treatments, but was asymptotic in the non-aerated ponds showing that lower dissolved oxygen levels dampened growth (Figure 8).

Hopkins (1992) recommends that in most cases, aquaculture growth data should be fitted to a curve which includes intermediate as well as initial and harvest data. If these data are linear, absolute growth is appropriate. If the data are exponential, the instantaneous growth rate should be used and if growth data are asymptotic, they should be fitted to the von Bertalanffy growth function:

$$Lt = L_\infty \ (1 - e^{-k(t - t_0)}) \tag{3}$$

where Lt = length at time t, L_∞ = asymptotic length, k is a growth coefficient and t_0 is a scaling constant. Values for k and L_∞ can be computed from the Gulland and Holt plot (Gulland and Holt, 1959) or from the Walford plot (Walford, 1946). The value of t_0 is estimated from the growth equation after estimates of k and L_∞ have been obtained. Fish growth among different populations can be compared with the values of k and L_∞. Since k is a slope, growth curves can be compared by analysis of covariance on the k value. The growth performance index, ϕ', combines k and L_∞ as follows:

$$\phi' = \log k + 2 \log L_\infty. \tag{4}$$

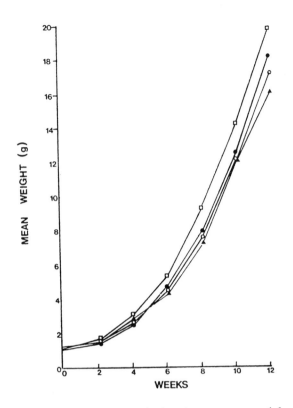

Figure 6. Graphic representation of fish growth showing an exponential pattern of weight gain over time. Data are from *Oreochromis mossambicus* X *O. niloticus* fed a 20% protein pelleted diet in aquaria. Experimental treatments were 6% (○), 12% (●), 18% (□), and 24% (■) dietary lipid. (Redrawn from De Silva, S. S., Gunasekera, R. M., and Shim, K. F., *Aquaculture*, 95, 305, 1991).

Mathews and Samuel (1990) reported that ϕ' is a flexible and precise estimator of growth performance in fishes. Pauly (1987) suggested that ϕ' may be useful for evaluating growth potentials under aquaculture conditions. Moreau et al. (1986) and Pauly et al. (1988) have used ϕ' to evaluate tilapia stocks for their aquaculture potential. Hopkins (1991) questioned the efficacy of using ϕ' to compare treatments in pond experiments noting that very different growth curves could have identical values of ϕ'.

Low-Intensity Aquacultures

Data on fish growth from the first three cycles of the CRSP Global Experiment were analyzed by Chang (1989). Mean daily growth ranged from 0.293 in Rwanda to 1.101 g in Thailand (Table 2) and was strongly correlated with solar radiation and air temperature. Generally fish growth was slower in rainy seasons than in dry seasons and the poor growth in Rwanda is largely explained by colder temperatures at that site.

Fish growth was also related to fertilization rate with an average growth rate ranging from 0.469 g/d in ponds receiving 125 kg/ha week of organic matter to 1.293 g/d in ponds fertilized at 1000 kg/ha week (Table 3).

Tilapia growth experiments in the southern United States are comparable to tropical locations because they are conducted during the summer months. Suffern et al. (1978) reported a growth rate of 1.07 g/d for the all-male hybrid, *O. mossambicus* (F) X *O. hornorum* (M), stocked in cages

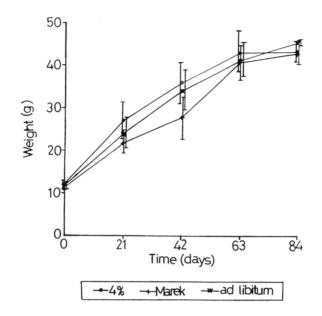

Figure 7. Graphic representation of fish growth showing an asymptotic pattern of weight gain over time. Data are from *Oreochromis niloticus* fed a 27% plant protein diet at three rates. 4% = 4% of fish biomass per day, Marek = 5.3% of biomass for fish < 10 g and 4.6% for fish > 10 g, ad libitum = unlimited food from a demand feeder. (From Pouamogne, V. and Mbongblang, J., Bamidgeh, 45, 147, 1993. With permission.)

suspended in sewage lagoons in Tennessee. Stickney and Hesby (1978) recorded growth rates of 1.10–1.29 g/d for male *O. aureus* stocked as fry in ponds fertilized with swine manure in Texas. The all-male hybrid, *O. niloticus* (F) X *O. hornorum* (M), grew 1.7 g/d when stocked at 29 g in Alabama in ponds fertilized with cattle manure (Collie and Smitherman, 1978). Green and Teichert-Coddington (1994) demonstrated the effect of stocking rate on the growth of male *O. niloticus* in Honduras in ponds fertilized with chicken litter. Growth rates were 2.26 g/d, 1.40 g/d and 0.80 g/d in ponds stocked at 2500/ha, 10,000/ha, and 20,000/ha, respectively.

In summary, tilapia growth rates in fertilized ponds in the tropics, stocked at 10,000 fish/ha ranged from 0.293–1.7 g/d. Growth rates as high as 2.26 g/d are possible with reduced stocking rates. Temperature, solar radiation, fertilization rate, and fish stocking size influenced fish growth in the studies cited here.

Intensive Aquacultures

Tal and Ziv (1978) recorded a growth rate of 2.1 g/d for *O. aureus* males stocked at 80,000/ha and fed a 25% protein pelleted diet in Israel. Aeration was required at this intensity level. Male *O. aureus* and male *O. niloticus*, fed a 36% protein diet in Alabama grew at rates of 2.74 g/d and 2.33 g/d, respectively (Anderson and Smitherman, 1978). In another study in Alabama, the all-male hybrid, *O. niloticus* (F) X *O. hornorum* (M), grew 2.8 g/d on a 36% protein diet (Collie and Smitherman, 1978). Fram and Pagan-Font (1978), working in Puerto Rico, recorded growth rates of 2.18 g/d for *O. niloticus* X *O. hornorum* stocked at an initial size of 15.9 g and fed a 30% protein diet. Fish stocked at an initial weight of 53.9 g grew 1.59 g/d in the same study. When mixed-sex *O. niloticus* were stocked at 8.3 g and fed a 30% protein pelleted ration, the growth rate was 0.8 g/d and 62% of the final crop was pond-spawned juveniles (Fram and Pagan-Font, 1978).

In summary, growth rates in the tropics can be substantially increased by the use of complete pelleted rations. The advantages of intensive methods can only be realized in mono-sex cultures.

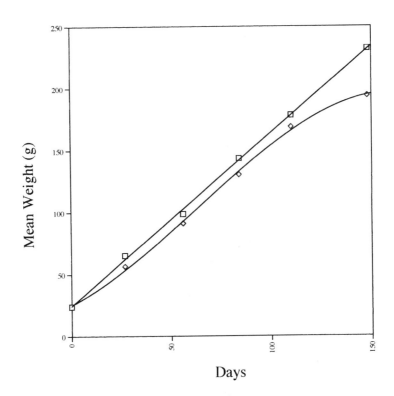

Figure 8.   Graphic representation of the growth of *Oreochromis niloticus*. The pattern of weight
            gain over time appeared linear in aerated treatments, but asymptotic in non-aerated
            ponds, showing that lower dissolved oxygen levels dampened growth. □ = Aerated;
            ◇ = non-aerated. (Redrawn from Teichert-Coddington, D., and Green, B. W., Aqua-
            culture, 118, 63, 1993.)

Table 2   Average Final Weight and
Daily Gain Of *Oreochromis niloticus*
Grown at Different Sites in the First
Three Cycles of the CRSP Global
Experiment

CRSP site	Final fish weight (g)	Daily growth (g)
Panama	62.378	0.596
Thailand	109.050	1.101
Indonesia	80.831	0.482
Honduras	88.876	0.854
Philippines	91.888	0.865
Rwanda	61.048	0.293

Note: Data from Chang, W. B., 1989.

Table 3 Relationship Between Rate of Fertilization with Organic Matter and Growth of *Oreochromis niloticus* for the First Three Cycles of the CRSP Global Experiment

Fertilization rate (kg/ha/week)	Daily growth (g)
125	0.469
250	0 705
500	0.817
1000	1.293

(Data From Chang, W. B., 1989).

REFERENCES

Abdalla, A. A. F., McNabb, C., Knud-Hansen, C., and Batterson, T., Growth of *Oreochromis niloticus* in the presence of un-ionized ammonia, in Egna, H. S., McNamara, M., and Weidner, N., Eds., *Ninth Annual Administrative Report,* Pond Dynamics/Aquaculture CRSP, 1991, Oregon State University, Corvallis, Oregon, 1992.

Anderson, C. E. and Smitherman, R. O., Production of normal male and androgen sex-reversed *Tilapia aurea* and *T. nilotica* fed a commercial catfish diet in ponds, in Smitherman, R. O., Shelton, W. L., Grover, J. H., Eds., *Culture of Exotic Fishes Symposium Proceedings.* Fish Culture Section, American Fisheries Society, Auburn, Alabama, 1978, 34–42.

Andrews, J. W., Murai, T., and Gibbons, G., The influence of dissolved oxygen on the growth of channel catfish, *Trans. Am. Fish. Soc.*, 102, 835–838, 1973.

Balarin, J. D., *Tilapia: A Guide to their Biology and Culture in Africa,* Unit of Aquatic Pathobiology, University of Stirling, Stirling, Scotland, 1979, 174.

Bertalanffy, L. von, A quantitative theory of organic growth, *Hum. Biol.*, 10, 181–213, 1938.

Boyd, C. E., *Water Quality in Ponds for Aquaculture,* Alabama Agricultural Experiment Station, Auburn University, Auburn, Alabama, 1990, 482.

Boyd, C. E., *Water Quality Management for Pond Fish Culture,* Elsevier Scientific Publishers, The Netherlands, 1982, 318.

Brett, J. R., Environmental factors and growth, in Hoar, W. S., Randall, D. J., and Brett, J. R., Eds., *Fish Physiology*, Volume VIII, *Bioenergetics and Growth*, Academic Press, Orlando, Florida, 1979, 599–677.

Carlson, A. R., Blocher, J., and Herman, L. J., Growth and survival of channel catfish and yellow perch exposed to lowered constant and diurnally fluctuating dissolved oxygen concentrations, *Progr. Fish Cult.*, 42, 73–78, 1980.

Chang, W. B., Fish production: data synthesis and model development, in Egna, H. S., and Horton, H. F., Eds., *Sixth Annual Administrative Report,* Pond Dynamics/Aquaculture CRSP, 1988, Oregon State University, Corvallis, Oregon, 41–49, 1989.

Collis, W. J. and Smitherman, R. O., Production of tilapia hybrids with cattle manure or a commercial diet, in Smitherman, R. O., Shelton, W. L., Grover, J. H., Eds., *Culture of Exotic Fishes Symposium Proceedings.* Fish Culture Section, American Fisheries Society, Auburn, Alabama, 43–54, 1978.

Colt, J. and Tchobanoglous, G., Chronic exposure of channel catfish, *Ictalurus punctatus*, to ammonia: effects on growth and survival, *Aquaculture*, 15, 353–372, 1978.

Davis, J. C., Minimal dissolved oxygen requirements of aquatic life with special emphasis on Canadian species: a review, *J. Fish. Res. Bd. Canada*, 32, 2295–2332, 1975.

De Silva, S. S., Gunasekera, R. M., and Shim, K. F., Interactions of varying dietary protein and lipid levels in young red tilapia: evidence of protein sparing, *Aquaculture*, 95, 305–318, 1991.

Downey, P. C. and Klontz, G. W., *Aquaculture techniques: Oxygen (PO₂) requirements for trout quality,* Idaho Water and Energy Resources Research Institute. University of Idaho, Moscow, Idaho, 1981, 42.

Fernandes, M. N. and Rantin, F. T., Relationships between oxygen availability and metabolic cost of breathing in Nile *tilapia* (*Oreochromis niloticus*): aquacultural consequences, *Aquaculture*, 127, 339–346, 1994.

Fram, M. J. and Pagan-Font, F. A., Monoculture yield trials of an all-male tilapia hybrid (F *Tilapia nilotica* x M *T. hornorum*) in small farm ponds in Puerto Rico, in Smitherman, R. O., Shelton, W. L., Grover, J. H., Eds., *Culture of Exotic Fishes Symposium Proceedings.* Fish Culture Section, American Fisheries Society, Auburn, Alabama, 55–64, 1978.

Fry, F. E. J., The aquatic respiration of fish, in Brown, M. E., Ed., *The Physiology of Fishes*, Volume I, Academic Press, Inc., New York, New York, 1957, 1–63.

Graham, J. M., Some effects of temperature and oxygen pressure on the metabolism and activity of speckled trout, *Salvelinus fontinalis, Can. J. Research*, 27, 270–288, 1949.

Green, B. W. and Teichert-Coddington, D. R., Growth of control and androgen-treated Nile tilapia, *Oreochromis niloticus* (L.), during treatment, nursery and grow-out phases in tropical fish ponds, *Aquaculture Fish Manage.*, 25, 613–621, 1994.

Grigg, G. C., The failure of oxygen transport in a fish at low levels of ambient oxygen. *Comp. Biochem. Physiol.*, 29, 1253–1257, 1969.

Gulland, J. A. and Holt, S. J., Estimation of growth parameters for data at unequal time intervals, *J. Cons. Cons. Int. Explor. Mer*, 25, 47–49, 1959.

Haskell, D. C., Trout growth in hatcheries. *New York Fish Game J.*, 6, 205–237, 1959.

Hopkins, K. D., ϕ′ not suited to compare growth in pond experiments, *Aquabyte*, 4, 6, 1991.

Hopkins, K. D., Reporting fish growth: a review of the basics. *J. World Mariculture Soc.*, 23, 173–179, 1992.

Itazawa, Y., Characteristics of respiration of fish considered from the arterio-venous difference of oxygen content, *Bull. Japanese Soc. Scien. Fish.*, 36, 571–577, 1970.

Jobling, M., *Fish Bioenergetics*, Chapman and Hall, Fish and Fisheries Series, London, 1994, 309.

Larmoyeaux, J. D. and Piper, R. G., The effects of water reuse on rainbow trout in hatcheries, *Progr. Fish Cult.*, 35, 2–8, 1973.

Legendre, M. and Albaret, J. J., Maximum observed length as an indicator of growth rate in tropical fishes, *Aquaculture*, 94, 327–341, 1991.

Liu, K. M. and Chang, W. Y. B., Bioenergetic modeling of effects of fertilization, stocking density, and spawning on growth of Nile tilapia, *Oreochromis niloticus* (L). *Aquaculture and Fish. Manage.*, 23, 291–301, 1992.

Mathews, C. P. and Samuel, M., Using the growth performance index ϕ′ to choose species for aquaculture: an example from Kuwait. *Aquabyte*, 3, 2–4, 1990.

Moreau, J., Bambino, C., and Pauly, D., A comparison of four indices of overall growth performance, based on 100 tilapia populations (Fam. Cichlidae), in Maclean, J. L., Dizon, L. B., and Hostilos, L. V., Eds., *The First Asian Fisheries Forum*, Asian Fisheries Society, Manila, Philippines, 1986, 201–206.

Moyle, P. B. and Cech, J. J., *Fishes: An Introduction to Ichthyology*, 3rd Edition, Prentice Hall, Upper Saddle River, New Jersey, 1996, 590.

Pauly, D., Application of information on age and growth to fishery management, in Summerfelt, R. C., and Hall, G. D., Eds., *The Age and Growth of Fish.* The Iowa University Press, Ames, Iowa, 1987, 495–506.

Pauly, D., Moreau, J., and Prein, M., Comparison of growth performance of tilapia in open waters and aquaculture, in Pullin, R. S. V., Bhukasawan, T., Tonguthai, K., and Maclean, J. L., Eds., *The Second International Symposium on Tilapia in Aquaculture*, ICLARM Conference Proceedings 15, Department of Fisheries, Bangkok, Thailand, and International Center for Living Aquatic Resources Management, Manila, Philippines, 1988, 469–479.

Platt, S. and Houser, W. J., Optimum temperature for feeding and growth of *Tilapia zillii, Progr. Fish Cult.*, 40, 105–107, 1978.

Piper, R. G., McElwain, J. B., Orme, L. E., McCraren, J. P., Fowler, L. G., and Leonard, J. R., *Fish hatchery management*, U. S. Fish and Wildlife Service, Washington, D. C., 1982, 517.

Pouomogne, V. and Mbongblang, J., Effect of feeding rate on the growth of tilapia (*Oreochromis niloticus*) in earthen ponds, *Bamidgeh*, 45, 147–153, 1993.

Rappaport, U., Sarig, S., and Marek, M., Results of tests of various aeration systems on the oxygen regime in the Genosaur experimental ponds and growth of fish there in 1975, *Bamidgeh*, 28, 35–49, 1976.

Robinette, H. R., Effects of selected sublethal levels of ammonia on the growth of channel catfish, *Ictalurus punctatus, Progr. Fish Cult.*, 38, 26–29, 1976.

Soderberg, R. W., Histopathology of rainbow trout, *Salmo gairdneri* (Richardson), exposed to diurnally fluctuating un-ionized ammonia levels in static-water ponds, *J. Fish Diseases*, 8, 57–64, 1985.

Soderberg, R. W., Temperature effects on the growth of blue tilapia in intensive aquaculture, *Progr. Fish Cult.*, 52, 155–157, 1990.

Soderberg, R. W., Linear fish growth models for intensive aquaculture, *Progr. Fish Cult.,* 54, 255–258, 1992.

Soderberg, R. W., *Flowing Water Fish Culture*, CRC Press, Inc., Boca Raton, Florida, 1995, 147.

Soderberg, R. W., Flynn, J. B., and Schmittou, H. R., Effects of ammonia on the growth and survival of rainbow trout in intensive static-water culture, *Trans. Am. Fish. Soc.*, 112, 448–451, 1983.

Soderberg, R. W., McGee, M. V., Grizzle, J. M., and Boyd, C. E., Comparative histology of rainbow trout and channel catfish in intensive static-water aquaculture, *Progr. Fish Cult.*, 46, 195–199, 1984a.

Soderberg, R. W., McGee, M. V., and C. E. Boyd, C. E., Histology of cultured channel catfish, *Ictalurus punctatus* (Rafinesque), *J. Fish Biol.*, 24, 683–690, 1984b.

Springborn, R. R., Jensen, A. L., Chang, W. Y. B., and Engle, C., Optimum harvest time in aquaculture: an application of economic principles to a Nile tilapia, *Oreochromis niloticus* (L.), growth model, *Aquaculture and Fish. Manage.*, 23, 639–647, 1992.

Stickney, R. R. and Hesby, J. H., Tilapia production in ponds receiving swine wastes, in Smitherman, R. O., Shelton, W. L., Grover, J. H., Eds., *Culture of Exotic Fishes Symposium Proceedings*. Fish Culture Section, American Fisheries Society, Auburn, Alabama, 1978, 90–101.

Suffern, J. S., Adams, S. M., Blaylock, B. C., Coutant, C. C., and Guthrie, C. A., Growth of mono-sex hybrid tilapia in the laboratory and sewage oxidation ponds, in Smitherman, R. O., Shelton, W. L., Grover, J. H., Eds., *Culture of Exotic Fishes Symposium Proceedings*. Fish Culture Section, American Fisheries Society, Auburn, Alabama, 1978, 65–81.

Tal, S. and I. Ziv, I., Culture of exotic fishes in Israel, in Smitherman, R. O., Shelton, W. L., Grover, J. H., Eds., *Culture of Exotic Fishes Symposium Proceedings*. Fish Culture Section, American Fisheries Society, Auburn, Alabama, 1978, 1–9.

Teichert-Coddington, D. and Green, B. W., Tilapia yield improvement through maintenance of minimal oxygen concentrations in experimental grow-out ponds in Honduras, *Aquaculture*, 118, 63–71, 1993.

Tucker, C. S. and Robinson, E. H., *Channel Catfish Farming Handbook*, Van Nostrand Reinhold, New York, New York, 1990, 454.

Tucker, L., Boyd, C. E., and McCoy, E. W., Effects of feeding rate on water quality, production of channel catfish and economic returns, *Trans. Am. Fish. Soc.*, 108, 389–396, 1979.

Walford, L. A., A new graphic method of describing the growth of animals, *Biol. Bull.*, 90(2), 141–147, 1946.

10   FRY AND FINGERLING PRODUCTION

Bartholomew W. Green, Karen L. Veverica, and Martin S. Fitzpatrick

INTRODUCTION

Tilapia production from aquaculture worldwide has grown rapidly, increasing from 241,681 metric tons (MT) in 1986 to 472,969 MT in 1992 (FAO, 1994). This growth has been facilitated in part through improvements in tilapia fry and fingerling production technology (e.g., mass production of monosex fingerlings through sex inversion) and dissemination of technology to fish farmers. Continued rapid expansion of tilapia aquaculture will require additional improvements in tilapia reproduction in order to ensure consistent, unlimited availability of high-quality fingerlings.

Mass production of tilapia fingerlings requires the successful implementation of a number of activities. The first step is the procurement of adequate numbers of good quality broodfish of known lineage and of the proper age and size. Adequate numbers of broodfish may be defined as the number of broodfish necessary to produce the number of eggs needed to give the desired number of advanced fingerlings for stocking into grow-out units. Good management practices are required to maintain the broodfish population in good physical and genetic condition. These management practices usually involve maintenance of broodfish in fertile earthen ponds and provision of a formulated supplemental feed. Original and replacement broodfish should be offspring from a large, randomly breeding population in order to avoid problems associated with greater potential for expression of deleterious recessive alleles as a result of inbreeding.

Spawning can be natural or artificial; artificial spawning involves the manual collection of eggs and milt from broodfish. As tilapia are sequential spawners, natural spawning predominates. Artificial spawning generally is limited to specific research projects. Spawning cycle duration will vary from days to months, depending on the management practice employed and on the size of fingerling produced. Segregation of broodfish by sex and intensive feeding of broodfish between spawning cycles (conditioning) may or may not be practiced. Broodfish are spawned in a variety of containers, and the eggs and sac-fry may or may not be removed for artificial incubation. A hatchery facility is necessary if eggs and sac-fry are removed from brooding females. Produced fry are mixed sex, or they can be treated with an androgen to produce monosex (all-male) populations. Finally, swim-up fry are nursed to advanced fingerling or stocker size for stocking into grow-out units.

Tilapia are classified as substrate incubators or mouth-brooders, the latter being either maternal or paternal/biparental (Trewavas, 1982; Chapter 8 in this volume). Both groups of tilapia provide their eggs and young with a high degree of parental care, thereby increasing offspring survival (Jalabert and Zohar, 1982; Philippart and Ruwet, 1982; Macintosh and Little, 1995). The discussion in this chapter primarily concentrates on the commercially important species, which include Nile tilapia (*Oreochromis niloticus*), blue tilapia (*O. aureus*), Mozambique tilapia (*O. mossambicus*), and red tilapia (*Oreochromis* sp.). *Oreochromis niloticus* is cultured most commonly worldwide, followed by *O. mossambicus* and *O. aureus* (FAO, 1994). The Egyptian and Ivory Coast strains of *O. niloticus* are the predominant aquacultural strains worldwide.

0-56670-274-7/97/$0.00+$.50
© 1997 by CRC Press LLC

TILAPIA BROODSTOCK

Tilapia broodfish should be procured from a reputable supplier who can document the purity and origin of the fish. Electrophoretic analysis can be used to verify species identification. Importation of tilapia should be done in accordance with the importing country's regulations pertaining to the introduction of exotic fish. In addition, fish should be certified disease-free prior to transfer to the farm, and once on the farm should undergo a quarantine period to ensure no diseases or parasites are introduced.

While tilapia broodfish can still be captured in their natural ranges in Africa (Philippart and Ruwet, 1982; Trewavas, 1982; Pullin, 1988; Pullin and Capili, 1988), most broodfish are obtained from hatcheries or fish farms. In collecting a founding stock from the wild, Tave (1993) recommends collecting a representative sample of the population from a wide geographic range and adjusting total numbers collected for future fish mortality and lack of spawning. Original and replacement broodfish should be offspring of a large, randomly breeding population when fish are obtained from a hatchery or produced on farm (Tave, 1993). Use of an inadequate effective breeding number results in inbreeding, which reduces viability, survival, and growth, and increases the number of abnormalities observed (Tave, 1993). Recommended minimum effective breeding numbers for fish farm populations range from 100 to 150 fish (Smitherman and Tave, 1987). Tave (1993) presents a thorough discussion of fish genetics and breeding programs, topics that exceed the scope of the present chapter.

Proper nutrition is an important aspect of broodfish management. While there are conflicting reports on the optimal dietary protein level required by tilapia broodfish (Santiago et al., 1985; Cissé, 1988; Wee and Tuan, 1988; De Silva and Radampola, 1990), supplemental dietary protein does increase tilapia seed production compared to a diet of natural foods only (Santiago et al., 1985). Brooding females, unable to eat, lose weight during the incubation period, and weight loss is correlated to duration of the incubation period (Little et al., 1993; Macintosh and Little, 1995). During the spawning period the daily feed ration is reduced to approximately 1.5% of the total fish biomass as brooding females are not consuming feed (Macintosh and Little, 1995). Following spawning, broodfish can be segregated by sex and stocked into ponds, tanks, or hapas for a conditioning period. Water in conditioning units should contain abundant natural food because of its assumed nutritional benefits (Macintosh and Little, 1995). Broodfish also should be fed a formulated ration ($\approx$25% protein) to satiation daily; the daily feed allowance can be given as two to three meals.

FACTORS AFFECTING TILAPIA FRY PRODUCTION

There is considerable plasticity in the age and size at which tilapia attain sexual maturity both in their native environments and under culture conditions (Rana, 1988). A number of genetic and environmental factors (e.g., condition factor, food availability, climatic variables, and intraspecific social interactions) are hypothesized to influence onset of sexual maturity in tilapia (Jalabert and Zohar, 1982; Rana, 1988; Brummett, 1995). In aquaculture situations, tilapia can attain sexual maturity and spawn at ages of three to six months and at sizes $\leq$20 grams (Jalabert and Zohar, 1982; De Silva and Radampola, 1990; Macintosh and Little, 1995). Tilapia reproductive biology and physiology have been the subjects of several thorough reviews (Jalabert and Zohar, 1982; Philippart and Ruwet, 1982; Rana, 1988; Brummett, 1995).

Age and Size

The size and number of eggs produced by tilapia is affected both by female age and size (Jalabert and Zohar, 1982; Peters, 1983; Rana, 1988). In general, larger females produce larger eggs (Peters, 1983; Rana, 1988), but female age, rather than weight, is thought to be more important in determining egg size (Rana, 1988). No correlation is observed between egg size and female weight within individual year classes for *O. niloticus* and *O. mossambicus*, although egg size is positively correlated with year class for both species (Rana and Macintosh, 1988). Hatchability of larger *O. niloticus* eggs is greater than smaller eggs, possibly because older females are more experienced egg brooders (Siraj et al., 1983; Mair et al., 1993). Larger eggs have been correlated with larger fry at hatch and with greater fry survival (Rana, 1988). In fact, the maximum fry length attained is positively correlated with mean egg size (Rana, 1985). *Oreochromis niloticus* and *O. mossambicus* fry that hatch from large eggs have larger yolk reserves, are able to grow to a larger size at the onset of feeding and survive starvation for longer periods than fry that hatch from smaller eggs (Rana, 1985; Rana and Macintosh, 1988). Growth of fry from larger eggs is faster during the first 60 d post-hatch under hatchery conditions (Rana and Macintosh, 1988). While egg size and weight and fry length increase significantly with older *O. niloticus* females, this advantage is no longer detectable at 20 d post-hatch in a green water system (Siraj et al., 1983).

Larger female tilapia not only produce larger eggs but also produce greater numbers of eggs than smaller females (Jalabert and Zohar, 1982; Galman et al., 1988; Rana, 1988). However, the relative fecundity (number of eggs/kg female) decreases with increased size (Siraj et al., 1983; Rana, 1988). The term "fecundity" is defined in many different ways in the aquacultural literature. Rana (1988) discusses some of the definitions (the total number of maturing oocytes in the ovaries prior to spawning, the total number of fry produced over a 12-month period) and proposes that fecundity in the aquacultural context be defined as the "number of eggs in a freshly spawned clutch." A more common usage of the term "fecundity" in tilapia fingerling production may be the number of seed per kilogram of female, with or without a time factor incorporated; seed, which includes eggs, sac-fry, and swim-up fry, must be defined clearly with first usage.

In tilapia, relative fecundity (seed/kg female) decreases with increased female age (Rana, 1986; Rana, 1988; Ridha and Cruz, 1989). Year class I *O. niloticus* and *O. spilurus* females have significantly greater relative fecundity than females from more advanced year classes (Hughes and Behrends, 1983; Siraj et al., 1983; Rana, 1988; Ridha and Cruz, 1989). But, relative fecundity of *O. niloticus* females within an individual year class can increase with successive spawns (Hughes and Behrends, 1983; Siraj et al., 1983) and with increasing female size (Rana, 1986). Although relative fecundity (seed/kg female) is greater in younger, smaller tilapia females, absolute productivity per spawn is greater in older, larger females (Hughes and Behrends, 1983; Rana, 1988), and older, larger *O. niloticus* females are more likely to spawn during each spawning cycle (Siraj et al., 1983).

Breeding Intensity

Frequency of spawning of tilapia in their natural environment depends on environmental variables, e. g., temperature, rainfall, light intensity, and duration, as well as on the fish's nutritional status (Jalabert and Zohar, 1982; Philippart and Ruwet, 1982; Rana, 1988; Macintosh and Little, 1995). While female tilapia are capable of spawning every 4 to 6 weeks under ideal conditions in their natural environment, synchrony of spawning among females is poor, and a continuous but low level of fry production results (Jalabert and Zohar, 1982; Little et al., 1993).

Efforts to increase breeding intensity of tilapia in aquacultural settings have concentrated on broodfish nutrition, reduction of interspawning interval, and synchronization of spawning. Results from broodfish nutrition work are inconclusive. Attempts to increase frequency of tilapia spawning through manipulation of dietary protein levels over the range of 20 to 50% protein generally do not produce significant improvements (Santiago et al., 1985; Cissé, 1988; Wee and Tuan, 1988; De Silva and Radampola, 1990). *Oreochromis niloticus* females produce significantly more seed when fed a 44% protein diet compared to when fed only natural foods (Santiago et al., 1985). Chang et al. (1988) reported a doubling of fry production when Taiwanese red tilapia brooders are fed a diet containing 44% protein compared to diets containing 22 to 24% protein. Diets that use soybean or corn oil as a dietary lipid source result in high tissue n-6 to n-3 fatty acid ratios, which generally enhances *O. niloticus* reproduction (Santiago and Reyes, 1993). It is clear that much research remains to be done on tilapia broodfish nutrition, particularly with respect to reproductive performance.

Interspawning interval is shortened and spawning frequency is increased by removal of eggs from mouth-brooding females (Dadzie, 1970a; Lee, 1979; Snow et al., 1983; Verdegem and McGinty, 1987; Mair et al., 1993). Removal of eggs and sac-fry at 5- to 10-d intervals significantly increases seed production, compared to natural egg incubation (Little et al., 1993) and to egg and sac-fry removal at 2- to 4-d intervals (Verdegem and McGinty, 1987). Interspawning interval also is shortened through broodfish exchange. For example, each time eggs are collected from mouth-brooding females, either spent females or all females are exchanged for broodfish that have been conditioned for 10 d (Little et al., 1993). In addition to reducing the interspawning interval, broodfish exchange increases spawning synchronicity, possibly through disruption of established social hierarchies (Little et al., 1993). While conditioning periods of 10 to 21 d have been reported for *O. niloticus* (Lovshin and Ibrahim, 1988; Little, 1989), Little (1989) reports lower fecundity (seed/kg female per day) for a 20-d compared to a 10-d conditioning period. Daily *O. niloticus* seed production increases from 31 seed/kg female when females hatch the eggs to 106 seed/kg female with removal of eggs from mouth-brooding females, to 160 seed/kg female when seed are removed and spawned females are exchanged, to 274 seed/kg female when seed are removed and all females are exchanged (Little et al., 1993).

Hypophysation has been attempted to improve spawning synchrony. Hypophysation involves injecting the hypophysis (i.e., the pituitary) as either a whole gland or extract preparation as a source of gonadotropin to stimulate spawning. Attempts to induce spawning using carp pituitary alone or luteinizing hormone alone did not improve spawning, but human chorionic gonadotropin (HCG) alone or in combination with carp pituitary were effective and consistent for inducing spawning in *O. aureus* (Dadzie, 1970b). However, hypophysation of *O. niloticus* with Chinese carp pituitary or HCG does not increase spawning frequency or synchronize spawning (Srisakultiew and Wee, 1988). Attempts to synchronize tilapia spawning or to increase spawning frequency through hypophysation only are justified for specialized research circumstances.

Water Temperature

Average water temperatures in excess of 20°C are an important stimulus for successful tilapia spawning (Philippart and Ruwet, 1982). Tilapia spawning can occur year-round in the tropics, as long as environmental cues (temperature and light) are adequate (Jalabert and Zohar, 1982; Philippart and Ruwet, 1982; Rana, 1988; Brummett, 1995). However, water temperature does decrease during the cooler season in the tropics, resulting in reductions or delays in tilapia spawning (Guerrero, 1986; Galman et al., 1988; Green and Teichert-Coddington, 1993; Maluwa and Costa-Pierce, 1993). The commercially important tilapia species spawn over a wide temperature range (20 to 35°C), although optimal spawning temperatures are 25 to 30°C for *O. niloticus* and *O. aureus*, and 20 to 35°C for *O. mossambicus* (Rothbard and Pruginin, 1975; Philippart and Ruwet, 1982; Rana, 1988).

Elevation affects water temperature significantly in the tropics. While earlier reports have stated that tilapia culture is not possible at elevations greater than 1500 m, this has been proven otherwise in Rwanda (Moehl et al., 1988). Pond culture of tilapia is possible at these elevations, although the reproductive capability of broodfish is diminished or inhibited. Persistent low water temperatures encountered at high elevations may prolong the time required for tilapia reproduction. Many people responsible for tilapia fingerling production in Rwanda, Burundi, and Kivu province of Zaire have noted repeatedly that 1- to 2-g fingerlings do not appear until 3 to 4 months after stocking ponds with broodfish (Veverica et al., in press). This is attributed in part to low growth rates following hatching. At elevations of approximately 1700 m in Rwanda, Nile tilapia are 6 to 9 months old when they are first able to spawn, and significant spawning activity is not observed for another two to three months (Hanson et al., 1988). At higher elevations (>2000 m), onset of reproduction is delayed until fish are 10 to 11 months old (Hanson et al., 1988).

There is circumstantial evidence that hatching is more sensitive to low water temperatures than is spawning. Rana (1988) notes that the temperature tolerance of tilapia eggs decreases as their developmental stage progresses. No recruitment is observed in ponds stocked with mixed-sex tilapia at high elevations (>2000 m) in Rwanda (Hanson et al., 1988; Rurangwa et al., 1992). During sampling, females are captured with eggs, but not sac-fry, in their mouths and only occasionally are a few fry captured from ponds. Hatching is assumed not to have occurred, probably due to the low water temperatures. Other incidences of "non-hatch" of tilapia eggs have occurred in Rwanda at 1625 m elevation, when water temperatures dip below 18°C for several hours in egg incubation jars.

Water temperature also seems to affect synchronization of spawning, with higher temperatures allowing for more synchronous spawns. Some authors (Hughes and Behrends, 1983; Srisakultiew and Wee, 1988) note that short exposure to cool temperatures ranging from 18 to 24°C followed by a return to optimum temperature helps induce spawning synchrony.

Salinity

Many species of tilapia exhibit a moderate degree of salinity tolerance where normal growth is not affected by salinity in the range of 15 to 25 g/L (Wohlfarth and Hulata, 1981; Philippart and Ruwet, 1982; Balarin, 1988). *Oreochromis niloticus* and *O. aureus* are reported to be capable of maintaining populations in habitats where salinity ranges from 11 to 29 g/L, and *O. mossambicus* can tolerate salinities as high as 120 g/L (Philippart and Ruwet, 1982). Rearing of tilapia in saline water affects their reproductive capability. *Oreochromis niloticus* spawn in brackish water ponds at salinities of 17 to 29 g/L, but not at salinities of 30 g/L and above (Fineman Kalio, 1988). In a laboratory study, *O. niloticus* have been spawned at 32 g/L (Watanabe and Kuo, 1985). *Oreochromis aureus* spawn in brackish water (17 g/L) ponds (Chervinski, 1966). However, the exact salinity at which *O. niloticus* or *O. aureus* cease to spawn is not known (Stickney, 1986). *Oreochromis mossambicus* are able to spawn at salinities greater than 30 g/L (Philippart and Ruwet, 1982) and even as high as 49 g/L (Payne, 1983). In fact, breeding populations of *O. mossambicus* have been reported in estuaries in Puerto Rico and Papua, New Guinea, in a brackish water pond in Hawaii, and at an atoll in the Pacific Ocean (Stickney, 1986; Nelson and Eldredge, 1991). Florida red tilapia (*O. urolepis hornorum* (F) × *O. mossambicus* (M)) spawn in full-strength seawater (Watanabe et al., 1989). In general, size of spawn, percent hatch, and fry survival of tilapia are reduced significantly under conditions of high salinity (Watanabe et al., 1989; Suresh and Lin, 1992).

Tilapia fingerling production is best accomplished in freshwater, although tilapia can spawn in brackish water. Spawning (eggs/g female body weight) of Florida red tilapia is significantly greater at 1 g/L than at higher salinities, and mean fertilization success (percent of eggs undergoing embryonic development) is significantly greater at less than 18 g/L (Watanabe et al., 1989). While Florida red tilapia eggs incubated at up to 36 g/L hatched, hatching success decreased as salinity increased (Watanabe et al., 1989). Ernst et al. (1991) also note that total production of Florida red tilapia seed at 5 g/L is greater than at 18 g/L. Total seed production of Florida red tilapia using the clutch-removal method is high at 12 g/L (Watanabe et al., 1992), possibly because the metabolic cost of osmoregulation is lowest at isosmotic salinity (Ferby and Lutz, 1987). While *O. niloticus* spawn at salinities ranging from freshwater to 32 g/L, hatching success of artificially incubated eggs from 2- to 3-year-old broodfish is greater in freshwater or at 5 g/L than for artificially incubated eggs from yearling broodfish spawned in freshwater, 10 g/L, or 15 g/L (Watanabe and Kuo, 1985).

Harvest Efficiency

Harvest efficiency is a measure of the effectiveness with which seed are recovered and often is evaluated in terms of fecundity (seed [recovered]/kg female) or in terms of productivity (e. g., seed/m²/d). Harvest efficiency depends on the type of container in which the fry are produced and is affected by the type of seed desired (e.g., eggs, sac-fry) and the equipment available for harvesting. Removal of eggs and sac-fry from mouth-brooding females results in the greatest harvest efficiency (Lee, 1979; Snow et al., 1983; Verdegem and McGinty, 1987; Little, 1989; Little et al., 1993; Mair et al., 1993). Fry production in net enclosures (hapas) allows 100% harvest efficiency but can be much lower if the people harvesting hapas are not conscientious. Circular tanks of various designs (e.g., with a lip to allow fry escapement) allow for highly efficient fry harvest. Harvest of swim-up and advanced fry is less efficient than harvest of eggs or sac-fry primarily because the older fry disperse throughout the water column.

Cannibalism can affect harvest efficiency significantly. High levels of cannibalism among tilapia fry are possible during the first 10 to 30 d following the onset of exogenous feeding, which occurs 6 to 7 d post-hatch (Macintosh and De Silva, 1984). The degree of cannibalism is inversely proportional to food availability (Macintosh and De Silva, 1984; Pantastico et al., 1988) and directly proportional to the size differential within the fry population (Pantastico et al., 1988).

Effects of cannibalism are more pronounced when swim-up or advanced fry are harvested, especially from earthen ponds. Any fry that escape harvest from hapas or tanks can prey upon the next group of fry produced, but these escapees can be controlled easily through management. In earthen pond production systems where partial harvests of swim-up or advanced fry are used, some fry will escape capture and grow to prey upon subsequent generations of fry. Where concrete tanks are used for fingerling production, the number of fry eluding harvest increases over time (Guerrero and Guerrero, 1985). Numbers of fingerlings captured decreases with consecutive partial harvests in earthen reproduction ponds; however, size of harvested fingerlings increases and biomass of fingerlings harvested remains constant (Broussard et al., 1983). Increase in the frequency of partial harvest from 30 to 7 d more than doubles fry yield (from 0.9 to 2.2 fry/m^2/d) and results in smaller, more uniform-sized fry (Green et al., 1994). Fry for sex inversion can be harvested from ponds weekly by edge seining the reproduction pond perimeter with a fine-mesh net (<1.6-mm mesh) (Verdegem and McGinty, 1989). The number of fry captured in each complete circuit of edge seining decreased from 51% of the estimated initial fry population in the first circuit to 10% of the estimated initial fry population in the third circuit (Verdegem and McGinty, 1989). Frequent harvest (six times a day) of fry from a 1740-m^2 reproduction pond resulted in 60% more fry than harvest three times per day (Little, 1989). Even with such frequent partial harvests, significant numbers of fry elude capture and can comprise 50 to 60% of the net yield (Little, 1989). While more frequent partial harvest likely acts to reduce cannibalism by narrowing the fry population size differential, harvest efficiency still is low compared to hapas and tanks.

Fry harvest efficiency from earthen ponds can be improved through frequent complete drain harvests of reproduction ponds (Rothbard et al., 1983; Popma and Green, 1990). Newly filled ponds are drained 16 to 21 d after being stocked with broodfish and uniform age and size fry are harvested. Prior to inundation for the subsequent fry production cycle, any puddles that remain on the pond bottom are poisoned to eliminate unharvested fry. Rothbard et al. (1983) estimate that 30% of total fry production remains unharvested as fry trapped in puddles or as eggs/sac-fry expelled prematurely from the female's mouth.

Measures of fry production efficiency are affected by the multiplicative effects of spawning rate, fecundity, hatchability, and fry survival (Siraj et al., 1983; Rana, 1986). For example, while year class I *O. niloticus* females produce 89% more eggs/kg than older fish, only 39% more fingerlings are produced from younger females because of reduced spawning rate, reduced hatchability, and reduced fry survival (Siraj et al., 1983). Management practices also influence measures of efficiency. Comparison of efficiencies for different fry production systems is shown in Table 1.

Incubation Success

Buccal incubation of eggs and sac-fry is the mechanism used by *Oreochromis* spp. to ensure the greatest survival of the relatively small number of eggs produced at each spawn. Survival can be increased further if eggs are removed from brooding females and incubated artificially. A mean of 9.7 *O. niloticus* fry/g female are recovered from artificial incubation compared to 6.5 fry/g female recovered from natural incubation (Mair et al., 1993). Artificial incubation, however, does not affect the post-hatch quality of *O. niloticus* fry compared to natural incubation as indicated by similar fry growth and survival during a 4-week grow-out period (Mair et al., 1993).

Tilapia eggs are negatively buoyant, and in the absence of a current to suspend them in the water column, eggs sink quickly and clump (Rana, 1986; Macintosh and Little, 1995). A variety of incubation systems have been used successfully for tilapia eggs. For research purposes, shaker tables rotating at 40 rpm are used to incubate relatively small numbers of eggs (Rothbard and Pruginin, 1975). Water in incubating vessels on the shaker table is changed daily, and dead eggs are removed daily to prevent spread of bacterial and fungal infections to all eggs (Rothbard and Pruginin, 1975). Rothbard and Hulata (1980) describe a recirculating incubation system that uses up-welling Zuger type hatching jars. This system replaced shaker tables because of the approximate 40% loss rate of egg batches incubated on shaker tables (Rothbard and Hulata, 1980). Zuger type (conical up-welling) incubators and shaker tables result in mechanical injury to and secondary infection of incubating eggs, and consequently, variable incubation success (Rana, 1986, 1988). Down-welling round-bottom incubators result in 17 to 22% greater hatchability and in improved overall survival (85%) from egg to 10-d-old swim-up fry when compared to overall survival (60%) for conical up-welling incubators (Rana, 1986, 1988).

Down-welling round-bottom incubators come in a variety of sizes and are relatively simple to construct from readily available materials. Two- and three-liter soft drink bottles make good incubators, as do 20-L plastic potable water bottles (Little, 1989; Macintosh and Little, 1995). Down-welling MacDonald jars also can be used to incubate tilapia eggs. Stocking rate for the smaller incubators is 650 to 1350 eggs/L, while in large incubators 4000 eggs/L are stocked (Little, 1989; Macintosh and Little, 1995). Water flow rates are 1 L/min in small incubators and 1 L/s per 10,000 eggs in large incubators (Little, 1989; Macintosh and Little, 1995).

Maintenance of water quality in incubators, especially for those in recirculating systems, is critical. Dissolved oxygen should be maintained as close to 100% saturation as possible. Hatchability of tilapia eggs is not affected significantly by 101 mg/L un-ionized ammonia or 140 mg/L total nitrate because of the relatively impermeable egg chorion (Rana, 1988; Subasinghe et al., 1990). However, fry are much less tolerant of poor water quality. An LC50 of 0.4 mg/L un-ionized ammonia is reported for 7- to 10-d-old tilapia fry at 28°C (Rana, 1988). A source of clean, good quality water is necessary for incubation facilities that use water only once, and a properly functioning biofilter is required to maintain low TAN and nitrite concentrations where incubation water is recirculated (Piper et al., 1982). Sand filtration or ultraviolet radiation can be used to reduce bacterial loads in water for incubators (Piper et al., 1982).

TILAPIA FINGERLING PRODUCTION SYSTEMS

A variety of production units are available to produce tilapia fingerlings: earthen ponds, concrete or fiberglass tanks, and hapas. Daily fry yield generally increases as the production unit progresses from earthen pond to tanks, primarily because of intensified management (see Table 1). In many respects the desired life stage to be produced, i.e., fertilized eggs, sac-fry, or swim-up fry, dictates the system employed. If uniform age and size fry must be collected for sex inversion, then the tank or hapa methods may be more advantageous, although small ponds with harvest basins have been used successfully. If mixed-sex fingerlings of a larger size (1 to 15 g) are desired, then more simple pond production methods may be sufficient. Systems that yield fertilized eggs and sac-fry, as opposed to swim-up fry, require incubation facilities, which in turn require a higher initial investment, increased operating costs, and trained hatchery technicians. Greater fingerling production is possible where hatchery facilities are available, but a favorable cost-benefit analysis is necessary before adoption of this level of technology.

Table 1 Management Parameters and Measures of Efficiency for Tilapia (*Oreochromis* sp.) Fry Production in Hapas, Tanks, or Earthen Ponds

System	Species[a]	Broodfish (No/m²)	Female biomass (kg/m²)	Female: male ratio	Daily yield		Source
					Seed/m²	Seed/g female	
HAPAS							
Harvest all seed at 10–18-d intervals	N	5.0	0.44	2:1	73.4	0.15	Hughes and Behrends, 1983
Harvest all seed after 21-d spawning period							
No broodfish exchange	N	2.7	0.16	2:1	34.7	0.15	Lovshin and Ibrahim, 1988
Female exchange	N	2.7	0.16	2:1	37.5	0.15	
Male and female exchange	N	2.7	0.16	2:1	43.3	0.18	
Harvest all seed at 5-d intervals with female exchange							
Females in spawning hapa only	N	2.0	0.09–0.14	1:1	86.0	0.61–0.96	Little, 1989
Females in spawning hapa and conditioning unit	N	4.0	0.27–0.42	3:1	86.0	0.20–0.32	
Harvest all seed at 5-d intervals with female exchange							
Females in spawning hapa only	N	1.5	0.12	1:1	64.7	0.54	Little, 1989
Females in spawning hapa and conditioning unit	N	3.0	1.53	3:1	64.7	0.04	
Harvest all seed at 7–13-d intervals	A	3.7	0.42	1:1	80.0	0.20	Behrends et al., 1993
Swim-up fry harvested daily							
Double net hapa	N	4.0	0.23	3:1	11.7	0.09	Guerrero and Garcia, 1983
Single net hapa	N	4.0	0.23	3:1	4.9	0.03	
Lower female:male ratio	N	4.0	0.65	3:1	4.6	0.01	
Higher female:male ratio	N	4.0	0.72	5:1	3.8	0.01	

TANKS

Fry dipped from edge of tanks 5–6 times daily

							Reference
Peak breeding period (day 20)	N	3.0	0.22	2:1	16.2	0.02	Guerrero and Guerrero, 1985
Total breeding period (60 d)	N	3.0	0.22	2:1	9.0	0.04	

Harvest all seed at 5-d intervals with female exchange

Females in spawning tank only	N	6.4	0.29–0.45	1:1	258.9	0.58–0.89	Little, 1989
Females in spawning tank and conditioning unit	N	12.7	0.57–0.89	3:1	258.9	0.29–0.45	

Harvest all seed at 5-d intervals with female exchange

Females in spawning tank only	N	8.0	0.50	1:1	302.9	0.61	Little, 1989
Females in spawning tank and conditioning unit	N	15.9	7.43	3:1	302.9	0.04	

Seed collected every 15–23 d

5 ppt salinity water	FL-R	8.2	1.61	4:1	77.7	0.05	Ernst et al., 1991
18 ppt salinity water	FL-R	8.2	1.75	4:1	47.4	0.03	
Harvest all seed at 8–9-d intervals	T-R	3.5	0.47–1.31	5:1	42.0–84.3	0.06–0.12	Chang et al., 1988

EARTHEN PONDS

Total harvest of swim-up fry and broodfish after 17–20 d	N	0.64	0.10	2:1	8.6–10.1	0.08–0.10	Green and Teichert-Coddington, 1993
Total harvest of swim-up fry and broodfish after 17–19 d	H	0.3–0.5	0.03–0.86	4:3	2.93	0.01–0.05	Rothbard et al., 1983
Daily harvest of swim-up fry from pond edge	N	0.41	0.02–0.03	1:1	2.46	0.1–0.16	Little, 1989
Twice daily harvest of swim-up fry from pond edge	N	4.0	0.32	3:1	2.4–6.5	0.01–0.03	Guerrero, 1986
Total harvest of swim-up fry and broodfish after 17–19 d	N	1.0	0.09	1.6:1	7.1	0.09	Green et al., 1995
	A	1.1	0.07	1.5:1	7.0	0.11	

[a] N = Nile tilapia (*Oreochromis niloticus*); A = blue tilapia (*O. aureus*); FL-R = Florida red tilapia (*O. urolepis hornorum × O. mossambica*); T-R = Taiwanese red tilapia (*O. mossambica × O. niloticus*); H = hybrid tilapia (*O. niloticus × O. aureus*).

Fingerling Production in Grow-Out Ponds

Direct stocking of broodfish in production ponds generally is used by tilapia farmers who do not have access to a consistent, reliable source of fingerlings or who, because of limited financial resources, are unable to purchase fingerlings to restock grow-out ponds. Direct stocking can be classified as either intentional or unintentional. Intentional stocking of broodfish in production ponds occurs in mixed sex tilapia production, as opposed to unintentional stocking of females that results from human error during segregation of fingerlings by sex, from unsuccessful sex inversion, or with fish that enter with supply water. If stocking of females is unintentional, then the fingerlings produced usually are not used for restocking grow-out ponds because the pond management system employed demands monosex fingerlings.

Most subsistence to small-scale tilapia farmers still rely on harvesting fingerlings with adult fish at the time of pond draining. Harvested fingerlings are held temporarily until the grow-out pond is reflooded for stocking. The number of fingerlings produced this way is highly variable and ranges from zero, in the presence of predator fish, to more than 50% of total fish yield by weight. Fingerling size also is highly variable. A major disadvantage to the use of fingerlings resulting from direct stocking is becoming apparent in Rwanda. Fingerlings range in size from 3 to 50 g at pond draining, depending on age and on sex. The first fingerlings produced (especially the males) during the grow-out cycle can attain sizes similar to stunted mature females from the original stock. Farmers often select the larger fingerlings for restocking their ponds, but because they are not trained to distinguish male fingerlings from female fingerlings, the fish population in ponds for the next production cycle is composed of a large proportion of mature females. This leads to early reproduction and increasingly larger numbers of offspring, which contributes to overpopulation and more stunting. Pond Dynamics/Aquaculture CRSP researchers in Rwanda have sampled several fish ponds in which greater than 80% of adult fish were stunted females that the farmers normally would identify as appropriate-sized fingerlings for restocking.

Earthen Spawning Ponds

Earthen ponds are used to produce more advanced fingerlings (0.5 to 5 g) or newly hatched fry for hormonal sex inversion. Ponds destined exclusively for fingerling production often are stocked with broodfish at a lower rate than are grow-out ponds, and stocking rates vary from about 3,000 to 10,000 fish/ha (150 to 2000 kg/ha total biomass) (Broussard et al., 1983; Mires, 1983; Rothbard et al., 1983; Little, 1989; Green and Teichert-Coddington, 1993; Green et al., 1995). Very low stocking rates (2000 fish/ha) are used in Rwanda because of the lack of a high quality feed and general scarcity of fertilizers. Broodfish sex ratio varies from 1 female:1 male to 4 females:1 male.

Advanced fingerlings are obtained by periodic, partial harvests of the spawning ponds using a seine net (1.0 to 6.2 mm^2 mesh). Several passes through the pond with the seine, whose length is about 25% greater than the width of the pond, are necessary to remove the majority of fingerlings. Experience in Honduras indicates that productivity of *O. niloticus* broodfish can be maintained for up to 6 months with weekly harvests of 1-g fingerlings. Productivity can decline if weekly harvests are not maintained as the larger fingerlings cannibalize smaller fry. Harvested mixed-sex fingerlings are nursed to sufficient size for identification of sex by external characteristics or are stocked directly in grow-out ponds. Biweekly harvests of spawning ponds in the Ivory Coast have been shown to yield the greatest number of small (1 to 2 g) fingerlings (Holl, 1983). In Rwanda, where temperatures are much cooler, monthly seining of *O. niloticus* fingerlings is more productive than seining every 2 weeks or every 2 months (Hishamunda and Moehl, 1989). Productivity of reproduction ponds in the Philippines subjected to partial harvests at 30-d intervals is 0.26 seed/m^2/d (Broussard et al., 1983). Intermittent seining of fingerlings can provide fish of more uniform age than a one-time draining. This is especially important if mixed-sex tilapia are used in grow-out ponds. If some of the fingerlings are old but small when stocked into production ponds, reproduction may occur very early in the production cycle, and overall fish growth can be retarded because of overpopulation.

Newly hatched fry (9 to 11 mm total length) suitable for hormonal sex inversion are harvested from ponds by frequent collection of swim-up fry with dipnets (Guerrero, 1986; Little, 1989; Verdegem and McGinty, 1989) or by complete drain harvest every 16 to 21 d (Rothbard et al., 1983; Popma and Green, 1990; Green and Teichert-Coddington, 1993). After release from their mother, swim-up fry school and prefer the warm, shallow waters near the pond edge (Philippart and Ruwet, 1982; Macintosh and Little, 1995), where they can be captured easily with hand-held dipnets. Edge seining the pond perimeter weekly using a 6.1-m long, 1.6-mm mesh seine yielded a total of 2.2 fry/m^2/d, of which >90% were of suitable size for sex inversion (Verdegem and McGinty, 1989). Verdegem and McGinty (1989) recommend that well-conditioned broodfish be stocked, that 1 to 2 circuits of the pond perimeter be conducted 1 to 3 times weekly, and that fingerlings that eluded capture be harvested every 2 weeks using a 6.4-mm mesh seine. Under more intensive management, fry are collected from two to six times daily using a 0.06- to 0.25-m^2 frame dip-net equipped with mosquito netting (Guerrero, 1986; Little, 1989). Production cycles vary from 30 to 40 d (Guerrero, 1986) to almost 120 d (Little, 1989). Ponds must be drained between cycles to eliminate recruits that have eluded capture, which can form a significant component of the harvest biomass (Little, 1989; Verdegem and McGinty, 1989).

Where fry suitable for hormonal sex inversion are obtained by complete drain harvest, ponds equipped with a harvest sump in the vicinity of the drain pipe are managed and harvested more easily. The sump, about 35-cm deep, can be constructed from concrete or stone, or excavated in compacted pond bottom. A water inlet located near the sump is recommended, but not required. Prior to pond inundation, a piece of nylon netting (1.25- to 2.54-cm square mesh) is draped over the harvest sump, and the edges of the netting are weighted down.

At harvest, the pond drain is covered with a fine-mesh screen (e.g., window screen) to retain all fry. The surface area of the drain screen should be large enough to reduce suction pressure during draining and clogging. The pond is drawn down until the water is level with the upper lip of the harvest sump. An adequate number of field workers are needed to lift the net placed in the harvest sump to remove all broodfish en masse. Broodfish are transferred by hauling tank to holding tanks for conditioning or restocking into spawning ponds. Fry are harvested from the harvest sump by means of 40- to 100-cm wide scoop net made of fine mesh (1.6-mm mesh nylon netting) or window screen/mosquito netting. Inlet water, if available, can be trickled into the harvest sump during fry harvest. Fry harvest must be done quickly to minimize stress; harvested fry are transferred to a nylon hapa (1.6 mm mesh) suspended in a hauling tank with clean water. Any puddles that remain in ponds after draining should be treated with a fish toxicant to eliminate fish that escaped harvest (Mires, 1983; Popma and Green, 1990). Females' mouths should be checked for eggs prior to restocking into spawning ponds (Popma and Green, 1990).

Rothbard et al. (1983) describe a system of complete drain harvest every 17 to 19 d that is based on assumed durations of the different steps in the reproductive process in ponds. Popma and Green (1990) base recommendations for duration of fry production cycles on water temperature. Because water temperature affects tilapia reproduction, the rule of thermal summation, or degree-days, can be used to predict temperature effects on tilapia fry production (Green and Teichert-Coddington, 1993). Degree-days are calculated as the mean daily water temperature minus the threshold temperature of 15°C, and the results are summed over the production period. Pond harvest is recommended between 195 to 220 cumulative degree-days (Green and Teichert-Coddington, 1993). At less than 195 degree-days, fry production still is increasing, and fry harvests will be sub-optimal; whereas at greater than 220 degree-days, the percentage of the fry population too large (≥13 mm total length) for sex inversion is increasing. Harvests, therefore, generally occur 14 to 20 d after stocking, with the shorter duration corresponding to the period of warmer water temperatures. Mass production of both *O. niloticus* and *O. aureus* has been accomplished using this technique (Green and Teichert-Coddington, 1993; Green et al., 1995).

Fry collected from ponds either with dip-nets or by complete drain harvest are accompanied by aquatic insects, particularly water boatmen (Corixidae) and backswimmers (Notonectidae), and suspended organic matter that must be separated from the fry before fry are transferred to treatment units for sex inversion. Little (1989) describes a method used in the Philippines and Thailand where a small quantity of gasoline is poured on the surface of the holding tank water to kill the air-breathing aquatic insects. He cautions that water quality must be maintained in order to prevent fry rising to the surface and becoming contaminated with gasoline. A different method is used in Honduras: harvested fry and accompanying by-catch are transferred quickly to a hapa suspended in a concrete tank supplied with a continuous flow of clean, well-aerated water for 6 to 24 h. Water depth in the hapa is 50 to 75 cm. The suspended organic matter and sediment are washed out by the water flow. After a period of several hours to overnight, the fry and aquatic insects have segregated into two different strata: schools of tilapia fry congregate at the bottom of the hapa, while the aquatic insects concentrated in the top 10 cm or so of the water column are removed easily with a fine mesh dip-net. A further benefit of this management technique is the separation of live fry from the handling mortality.

Hapas

Hapas that range in size from 2 to 100 m^2 are used for tilapia seed production. Hapas are constructed from fine mesh netting (usually 1.6 mm mesh size) that retains both eggs and newly released fry. Hapas are suspended in tanks, ponds, lakes, or coastal waters and allow for easy management of broodfish (Figure 1). Advantages to the use of hapas for tilapia reproduction include a small initial investment, ease of harvest, and low fry mortality during harvest. Disadvantages include the short (6 to 12 months) usable life of untreated hapas, deterioration of water quality because fine mesh clogs with algae, and the need for incubation facilities for eggs spit out by females during fry harvest (Lovshin and Ibrahim, 1988).

Broodfish stocking rates range from less than one brooder/m^2 to 12.5 brooders/m^2, but averages about 4 brooders/m^2 (Guerrero and Garcia, 1983; Hughes and Behrends, 1983; Siraj et al., 1983; Bautista et al., 1988; Lovshin and Ibrahim, 1988; Little, 1989; Behrends et al., 1993; Mair et al., 1993; Maluwa and Costa-Pierce, 1993; Costa-Pierce and Hadikusumah, 1995). Bautista et al. (1988) report no differences in mean daily seed productions at stocking rates of 4, 7, or 10 *O. niloticus* females/m^2. Broodfish sex ratios also vary considerably, from 1 female:1 male to 10 females:1 male, while ratios of one to three females per male are most common (Guerrero and Garcia, 1983; Hughes and Behrends, 1983; Siraj et al., 1983; Bautista et al., 1988; Lovshin and Ibrahim, 1988; Little, 1989; Behrends et al., 1993; Mair et al., 1993; Maluwa and Costa-Pierce, 1993; Costa-Pierce and Hadikusumah, 1995). No differences in daily seed production are reported for broodfish sex ratios (female:male) of 5:1 and 3:1 (Guerrero and Garcia, 1983), or 4:1, 7:1, and 10:1 (Bautista et al., 1988).

Tilapia seed yields from hapas range from two to more than 80 seed/m^2/d (Guerrero and Garcia, 1983; Hughes and Behrends, 1983; Lovshin and Ibrahim, 1988; Little, 1989; Behrends et al., 1993; Maluwa and Costa-Pierce, 1993). Harvest frequency of seed from hapas ranges from daily to 21- to 32-d intervals. Developmental stage of seed from intensively harvested hapas ranges from newly fertilized eggs with no pigmentation to yolk-sac mostly absorbed, actively feeding swim-up fry (Behrends et al., 1993). Swim-up fry are harvested from the less intensively harvested, less productive systems. Swim-up fry less than 1 week old are harvested easily using dip-nets, while older fry are more difficult to capture with dip-nets because they congregate near the bottom of the hapa. Single-net hapas generally are used for tilapia reproduction, although a double-net hapa has been used in the Philippines and Indonesia, and is particularly well-suited for lacustrine aquaculture (Guerrero and Garcia, 1983; Costa-Pierce and Hadikusumah, 1995). The double-net hapa system involves a smaller hapa suspended inside of a larger hapa. The mesh size of the inner hapa is 30 mm on the sides and about 2 mm on the bottom, while the outer hapa is constructed entirely of 2 mm mesh. Broodfish are confined to the inner hapa, and swim-up fry are able to pass through to the outer hapa. Daily seed production ranges from 6.8 to 12.2 seed/m^2 (Guerrero and Garcia, 1983; Costa-Pierce and Hadikusumah, 1995). Total seed yield from a single-net hapa (48 m^2) is the same as from a double-net hapa (48-m^2 outer hapa, 20-m^2 inner hapa), but because fewer broodfish are used in the double-net hapa, its productivity is more than double (11.6 seed/m^2/d) that of the single-net hapa (4.9 seed/m^2/d) (Guerrero and Garcia, 1983).

Concrete or Fiberglass Tanks

Tilapia seed, ranging from fertilized eggs to swim-up fry, are produced in 4 to 100 m^2 tanks that are 1 to 1.5 m deep. Stocking rate of broodfish varies from 3 to 10 fish/m^2, and broodfish sex ratio varies from 1 female:1 male to 10 females:1 male (Guerrero and Guerrero, 1985, 1988; Bautista et al., 1988; Chang et al., 1988; Galman et al., 1988; Ernst et al., 1991; Little et al., 1993). Stocking 4, 7, or 10 females/m^2, at a 4 female to 1 male ratio, does not significantly affect daily seed production per *O. niloticus* spawner, although the highest observed production is with 4 females/m^2 (Bautista et al., 1988). Daily seed production per spawner is not affected significantly when the female to male *O. niloticus* broodfish ratio is increased from 4:1 to 10:1, however the highest value is at the 4 females to 1 male ratio (Bautista et al., 1988). Seed productivity in tanks increases as earlier life stages are harvested, and ranges from 8 to 16 seed/m^2/d where swim-up fry are harvested (Guerrero and Guerrero, 1985, 1988) to 42 to 84 seed/m^2/d where sac-fry and swim-up fry are harvested (Chang et al., 1988; Little, 1989; Ernst et al., 1991) to 239 seed/m^2/d where eggs and sac-fry are harvested (Little, 1989).

PRODUCTION OF MONOSEX FINGERLINGS

In recent years, the use of monosex populations of fish in aquaculture has become common. Monosex populations offer several benefits in aquaculture, including faster growth (if one sex grows faster and larger) and prevention of unwanted reproduction. Unwanted reproduction carries two potentially negative impacts: fish divert energy to gametic rather than somatic growth, and, if reproduction occurs, a significant part of the harvest may be unmarketable juvenile fish. In tilapia aquaculture, the rationale for using all-male populations arises from the prolific reproductive potential of these species and from the faster and larger growth potential of males.

Four general methods exist for culturing all-male populations of tilapia: (1) census the entire juvenile population in the pond and select the males based on secondary sex characteristics (the difference in appearance of the genital papilla); (2) produce all-male populations through interspecific hybridization; (3) produce all-male populations by treating young fry with androgens; and (4) produce all-male populations by using YY males to sire the offspring. The first method is less desirable in most instances because of its inefficiency (i.e., half the population is discarded) and labor intensity. The last method still is under development and offers the advantage that hormone is used only in broodstock. The use of androgens to masculinize tilapia directly is by far the most common practice for many aquaculturists the world over.

Production of all-male populations of tilapia through interspecific hybridization was practiced widely during the early stages of development of tilapia aquaculture and still is practiced to a limited extent. Two mechanisms of sex determination have been proposed for tilapia: homogametic female (XX)-heterogametic male (XY), e.g., *O. mossambicus*, *O. niloticus*, and heterogametic female (WZ)-homogametic male (ZZ), e.g., *O. urolepis hornorum*, *O. macrochir*, *O. aureus* (Wohlfarth and Hulata, 1981; Lovshin, 1982; Trombka and Avtalion, 1993). Interspecific hybridization between a homogametic female and homogametic male results in all-male offspring (Pruginin et al., 1975; Wohlfarth and Hulata, 1981; Lovshin, 1982; Trombka and Avtalion, 1993). Often, fewer fry are produced by interspecific hybridization than are produced by pure line crosses of the parental species (Lovshin, 1980; Mires, 1983, 1985). This reduction in productivity is attributed to

A

B

Figure 1.   A hapa is an open-top cage constructed from netting. Hapas used for tilapia egg or fry production are constructed from fine mesh (usually 1.6 mm mesh) to retain eggs and sac-fry (Photo A). Production of tilapia eggs/fry can be conducted in large (40 to 100 m^2) hapas as in Thailand (Photo B) or in small (~1 m^2) hapas as in the Philippines (Photo C). In Honduras, sex reversal is accomplished in hapas (2 to 4 m^2) suspended in earthen ponds (Photo D).

reproductive incompatibilities between the two parent species (Lovshin, 1980; Mires, 1983, 1985). Hulata et al. (1985) reported that Ghana strain of *O. niloticus* females were able to hybridize more easily with *O. urolepis hornorum* males than were *O. niloticus* Ivory Coast females, and suggested that production of hybrid fry may be increased through use of suitable broodstock. Maintenance of broodfish purity is of utmost importance for successful production of all-male hybrid tilapia. Hybrid tilapia are reproductively viable, will breed readily with parent species, and often are difficult to distinguish visually from parent species. Thus, management of the fingerling production facility

C

D

Figure 1 (continued)

requires strict safeguards to ensure maintenance of broodfish purity. Commercial production of hybrid tilapia requires selection of broodfish based on its ability to consistently produce offspring with a stable sex ratio, maintenance of broodfish purity through electrophoretic examination of blood serum protein, and elimination of broodfish of doubtful purity (Mires, 1983). Perhaps for these reasons, sex inversion, rather than interspecific hybridization, has been adopted as the method of choice for commercial mass production of all-male tilapia.

Two methods for inducing one sex in place of another, i.e., sex inversion and production of YY males, take advantage of the extremely plastic gonadal development of most fishes. Although sex seems to have a genetic basis in most fish species, gonadal differentiation can be directed to the gender opposite to that which would have occurred under normal circumstances. The process of gender manipulation through treatment with androgen or estrogen has been called sex reversal or sex inversion. The latter term refers to those cases where the undifferentiated gonad is directed to a particular sex, and the former term is properly applied when referring to fish where the already differentiated gonad is induced to become that of the opposite sex. However, review of the tilapia literature shows that both terms are used interchangeably, even by the same author. A number of androgens have been shown to masculinize newly hatched tilapia fry of the commercially important species (see reviews by Schreck, 1974; Shelton et al., 1978; Hunter and Donaldson, 1983; Pandian and Sheela, 1995). A list of the androgens used to masculinize tilapia in recent years includes 17 α-methyltestosterone (Shelton et al., 1978; Tayamen and Shelton, 1978; Guerrero and Guerrero, 1988); mibolerone (Meriwether and Torrans, 1986; Torrans et al., 1988); fluoxymesterone (Phelps et al., 1992); norethisterone acetate (Varadaraj, 1990); 17 α-ethynyltestosterone (Shelton et al., 1981; Berger and Rothbard, 1987; Watanabe et al., 1993); testosterone propionate (Nsengiyumva et al., 1990); and 17 α-methylandrostendiol (Varadaraj and Pandian, 1987). All of these androgens share common characteristics: they are synthetic compounds alkylated at the 17-α position.

Sex Inversion

To induce masculinization, treatment of tilapia fry with androgen should be initiated before onset of gonadal differentiation. Newly hatched fry (9 to 11 mm total length) are preferred for sex inversion because the gonadal tissue of tilapia is presumed undifferentiated (Shelton et al., 1978). Gonadal differentiation seems to occur between 8 to 25 d post-hatch, with environmental conditions likely responsible for the observed variation (Eckstein and Spira, 1965; Nakamura and Takahashi, 1973; Alvendia-Casauay and Cariño, 1988). In pond aquaculture of tilapia, because monitoring individual broods and the exact time of the onset of exogenous feeding is impractical, fry of uniform size and age must be selected from the population. This is achieved using fry production techniques described above. Fry harvested from ponds should be graded through a grader fitted with 3.2-mm square mesh netting to remove fry larger than 13 mm total length, as these fry are considered too large for successful sex inversion (Popma and Green, 1990). Fry that pass through the grader are stocked into treatment units for sex inversion.

Fish exposure to androgens usually occurs through dietary treatment. The most commonly used androgen is 17 α-methyltestosterone (MT). Feed preparation has been described in detail by Popma and Green (1990). Feeds that contain 25 to 45% protein are recommended, although a 20% protein feed has been used successfully (Popma and Green, 1990). Androgen-treated feed is prepared by (1) thoroughly mixing androgen that has been dissolved in solvent (usually 80 to 95% ethanol) with finely ground (approximately 0.5- to 1.0-mm diameter particle size) feed or (2) spraying the dissolved androgen and solvent onto the feed. The advantage of the first method is the minimal atomization of androgen solution, which reduces the chance for contamination of workers or workplace; the second method is advantageous in that the volume of solvent can be reduced. In both cases, the alcohol is allowed to evaporate and the dried feed stored in a cool, well-ventilated area or refrigerated until used; good air circulation around the complete feed container helps maintain feed quality during storage. Because MT is photosensitive, pure MT should be protected from sunlight, and treated feed should not be dried or stored exposed to direct sunlight. Feed prepared with MT that has been stored exposed to light and air, and MT-treated feed stored exposed to light and air were ineffective in sex inversion of *O. mossambicus* (Varadaraj et al., 1994). Otherwise, MT is quite stable within conditions normally encountered on fish farms. Refrigeration, however, is recommended for treated feed to ensure feed quality. It is not necessary to use reagent or USP grade ethanol as a solvent for MT. In fact, in Honduras denatured ethanol (denatured with glycerin at 0.5% by volume) is obtained from a local distillery at low cost and used successfully.

Oral administration of MT-treated feed (30 to 60 mg MT/kg feed) to tilapia fry during a 3- to 4-week period yields populations composed of ≥95% males (Shelton et al., 1978; Tayamen and Shelton, 1978; Owusu-Frimpong and Nijjhar, 1981; Nakamura and Iwahashi, 1982; Obi and Shelton, 1983; Chambers, 1984; Shepperd, 1984; Nakamura and Takahashi, 1985; McGeachin et al., 1987; Popma, 1987; Guerrero and Guerrero, 1988; Jo et al., 1988; Pandian and Varadaraj, 1988;

Varadaraj and Pandian, 1989; Green and Lopez, 1990; Hiott and Phelps, 1993; Meyer, 1990; Green and Teichert-Coddington, 1991; Phelps et al., 1992). The efficacy of MT in sex inversion of tilapia fry has been demonstrated both in the laboratory (e.g., Tayamen and Shelton, 1978; McGeachin et al., 1987; Pandian and Varadaraj, 1988; Varadaraj and Pandian, 1989) and in large-scale, prolonged field trials (Popma, 1987; Guerrero and Guerrero, 1988; Green and Lopez, 1990). Treatment duration of 3 to 4 weeks consistently produces tilapia populations composed of ≥95% phenotypic males, and treatment periods that exceed 4 weeks do not improve the efficacy of the treatment (Tayamen and Shelton, 1978; Owusu-Frimpong and Nijjhar, 1981; Nakamura and Takahashi, 1985). However, at cooler water temperatures (20 ± 2°C), increasing treatment (60 mg MT/kg feed) duration from 20 to 40 d resulted in an increase in efficacy from 69 to 95% of the population being composed of males (Mbarererehe, 1992). Incorporation of 90, 100, or 120 mg MT/kg feed did not enhance, accelerate, or depress sex inversion of tilapia fry when compared to results obtained with 30 to 60 mg MT/kg feed (Nakamura and Iwahashi, 1982; McGeachin et al., 1987). Treatment with higher dosages of MT (240, 480, 600, 1000, or 1200 mg MT/kg feed) for 19 or 28 d do not result in successful sex inversion (Nakamura, 1975; Okoko and Phelps, 1995).

Oral administration of 30 to 60 mg MT/kg feed to tilapia fry does not induce toxic effects. Growth and survival of tilapia fry fed an MT-treated (60 mg MT/kg feed) diet for sex inversion were similar to fry fed androgen-free diet (Green and Teichert-Coddington, 1994). In fact, no significant differences in growth or survival were noted between MT-treated and untreated fish during an additional 244 d of growth. Ridha and Lone (1990) reported no significant differences in tilapia, fry survival, and final weights after a 38-d trial where fry were fed androgen-free diet or diets that contained 30, 50, or 70 mg MT/kg. Tilapia fry offered a diet that contains 0, 60, 90, or 120 mg MT/kg feed for 22 d had similar survival rates (McGeachin et al., 1987). In an earlier study, Guerrero (1975) did not observe significant differences in survival or total length between control fry and MT-treated (0, 15, 30, or 60 mg MT/kg of feed) fry after a 21-d treatment period. Results of other studies also showed no significant difference in survival between MT-treated and control tilapia (Nakamura, 1975; Tayamen and Shelton, 1978; Jo et al., 1988; Phelps et al., 1992). No gross abnormalities were observed in any of the treated fish in any of the studies. Thus, available evidence indicates that oral administration of 30 to 60 mg MT/kg feed for a 30-d treatment period does not produce deleterious effects in tilapia fry.

The presence of natural food in the treatment unit does not affect efficacy of androgen treatment (Buddle, 1984; Chambers, 1984; Berger and Rothbard, 1987; Phelps and Cerezo, 1992; Phelps et al., 1995). Successful sex inversion of tilapia has been accomplished in hapas suspended in fertile earthen ponds (Berger and Rothbard, 1987; Popma and Green, 1990; Green and Teichert-Codding-ton, 1991; Macintosh and Little, 1995), in hapas suspended in concrete tanks (Jo et al., 1988; Phelps and Cerezo, 1992; Argue and Phelps, 1996), stocked free in tanks (Rothbard et al., 1983; Guerrero and Guerrero, 1988; Macintosh et al., 1988; Watanabe et al., 1993; Argue and Phelps, 1996), or stocked free in earthen ponds (Phelps et al., 1995). Stocking rates for MT treatment vary from 2000 to 6000 fry/m^2 in hapas suspended in either ponds or tanks, from 150 to 750 fry/m^2 free in tanks, from 75 to 260 fry/m^2 free in ponds, from 6000 to 12,000 fry/m^3 in recirculating tank systems, and from 8000 to 17,000 fry/m^2 in tanks with continuous water exchange. Uneven exposure of fry to dietary androgens can occur because feeding hierarchies become established, particularly when treatment units are stocked with high numbers of fry. Therefore, to ensure efficacy of androgen treatment, the daily ration should be divided into at least four meals evenly spaced throughout daylight hours (Popma and Green, 1990; Bocek et al., 1992).

Immersion of tilapia fry in androgen solutions may be an alternative to oral administration of androgen. This technique is well developed in salmonid aquaculture (Piferrer and Donaldson, 1989); however, it remains largely experimental in tilapia culture. For *O. aureus*, immersion of fry in mibolerone at 0.6 mg/L for 5 weeks resulted in populations that were 82% male (the remaining fish had ovo-testes), while 0.3 mg/L mibolerone immersion for 5 weeks resulted in less than 1% functional females (Torrans et al., 1988). Immersion of *O. mossambicus* in

17 α-methylandrostendiol at 5 µg/L for 11 d beginning at 7 or 10 d post-hatching caused 100% masculinization (Varadaraj and Pandian, 1987). Immersion of *O. niloticus* in 17 α-methyldihydro-testosterone at 0.5 mg/L at 10 and 13 d post-fertilization for 3 h resulted in populations that were between 93 and 100% males (Gale et al., 1995).

Induction of masculinization by feeding has several advantages over immersion: treatment can be accomplished within pens in ponds, and effective doses and protocols have been established, particularly for MT. However, feeding androgen carries some potential disadvantages: inefficiency in masculinization due to unequal exposure, increased risk of exposure to workers during food preparation and feeding unless proper safety measures are observed, and unknown effects of androgen on untargeted elements of the pond ecosystem.

Induction of masculinization by immersion has several advantages over feeding: all fish are exposed to the same dose, a shorter exposure period is required, and hatchery workers have greater control over the process. Potential disadvantages of immersion treatment include the need for an enclosed tank for treatment; however, this need also results in the androgen being contained for easy filtration (e.g., carbon filtration) and removal. The primary impediment to the adoption of this technique is that minimum effective treatments have yet to be established for all species.

Immersion as a viable alternative to dietary treatment with steroids awaits demonstration that it can be scaled to production levels. The following examples illustrate the total amounts of androgen required for masculinization by dietary treatment or immersion. *Oreochromis niloticus* fry were masculinized by dietary treatment with 60 mg MT/kg feed for 21 d (Green and Teichert-Coddington, 1991). Based on the beginning and ending body weights and the stated feeding rate, the total MT used to masculinize 6000 fry was 30 to 35 mg. Similarly, Goudie et al. (1986a), starting with larger fish and treating at 30 mg MT/kg feed, calculated that each fish received 10 µg MT during 21 d of dietary treatment, which if scaled to 6000 fry treated would be 60 mg. *Oreochromis aureus* fry were masculinized by immersion in 0.6 mg/L mibolerone for 5 weeks (Torrans et al., 1988). Based on the frequency of water changes and the volume of the treatment tank, the total mibolerone used to treat 250 fry was 175 mg. Both the large-scale feeding study and the immersion study raised tilapia at about the same density (6 to 7 fish/L). The first study was conducted on fry fed in 1-m^3 hapas; therefore, to be as efficient, immersion would have to be effective at 35 µg/L, or 20% of the dose used by Torrans et al. (1988). Varadaraj and Pandian (1987) reported a minimum effective immersion dose of 5 µg/L for methylandrostendiol in *O. mossambicus*; however, it is unknown if and how often the immersion solution was changed and therefore the total dose of steroid cannot be calculated. Although immersion as an alternative treatment does show potential, additional research is needed to demonstrate effective treatment parameters and scaling to production levels.

Human food safety is an important consideration with food animals that have been exposed to synthetic androgens during the production process. To this end, studies have been undertaken on the disposition of hormone in tilapia fry fed MT for masculinization. Ingested 17 α-methyltestosterone is rapidly metabolized into active and inactive metabolites, and rapidly excreted from the fish. In a study in which sexually undifferentiated tilapia fry were fed radio-labeled (^{3}H–) MT for 30 d, 97% of the ^{3}H–MT residues were polar metabolites 1 d after dosing; 10 d after androgen withdrawal, only trace amounts of ^{3}H–MT metabolites were detected (Curtis et al., 1991). Biological half-lives for polar and chloroform-extractable MT metabolites are 1.1 and 2.2 d, respectively. Goudie et al. (1986a) report that 21 d after androgen withdrawal in tilapia fry treated for sex inversion, less than 1% of the original radioactivity (^{3}H–MT and ^{14}C–MT) remained (equivalent to a total of 0.5 ng MT and metabolites/100 mg tissue, or 0.3 ng MT/100 mg viscera and 0.2 ng MT/100 mg carcass). Johnstone et al. (1983) report that >95% of radioactivity detected in tilapia fry fed ^{3}H–MT-treated feed was associated with the viscera. Whole-body levels of radioactivity had fallen to <1% of initial values within 100 h of hormone withdrawal.

Elimination pathways of radio-labeled MT have been identified from studies on larger fish. Research protocol involves either feeding adult fish a single dose of radio-labeled MT (^{3}H– and/or ^{14}C–) or feeding ^{3}H–MT-treated diet for 3 to 12 d, and following elimination of radioactivity from the different organ systems for 10 to 21 d after androgen withdrawal. In tilapia, most of the radioactivity is detected in the gall bladder and liver, which supports a hepatobiliary route for MT metabolism (Goudie et al., 1986b; Curtis et al., 1991). Studies on coho salmon, rainbow trout, and common carp also show that MT metabolism is via the hepatobiliary route (Fagerlund and McBride, 1978; Fagerlund and Dye, 1979; Lone and Matty, 1981; Cravedi et al., 1989).

If it is assumed that the 0.5 ng MT and metabolites/100 mg tissue residue persist and the distribution between viscera and carcass remains the same, then a marketable-size tilapia (250 to 300 g) would have only 10 pg MT and metabolites/g of carcass (Goudie et al., 1986a). If it is assumed further that this residue is comprised wholly of MT, then a person would have to consume 1000 kg of fish flesh in 1 d to have consumed the minimum daily MT dosage for human medicine. Dosage rates of MT in human medicine range from 10 to 200 mg daily (Wilson, 1988; U.S. Pharmacopeial Convention, 1991; Drug Facts and Comparisons, 1994).

Very little information is available on adverse effects to aquaculturists handling MT. However, the potential risks of long-term exposure can be summarized based on the warnings listed for medical use of androgens, which include carcinogenesis, liver and heart damage that may be life-threatening, as well as other effects, such as changes in fertility, hirsutism, and pattern baldness (Drug Facts and Comparisons, 1994). The amounts of MT that may be handled during preparation of treated food and actual feeding at a large-scale aquaculture facility are within the dosage rate range used for human medicine. Androgens can be absorbed readily through the skin and should be handled with extreme care in all forms (powder, liquid, and in feed). A respirator and eye protection should be used while handling pure androgen, and gloves should be worn during all phases of treatment.

Production of YY Males

In an effort to limit the use of androgens in production of monosex tilapia, another method of masculinization is being developed. This method is the use of YY males that sire only male offspring (Scott et al., 1989). Since the phenotypic sex of tilapia can be inverted, then it is possible, in the case of *O. niloticus* and *O. mossambicus*, to feminize genetic males (XY) by treatment with estrogen. When mated with a normal male (XY), the feminized genetic male will produce offspring with ratios of the following genotypes: 1 XX female: 2 XY males: 1 YY male, provided that this last genotype is viable. YY males have been identified in some studies (Mair et al., 1991a). Theoretically, only male offspring should result from mating normal females (XX) with the YY male. Similarly, feminization of *O. aureus* genetic males (ZZ female) and subsequent mating with normal males result in all-male offspring (Lahav, 1993).

Despite the potential for production of all male tilapia with the use of YY males (or ZZ females), the technique has several limitations. The creation of a YY broodstock requires several generations because of the need for progeny testing (Scott et al., 1989). To circumvent the need for constant progeny testing, some researchers have suggested the creation of YY females through feminization with estrogen; such animals could then be mated with YY males to establish a YY broodstock that no longer requires progeny testing (Scott et al., 1989). This system relies on the simple monofactorial XX female to XY male model to explain sex determination. However, the small but significant deviations from expected sex ratios suggest that other factors may be involved in sex determination (Shelton et al., 1983; Wolhfarth and Hulata, 1989; Mair et al., 1991a; Trombka and Avtalion 1993). Not all YY males produce 100% male progeny (Scott et al., 1989); therefore, the use of YY males must be monitored to ensure the desired results. Furthermore, unless genetic engineering or chromosome set manipulation techniques evolve to the point of practical use, generation of YY males will still require the use of hormones for sex inversion, albeit not in food fish. Several studies have succeeded in feminizing tilapia through dietary treatment (Hopkins et al., 1979; Scott et al., 1989; Lahav, 1993), although only with partial success. Therefore, feminization methodology requires further development to optimize success.

Experimental Techniques

Most experimental evidence only partially supports a monofactorial model for sex determination in tilapia; autosomal and/or environmental factors also are thought to influence sex determination (Wohlfarth and Hulata, 1981; Majumdar and McAndrew, 1983; Shelton et al., 1983; Mair et al., 1991a; Mair et al., 1991b; Trombka and Avtalion, 1993). Temperature is the environmental factor thought most likely to affect sex determination. In a study using *O. vulcani* and *O. niloticus*, Mires (1974) concluded that exposure to cooler temperatures (21 to 27°C) during incubation and early fry rearing did not affect sex ratios significantly compared to incubation and rearing at 27 to 29°C. Rearing of *O. niloticus* (Ivory Coast strain) fry less than 12-d old at 21.4°C or at 33.6°C for a 29-d period, corresponding to the time gonadal differentiation occurs, did not result in significant deviations from the expected 1 male to 1 female sex ratio (Argue and Phelps, 1995). However, Baroiller et al. (1995) reported that a significantly increased proportion of males were obtained when 9- to 13-d post-fertilization *O. niloticus* (Ivory Coast strain) fry were reared for 21 d at 32 to 34°C compared to 27 to 29°C. In a series of experiments conducted over the 19 to 32°C temperature range, first-feeding fry were maintained at constant temperature for 1568 degree-days to assess the effect of temperature on sex ratio (Mair et al., 1990). Temperature did not affect sex ratios of *O. niloticus*, while some results for *O. aureus* indicated that a higher percentage of males may be obtained at lower temperatures (Mair et al., 1990). A significantly greater proportion of males was obtained when *O. mossambicus* fry were reared at low (<25°C) temperatures (Mair et al., 1990). Additional research is needed in the area of temperature effects on tilapia sex determination.

POST-HATCHERY FINGERLING REARING

Fingerlings generally must be reared to advanced fingerling or stocker size for stocking into grow-out units. Tilapia fingerlings average 0.1 to 1.0 g each following completion of sex inversion or at partial harvest from mixed-sex fingerling production ponds. Direct stocking of these small fingerlings into ponds for grow-out is an inefficient use of pond facilities because of low fish biomass during the first months of production. Nursery rearing of tilapia fingerlings can be accomplished in hapas, tanks, or ponds, and can involve one or more phases in which stocking rates are reduced periodically to reduce fish standing stock in order to maintain fast growth of fingerlings. Nursery pond stocking rates vary from 15 to 265 fry/m^2, nursery phase duration varies from 21 to 90 d, final weight varies from 0.5 to 70 g, and survival ranges from 60 to 95%, with an average of 75% (Sarig and Marek, 1974; Broussard et al., 1983; Snow et al., 1983; Barash, 1984; Guerrero, 1986; Watanabe et al., 1990; Perez, 1995). For example, in Central America, one nursery pond management strategy involved stocking fry from sex inversion treatment at 110 fry/m^2 for growth to 20 g, followed by a reduction in stocking rate to 28 fingerlings/m^2 for growth from 20 to 50 g (Perez, 1995). Fry size at stocking affects survival in nursery ponds: in earthen nursery ponds, survival rate for 0.01-g swim-up fry is 57.3%, compared to a 72.4% survival rate for 0.05-g swim-up fry (Guerrero, 1986). However, survival during the nursery phase did not differ when 4- or 7-d post-swim-up fry were stocked (Snow et al., 1983). Fry must be large enough to escape both cannibalism by conspecifics and predation by aquatic insects.

Growth of fry during the nursery phase also is affected by availability of food. Natural food is sufficient to maintain rapid growth of fingerlings until a standing crop of 800 to 1200 kg/ha is achieved, at which point supplemental feed is required to maintain fast growth (Green, 1992). Snow et al. (1983) reported standing crops of 511 to 605 kg/ha for chemically fertilized nursery ponds, 2873 kg/ha for ponds with supplemental feeding only, and 5450 kg/ha in nursery ponds managed with fertilization and supplemental feeding. Nursery pond standing crops of 7000 to 14,000 kg/ha have been reported for intensively managed ponds (Sarig and Marek, 1974; Perez, 1995).

Supplemental feeds offered during nursing vary from 24 to 54% protein (Sarig and Marek, 1974; Snow et al., 1983; Barash, 1984; Siraj et al., 1988; Shiau et al., 1989; Dambo and Rana, 1992), but the optimal dietary protein level for tilapia during nursery rearing is not yet identified.

Nursery rearing may not be possible always because of inadequate facilities or because of growing season length. Young-of-year culture can be a viable alternative under these circumstances. Fingerlings are produced early in the season and may or may not be nursed to advanced fingerling stages prior to stocking into grow-out units. In Alabama, 5-g young-of-year mixed-sex tilapia stocked at 5000 or 10,000/ha are reared to 150 g (700 to 1200 kg/ha standing crop) in 104 d (Lovshin et al., 1990). Standing crops of 1400 to 3725 kg/ha are attained when 0.5- to 3-g young-of-year tilapia are stocked for a 145-d grow-out period (Green et al., 1995; Green et al., 1996). Mixed-sex populations of tilapia can be used for young-of-year culture because females can grow to 125 to 200 g in size before significant production of offspring occurs (Lovshin et al., 1990; Green et al., 1995); however, population size variation is reduced when monosex fish are stocked (Green et al., 1996).

REFERENCES

Alvendia-Casauay, A. and Cariño, V. S., Gonadal sex differentiation in *Oreochromis niloticus,* in Pullin, R. S. V., Bhukaswan, T., Tonguthai, K., and Maclean, J. L., Eds., *The Second International Symposium on Tilapia in Aquaculture,* ICLARM Conference Proceedings 15, Department of Fisheries, Bangkok, Thailand, and International Center for Living Aquatic Resources Management, Manila, Philippines, 1988, 121–124.

Argue, B. J. and Phelps, R. P., Temperature effect on sex ratios in *Oreochromis niloticus, J. Appl. Ichthyol.,* 11, 126–128, 1995.

Argue, B. J. and Phelps, R. P., Evaluation of techniques for producing hormone sex-reversed *Oreochromis niloticus* fry. *J. Aquaculture Tropics,* 11, 153–159, 1996.

Balarin, J. D., Development planning for tilapia farming in Africa, in Pullin, R. S. V., Bhukaswan, T., Tonguthai, K., and Maclean, J. L., Eds., *The Second International Symposium on Tilapia in Aquaculture,* ICLARM Conference Proceedings 15, Department of Fisheries, Bangkok, Thailand, and International Center for Living Aquatic Resources Management, Manila, Philippines, 1988, 531–538.

Barash, H., Growth rates of young tilapia fingerlings fed on commercial eel and trout diets, *Bamidgeh,* 36, 70–79, 1984.

Baroiller, J. F., Chourrout, D., Fostier, A., and Jalabert, B., Temperature and sex chromosomes govern sex ratios of the mouthbrooding cichlid fish *Oreochromis niloticus, J. Exp. Zool.,* 273, 216–223, 1995.

Bautista, A. M., Carlos, M. H., and San Antonio, A. I., Hatchery production of *Oreochromis niloticus* L. at different sex ratios and stocking densities, *Aquaculture,* 73, 85–95, 1988.

Behrends, L. L., Kingsley, J. B., and Price, A. H., III, Hatchery production of blue tilapia, *Oreochromis aureus* (Steindachner), in small suspended net hapas, *Aquaculture Fish. Manage.,* 24, 237–243, 1993.

Berger, A. and Rothbard, S., Androgen induced sex-reversal of red tilapia fry stocked in cages within ponds, *Bamidgeh,* 39, 49–57, 1987.

Bocek, A., Phelps, R. P., and Popma, T. J., Effect of feeding frequency on sex reversal and on growth of Nile tilapia, *Oreochromis niloticus, J. Aquaculture Tropics,* 1, 97–103, 1992.

Broussard, M. C., Reyes, R., and Raguindin, F., Evaluation of hatchery management schemes for large scale production of *Oreochromis niloticus* fingerlings in Central Luzon, Philippines, in Fishelson, L. and Yaron, Z., compilers, *Proc. Int. Symposium on Tilapia in Aquaculture, Nazareth, Israel, 8–13 May 1983,* Tel Aviv University, Israel, 1983, 414–424.

Brummett, R. E., Environmental regulation of sexual maturation and reproduction in tilapia, *Rev. Fish. Sci.,* 3, 231–248, 1995.

Buddle, C. R., Androgen inducement of sex inversion of *Oreochromis* (Trewavas) hybrid fry stocked in cages standing in an earthen pond, *Aquaculture,* 40, 233–239, 1984.

Chambers, S. A, Sex Reversal of Nile Tilapia in the Presence of Natural Food, M.S. thesis, Auburn University, Auburn, AL, 1984.

Chang, S.-L., Huang, C.-M., and Liao, I.-C., The effect of various feeds on seed production by Taiwanese red tilapia, in Pullin, R. S. V., Bhukaswan, T., Tonguthai, K., and Maclean, J. L., Eds., *The Second International Symposium on Tilapia in Aquaculture,* ICLARM Conference Proceedings 15, Department of Fisheries, Bangkok, Thailand, and International Center for Living Aquatic Resources Management, Manila, Philippines, 1988, 319–322.

Chervinski, J., Growth of *Tilapia aurea* in brackish water ponds, *Bamidgeh,* 18, 81–83, 1966.

Cissé, A., Effects of varying protein levels on spawning frequency and growth of *Sarotherodon melanotheron,* in Pullin, R. S. V., Bhukaswan, T., Tonguthai, K., and Maclean, J. L., Eds., *The Second International Symposium on Tilapia in Aquaculture,* ICLARM Conference Proceedings 15, Department of Fisheries, Bangkok, Thailand, and International Center for Living Aquatic Resources Management, Manila, Philippines, 1988, 329–333.

Costa-Pierce, B. A. and Hadikusumah, H., Production management of double-net tilapia *Oreochromis* spp. hatcheries in a eutrophic tropical reservoir, *J. World Aquaculture Soc.,* 26, 453–459, 1995.

Cravedi, J. P., Delous, G., and Rao, D., Disposition and elimination routes of 17 α-methyltestosterone in rainbow trout (*Salmo gairdneri*), *Can. J. Fish. Aquatic Sci.,* 46, 159–165, 1989.

Curtis, L. R., Diren, F. T., Hurley, M. D., Seim, W. D., and Tubb, R. A., Disposition and elimination of 17 α-methyltestosterone in Nile tilapia (*Oreochromis niloticus*), *Aquaculture,* 99, 193–201, 1991.

Dadzie, S., Laboratory experiment on the fecundity and frequency of spawning in *Tilapia aurea, Bamidgeh,* 22, 14–18, 1970a.

Dadzie, S., Preliminary report on induced spawning of *Tilapia aurea, Bamidgeh,* 22, 9–13, 1970b.

Dambo, W. B. and Rana, K. J., Effect of stocking density on growth and survival of *Oreochromis niloticus* (L.) fry in the hatchery, *Aquaculture Fish. Manage.,* 23, 71–80, 1992.

De Silva, S. S. and Radampola, K., Effect of dietary protein level on the reproductive performance of *Oreochromis niloticus,* in Hirano, R. and Hanyu, I., Eds., *The Second Asian Fisheries Forum,* Asian Fisheries Society, Manila, Philippines, 1990, 559–563.

Drug Facts and Comparisons, Androgens, in *Drugs Facts and Comparisons,* 48th ed., Facts and Comparisons Division, J. B. Lippincott, St. Louis, MO, 1994, 397–430.

Eckstein, B. and Spira, M., Effect of sex hormone on gonadal differentiation in a cichlid, *Tilapia aurea, Biol. Bull.,* 129, 482–489, 1965.

Ernst, D. H., Watanabe, W. O., Ellingson, L. J., Wicklund, R. I., and Olla, B. L., Commercial-scale production of red tilapia seed in low- and brackish-salinity tanks, *J. World Aquaculture Soc.,* 22, 36–44, 1991.

Fagerlund, U. H. M. and McBride, J. R., Distribution and disappearance of radioactivity in blood and tissues of coho salmon (*Oncorhynchus kisutch*) after oral administration of ^{3}H-testosterone, *J. Fish. Res. Bd. Can.,* 35, 893–900, 1978.

Fagerlund, U. H. M. and Dye, H. M., Depletion of radioactivity from yearling coho salmon (*Oncorhynchus kisutch*) after extended ingestion of anabolically effective doses of 17 α-methyltestosterone-1,2-^{3}H, *Aquaculture,* 18, 303–315, 1979.

FAO (Food and Agriculture Organization of the United Nations), Aquaculture Production 1986–1992, FAO Fisheries Circular No. 815, Revision 6, Food and Agriculture Organization of the United Nations, Rome, Italy, 1994.

Ferby, R. and Lutz, P., Energy partitioning in fish: the activity-related cost of osmoregulation in a euryhaline cichlid, *J. Exp. Biol.,* 128, 63–85, 1987.

Fineman Kalio, A. S., Preliminary observations on the effect of salinity on the reproduction and growth of freshwater Nile tilapia, *Oreochromis niloticus* (L.), cultured in brackish water ponds, *Aquaculture Fish. Manage.,* 19, 313–320, 1988.

Gale, W. L., Fitzpatrick, M. S., and Schreck, C. B., Immersion of Nile tilapia (*Oreochromis niloticus*) in 17-α methyltestosterone and mestanolone for the production of all-male populations, in Goetze, F. W. and Thomas, P., Eds., *Proceedings of the Fifth International Symposium on Reproductive Physiology of Fish, Fish Symposium 95,* Austin, TX, 1995, 117.

Galman, O. R., Moreau, J., and Avtalion, R. R., Breeding characteristics and growth performance of Philippine red tilapia, in Pullin, R. S. V., Bhukaswan, T., Tonguthai, K., and Maclean, J. L., Eds., *The Second International Symposium on Tilapia in Aquaculture,* ICLARM Conference Proceedings 15, Department of Fisheries, Bangkok, Thailand, and International Center for Living Aquatic Resources Management, Manila, Philippines, 1988, 169–175.

Goudie, C. A., Shelton, W. L., and Parker, N. C., Tissue distribution and elimination of radio-labeled methyltestosterone fed to sexually undifferentiated blue tilapia, *Aquaculture,* 58, 215–226, 1986a.

Goudie, C. A., Shelton, W. L., and Parker, N. C., Tissue distribution and elimination of radio-labeled methyltestosterone fed to adult blue tilapia, *Aquaculture,* 58, 227–240, 1986b.

Green, B. W., Substitution of organic matter for pelleted feed in tilapia production, *Aquaculture,* 101, 213–222, 1992.

Green, B. W., El Nagdy, Z., Kenawy, D. A. R., Shaker, I., and El Gamal, A. R., Yield characteristics of two species of tilapia under two different pond environments, in Goetze, B., Berkman, H., and Egna, H. Eds., *Egypt Project Final Report, October 1992–March 1995,* Pond Dynamics/Aquaculture CRSP, Office of International Research and Development, Oregon State University, Corvallis, Oregon, 1995, 10–12.

Green, B. W. and Lopez, L. A., Factibilidad de la produccion masiva de alevines machos de *Tilapia nilotica* a través de la inversión hormonal de sexo en Honduras, *Agronomia Mesoamericana,* 1, 21–25, 1990.

Green, B. W., Rizkalla, E. H., and El Gamal, A. R., Mass production of Nile (*Oreochromis niloticus*) and blue (*O. aureus*) tilapia fry, in Egna, H. S., Bowman, J., Goetze, B., and Weidner, N., Eds., *Twelfth Annual Administrative Report,* Pond Dynamics/Aquaculture CRSP, 1994, Oregon State University, Corvallis, 1995, 192–195.

Green, B. W. and Teichert-Coddington, D. R., Effects of fry stocking rate, hormone treatment, and temperature on the production of sex-reversed *Oreochromis niloticus,* in Egna, H. S., Bowman, J., and McNamara, M., Eds., *Eighth Annual Administrative Report,* Pond Dynamics/Aquaculture CRSP, 1990, Oregon State University, Corvallis, 1991, 22–25.

Green, B. W. and Teichert-Coddington, D. R., Production of *Oreochromis niloticus* fry for hormonal sex reversal in relation to water temperature, *J. Appl. Ichthyol.,* 9, 230–236, 1993.

Green, B. W. and Teichert-Coddington, D. R., Growth of control and androgen-treated Nile tilapia during treatment, nursery and grow-out phases in tropical fish ponds, *Aquaculture Fish. Manage.,* 25, 613–621, 1994.

Green, B. W., Teichert-Coddington, D. R., and Hanson, T. R., Development of Semi-Intensive Aquaculture Technologies in Honduras, International Center for Aquaculture and Aquatic Environments Research and Development Series No. 39, Auburn University, Auburn, AL, 1994.

Guerrero, R. D., Use of androgens for the production of all-male *Tilapia aurea* (Steindachner), *Trans. Am. Fish. Soc.,* 104, 342–348, 1975.

Guerrero, R. D., III, Production of Nile tilapia fry and fingerlings in earthen ponds at Pila, Laguna, Philippines, in Maclean, J. L., Dizon, L. B., and Hosillos, L. V., Eds., *The First Asian Fisheries Forum,* Asian Fisheries Society, Manila, Philippines, 1986, 49–52.

Guerrero, R. D., III and Garcia, A. M., Studies on the fry production of *Sarotherodon niloticus* in a lake-based hatchery, in Fishelson, L. and Yaron, Z., compilers, *Proc. Int. Symp. on Tilapia in Aquaculture, Nazareth, Israel, 8–13 May 1983,* Tel Aviv University, Israel, 1983, 386–393.

Guerrero, R. D., III and Guerrero, L. A., Further observations on the fry production of *Oreochromis niloticus* in concrete tanks, *Aquaculture,* 47, 257–261, 1985.

Guerrero, R. D., III and Guerrero, L. A., Feasibility of commercial production of sex-reversed Nile tilapia fingerlings in the Philippines, in Pullin, R. S. V., Bhukaswan, T., Tonguthai, K., and Maclean, J. L., Eds., *The Second International Symposium on Tilapia in Aquaculture,* ICLARM Conference Proceedings 15, Department of Fisheries, Bangkok, Thailand, and International Center for Living Aquatic Resources Management, Manila, Philippines, 1988, 183–186.

Hanson, B. J., Moehl, J. F., Jr., Veverica, K. L., Rwangano, F., and Van Speybroeck, M., Pond culture of tilapia in Rwanda, a high altitude equatorial African country, in Pullin, R. S. V., Bhukaswan, T., Tonguthai, K., and Maclean, J. L., Eds., *The Second International Symposium on Tilapia in Aquaculture,* ICLARM Conference Proceedings 15, Department of Fisheries, Bangkok, Thailand, and International Center for Living Aquatic Resources Management, Manila, Philippines, 1988, 553–559.

Hiott, A. E. and Phelps, R. P., Effects of initial age and size on sex reversal of *Oreochromis niloticus* fry using methyltestosterone, *Aquaculture,* 112, 301–308, 1993.

Hishamunda, N. and Moehl, J. F., Jr., Rwanda National Fish Culture Project, International Center for Aquaculture and Aquatic Environments Research and Development Series No. 34, Auburn University, Auburn, AL, 1989.

Holl, M., Production d'alvins de Tilapia nilotica en Station Domaniale, Project PNUD/FAO/IVC/77/003, Developpement de la pisciculture en Eaux continentales en Côte d'Ivoire, Document Technique No. 10, 1983.

Hopkins, K. D., Shelton, W. L., and Engle, C. R., Estrogen sex-reversal of *Tilapia aurea, Aquaculture,* 18, 263–268, 1979.

Hughes, D. G. and Behrends, L. L., Mass production of *Tilapia nilotica* seed in suspended net enclosures, in Fishelson, L. and Yaron, Z., compilers, *Proc. Int. Symp. on Tilapia in Aquaculture, Nazareth, Israel, 8–13 May 1983,* Tel Aviv University, Israel, 1983, 394–401.

Hulata, G., Rothbard, S., Itzkovich, J., Wohlfarth, G., and Halevy, A., Differences in hybrid fry production between two strains of the Nile tilapia, *Progr. Fish Cult.,* 47, 42–49, 1985.

Hunter, G. A. and Donaldson, E. M., Hormonal sex control and its application to fish culture, in Hoar, W. S., Randall, D. J., and Donaldson, E. M., Eds., *Fish Physiology,* Vol. 9B, Academic Press, New York, 1983, 223–303.

Jalabert, B. and Zohar, Y., Reproductive physiology in cichlid fishes, with particular reference to *Tilapia* and *Sarotherodon,* in Pullin, R. S. V. and Lowe-McConnell, R. H., Eds., *The Biology and Culture of Tilapias,* ICLARM Conference Proceedings 7, International Center for Living Aquatic Resources Management, Manila, Philippines, 1982, 129–140.

Jo, J.-Y., Smitherman, R. O., and Behrends, L. L., Effects of dietary 17-α methyltestosterone on sex reversal and growth of *Oreochromis aureus,* in Pullin, R. S. V., Bhukaswan, T., Tonguthai, K., and Maclean, J. L., Eds., *The Second International Symposium on Tilapia in Aquaculture,* ICLARM Conference Proceedings 15, Department of Fisheries, Bangkok, Thailand, and International Center for Living Aquatic Resources Management, Manila, Philippines, 1988, 203–207.

Johnstone, R., Macintosh, D. J., and Wright, R. S., Elimination of orally administered 17 α-methyltestosterone by *Oreochromis mossambicus* (Tilapia) and *Salmo gairdneri* (Rainbow trout) juveniles, *Aquaculture,* 35, 249–257, 1983.

Lahav, E., Use of sex-reversed females to produce all-male tilapia (*Oreochromis aureus*) fry, *Israeli J. Aquaculture Bamidgeh,* 45, 131–136, 1993.

Lee, J. C., Reproduction and Hybridization of Three Cichlid Fishes, *Tilapia aurea, T. hornorum,* and *T. nilotica* in Aquaria and Plastic Pools, Ph.D. dissertation, Auburn University, Auburn, AL, 1979.

Little, D. C., An Evaluation of Strategies for Production of Nile Tilapia (*Oreochromis niloticus* L.) Fry Suitable for Hormonal Treatment, Ph.D. dissertation, University of Stirling, Stirling, Scotland, 1989.

Little, D. C., Macintosh, D. J., and Edwards, P., Improving spawning synchrony in the Nile tilapia, *Oreochromis niloticus* (L.), *Aquaculture Fish. Manage.,* 24, 399–405, 1993.

Lone, K. P. and Matty, A. J., Uptake and disappearance of radioactivity in blood and tissues of carp (Cyprinus carpio) after feeding ^{3}H-testosterone, *Aquaculture,* 24, 315–326, 1981.

Lovshin, L. L., Progress Report on Fisheries Development in Northeast Brazil, International Center for Aquaculture and Aquatic Environments Research and Development Series No. 26, Auburn University, Auburn, AL, 1980.

Lovshin, L. L., Tilapia hybridization, in Pullin, R. S. V. and Lowe-McConnell, R. H., Eds., *The Biology and Culture of Tilapias,* ICLARM Conference Proceedings 7, International Center for Living Aquatic Resources Management, Manila, Philippines, 1982, 279–308.

Lovshin, L. L. and Ibrahim, H. H., Effects of broodstock exchange on *Oreochromis niloticus* egg and fry production in net enclosures, in Pullin, R. S. V., Bhukaswan, T., Tonguthai, K., and Maclean, J. L., Eds., *The Second International Symposium on Tilapia in Aquaculture,* ICLARM Conference Proceedings 15, Department of Fisheries, Bangkok, Thailand, and International Center for Living Aquatic Resources Management, Manila, Philippines, 1988, 231–236.

Lovshin, L. L., Tave, D., and Lieutaud, A. O., Growth and yield of mixed-sex, young-of-the-year *Oreochromis niloticus* raised at two densities in earthen ponds in Alabama, USA, *Aquaculture,* 89, 21–26, 1990.

Macintosh, D. J. and De Silva, S. S., The influence of stocking density and food ration on fry survival and growth in *Oreochromis mossambicus* and *O. niloticus* female x *O. aureus* male hybrids reared in a closed circulated system, *Aquaculture,* 41, 345–358, 1984.

Macintosh, D. J. and Little, D. C., Nile tilapia (*Oreochromis niloticus*), in Bromage, N. R. and Roberts, R. J., Eds., *Broodstock Management and Egg and Larval Quality,* Blackwell Science Ltd., Oxford, 1995, 277–320.

Macintosh, D. J., Singh, T. B., Little, D. C., and Edwards, P., Growth and sexual development of 17 α-methyltestosterone- and progesterone-treated Nile tilapia (*Oreochromis niloticus*) reared in earthen ponds, in Pullin, R. S. V., Bhukaswan, T., Tonguthai, K., and Maclean, J. L., Eds., *The Second International Symposium on Tilapia in Aquaculture,* ICLARM Conference Proceedings 15, Department of Fisheries, Bangkok, Thailand, and International Center for Living Aquatic Resources Management, Manila, Philippines, 1988, 457–463.

Mair, G. C., Beardmore, J. A., and Skibinski, D. O. F., Experimental evidence for environmental sex determination in *Oreochromis* species, in Hirano, R. and Hanyu, I., Eds., *The Second Asian Fisheries Forum,* Asian Fisheries Society, Manila, Philippines, 1990.

Mair, G. C., Estabillo, C. C., Sevilleja, R. C., and Recometa, R. D., Small-scale fry production systems for Nile tilapia, *Oreochromis niloticus* (L.), *Aquaculture Fish. Manage.,* 24, 229–235, 1993.

Mair, G. C., Scott, A. G., Penmann, D. G., Beardmore, J. A., and Skibinski, D. O. F., Sex determination in the genus *Oreochromis.* 1. Sex reversal, gynogenesis and triploidy in *O. niloticus* (L.), *Theor. Appl. Genet.,* 82, 144–152, 1991a.

Mair, G. C., Scott, A. G., Penmann, D. J., Skibinski, D. O. F., and Beardmore, J. A., Sex determination in the genus *Oreochromis.* 2. Sex reversal, gynogenesis and triploidy in *O. aureus* Steindachner, *Theor. Appl. Genet.,* 82, 153–160, 1991b.

Majumdar, K. C. and McAndrew, B. J., Sex ratios from interspecific crosses within the tilapias, in Fishelson, L. and Yaron, Z., compilers, *Proc. Int. Symp. on Tilapia in Aquaculture, Nazareth, Israel, 8–13 May 1983,* Tel Aviv University, Israel, 1983, 261–269.

Maluwa, A. O. and Costa-Pierce, B. A., Effect of broodstock density on *Oreochromis shiranus* fry production in hapas, *J. Appl. Aquaculture,* 2, 63–76, 1993.

Mbarererehe, F., Contribution à l'étude de l'influencede la température et la durée de traitement sur la production des alevins monosexes du *Tilapia nilotica.* Mémoire présenté en vue de l'obtention du diplôme d'ingénieur technicien. Institut Supérieur d'Agriculture et d'Elevage de Busogo, Ruhengeri, Rwanda, 1992.

McGeachin, R. B., Robinson, E. H., and Neil, W. H., Effect of feeding high levels of androgens on the sex ratio of *Oreochromis aureus, Aquaculture,* 61, 317–321, 1987.

Meriwether, F. H. and Torrans, E. L., Evaluation of a new androgen (Mibolerone) and procedure to induce functional sex reversal in tilapia, in Maclean, J. L., Dizon, L. B., and Hosillos, L. V., Eds., *The First Asian Fisheries Forum,* Asian Fisheries Society, Manila, Philippines, 1986, 675–678.

Meyer, D. E., Growth, Survival and Sex Ratios of *Tilapia hornorum, Tilapia nilotica* and Their Hybrid (*T. nilotica* female x *T. hornorum* male) Treated with 17 Alpha-Methyltestosterone, Ph.D. dissertation, Auburn University, Auburn, AL, 1990.

Mires, D., On the high percent of tilapia males encountered in captive spawnings and the effect of temperature on this phenomenon, *Bamidgeh,* 26, 3–11, 1974.

Mires, D., Current techniques for the mass production of tilapia hybrids as practiced at Ein Hamifratz fish hatchery, *Bamidgeh,* 35, 3–8, 1983.

Mires, D., Genetic problems concerning the production of tilapia in Israel, *Bamidgeh,* 37, 51–54, 1985.

Moehl, J. F., Jr., Veverica, K. L., Hanson, B. J., and Hishamunda, N., Development of appropriate pond management techniques for use by rural Rwandan farmers, in Pullin, R. S. V., Bhukaswan, T., Tonguthai, K., and Maclean, J. L., Eds., *The Second International Symposium on Tilapia in Aquaculture,* ICLARM Conference Proceedings 15, Department of Fisheries, Bangkok, Thailand, and International Center for Living Aquatic Resources Management, Manila, Philippines, 1988, 561–568.

Nakamura, M., Dosage-dependent changes in the effect of oral administration of methyl-testosterone on gonadal sex differentiation in *Tilapia mossambica, Bull. Fac. Fish. Hokkaido Univ.,* 26, 99–108, 1975.

Nakamura M. and Iwahashi, M., Studies on the practical masculinization in *Tilapia nilotica* by oral administration of androgen, *Bull. Jpn. Soc. Sci. Fish.,* 48, 763–769, 1982.

Nakamura, M. and Takahashi, H., Gonadal sex differentiation in *Tilapia mossambica,* with special reference to the time of estrogen treatment effective in inducing complete feminization of genetic males, *Bull. Fac. Fish. Hokkaido Univ.,* 24, 1–13, 1973.

Nakamura M. and Takahashi, H., Sex control in cultured tilapia (*Tilapia mossambica*) and salmon (*Onchorhynchus masou*), in Lofts, B. and Holmes, W. N., Eds., *Current Trends in Comparative Endocrinology,* Hong Kong University Press, Hong Kong, 1985, 1255–1260.

Nelson, S. G. and Eldredge, L. G., Distribution and status of introduced cichlid fishes of the genera *Oreochromis* and *Tilapia* in the islands of the South Pacific and Micronesia, *Asian Fish. Sci.,* 4, 11–22, 1991.

Nsengiyumva, V., Rurangwa, E., and Veverica, K. L., Sex reversal of *Oreochromis niloticus* fry in a relatively cool environment using two different hormones, in Egna, H. S., Bowman, J., Goetze, B., and Weidner, N., Eds., *Seventh Annual Administrative Report,* Pond Dynamics/Aquaculture CRSP, 1989, Oregon State University, Corvallis, 1990, 26–28.

Obi, A. and Shelton, W. L., Androgen and estrogen sex reversal in *Tilapia hornorum,* in Fishelson, L. and Yaron, Z., compilers, *Proc. Int. Symp. on Tilapia in Aquaculture, Nazareth, Israel, 8–13 May 1983,* Tel Aviv University, Israel, 1983, 165–173.

Okoko, M. and Phelps, R. P., Effect of methyltestosterone concentration on sex ratio, growth and development of Nile tilapia, in Goetze, F. W. and Thomas, P., Eds., *Proc. Fifth International Symposium on Reproductive Physiology of Fish,* Fish Symposium 95, Austin, TX, 1995.

Owusu-Frimpong, M. and Nijjhar, B., Induced sex reversal in *Tilapia nilotica* (Cichlidae) with methyltestosterone, *Hydrobiologia,* 78, 157–160, 1981.

Pandian, T. J. and Sheela, S. G., Hormonal induction of sex reversal in fish, *Aquaculture,* 138, 1–22, 1995.

Pandian, T. J. and Varadaraj, K., Techniques for producing all-male and all-triploid *Oreochromis mossambicus,* in Pullin, R. S. V., Bhukaswan, T., Tonguthai, K., and Maclean, J. L., Eds., *The Second International Symposium on Tilapia in Aquaculture,* ICLARM Conference Proceedings 15, Department of Fisheries, Bangkok, Thailand, and International Center for Living Aquatic Resources Management, Manila, Philippines, 1988, 243–249.

Pantastico, J. B., Dangilan, M. M. A., and Eguia, R. V., Cannibalism among different sizes of tilapia (*Oreochromis niloticus*) fry/fingerlings and the effect of natural food, in Pullin, R. S. V., Bhukaswan, T., Tonguthai, K., and Maclean, J. L., Eds., *The Second International Symposium on Tilapia in Aquaculture,* ICLARM Conference Proceedings 15, Department of Fisheries, Bangkok, Thailand, and International Center for Living Aquatic Resources Management, Manila, Philippines, 1988, 465–468.

Payne, A. I., Estuarine and salt tolerant tilapias, in Fishelson, L. and Yaron, Z., compilers, *Proc. Int. Symp. on Tilapia in Aquaculture, Nazareth, Israel, 8–13 May 1983,* Tel Aviv University, Israel, 1983, 534–543.

Perez, H., Cultivo de tilapia en Centroamerica, First National Meeting on Tilapia Culture in Nicaragua, PRADEPESCA, Managua, Nicaragua, 1995.

Peters, H. M., Fecundity, egg weight and oocyte development in tilapias (Cichlidae, Teleostei), ICLARM Translations 2, International Center for Living Aquatic Resources Management, Manila, Philippines, 1983.

Phelps, R. P. and Cerezo, G., The effect of confinement in hapas on sex reversal and growth of *Oreochromis niloticus,* *J. Appl. Aquaculture,* 1, 73–81, 1992.

Phelps, R. P., Cole, W., and Katz, T., Effect of fluoxymesterone on sex ratio and growth of Nile tilapia, *Oreochromis niloticus* (L.), *Aquaculture Fish. Manage.,* 23, 405–410, 1992.

Phelps, R. P., Conterras Salazar, G., Abe, V., and Argue, B., Sex reversal and nursery growth of Nile tilapia, *Oreochromis niloticus* (L.), free-swimming in earthen ponds, *Aquaculture Res.,* 26, 293–295, 1995.

Philippart, J.-Cl. and Ruwet, J.-Cl., Ecology and distribution of tilapias, in Pullin, R. S. V. and Lowe-McConnell, R. H., Eds., *The Biology and Culture of Tilapias,* ICLARM Conference Proceedings 7, International Center for Living Aquatic Resources Management, Manila, Philippines, 1982, 15–59.

Piferrer, F. and Donaldson, E. M., Gonadal differentiation in coho salmon, *Oncorhynchus kisutch,* after a single treatment with androgen or estrogen at different stages during ontogenesis, *Aquaculture,* 77, 251–262, 1989.

Piper, R. G., McElwain, I. B., Orme, L. E., McCraren, J. P., Fowler, L. G., and Leonard, J. R., *Fish Hatchery Management,* United States Department of the Interior, Fish and Wildlife Service, Washington, D.C., 1982.

Popma, T. J., Freshwater Fish Culture Development Project, ESPOL, Guayaquil, Ecuador: Final Technical Report, Department of Fisheries and Allied Aquacultures, Auburn University, Auburn, AL, 1987.

Popma, T. J. and Green, B. W., Sex Reversal of Tilapia in Earthen Ponds, International Center for Aquaculture and Aquatic Environments Research and Development Series No. 35, Auburn University, Auburn, AL, 1990.

Pruginin, Y., Rothbard, S., Wohlfarth, G., Halevy, A., Moav, R., and Hulata, G., All-male broods of *Tilapia nilotica* x *T. aurea* hybrids, *Aquaculture,* 6, 11–21, 1975.

Pullin, R. S. V., Ed., Tilapia genetic resources for aquaculture, in ICLARM Conference Proceedings 16, International Center for Living Aquatic Resources Management, Manila, Philippines, 1988.

Pullin, R. S. V. and Capili, J. B., Genetic improvements of tilapias: problems and prospects, in Pullin, R. S. V., Bhukaswan, T., Tonguthai, K., and Maclean, J. L., Eds., *The Second International Symposium on Tilapia in Aquaculture,* ICLARM Conference Proceedings 15, Department of Fisheries, Bangkok, Thailand, and International Center for Living Aquatic Resources Management, Manila, Philippines, 1988, 259–266.

Rana, K. J., Influence of egg size on the growth, onset of feeding, point-of-no-return, and survival of unfed *Oreochromis mossambicus* fry, *Aquaculture,* 46, 119–131, 1985.

Rana, K. J., An evaluation of two types of containers for the artificial incubation of *Oreochromis* eggs, *Aquaculture Fish. Manage.,* 17, 139–145, 1986.

Rana, K. J., Reproductive biology and hatchery rearing of tilapia eggs and fry, in Muir, J. F. and Roberts, R. J., Eds., *Recent Advances in Aquaculture,* Vol. 3, Croom Helm Ltd., London, 1988, 343–407.

Rana, K. J. and Macintosh, D. J., A comparison of the quality of hatchery-reared *Oreochromis niloticus* and *O. mossambicus* fry, in Pullin, R. S. V., Bhukaswan, T., Tonguthai, K., and Maclean, J. L., Eds., *The Second International Symposium on Tilapia in Aquaculture,* ICLARM Conference Proceedings 15, Department of Fisheries, Bangkok, Thailand, and International Center for Living Aquatic Resources Management, Manila, Philippines, 1988, 497–502.

Ridha, M. and Cruz, E. M., Effect of age on the fecundity of the tilapia *Oreochromis spilurus*, *Asian Fish. Sci.,* 2, 239–247, 1989.

Ridha, M. T. and Lone, K. P., Effect of oral administration of different levels of 17-α methyltestosterone on the sex reversal, growth, and food conversion efficiency of the tilapia *Oreochromis spilurus* (Günther) in brackish water, *Aquaculture Fish. Manage.,* 21, 391–397, 1990.

Rothbard, S. and Hulata, G., Closed-system incubator for cichlid eggs, *Progr. Fish-Cult.,* 42, 203–204, 1980.

Rothbard, S. and Pruginin, Y., Induced spawning and artificial incubation of Tilapia, *Aquaculture,* 5, 315–321, 1975.

Rothbard, S., Solnik, E., Shabbath, S., Amado, R., and Grabie, I., The technology of mass production of hormonally sex-inversed all-male tilapias, in Fishelson, L. and Yaron, Z., compilers, *Proc. Int. Symp. on Tilapia in Aquaculture, Nazareth, Israel, 8–13 May 1983,* Tel Aviv University, Israel, 1983, 425–434.

Rurangwa, E., Veverica, K. L., Seim, W. K., and Popma, T. J., On-farm production of mixed sex *Oreochromis niloticus* at different elevations (1370 to 2230 m), in Egna, H. S., McNamara, M., and Weidner, N., Eds., *Ninth Annual Administrative Report,* Pond Dynamics/Aquaculture CRSP, 1991, Oregon State University, Corvallis, 1992, 35–40.

Santiago, C. B., Aldaba, M. B., Abuan, E. F., and Laron, M. A., The effects of artificial diets on fry production and growth of *Oreochromis niloticus* breeders, *Aquaculture,* 47, 193–203, 1985.

Santiago, C. B. and Reyes, O. S., Effects of dietary lipid source in reproductive performance and tissue lipid levels on Nile tilapia *Oreochromis niloticus* (Linnaeus) broodstock, *J. Appl. Ichthyol.,* 9, 33–40, 1993.

Sarig, S. and Marek, M., Results of intensive and semi-intensive fish breeding techniques in Israel 1971–1973, *Bamidgeh,* 26, 28–48, 1974.

Schreck, C. B., Hormone treatment and sex manipulation in fishes, in Schreck, C. B., Ed., *Control of Sex in Fishes,* Virginia Polytechnic Institute and State University, Blacksburg, VA, 1974, 84–106.

Scott, A. G., Penman, D. J., Beardmore, J. A., and Skibinski, D. O. F., The "YY" supermale in *Oreochromis niloticus* (L.) and its potential in aquaculture, *Aquaculture,* 78, 237–251, 1989.

Shelton, W. L., Hopkins, K. D., and Jensen, G. L., Use of hormones to produce monosex tilapia, in Smitherman, R. O., Shelton, W. L., and Grover, J. L., Eds., *Culture of Exotic Fishes Symposium Proceedings,* Fish Culture Section, American Fisheries Society, Auburn, AL, 1978, 10–33.

Shelton, W. L., Meriwether, F. H., Semmens, K. J., and Calhoun, W. E., Progeny sex ratios from intraspecific pair spawnings of *Tilapia aurea* and *Tilapia nilotica*, in Fishelson, L. and Yaron, Z., compilers, *Proc. Int. Symp. on Tilapia in Aquaculture, Nazareth, Israel, 8–13 May 1983,* Tel Aviv University, Israel, 1983, 270–280.

Shelton, W. L., Rodriguez-Guerrero, D., and Lopez-Macias, J., Factors affecting androgen sex reversal of *Tilapia aurea*, *Aquaculture,* 25, 59–65, 1981.

Shepperd, V. D., Androgen Sex Inversion and Subsequent Growth of Red Tilapia and Nile Tilapia, M.S. thesis, Auburn University, Auburn, AL, 1984.

Shiau, S.-Y., Kwok, C.-C., Hwang, J.-Y., Chen, C.-M., and Lee, C.-M., Replacement of fishmeal with soybean meal in male tilapia (*Oreochromis niloticus* x *O. aureus*) fingerling diets at a suboptimal protein level, *J. World Aquaculture Soc.,* 20, 230–235, 1989.

Siraj, S. S., Kamaruddin, Z., Satar, M. K. A., and Kamarudin, M. S., Effects of feeding frequency on growth, food conversion and survival of red tilapia (*Oreochromis mossambicus/O. niloticus*) hybrid fry, in Pullin, R. S. V., Bhukaswan, T., Tonguthai, K., and Maclean, J. L., Eds., *The Second International Symposium on Tilapia in Aquaculture,* ICLARM Conference Proceedings 15, Department of Fisheries, Bangkok, Thailand, and International Center for Living Aquatic Resources Management, Manila, Philippines, 1988, 383–386.

Siraj, S. S., Smitherman, R. O., Castillo-Gallusser, S., and Dunham, R. A., Reproductive traits for three year classes of *Tilapia nilotica* and maternal effects on their progeny, in Fishelson, L. and Yaron, Z., compilers, *Proc. Int. Symp. on Tilapia in Aquaculture, Nazareth, Israel, 8–13 May 1983,* Tel Aviv University, Israel, 1983, 210–218.

Smitherman, R. O. and Tave, D., Maintenance of genetic quality in cultured tilapia, *Asian Fish. Sci.,* 1, 75–82, 1987.

Snow, J. R., Berrios-Hernandez, J. M., and Ye, H. Y., A modular system for producing tilapia seed using simple facilities, in Fishelson, L. and Yaron, Z., compilers, *Proc. Int. Symp. on Tilapia in Aquaculture, Nazareth, Israel, 8–13 May 1983,* Tel Aviv University, Israel, 1983, 402–413.

Srisakultiew, P. and Wee, K. L., Synchronous spawning of Nile tilapia through hypophysation and temperature manipulation, in Pullin, R. S. V., Bhukaswan, T., Tonguthai, K., and Maclean, J. L., Eds., *The Second International Symposium on Tilapia in Aquaculture,* ICLARM Conference Proceedings 15, Department of Fisheries, Bangkok, Thailand, and International Center for Living Aquatic Resources Management, Manila, Philippines, 1988, 275–284.

Stickney, R. R., Tilapia tolerance of saline waters: a review, *Progr. Fish-Cult.,* 48, 161–167, 1986.

Subasinghe, R. P., Sommerville, C., and Rana, K. J., Effect of nitrite-nitrogen (NO_2-N) on the eggs and sac fry of *Oreochromis mossambicus* (Peters), in Hirano, R. and Hanyu, I., Eds., *The Second Asian Fisheries Forum,* Asian Fisheries Society, Manila, Philippines, 1990, 69–72.

Suresh, A. V. and Lin, C. K., Tilapia culture in saline waters: a review, *Aquaculture,* 106, 201–226, 1992.

Tave, D., Genetics and breeding of tilapias: a review, in Pullin, R. S. V., Bhukaswan, T., Tonguthai, K., and Maclean, J. L., Eds., *The Second International Symposium on Tilapia in Aquaculture,* ICLARM Conference Proceedings 15, Department of Fisheries, Bangkok, Thailand, and International Center for Living Aquatic Resources Management, Manila, Philippines, 1988, 285–293.

Tave, D., *Genetics for Fish Hatchery Managers,* 2nd ed., Van Nostrand Reinhold, New York, 1993.

Tayamen, M. M. and Shelton, W. L., Inducement of sex reversal in *Sarotherodon niloticus* (Linnaeus), *Aquaculture,* 14, 349–354, 1978.

Torrans, L., Meriwether, F., Lowell, F., Wyatt, B., and Gwinup, P. D., Sex-reversal of *Oreochromis aureus* by immersion in milbolerone, a synthetic steroid, *J. World Aquaculture Soc.,* 19, 97–102, 1988.

Trewavas, E., Tilapias: Taxonomy and speciation, in Pullin, R. S. V. and Lowe-McConnell, R. H., Eds., *The Biology and Culture of Tilapias,* ICLARM Conference Proceedings 7, International Center for Living Aquatic Resources Management, Manila, Philippines, 1982, 3–13.

Trombka, D. and Avtalion, R., Sex determination in tilapia — a review, *Israeli J. Aquaculture Bamidgeh,* 45, 26–37, 1993.

United States Pharmacopeial Convention, USP Dispensing Information (USP DI), Vol. I, *Drug Information for the Health Care Professional,* 11th ed., United States Pharmacopeial Convention, Rockville, MD, 1991.

Varadaraj, K., Production of monosex male *Oreochromis mossambicus* (Peters) by administering 19-norethisterone acetate, *Aquaculture Fish. Manage.,* 21, 133–135, 1990.

Varadaraj, K., Kumari, S. S., and Pandian, T. J., Comparison of conditions for hormonal sex reversal of Mozambique tilapias, *Progr. Fish-Cult.,* 56, 81–90, 1994.

Varadaraj, K. and Pandian, T. J., Masculinization of *Oreochromis mossambicus* by administration of 17a-methyl-5-androsten-3b-17b-diol through rearing water, *Curr. Sci.,* 56, 412–413, 1987.

Varadaraj, K. and Pandian, T. J., Monosex male broods of *Oreochromis mossambicus* produced through artificial sex reversal with 17α methyl-4 androsten-17β-ol-3-one, *Curr. Trends Life Sci.,* 15, 169–173, 1989.

Verdegem, M. C. and McGinty, A. S., Effects of frequency of egg and fry removal on spawning by *Tilapia nilotica* in hapas, *Progr. Fish-Cult.,* 49, 129–131, 1987.

Verdegem, M. C. and McGinty, A. S., Evaluation of edge seining for harvesting *Oreochromis niloticus* fry from spawning ponds, *Aquaculture,* 80, 195–200, 1989.

Veverica, K. L., Rurangwa, E., Lund, J., and Henderson, C., *Proc. Third Conference on High Altitude Tilapia Culture in Africa,* Pond Dynamics/Aquaculture CRSP Research Report, Oregon State University, Corvallis, in press.

Watanabe, W. O., Burnett, K. M., Olla, B. L., and Wicklund, R. I., The effects of salinity on reproductive performance of Florida red tilapia, *J. World Aquaculture Soc.,* 20, 223–229, 1989.

Watanabe, W. O., Clark, J. H., Dunham, J. B., Wicklund, R. I., and Olla, B. L., Production of fingerling Florida red tilapia (*Tilapia hornorum* x *T. mossambica*) in floating marine cages, *Progr. Fish-Cult.,* 52, 158–161, 1990.

Watanabe, W. O. and Kuo, C. M., Observations on the reproductive performance of Nile tilapia (*Oreochromis niloticus*) in laboratory aquaria at various salinities, *Aquaculture,* 49, 315–323, 1985.

Watanabe, W. O., Mueller, K. W., Head, W. D., and Ellis, S. C., Sex reversal of Florida red tilapia in brackish water tanks under different treatment durations of 17 a-ethynyltestosterone administered in feed, *J. Appl. Aquaculture,* 2, 29–41, 1993.

Watanabe, W. O., Smith, S. J., Wicklund, R. I., and Olla, B. L., Hatchery production of Florida red tilapia seed in brackish water tanks under natural- and clutch-removal methods, *Aquaculture,* 102, 77–88, 1992.

Wee, K. L. and Tuan, N. A., Effects of dietary protein level on growth and reproduction in Nile tilapia (*Oreochromis niloticus*), in Pullin, R. S. V., Bhukaswan, T., Tonguthai, K., and Maclean, J. L., Eds., *The Second International Symposium on Tilapia in Aquaculture,* ICLARM Conference Proceedings 15, Department of Fisheries, Bangkok, Thailand, and International Center for Living Aquatic Resources Management, Manila, Philippines, 1988, 401–410.

Wilson, J. D., Androgen abuse by athletes, *Endocrine Rev.,* 9, 181–199, 1988.

Wohlfarth, G. W. and Hulata, G. I., Applied Genetics of Tilapia, *ICLARM Studies and Reviews 6,* International Center for Living Aquatic Resources Management, Manila, Philippines, 1981.

Wohlfarth, G. W. and Hulata, G. I., Selective breeding of cultivated fish, in Shilo, M. and Sarig, S., Eds., *Fish Culture in Warm Water Systems: Problems and Trends,* CRC Press, Boca Raton, FL, 1989, 21–63.

11   FEEDING STRATEGIES

James S. Diana

INTRODUCTION

Aquaculture of tilapia originally developed as an extensive to semi-intensive system for local consumption (Pillay, 1990). While this sort of aquaculture continues today, high demand in the U.S. and elsewhere has made intensive production and export of tilapia more feasible (Anonymous, 1995). Such culture systems usually include provision of formulated or complete feed. Utilization of such feed may be made more efficient by certain feeding practices or techniques. It may also be more cost effective because it utilizes locally available items rather than relying entirely on imported complete feeds. The purpose of this chapter is to review CRSP experiments related to supplemental feeding of Nile tilapia and to put these into context as they relate to water quality and to pond fertilization.

PRINCIPLES OF FEEDING

Feeding is done in aquaculture situations for a variety of reasons and with differing materials. As aquaculture intensifies from extensive through intensive, more complete feeds are applied at a higher rate until all fish nutrition originates from supplied materials. For this chapter, natural food is defined as organisms produced in the pond system that are consumed by the tilapia. Complete feed is defined as materials added to the pond for consumption which contain all nutrient requirements of the fish. Supplemental feed includes materials added to the pond to supplement consumption but that do not take care of all nutrient requirements of the fish. The Pond Dynamics/Aquaculture CRSP has focused on the semi-intensive culture of Nile tilapia. Production of natural food in the pond is maximized under this type of management, while supplemental feeding may be used as a technique to enhance carrying capacity of the pond or to grow fish to a larger size than is possible with natural food.

Critical Standing Crop and Carrying Capacity

Hepher (1978) formalized the role of supplemental feeding and other management practices into two major aquaculture concepts: critical standing crop and carrying capacity. These concepts were based on aquacultural and ecological theory related to maximum growth rates and intraspecific competition. Hepher theorized that for any species, maximum growth is physiologically constrained by fish size and food availability, ideas that are well supported in the ecological and physiological literature. For aquaculture systems, where maximum growth is a common goal, a crop of young fish will likely grow at a near maximum rate until food or other environmental conditions become limiting. This point is termed the critical standing crop (CSC), which is the biomass of fish in any aquaculture system that results in growth reductions for each individual. Even though growth is reduced at CSC, biomass continues to increase once fish exceed CSC until the population reaches

0-56670-274-7/97/$0.00+$.50
© 1997 by CRC Press LLC

carrying capacity (K). At K, density effects of the population are so strong that growth reaches zero, and biomass remains stable.

Hepher used these concepts to better understand management of fish ponds. Obviously, CSC and K can be exceeded either by increasing the number of fish beyond the biomass at CSC and K or by keeping the number of fish constant but allowing them to grow until biomass exceeds CSC or reaches K. The latter practice is more common in aquaculture, although stocking densities may be altered by harvest or other practices to keep density in line with CSC and K. Once a density is set for a pond and fish are allowed to grow, the biomass will ultimately increase until K is reached or fish are harvested.

Hepher graphically evaluated CSC and K for carp ponds in Israel (Figure 1). His overall results showed that maximum growth could be related to fish size by the equation

$$G = 0.176W^{0.66} \tag{1}$$

where G is growth (g/d) and W is body weight (g). He also reviewed literature indicating that the weight exponent and intercept (0.66 in Equation 1) varied among species, with weight exponents ranging from 0.5 to 0.83. This equation allows one to predict maximum growth rate of a species for any size of fish cultured under optimal temperatures and other environmental conditions.

According to Hepher, if fish are stocked in ponds at low density and natural foods are abundant, they will grow at a maximum rate for that temperature. Addition of supplemental feed at this point should have no effect, because food is not limiting. However, once CSC is reached, food becomes limiting. Growth then becomes reduced unless management is intensified. If natural food production can be enhanced by fertilization, growth should again increase until a new CSC is reached. At that point, supplemental feed may be required to increase growth to maximum. Again, another CSC will be reached, until food quality (e.g., essential amino acids) or water quality limits fish growth. This progression of CSC under different management schemes was demonstrated by Hepher (1978; Figure 1).

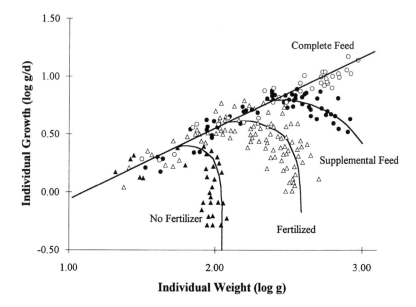

Figure 1.   The relationship between growth rate and size of carp in ponds with differing inputs. CSC is the point where growth begins to decline in a treatment, K is when growth reaches zero. (Redrawn from Hepher, B., 1978.)

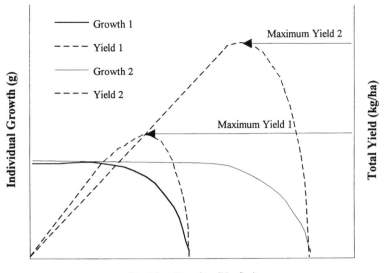

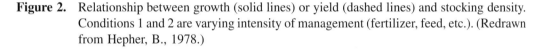

Figure 2. Relationship between growth (solid lines) or yield (dashed lines) and stocking density. Conditions 1 and 2 are varying intensity of management (fertilizer, feed, etc.). (Redrawn from Hepher, B., 1978.)

The concepts of CSC and K are important in understanding feeding strategies, because they indicate when feeding may be most efficiently done but not how much food should be added. The concepts are also important in developing harvest strategies. Hepher developed relationships between CSC, K, and yield, which indicated that maximum yield can be harvested under conditions between CSC and K (Figure 2).

How well do these concepts apply to tilapia culture data for the CRSP experiments? Data from Thailand for tilapia stocked at 1 or 2 fish/m² show considerable variability in growth rates (Figure 3). While these data roughly compare to the carp ponds shown earlier, it appears that most CRSP experiments with tilapia were mainly done at sizes and densities exceeding CSC. Furthermore, these data indicate that maximum growth in ponds is related to fish weight by

$$G = 0.836W^{0.287} \tag{2}$$

where ($r^2 = 0.74$, $P < 0.001$), which has a considerably different exponent and intercept than Equation 1. However, until cycle 4 of our experiments, we stocked fish at 30 to 40 g in size, which eliminated data from very small fish. Consequent grow-outs with smaller fish have provided results similar to Figure 3. Since Figure 3 is a complex curve based on expectations from Figure 1, the curving sections are not fitted by any statistical techniques but rather by eye.

Data from Figure 3 indicate CSC occurred at an individual fish weight of about 35.4 g in well-fertilized ponds, with a biomass of about 319 kg/ha. Carrying capacity was not exceeded in any of these ponds, but it appears that K would occur at a weight of about 354 g or a biomass of about 3190 kg/ha. These well-fertilized ponds were given large doses of chicken manure. Nitrogen was likely limiting primary production at times in these ponds (Diana et al., 1991; Knud-Hansen et al., 1991a).

Similar data from Honduras can be used to extrapolate CSC and K for conditions there (Figure 4), although less growth data are available in Honduras for small fish. Overlying the Thailand maximum growth curve on these data indicates that maximum growth in Thailand

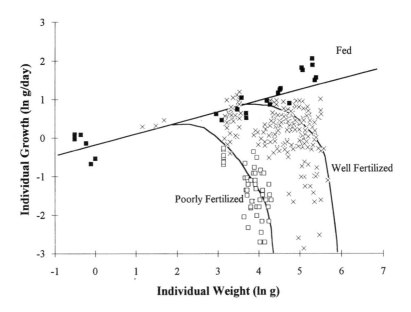

Figure 3. The relationship between growth and weight for tilapia stocked at 1 to 2 fish/m^2 and held in poorly fertilized (open squares), well-fertilized (x), or fed (closed squares) ponds in Thailand.

appears similar to Honduras. For this curve, CSC and K occur at similar values to Figure 3. The similarity of the two growth relationships may indicate some validity for these curves, but both are tentative.

These data, while preliminary, point out potential feeding strategies for tilapia, as well as differences in management that may be necessary among sites. Tilapia stocked at small size and low density may be grown very well on natural feeds, and fertilization of the ponds is probably

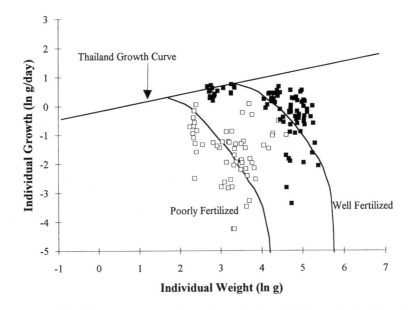

Figure 4. The relationship between growth and weight for tilapia stocked at 1 to 2 fish/m^2 in ponds in Honduras. Symbols as in Figure 3.

the best management strategy. Once the fish exceed 40 to 80 g in individual weight (at a density of 1 to 2/m^2), growth begins to decline and then will decline dramatically as individual weights approach 300 g. Thus, harvest at sizes of about 150 g (between CSC and K) would likely produce optimal maximum yields in fertilized ponds, while efforts to grow larger tilapia would require supplemental feeding. Such feeding could be staged to begin at some advanced fish size (at CSC around 35 g in size). This strategy would be successful in biological terms, but economic evaluations on these trade-offs are also important.

Nutritional Requirements of Tilapia

Tilapias in general are herbivores and detritivores, although they show ontogenetic shifts from zooplankton at young ages to phytoplankton, macrophytes, and detritus at advanced ages (Bowen, 1982). While tilapia can persist on low protein diets (Bowen, 1979, 1982) optimal growth requires protein levels similar to other fish (Bowen et al., 1995). Wee and Tuan (1988) demonstrated that Nile tilapia grew well in tank systems when dietary protein levels were varied from 20 to 50% in pellets, and growth was not significantly different among fish at different protein levels. They proposed optimal dietary protein levels of 27.5 to 35% based on these results; similar numbers have been generated for other tilapia species (Cissé, 1988).

Protein sources for fish diets can include either fish meal or plant proteins (Stickney, 1979). Research on substitution of plant protein sources for fish meal has had mixed results. Stickney (1993) proposed that up to 25% replacement of animal protein with plant sources is acceptable. Complete substitution of plant proteins into tilapia diets produced excellent growth when these diets were supplemented with minerals and vitamins (Hughes and Handwerker, 1993, cited in Stickney, 1995). Partial replacement of animal protein with plant protein in the diet is possible for some plant sources, but not all (El Sayed, 1992; Keembiyehetty and de Silva, 1993; Omoregie and Ogbemudia, 1993). Due to cost constraints and limited availability of fish meal in some areas, local research on plant protein substitution is probably still warranted. Currently, fish protein sources are still most reliable in producing rapid growth of tilapia on complete feeds.

Amino acid requirements have been most thoroughly studied for *Oreochromis mossambicus*. These requirements have been synthesized by Stickney (1993) and include ten amino acids, similar for most fish and mammals (Landau, 1992). Probably all tilapia share similar amino acid requirements.

Bowen (1982) and Bowen et al. (1995) also determined that tilapias are particularly adept at assimilating energy from algal chlorophyll and other plant sources. This ability is based on the low pH generated in tilapia stomachs (as low as 1 to 1.25), compared to moderate values in other animals (2 to 2.2). Coupled with low stomach pH is a long intestine, which may reach from 7 to 10 times the fish's length. Both of these adaptations make tilapia particularly good at extracting energy from plant matter and allow inexpensive supplemental feeds to be useful in tilapia culture. Certainly, tilapia in nature appear capable of growing on relatively low quality foods such as benthic detrital aggregate (Bowen, 1982).

Stickney (1993) provides an excellent review of nutritional studies on tilapia, including lipids, vitamins, and minerals. Lipid requirements of tilapia appear similar to other fishes. Vitamin and mineral contents of diets developed for other species produce excellent growth of tilapia as well. Tilapia probably differ little from other fish species in these details of their nutrition.

POND PRODUCTION OF TILAPIA

Natural Feeds

Most CRSP experimentation has concentrated on semi-intensive culture of tilapia in ponds. Optimal fertilization strategies vary for such feed production depending on site, nutrient balance in fertilizers, and water quality (refer to Chapter 4 of this book). In this chapter, the kinds of natural foods consumed by Nile tilapia will be the main emphasis.

Tilapia as a group are usually herbivores and detritivores, although some feed on zooplankton as well (Bowen, 1982). Some of our pond experiments have evaluated plankton communities in the ponds, as well as food habits of sampled tilapia. Tilapia less than 35 g appear to be particulate feeders, selecting individual plankton, especially crustaceans, from the water column (Bowen, 1982). At about 35 g, tilapia make a shift to filter feeding and utilize mainly phytoplankton and smaller zooplankton, such as rotifers (Bowen, 1982; Diana et al., 1991). At this stage, large zooplankton may be capable of avoiding their filter mechanism, or at least they are uncommon prey. Diana et al. (1991) found that tilapia growth declined with density in fertilized ponds, indicating that food became limited as density increased. The ecosystem in each pond did not change dramatically with tilapia density; indeed, the only significant difference in food organisms among ponds with different density treatments was that small zooplankton (mainly rotifers) were much more abundant in fishless ponds than in ponds with fish. Thus, tilapia appeared either to crop rotifers effectively or to reduce their abundance through competition, and while they also foraged on phytoplankton, this foraging did not effectively reduce phytoplankton biomass.

The previous results contrast somewhat with one accepted role of tilapia in aquaculture systems — that of a biological filter for phytoplankton. Landau (1992) and Bardach et al. (1972) review this role of tilapia, stocked into reservoir ponds, in reducing algal growth and improving water quality, much like the silver carp in Chinese polyculture (Lin, 1982). Diana et al. (1991) and Boyd (1990) found no effects of tilapia on phytoplankton populations in fertilized fish ponds. Boyd (1990) reviewed polyculture experiments, including tilapia, and found limited evidence of tilapia controlling phytoplankton production. In fact, data from Perschbacher (1975) indicate that phytoplankton production is stimulated by the grazing of *Oreochromis aureus*. There seems to be little doubt that Nile tilapia consume phytoplankton, but their role in phytoplankton control is questionable and probably depends on the relative biomass and production of tilapia and their prey.

Natural food production can be greatly enhanced in tilapia ponds by fertilization (Chapter 4). Organic and inorganic fertilizers can both result in high levels of primary production. However, organic fertilizers often result in higher tilapia growth and yield (Figure 5), apparently due to increased production of heterotrophic organisms (Diana et al., 1991). Schroeder (1978) has estimated that more than half of the fish production in organically fertilized ponds is due to increased heterotrophic production caused by organic fertilization. In addition, manures may contain materials (such as chicken feed) which can be directly consumed by tilapia.

Supplemental Feeds

Due to the diverse food habits of tilapia, a large variety of supplemental feeds has been used in tilapia production. Many materials, including soybean, corn, peanut, or cottonseed meal, as well as rice bran or other cereal brans, are commonly fed to tilapia (Landau, 1992; Stickney, 1993). Many feeding programs grow tilapia on waste materials from farm or sewage operations (Landau, 1992). Micha et al. (1988) fed *Azolla* to tilapia and measured positive but reduced growth compared to culture on complete feeds. Similar results were found for tilapia fed algae (*Cladophora*; Appler and Jaucey, 1983), pito brewery waste (a byproduct of sorghum fermentation; Odura-Boateny and Bart-Plange, 1988), waste vegetables (McLarney, 1987), and duck weed (*Lemna*; Hassan and Edwards, 1992), to name just a few. McLarney (1987) emphasizes techniques to use waste products as feed and to attract natural organisms to ponds by use of bug lights or other devices.

Tilapia have been the subject of considerable physiological investigation in recent years, and a variety of studies have been done on various feeds and their effects on body composition and growth (see Stickney, 1993). These studies indicate a broad dietary tolerance for different feeds.

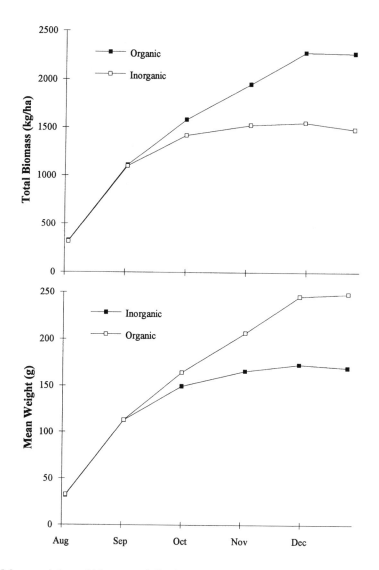

Figure 5. Mean weight and biomass of tilapia stocked at 1 fish/m² in ponds fertilized with chicken manure or inorganic fertilizer at similar loading rates for N and P. (Data from Diana, J. S., Lin, C. K., and Schneeberger, P. J., 1991.)

Both prepared and other supplemental feeds listed above can prove economical for rearing tilapia, so it is useful to experiment on locally available food stuffs for supplemental feeding.

Complete Feeds

Tilapia are often raised in intensive systems throughout the world, utilizing complete feed. Stickney (1993) summarized the essential amino acid composition of complete tilapia feeds. Protein requirements vary from 30 to 40% for rearing tilapia, while high protein diets (44%) are considered optimal for brooders. Plant proteins may be substituted for up to 25% of the total protein requirement without a reduction in performance, although higher rates of substitution may be acceptable (Stickney, 1993).

FEEDING TECHNIQUES

Feeding Alone

Nile tilapia are capable of rapid growth utilizing natural feeds, complete feeds, or combinations of natural and supplemental feeds. Complete feeding of tilapias can reach feed conversion ratios approaching 1.0 (Coche, 1982; Hepher and Pruginin, 1982; Diana et al., 1994). Complete feeding in ponds may include use of natural foods and can produce feed conversion ratios of less than 1.0. Tilapia may easily be trained to use demand feeders, particularly in intensive culture. Feed requirements are known to vary with temperature and fish mass (Ross et al., 1988), as is the case for other teleosts. Due to the continual feeding pattern of tilapia, several meals per day may produce higher or more efficient growth, particularly for young fish (Siraj et al., 1988).

Fertilization and Feeding

The Pond Dynamics/Aquaculture CRSP has emphasized semi-intensive systems for tilapia production. Few experiments have been done on feeding alone under controlled conditions, while a number have been done on feed and fertilizer combinations. Such combinations may be very effective because fertilization rates can be reduced due to enrichment gained from fish excreta. Combinations of fertilization and feeding can include addition of both simultaneously, staged addition of feed, or feeding of caged fish within fertilized ponds.

Diana et al. (1994) conducted two experiments to provide complete feeds at various rates into fertilized fish ponds. Fertilization was done using inorganic materials at proposed optimum rates (Knud-Hansen et al., 1991b). These rates were defined as inputs of N and P at 4:1 ratios and provision of 1 kg/ha/d of P. Fertilizers were applied weekly, and included triple superphosphate and urea. The first experiment compared fertilization alone, satiation feeding alone, or fertilization combined with satiation feeding in replicated pond studies. It is important to realize that satiation feeding in such cases is done *in situ*, meaning that the consumption of feed is affected by natural food production in the ponds and may not equal maximum consumption rates in the lab. In the first experiment, weight at harvest ranged from an average of 184 g for fertilization alone to 325 g for feed alone or feed and fertilizer after 162 d of culture (Figure 6). Biomass in the pond also ranged from 3180 to 5850 kg/ha, respectively. Fertilization alone led to low alkalinity in some ponds due to high rates of primary productivity and low carbon input from inorganic fertilizers. Feeding sometimes resulted in high inorganic nitrogen (ammonia, nitrate, and nitrite) in ponds, which significantly lowered fish production. However, there was no significant decline in fish production due to specific components of inorganic nitrogen such as ammonia or nitrate.

This first experiment provided one particularly interesting result. Differences in tilapia growth among ponds and treatments occurred from the first month. Hepher's (1978) concepts would predict that growth in all treatments should remain similar until critical standing crop was reached. For this experiment, critical standing crop apparently occurred at a fish size around 30 g (size at first month sampled). Earlier analyses (Figure 3) indicated CSC occurred in well-fertilized ponds in Thailand at an individual fish weight of about 35.4 g and a biomass of 319 kg/ha.

The second experiment in Thailand on simultaneous addition of feed and fertilizer used a modified approach with five treatments: satiation feeding alone, fertilizer alone, 0.75 satiation feeding plus fertilizer, 0.5 satiation feeding plus fertilizer, and 0.25 satiation feeding plus fertilizer. Again, these five treatments resulted in differential growth and standing crop over the 155 d of culture (Figure 7). In this experiment, the three highest feed treatments had statistically similar growth and biomass; the 0.25 feeding treatment was intermediate, and the fertilizer only treatment had the lowest values. This indicates that feeding at 0.5 satiation ration, plus fertilization, could produce similar growth to 100% feeding with reduced food input (42% less food added) and increased conversion efficiency (61% better; Table 1).

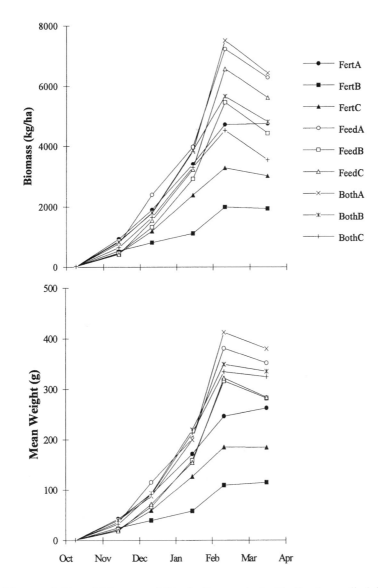

Figure 6. Mean weight and biomass of tilapia for ponds in Thailand supplied fertilizer alone (Fert), feed alone (Feed), or fertilizer and feed (Both). (Redrawn from Diana, J. S., Lin, C. K., and Jaiyen, K., 1994.)

There were some interesting outcomes for this experiment and the first one combined. Utilizing multiple regression, the relationships between growth or yield and the treatment variables, as well as water quality variables, could be evaluated. In experiment 1, variation among ponds was high, and multiple regression showed no significant effects of feeding or fertilization on growth (R^2 = 0.465), but significant effects on yield (R^2 = 0.519), indicating that probably both growth and survival varied among ponds. For experiment 2, variation among ponds was less, and the treatment effects on growth and yield were highly significant (R^2 = 0.873 and 0.894, respectively). Combining both experiments, the effects of water quality variables could also be assessed. Feed input, fertilizer input, alkalinity, and total inorganic nitrogen were significantly correlated to fish growth (R^2 = 0.900), with nitrogen having a negative correlation. Feed input and the number of low dissolved

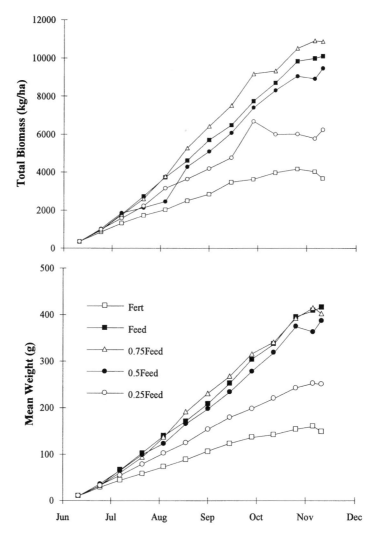

Figure 7. Mean weight and biomass of tilapia for ponds in Thailand, either fertilized or fertilized and fed at different fractions of maximum ration. (Redrawn from Diana, J. S., Lin, C. K., and Jaiyen, K., 1994.)

oxygen events were significantly correlated to yield (R^2 = 0.888), with number of low dissolved oxygen (DO) events having a negative coefficient. This indicates that primary production, feed input, and water quality all combined to produce the observed differences in growth and yield. Feed and fertilizer inputs were constant across replicate ponds in a treatment, while water quality variables differed among replicates and treatments. This also agrees with data from Teichert-Coddington and Green (1993) that DO concentrations had to be maintained above 10% satiation or growth declined in Nile tilapia.

Another interpretation of experiment 2 was related to CSC. Once again, treatments differed significantly in growth from the beginning of the experiment. These differences were clear by the second biweekly measurement period (Figure 7), and indicated again that CSC had to occur at a size less than 50 g.

Combinations of feed and fertilizer can produce rapid growth rates of tilapia and larger size at harvest than fertilizer alone. In areas where large fish fetch a higher price, this may be advantageous. In the first combination experiment, size at harvest did not exceed 350 g, due to an apparent

Table 1   Growth (g/d) and Yield (kg/d) for Fed and Fertilizer-Only Stages of Tilapia Production for Experiment 2

Treatment	Growth (g/d)	Yield (kg/d)
Fed Stage		
50 g	2.75	1.75
100 g	3.32	2.09
150 g	3.17	1.90
200 g	3.27	1.62
250 g	2.71	1.72
Fertilized Stage		
50 g	1.14	1.06
100 g	1.14	0.80
150 g	1.21	0.81
200 g	1.26	0.77
250 g	1.03	0.71
Overall		
50 g	2.51	1.63
100 g	2.53	1.62
150 g	2.00	1.25
200 g	2.00	1.10
250 g	1.47	0.97

decline in growth over the last month. However, this decline in growth may also be due to biased seine sampling for monthly data (Hopkins and Yakupitiyage, 1991). Growth rates in experiment 2 remained linear, and in the three highest feeding rates, fish averaged over 400 g at harvest. Since these treatments differed significantly in the amount of feed applied, the feed conversion rates in each treatment also differed. The FCR was 1.42 in satiation ponds, 1.1 in 75% satiation ponds, and 0.88 in 50% satiation ponds (Table 1). Feed is usually more expensive than fertilizer, so economic efficiency might be highest using a 50% satiation ration plus fertilizer. This treatment also resulted in rapid growth and minimal deterioration of water quality.

Similar experiments on combined fertilization and feeding were conducted in Honduras. Green (1992) reported that fed and fertilized ponds had similar growth and production of tilapia compared to feed only ponds. FCR was significantly better in the former treatment (0.95 compared to 1.83), and economic returns were best in this treatment as well. While he did not report on water quality in these ponds, mortality did not differ significantly among treatments.

Staged Feeding

Feeding strategies can be fine-tuned further to increase efficiency using CSC and K concepts. Because growth is ultimately limited by food availability, there should be time periods early in a grow-out when growth on natural foods occurs at a high rate. In this case, supplemental feed applied then would only supplant natural food, with little increase in growth; indeed, supplemental feed might even reduce growth if food quality is low. Using Hepher's (1978) concepts, supplemental feeding prior to a pond reaching CSC would be wasteful.

In an effort to further analyze CSC, experiments were conducted in Thailand on staged feeding (Diana et al., 1996). In this study, all ponds were fertilized at rates defined previously. Five treatments were used, each differing in the average size of fish when supplemental feed was first added. Sizes at onset of first feeding were the following: treatment A, 50 g; B, 100 g; C, 150 g;

D, 200 g; and E, 250 g. Fish size was measured every two weeks, and feeding (at 50% satiation) commenced the first week that measured size exceeded the target value. Feeding continued until size at harvest was at least 500 g.

It took tilapia 38 d to reach 50 g and 234 d to reach 200 g on fertilizer alone (Figure 8). Total times to harvest varied among treatments from 236 to 328 d. Growth rates (g/d) and yield rates (kg/ha/d) were not significantly different among treatments in a stage (fed or unfed) but were significantly different between stages (Table 1). Interestingly, growth rates appeared linear in all treatments during the fertilization stage and then increased to a new and similar slope once feeding commenced (Figure 8). Fish fed at a later age and larger size might have been expected to show slower growth rates when feed was finally applied, compared to fish fed at a smaller size.

One other intriguing result occurred among treatments A, B, and C. Treatment A took 38 d to onset of feeding and 236 to harvest. Treatment B took 80 d to onset of feeding, but still only 236 to harvest. Treatment C took 153 d to first feeding, but only 265 to harvest. While time to onset

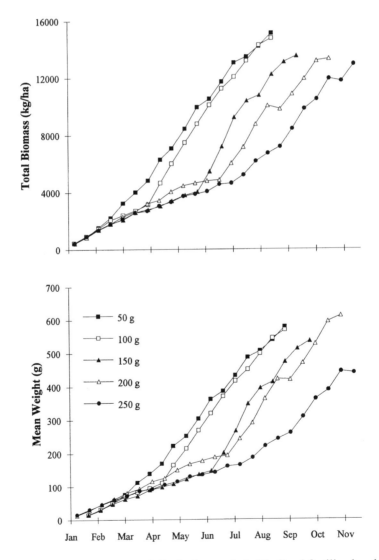

Figure 8.   Mean weight and biomass of tilapia for ponds in Thailand fertilized and then fed 50% satiation rations with different initial weights for first feeding.

of feeding approximately doubled in each treatment compared to the previous treatment, total time to harvest increased only 29 d. From this result, treatment C could be most advantageous because it resulted in less feeding time than A or B (112 d compared to 198 and 156, respectively) yet only took 29 more days to reach harvest size. However, fish in treatment C actually ate more total feed than A or B during the feeding stage (Table 2). Feed conversion rates averaged 1.033 for all treatments during feeding and did not differ among treatments.

Table 2 Evaluation of Simultaneous Fertilizer and Feed Experiments, Including Feed Added, Feed Conversion Rate (FCR), Yield, Time to Harvest, and Size at Harvest

Treatment	Feed added (kg/ha/yr)	FCR	Yield (kg/ha/yr)	Time to harvest (d)	Mean size at harvest (g)
Continual Feeding					
0.5	18,838	0.87	21,552	155	402
0.75	26,845	1.08	24,777	155	416
1.0	32,685	1.42	22,996	155	388
Staged Feeding					
at 50 g	22,147	0.96	23,131	236	593
at 100 g	20,353	0.88	23,117	236	596
at 150 g	24,131	1.35	17,861	265	534

Multiple regression was used to sort out treatment effects and water quality effects. Treatment effects (feed input and days of culture) were both significant ($R^2 = 0.738$). In addition, chlorophyll *a* content was included in the final correlation (final $R^2 = 0.889$). Differences among ponds in primary productivity affected replicates and treatments. Interestingly, this feeding scheme did not appear to produce negative effects of water quality as did previous experiments, probably because feed input rate was only at 50% satiation due to results of the previous experiment.

Growth increased similarly and linearly once food was added, regardless of fish size (Figure 8). CSC appeared to occur at a size less than 50 g. Growth on fertilizer alone did not decrease much from 50 to 250 g, indicating a wide range of sizes beyond CSC that still had good growth. These sizes still fall well below the predicted size at K of 354. Tilapia show obvious ontogenetic trends in food consumption, behavior, and growth. These ontogenetic trends may make it possible to use natural and supplemental feeds to optimize growth of different life stages.

The staged feeding experiment indicated that significant reductions in feed requirements could be made using such techniques. One interesting outcome is to compare the overall practices across the three feeding experiments (Table 2). Highest yields were achieved either in staged feeding at 50 or 100 g or by 75% or satiation feeding continually. Most efficient feeding (measured by overall FCR) occurred by 50 to 75% continual feeding or by staged feeding at 50 to 100 g. The best overall practices appeared to be continual feeding at 50 to 75% maximum ration or staged feeding after fish reached 50 to 100 g. More fine-tuning of this system is desirable.

Caged Feeding in Ponds

Feeding of tilapia can result in rapid growth, particularly of larger fish. However, water quality can deteriorate during feeding, although fertilization may facilitate primary production and algal uptake of fish excretory products by balancing nutrient levels in ponds. Discharge water from completely fed ponds may add considerable sediment, nitrogen, and phosphorous loads to receiving waters (Pillay, 1992). One means to reverse this trend would be to treat this water in a settling pond prior to discharge, which may also include some conditioning of the water by organisms in the settling pond. Another means might be to feed fish in cages within a pond stocked with fish. Such a system could significantly improve discharge water quality through primary production and use of nutrients, while also harvesting an additional crop. CRSP research has targeted two such systems, *Clarias* in cages within tilapia ponds and large tilapia in cages in tilapia ponds.

Growing *Clarias* on prepared feeds is a common practice in Thailand. Usually, these fish can be grown either in concrete tanks flushed with water from a reservoir or in earthen ponds with no water replacement. The fish can tolerate low oxygen in water due to their ability to breathe air. However, water quality in these rearing systems does appear to influence growth and survival (Diana et al., 1988). Water quality in circulating systems can also be improved by flushing, by lowering stocking density in tanks, or by lowering feeding rates. An alternative might be to combine *Clarias* in cages with tilapia at large and allow primary production to occur in the pond waters to utilize surplus nutrients. Tilapia may harvest phytoplankton as food and further improve water quality. (However, see previous section on natural feeds.) Silver carp or other phytoplanktivores are also believed to reduce phytoplankton abundance (Lin, 1982). This system would combine a fed, grow-out pond and a settling pond into one system and harvest two crops.

Preliminary experiments with such a system have been very successful (Lin and Diana, 1995). *Clarias* have grown well on pelleted feed. Tilapia growth in cage ponds has been higher than in fertilized ponds, undoubtedly due to their utilization of uneaten feed as well as natural food produced by assimilation of *Clarias* excretory products. Water quality on discharge is significantly improved over normal *Clarias* culture. This system has been adopted since 1992 on a commercial scale in northeast Thailand.

The commercial system utilizes 600-m^2 ponds. Each pond includes 15 cages, 1 m^3 in size, with 1000 *Clarias* per cage. The *Clarias* grow from 5 to 100 g (market size) in 1.5 months, giving about three *Clarias* crops in one tilapia grow-out period. Production approaches 9000 kg of catfish and 1000 kg of tilapia in a grow-out. This equates to 45,000 kg/ha/yr of catfish, and 5000 kg/ha/yr of tilapia. Pelleted feed can be used for *Clarias*, while fertilizer balancing (adding phosphorous or nitrogen to balance the N:P ratio) may be used to improve tilapia growth. In an interesting twist on this system, a commercial system in Thailand uses pig dung in bins to attract flies for maggot production. The maggots are then fed to the *Clarias*, replacing pellets. One such system can produce 200 kg of maggots per day! This provides opportunity to replace expensive pellets with inexpensive feeds for the intensive portion of the system.

A system with caged tilapia is still in development. As mentioned earlier, the best tilapia production on fertilizer probably results in 6000 kg/ha/yr of fish grown for about 150 d to a size of 180 g or so. In many locations, larger fish may be preferable but may take too long to grow extensively. Rather than staged feeding at 3/m^2 in ponds at large, caged feeding could improve growth of larger fish while simultaneously growing a crop of smaller ones. This would require fewer facilities than the two separate systems.

EFFECTS OF FEEDING ON WATER QUALITY

Dissolved Oxygen

Feeding systems can be used to increase stocking density and growth rate. Both of these variables put a higher oxygen demand on a pond, mainly through increased inputs required for the larger fish biomass. The increased oxygen demand can lead to decreased growth or increased mortality of the fish. CRSP experiments generally have focused on semi-intensive systems, which have not used aerators to improve water quality. Rather the CRSP has attempted to utilize fertilization schemes to increase primary productivity and thereby improve oxygen conditions. Most of our stocking densities have been 1 to 3 fish/m^2, again limiting oxygen demand.

In some CRSP experimental treatments, oxygen conditions have declined to low levels. CRSP data show frequent occurrences of oxygen conditions at sunrise that were less than 1 mg/L but were accompanied by little or no mortality. Tilapia appear capable of tolerating short-term conditions near 0 mg/L of oxygen without lethal effects. However, their growth appears to decline when conditions reach such levels.

Analysis of sporadically occurring low oxygen events is difficult. Diana et al. (1994) found occasional low oxygen events in supplemental feeding experiments and also found that such events were significantly correlated with reduced yields in these ponds. Surprisingly, events of low DO (less than 1 mg/L at dawn) were not correlated to total inputs in ponds (total amounts of fertilizer and feed added) or to phytoplankton biomass, but rather occurred sporadically in ponds spanning the range of lowest to highest feeding rates.

Teichert-Coddington and Green (1993) evaluated the effects of aeration and DO regulation on growth and yield of fish from CRSP ponds in Honduras. They stocked *Oreochromis niloticus*, *Cichlasoma maculicauda*, and *C. managuense*, and fertilized the ponds for 2 months with chicken manure, then fed the fish a commercial shrimp ration at 3% body weight per day. They had three treatments: no aeration, aeration starting when DO reached 10% saturation, and aeration at 30% saturation. Tilapia yield and growth were higher in aerated treatments than in controls. There were no treatment effects on survival.

While low DO limits growth, CRSP experiments with feeding rates up to satiation and stocking densities up to 3 fish/m^2 have not had consistent effects on the number of low DO events. Apparently, fertilizer balance, wind, and primary productivity also affect the added oxygen demand of the fish and feed in most ponds, even at highest loading rates.

Nitrogen

High levels of nitrogen in the form of un-ionized ammonia or nitrite can be toxic to fish. Such conditions may commonly be produced under high loading rates of fed fish. Nitrite and un-ionized ammonia are relatively higher at high pH, so there may be an interactive effect of ammonia and pH on fish survival and growth. Boyd (1990) reviewed the toxic and sublethal effects of these compounds in pond aquaculture.

High stocking rates and feeding rates of fish can produce high levels of ammonia loading. Ammonia is the most common nitrogenous excretory product of fishes; it can also be released during decomposition of feed and feces. Thus, ammonia loading can be most dramatic in intensive culture. Again, Diana et al. (1994) evaluated the frequency of high ammonia events (ammonia measures greater than 1 mg/L) in waters from tilapia ponds with intensive feeding. While such events occurred sporadically in many ponds, there was no relationship between the frequency of such events and fish growth or yield. However, there was a negative correlation between total inorganic nitrogen concentration (total of ammonia, nitrite, and nitrate) and fish growth. This may indicate a sublethal response to the effects of ammonia and/or nitrite, but neither component of dissolved inorganic nitrogen (DIN) was significantly correlated to growth.

Primary Production

Intense feeding of ponds produces added nutrients from decomposition of feed and increased fish excretion. These nutrients may affect primary production differently than fertilizers alone. Diana et al. (1994) evaluated such potential changes in ponds with fertilizer alone, feed alone, or both fertilizer and feed. This experiment utilized inorganic fertilizers, and fertilization treatments had lower alkalinity than the other treatments, probably due to carbon utilization (without replacement) in photosynthesis or destruction of alkalinity as a result of transformation from NH_3 to NO_3. Treatments with fertilizer (alone or with feed) developed higher rates of primary productivity than feed only ponds. These same ponds also developed stronger oxygen stratification than feed only ponds. Fed ponds produced some nutrients, but primary production remained highest in fertilized ponds. However, ponds with feed and fertilizer showed similar trends to ponds with fertilizer alone. Nutrient balance remains a question in these ponds. Comparisons of nutrients in each treatment indicate that total phosphorous and total inorganic nitrogen were significantly higher in fertilizer treatments than in feed only ponds. The resultant N:P ratio varied significantly at 1.71 for fertilizer only ponds, 0.69 for fed only ponds, and 1.00 for fed and fertilized ponds. Apparently, phosphorous was produced at a high level in fed treatments, and further nitrogen supplementation in fed ponds might drive still higher primary production. An alternative explanation may be that more N was retained in fish in the fed ponds.

SUMMARY

Nile tilapia in fertilized ponds appear to reach CSC at 35 g (354 kg/ha) and K at 354 g (3190 kg/ha). Most CRSP experiments have either stocked fish beyond CSC or the fish have reached CSC within the first month of culture.

Feeding experiments in Thailand and Honduras indicate that combined fertilization and feeding moderate water quality and produce rapid growth. Experiments in Honduras indicate that mainte-nance of DO in excess of 10% saturation, in this case through aeration, produces more rapid growth. Thus, feeding at moderate levels or moderate densities may provide a better environment to grow tilapia. This was also illustrated by experiments in Thailand, where feeding at 0.5 satiation rates, coupled with fertilization, gave comparable growth and yield to full satiation feeding and provided better water quality.

Staged feeding of tilapia can increase efficiency by providing food only after food becomes limiting to growth. Experiments in Thailand indicate that onset of feeding at 50-, 100-, or 150-g size resulted in similar yields. Initiation of feeding at 100 g appeared to be the best system, considering yield, feed application rate, and feed conversion efficiency.

Combining these results indicated that the best system for feeding tilapia in Thailand is to combine fertilization and feeding. Early growth is rapid under fertilization alone, and feeding can be delayed until the fish approach 100 g. From that time, feeding should continue at the 50% satiation level. Results from Honduras may be comparable to these. This response will be lower in Rwanda, because lower temperatures will probably result in slower growth rates and lower consumption rates.

REFERENCES

Anonymous, Annual tilapia situation and outlook report, *Aquaculture Mag.*, 21(5), 6, 1995.

Appler, H. N. and Jaucey, K., The utilization of a filamentous green alga (*Cladophora glomerata* (L.) Kutzin) as a protein source in pelleted feeds for *Sarotherodon* (*Tilapia*) *nilotica* fingerlings, *Aquaculture,* 30, 31, 1983.

Bardach, J. E., Ryther, J. H., and McLarney, W. O., *Aquaculture: The Farming and Husbandry of Freshwater and Marine Organisms,* John Wiley & Sons, New York, 1972, 868.

Bowen, S. H., A nutritional constraint in detritivory by fishes: the stunted population of *Sartherodon mossa-mbicus* in Lake Sibaya, South Africa, *Ecol. Monogr.,* 49, 17, 1979.

Bowen, S. H., Feeding, digestion and growth, qualitative considerations, in *The Biology and Culture of Tilapias,* Pullin, R. S. V. and Lowe-McConnell, R. H., Eds., International Center for Living Aquatic Resources Management, Manila, 1982, 141.

Bowen, S. H., Lutz, E. V., and Ahlgren, M. O., Dietary protein and energy as determinants of food quality: trophic strategies compared, *Ecology,* 76, 899, 1995.

Boyd, C. E., *Water Quality in Ponds for Aquaculture,* Agriculture Experiment Station, Auburn University, Auburn, AL, 1990, 482.

Cissé, A., Effects of varying protein levels on spawning frequency and growth of *Sartherodon melanotheron* young breeders, in *The Second International Symposium on Tilapia in Aquaculture*, Pullin, R. S. V., Bhukaswan, T., Tonguthai, K., and Maclean, J. L., Eds., International Center for Living Aquatic Resources Management, Manila, 1988, 329.

Coche, A. G., Cage culture of tilapias, in *The Biology and Culture of Tilapias*, Pullin, R. S. V. and Lowe-McConnell, R. H., Eds., International Center for Living Aquatic Resources Management, Manila, 1982, 205.

Diana, J. S., Kohler, S. L., and Ottey, D. R., A yield model for walking catfish production in aquaculture systems, *Aquaculture*, 71, 23, 1988.

Diana, J. S., Lin, C. K., and Yi, Y., Timing of supplemental feeding for tilapia in production, *J. World Aquaculture Soc.*, 27, 410, 1996.

Diana, J. S., Lin, C. K., and Schneeberger, P. J., Relationships among nutrient inputs, water nutrient concentrations, primary production, and yield of *Oreochromis niloticus* in ponds, *Aquaculture*, 92, 323, 1991.

Diana, J. S., Lin, C. K., and Jaiyen, K., Supplemental feeding of tilapia in fertilized ponds, *J. World Aquaculture Soc.*, 25, 497, 1994.

El-Sayed, A. F., Effects of substituting fish meal with *Azolla pinnata* in practical diets for fingerling and adult Nile tilapia, *Oreochromis niloticus* (L.), *Aquaculture Fish. Manage.*, 23, 167, 1992.

Green, B. W., Substitution of organic manure for pelleted feed in tilapia production, *Aquaculture*, 101, 213, 1992.

Hassan, M. S. and Edwards, P., Evaluation of duckweed (*Lemna perpusilla* and *Spirodela polyrrhiza*) as feed for Nile tilapia (*Oreochromis niloticus*), *Aquaculture*, 104, 315, 1992.

Hepher, B., Ecological aspects of warmwater fishpond management, in Gerking, S. D., Ed., *Ecology of Freshwater Fish Production*, John Wiley & Sons, New York, 1978, chap. 18.

Hepher, B. and Pruginin, Y., Tilapia culture in ponds under controlled conditions, in *The Biology and Culture of Tilapias*, Pullin, R. S. V. and Lowe-McConnell, R. H., Eds., International Center for Living Aquatic Resources Management, Manila, 1982, 185.

Hopkins, K. D. and Yakupitiyage, A., Bias in seine sampling of tilapia, *J. World Aquaculture Soc.*, 23, 173, 1991.

Keembiyehetty, C. N. and de Silva, S. S., Performance of juvenile *Oreochromis niloticus* (L.) reared on diets containing cowpea, *Vigna catiang* and black gram, *Phaseolus mungo*, seeds, *Aquaculture*, 112, 207, 1993.

Knud-Hansen, C. F., Batterson, T. R., McNabb, C. D., Harahat, I. S., Sumantadinata, K., and Eidman, H. M., Nitrogen input, primary production and fish yield in fertilized freshwater ponds in Indonesia, *Aquaculture*, 94, 49, 1991b.

Knud-Hansen, C. F., McNabb, C. D., and Batterson, T. R., Application of limnology for efficient nutrient utilization in tropical pond aquaculture, *Proc. Int. Assoc. Theor. Appl. Limnol.*, 24, 2541, 1991a.

Landau, M., *Introduction to Aquaculture*, John Wiley & Sons, New York, 1992, 440.

Lin, C. K. and Diana, J. S., Co-culture of catfish (*Clarias maciocephalus* x *C. gariepinus*) and tilapia (*Oreochromis niloticus*) in ponds. *Aquat. Living Resour.*, 8, 449, 1995.

Lin, H.-R., Polycultural system of freshwater fish in China, *Can. J. Fish. Aquatic Sci.*, 39, 143, 1982.

McLarney, W., *The Freshwater Aquaculture Book*, Hartley and Marks Publishers, Point Roberts, WA, 1987, 583.

Micha, J.-C., Antoine, T., Wery, P., and Van Hove, C., Growth, ingestion capacity, comparative appetency and biochemical composition of *Oreochromis niloticus* and *Tilapia rendalli* fed with *Azolla*, in *The Second International Symposium on Tilapia in Aquaculture*, Pullin, R. S. V., Bhukaswan, T., Tonguthai, K., and Maclean, J. L., Eds., International Center for Living Aquatic Resources Management, Manila, 1988, 83.

Odura-Boateny, F. and Bart-Plange, A., Pito brewery waste as an alternative protein source to fish meal in feeds for *Tilapia busumana*, in *The Second International Symposium on Tilapia in Aquaculture*, Pullin, R. S. V., Bhukaswan, T., Tonguthai, K., and Maclean, J. L., Eds., International Center for Living Aquatic Resources Management, Manila, 1988, 357.

Omoregie, E. and Ogbemudia, F. J., Effect of substituting fishmeal with palm kernel meal on growth and food utilization of the Nile tilapia, *Oreochromis niloticus*, *Bamidgeh*, 45, 113, 1993.

Perschbacher, P. W., The Effect of an Herbivorous Fish, *Tilapia aurea* (Steindachner), on the Phytoplankton Community of Fertilized Ponds, M.S. thesis, Auburn University, Auburn, AL, 1975, 54.

Pillay, T. V. R., *Aquaculture Principles and Practices*, Fishing Book News, London, 1990, 575.

Pillay, T. V. R., *Aquaculture and the Environment*, Fishing Book News, London, 1992, 189.

Ross, L. G., McKinney, R. W., and Ross, B., Energy budgets for cultured tilapias, in *The Second International Symposium on Tilapia in Aquaculture*, Pullin, R. S. V., Bhukaswan, T., Tonguthai, K., and Maclean, J. L., Eds., International Center for Living Aquatic Resources Management, Manila, 1988, 83.

Schroeder, G. L., Autrotrophic and heterotrophic production of micro organisms in intensely-manured fish ponds, and related fish yields, *Aquaculture*, 14, 303, 1978.

Siraj, S. S., Kamaruddin, A., Satlar, M. K. A., and Kamarudin, M. S., Effects of feeding frequency on growth, food conversion and survival of red tilapia (*Oreochromis mossambicus/Oreochromis niloticus*) hybrid fry, in *The Second International Symposium on Tilapia in Aquaculture*, Pullin, R. S. V., Bhukaswan, T., Tonguthai, K., and Maclean, J. L., Eds., International Center for Living Aquatic Resources Management, Manila, 1988, 387.

Stickney, R. R., *Principles of Warmwater Aquaculture*, John Wiley & Sons, New York, 1979, 375.

Stickney, R. R., Tilapia, in Stickney, R. R., Ed., *Culture of Nonsalmonid Freshwater Fishes*, CRC Press, Boca Raton, FL, 1993, chap. 3.

Stickney, R. R., Tilapia update, *World Aquaculture*, 25(3) 14, 1995.

Teichert-Coddington, D. and Green, B. W., Tilapia yield improvement through maintenance of minimal oxygen concentrations in experimental grow-out ponds in Honduras, *Aquaculture*, 118, 63, 1993.

Wee, K. L. and Tuan, N. A., Effects of dietary protein level on growth and reproduction of Nile tilapia (*Oreochromis niloticus*), in *The Second International Symposium on Tilapia in Aquaculture,* Pullin, R. S. V., Bhukaswan, T., Tonguthai, K., and Maclean, J. L., Eds., International Center for Living Aquatic Resources Management, Manila, 1988, 401.

12

DISEASES OF TILAPIA

Kamonporn Tonguthai and Supranee Chinabut

INTRODUCTION

Tilapias originated in Africa. They have a rapid growth rate and relatively few diseases, probably because of the nature of the environment within which they have evolved, with its regular droughts and other stresses. They have become one of the most economically important group of cultured species. Tilapia farms are widespread in the tropics and subtropics. The fish are reared in ponds, cages, or pens, and they grow well in freshwater and brackishwater environments. The high fecundity of the fish, its rapid growth rate, its few disease problems, and the ready availability of tilapia fry have resulted in intensification of production. Papers on diseases of tilapia were first published early this century.

Under the original extensive or semi-intensive culture systems, tilapias were more resistant to disease than many other fish species (Roberts and Sommerville, 1982). However, the intensification of culture systems and resultant deterioration in the environment have been associated with an increase in parasitic and infectious disease problems.

Formerly, parasitic diseases appeared to be more significant than other forms of infection, but the incidence of nonparasitic infections appears to be increasing. Consequently, although the literature on infectious diseases of tilapia is increasing rapidly, there has only been a slight increase in the reports of parasitic problems (Vega, 1988).

This chapter reviews the work on diseases of both wild and cultured tilapias.

PARASITIC DISEASES OF TILAPIA

The parasitology of wild and cultured cichlids has been reviewed by Sarig (1975), Fryer and Iles (1972), Balarin and Hatton (1979), Roberts and Sommerville (1982), Paperna (1980a), Paperna et al. (1983), Natividad et al. (1986), and others. There are several species of internal and external parasites that affect tilapias, but the literature focuses on describing the parasites themselves rather than the effects of those parasites on the health and productivity of the fish. Parasites generally do not severely damage the fish unless they are present in large numbers and attack the vital organs. However, blood feeding parasites do cause severe damage, even if they are only present in small numbers. Their detrimental effects are particularly severe in small fish.

External Parasites

External parasites are the most frequent problem in cultured tilapias. In the wild, external parasites are commonly found but in relatively low numbers on individual fish. Many species of external parasites can be removed easily from the skin. However, some that penetrate through the skin or have attachment organs firmly embedded into the flesh are much more difficult to dislodge. Many of these parasites may predispose fish to bacterial infection.

0-56670-274-7/97/$0.00+$.50
© 1997 by CRC Press LLC

Protozoa

Protozoa are unicellular microscopic organisms. They have the ability to multiply on or within the host. Harmful effects of parasitic protozoa to the host depend on their incidence, site of infection, and feeding habits. Protozoa are often present in low numbers on or in fishes in their natural environment. Alteration of the environment may be to the advantage of the parasite. Intensive culture practice favors the protozoan populations.

There are many different species of protozoan fish parasites, but tilapias appear to have fewer parasitic protozoa problems compared to other cultured species, such as carps, catfishes, or snakehead. Protozoa are, however, the most commonly reported group of external parasites on tilapia. External protozoa have a direct life cycle and reproduce by binary fission on the skin and gills of their hosts. The rate of reproduction of most protozoa depends on water quality and temperature. Some protozoa can survive for a short time without a host, but in a confined area such as a pond with large numbers of potential hosts can readily reinfest another host.

Heavy infestation with protozoa may provoke a copious mucous exudate. The epithelial cells of the skin subsequently degenerate and die. Affected fish exhibit abnormal behavior or coloration. Severe damage to gill tissues disrupts normal function of the gills. Damage caused by protozoa infestations may also predispose fish to secondary bacterial infections. Severe effects of parasitic protozoa may be directly responsible for death of fry through osmotic imbalance.

Three groups of parasitic protozoa, e.g., ciliates, flagellates, and sporozoa, have been reported on tilapias.

Parasitic ciliate protozoa are mainly ectoparasitic, though some burrow deep into the skin. Several genera have been involved in disease epizootics in tilapia. Paperna and Van As (1984) reported *Chilodonella hexasticha* in high number on *Oreochromis mossambicus* stressed by high temperature. *Chilodonella* sp. are heart-shaped, motile, 30 to 70 μ in length and 21 to 40 μ in width (Figure 1). They are generally found on the skin, fins, and gills and feed on epithelial cells. Fish infested with *Chilodonella* sp. usually have a bluish-gray mucous investing the infected skin. The gills may be badly damaged, resulting in tissue hyperplasia, which causes the respiratory surface to become nonfunctional. Van As and Basson (1984) reported that during winter, parasite levels were lower but appeared to remain on the fish stocks.

Trichodinids were also reported to be predominant ciliate protozoa on tilapia (Figure 2). They were reported to cause heavy losses in small tilapia in Ghana (Sarig, 1975); Paperna and Van As (1984) referred to trichodinids as being significant parasites in cultured tilapias. Trichodinid on fingerling *O. mossambicus* increased dramatically during winter at Harbeesport Dam, South Africa (Oldewage and Van As, 1987). *Trichodina acuta* was reported on *Tilapia zilii* and *O. mossambicus* (Duncan, 1977). Five species of *Tripartiella* (Figure 3) were observed on the gills of freshwater

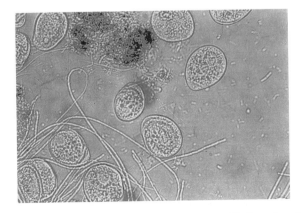

Figure 1.   *Chilodonella* sp. Heart-shaped ciliate protozoa generally found on the skin, fins, and gills.

Figure 2. *Trichodinids* sp. commonly found on the gills and skin. (Courtesy of M. G. Bondad-Reantaso.)

and brackishwater tilapia in the Philippines (Bondad-Reantaso, 1989). Trichodinids are considered to be among the most highly complex of the protozoa commonly found on the skin and gills of fishes. The body is saucer or bell-shaped. Embedded within the parasite under the attachment organ or sucker is a ring of denticles or tooth-like structures. The harmful effects of trichodinids are aggravated by their mobility so that each individual extends its influence over a large area. They can infest the fry by invading the mouth of broodstock during the mouth-brooding period, which is a feature of many tilapia life cycles (Fryer and Iles, 1972). Infested fish may exhibit abnormal behavior or coloration. Hyperplasia, degeneration, and necrosis of the epithelial cells occur, accompanied by proliferation of mucous cells. Secondary bacterial infections are a common sequela.

Ichthyophthiriasis is one of the most prevalent diseases of freshwater fishes, tilapias included. The disease is caused by the ciliate *Ichthyophthirius multifiliis*, which is the largest protozoan found on fishes. The body of the parasite is subspherical to ovoid, with a diameter of 50 to 100 μ. It is uniformly covered with cilia (Figure 4). Signs of infestation are irritation and white spots on the skin and gills. Penetration into the epithelium by the tomites of the parasite results in extensive changes in the surrounding integumental tissues. As the parasite grows, the epithelium is pushed outward to form the white swelling. When the parasite breaks out of the skin, the epithelium may be completely destroyed and the dermis exposed to bacterial or fungal infection. The gill epithelium is similarly affected. In severe cases, especially in small fish, mortalities may occur due to the

Figure 3. *Tripartiella* sp. were observed on the gills and skin of tilapia. (Courtesy of M. G. Bondad-Reantaso.)

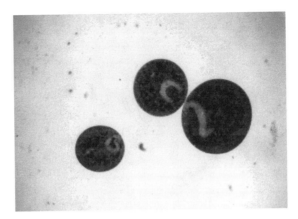

Figure 4. *Ichthyophthirius multifiliis*, causing white spot disease on the skin and gills.

direct osmotic effects of the parasite. In a review of diseases affecting cichlids by Paperna and Van As (1984), *I. multifiliis* has been recorded in high numbers on temperature-stressed fish. A similar case has been observed on tilapia fry at 25 to 27°C in Alabama (R. Phelps, personal communication, 1995). The parasite was observed in *O. mossambicus* imported from Singapore to Hawaii (Brock and Takata, 1955), and Paperna (1970) has reported it on wild tilapia in East Africa.

Epistylis sp. (Figure 5), *Scyphidia* sp. (Figure 6), and *Apisoma* sp. (Figure 7) are also ciliated protozoa. These parasites are distinguished by the circular or subcircular arrangement of cilia near the oral end of the cell. They possess more or less definite stalks by which they attach to substrata. *Epistylis* sp. form colonies. They are found on skin, fins, and gills. They favor soft skin, devoid of large scales. They are, therefore, most common on juvenile fish. They have been recorded in high numbers on temperature-stressed tilapia (Paperna and Van As, 1984). Heavy infestation with this parasite provokes a copious exudate of mucous, the skin becomes hyperaemic, and the scales may bristle. Mass colonization on the opercula can disturb their normal movement and affect respiration.

Few genera of flagellate protozoa have been reported to infest tilapia. Among them are *Ichthyobodo* sp. and *Amyloodinium* sp. Paperna and Van As (1981) reported several species of *Ichthyobodo* on tilapia, including *Ichthyobodo necator*. Sarig (1971) reported that infestation of this parasite was common in late autumn and winter in Israel.

The parasitic attaching stage of *Ichthyobodo* (Figure 8) is usually spherical or pyriform and attaches to the host by a set of root-like cytoplasmic structures or rhizoids. They have four flagella, 10 to 15 μ in diameter. These parasites are most commonly found in the scale pockets. In severe

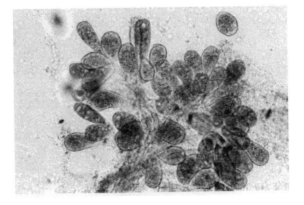

Figure 5. *Epistylis* sp. attached to the skin, fins, and gills, commonly found on juvenile fish.

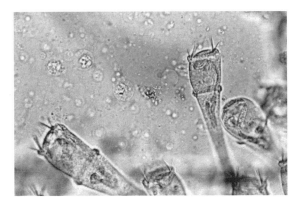

Figure 6. *Scyphidia* sp. Ciliated protozoa commonly found on skin and fins.

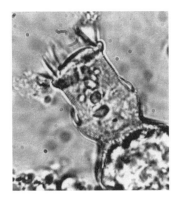

Figure 7. *Apiosoma* sp. Ciliated protozoa on skin and fins. (Courtesy of M. G. Bondad-Reantaso.)

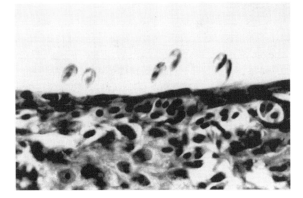

Figure 8. The parasitic attacking stage of *Ichthyobodo* sp. They are most commonly found in scale pockets. In severe infestation, they may spread to the gills.

infection they may spread onto gills. The damaging effects are the result of the functional disorder of the epithelium caused by the attachment and feeding of the parasite (Schubert, 1968). *Amyloodinium* sp. (Figure 9) are similar to *Ichthyobodo*, but they are a dinoflagellate protozoa. Paperna (1980b) reported *Amyloodinium ocellatum* on *O. mossambicus* acclimated in seawater. It is interesting to learn that fish that had recovered from epizootic infestation of *Amyloodinium* sp. were not reinfested (Paperna, 1991).

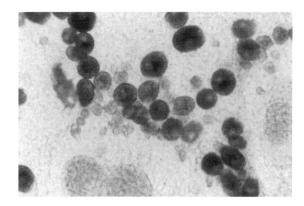

Figure 9. *Amyloodinium* sp. A dinoflagellate protozoan commonly found on the gills.

Monogenea

Monogeneans are large ectoparasites usually found on the skin and gills of fish. Some inhabit the mouth cavity. They attach to the host by means of an organ called the opisthaptor. Hooks on the opisthaptor are responsible for much of the damage to the host cells. They penetrate into the surface layers of the skin and gills. Monogenea may be oviparous or viviparous and can build up large populations within a short time. Light infestation is considered harmless, but a light infestation frequently becomes heavy because its reproduction potential is great.

In case of heavy infestation, the parasites may be visible, and the skin often appears paler than normal. Abnormally copious mucous production is common. Small red spots may mark the site of local lesions. Large portions of the skin may slough off, with accompanying scale loss. This results in respiratory and osmoregulatory difficulties, especially in young or small fish. A large number of monogenean species has been recorded on fish, but only a few species have been found on tilapia.

Gyrodactylus sp. and *Dactylogyrus* sp. are the two significant genera in cultured tilapia. The major characteristic of *Gyrodactylus* (Figure 10) is the opisthaptor, which has 16 marginal hooks and one pair of anchors. Eye spots are absent. The uterus contains a single embryo that, in turn, contains an embryo of the next generation. *Dactylogyrus* has only 14 marginal hooks with two pairs of eye spots and usually has four head lobes. They commonly occur on gill filaments but when present in large numbers can be distributed all over the body of the host (Amlacher, 1970).

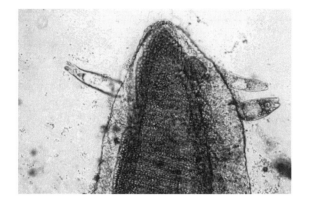

Figure 10. *Gyrodactylus* sp. A monogenean in cultured tilapia, one of the significant genera in cultured tilapia.

Gyrodactylus sp. have been reported in various countries in South East Asia and in various species of fish. However, there are very few reports of *Gyrodactylus* sp. in tilapias. Natividad et al. (1986) reported *Gyrodactylus* sp. in *O. niloticus* cultured in both brackish water and freshwater in the Philippines. *Gyrodactylus cichlidarum* has been reported from wild tilapia in Israel (Paperna and Lahav, 1971), and the records showed that this parasite was found to have heavily infested blue tilapia in Puerto Rico and moderately infested Mozambique tilapia (Williams and Williams, 1994). Nguenga (1988) observed *Dactylogyrus* sp. on gills of *O. niloticus* kept in tanks in Cameroon. Fryer and Iles (1972) recorded an outbreak of gyrodactyliasis in pond-reared tilapias in Uganda, where they were associated with corneal damage. Two new species of Gyrodactylus were described from the body surface of Nile tilapia from fish farms in the Philippines (Cone et al., 1995). They are *G. shariffi* sp. n occurring on fish in brackishwater ponds and *G. niloticus* sp. n on fish cultured in freshwater ponds.

Infestation of the gill by the above monogeneans causes severe hyperplasia of the epithelium. Extensive proliferation of the gill epithelium interferes with respiratory function and may be a direct cause of death, particularly in small fish. Paperna (1963b) showed that large fish can tolerate heavy infestation of *Dactylogyrus* sp. better than the small fish. He reported that mortality was not observed in fish longer than 32 to 35 mm, and fish more than 35 mm long were able to tolerate a heavy infestation of up to 300 parasites per fish.

In addition to *Gyrodactylus* sp. and *Dactylogyrus* sp., *Cichlidogyrus* sp. appear to be widespread in Africa as well as East Asia. A high incidence of five species of *Cichlidogyrus* — *C. sclerosus, C. tilapiae, C. tiberianus, C. lonicornis,* and *C. branchialis* — was reported in cultured tilapia in Africa (Paperna, 1960; Paperna and Thurston, 1969). *C. tilapiae* was reported in tilapia in Puerto Rico; it is only gill worm on these hosts (Williams and Williams, 1994). Duncan (1973) has also reported *C. sclerosus* on *O. mossambicus* in the Philippines. This is probably the first record of this species outside Africa and the Middle East. These parasites are believed to have been introduced through the importation of tilapia to the Philippines (Natividad et al., 1986; Bondad-Reantaso and Arthur, 1990).

Enterogyrus cichlidonum was reported on *O. niloticus* in the Philippines (Bondad-Reantaso and Arthur, 1990). A heavy infestation of *Neobenedenia melleni,* a marine monogenean (Figures 11 and 12), was reported on *O. mossambicus* in saltwater cages in Kaneohe Bay, Hawaii and was found to be synomized with *N. girellae* (Ogawa et al., 1995). Up to 400 parasites were found on individual fish, attached over nearly the entire body surface (Kaneko et al., 1988). *Benedenia monticelli* was found in *O. aurea* associated with the stress of gradual acclimatization to seawater (Paperna, 1975; Paperna and Overstreet, 1981; Paperna et al., 1984).

Figure 11. *Neobenedenia girellae:* a monogene on the skin of tilapia cultured in saltwater. (Courtesy of M. G. Bondad-Reantaso.)

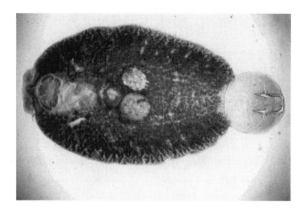

Figure 12. *Neobenedenia girellae.* (Courtesy of M. G. Bondad-Reantaso.)

Transversotrema sp., (Figure 13) an ectoparasitic digene, was found on the skin of cultured tilapia in Thailand (Sirikan, 1983) and in the Philippines (Natividad et al., 1986).

Crustacea

The most common species of parasitic crustaceans on tilapia are *Argulus* sp., *Ergasilus* sp., and *Lernaea* sp. Some reports refer to severe damage caused by these parasites, but some refer only to finding them on the fish.

Crustacean parasites damage the fish by attachment and feeding. The pressure of the attached parasite can cause tissue atrophy. If the parasites are present in the gill cavity, they can seriously reduce the respiratory area of the fish. Mode of feeding of some crustacean parasites involves secretion and injection of relatively large quantities of digestive fluids and the tearing of tissues by the buccal apparatus and attachment appendages. If the parasite feeds on erythrocytes, they can cause a serious and prolonged energy drain.

Several species of *Argulus* (Figure 14) have been reported on tilapia (Paperna, 1969). *A. indicus* and *A. foliaceus* caused problems in *O. mossambicus* and *O. niloticus* in Indonesia and Thailand (Kabata, 1985). *A. japonicus* was reported in tilapia in Puerto Rico (Williams and Williams, 1994). The problem mostly occurred in standing water. Fish are damaged by *Argulus* from attachment of the cup-like stalked suckers and by a sharp piercing organ or stylet. *Argulus* sp. feeds on plasma, but the diameter of the stylet is too small to pass erythrocytes. During feeding, toxic substances are released, which may cause a severe inflammatory response. The tiny wounds caused by feeding

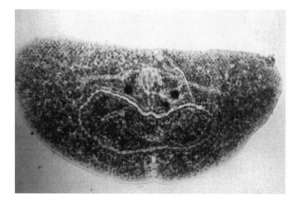

Figure 13. *Transvosotrema* sp.: an ectoparasitic digene on the skin of cultured Nile tilapia. (Courtesy of M. G Bondad-Reantaso.)

Figure 14.   *Argulus* sp.: one of the most common species of parasitic crustaceans on tilapia.

may also lead to bacterial infection. *Argulus* sp. are large parasites; if they remain in one spot for a long time, the skin develops a depression, and eventually all the mucous cells of the epidermis under the parasite may disappear.

Dolops ranarum, an opportunistic argulid parasite, was reported to occur predominantly on the buccal and branchial integument of *O. mossambicus* in Southern Africa, with a prevalence of 26% (Avenant and Van As, 1986).

Ergasilus sp. has been reported to cause severe damage to fish in Israeli polyculture systems (Paperna, 1964b; Paperna and Lahav, 1971). Williams and Williams (1994) reported *E. caerulae* as one of the parasites of Puerto Rico freshwater sport fishes; Puerto Rico may be the only exotic location of this parasite. Gills are the main habitat of *Ergasilus* sp. In severe infestations, they may also be found on the skin and fins. They attach by powerful claws developed from second antennae, resulting in responsive hypertrophy and fusion of the filaments. The damage to the gill can reduce the respiratory area; *Ergasilus* sp. feed on integumental cells, which can also cause reduction of the respiratory area. Severe infestation of this parasite on tilapia in high stocking densities may lead to high mortality.

High tilapia mortalities have also been reported due to *Lernaea* (Figure 15) in Nigerian ponds (Fryer, 1961, 1968) and Israel (Paperna, 1969). In Southeast Asia, at least five species of Lernaea — *L. cyprinacea, L. arcuata, L. polymorpha, L. oryzophila,* and *L. lophiara* — have been recorded on *O. mossambicus* and *O. niloticus* (Kabata, 1985; Shariff and Sommerville, 1986). *L. hardingi* was found on *O. niloticus, Sarotherodon galilaea, T. heudeloti,* and *T. zillii* in reservoirs and ponds in Ghana (Paperna, 1969).

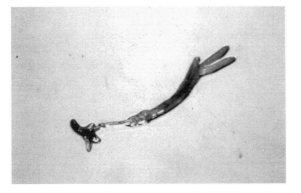

Figure 15.   *Lernaea* sp.: one of the parasitic crustaceans that cause severe damage to tilapia.

Lernaea sp. embed their anterior end into the tissue of the host and then develop finger-like shaped holdfasts structure. The span of this holdfast may reach several centimeters. Sites of attachment may be inflamed, with the wound remaining erythemic and hemorrhagic after the parasites have fallen off. *Lernaea* sp. feed on tissue debris and erythrocytes, causing the area to become ulcerated, and secondary bacterial infections are common. Epithelial mucous cells in adjacent areas increase in number, with copious mucous secretion. The life cycle of these parasites is completed within 20 to 25 d or less depending on the temperature. Thus, once they get into a pond, the population increases in a relatively short time.

Caligus sp. and *Lamproglena* sp. have been observed on tilapia. Natividad et al. (1986) reported an abundant infestation of *Caligus* sp. on the skin of Nile tilapia. It was a severe case as the infestation reached a maximum mean intensity of 111.7 parasites per fish. *Caligus* sp. live on the skin and fins, as well as in the buccal and branchial cavities. This parasite feeds on the epithelial cell around the attachment area. Destruction of the skin and bleeding wounds may be observed, but severity of the damage depends on the area of attachment and the number of parasites present. *Lamproglena* sp. has recently been reported in gills of *O. niloticus* cultured in cages in the Philippines (A. V. Yambot, personal communication, 1994). Generally it does not occur in large populations. In this case, however, high numbers of parasites were present on the gill. This parasite can cause tissue hypertrophy and local degeneration of the capillaries of the gill filament. Additional damage is caused by feeding activities. When they occurred in large number on *O. niloticus* in the Philippines, they caused severe injury to fish.

Alitropus typus is a very common isopod in freshwater, which has been observed on the body and gill of tilapia in reservoirs in Thailand (S. Kanchanakan, personal communication, 1993). The harmful effects of isopods are similar to the copepods, mainly due to attachment and feeding activity, although the effects may be more severe, especially in small fish, if the parasite feeds on blood. Fish mortalities may occur within a relatively short period of time.

Internal Parasites

Internal parasites can be found in various organs. Immature forms of parasites are generally found in the gill, muscle, and mesentery, but mature forms are generally found in the digestive tract. Damage to fish depends on site of infection.

Protozoa

On tilapia, sporozoans are likely to be found more often than other endoparasitic protozoa. Among them, *Myxobolus* and *Myxosoma* sp. have been recorded from wild tilapias (Baker, 1963). In farmed cichlids in Malawi, large subcutaneous cysts in juvenile *O. mossambicus* caused by myxosporean infestation were observed (Paperna, unpublished). Roberts and Sommerville (1982) reported that these parasites are more likely to be a problem in countries where the height of the water table does not allow seasonal drainage and drying of ponds.

The effect of these parasites depends on the number and location of the cysts. Heavy infection in the gill may lead to occlusion of branchial circulation, necrosis, and wide-scale respiratory dysfunction. Those found in muscle have little effect on the host. To date there are no reports of high mortality caused by these parasites in cultured tilapias.

Eimeria sp. (Figure 16) are other sporozoans that have also been reported on tilapias. *Eimeria vanasi* was reported in intestines of wild and farmed cichlids in Israel and South Africa (Lansberg and Paperna, 1987). This species was also found in juvenile *O. niloticus* in Thailand (Paperna, 1991). The key feature of *Eimeria* sp. is an oocyst that consists of four sporocysts, each with two sporozoites. Some eimerian species release sporocysts from the oocyst, either inside the host or soon after excretion. The sporocyst is ingested by a new host where sporozoites are released and enter cells of the suitable organ of the new host. The severity of infection can be judged by the

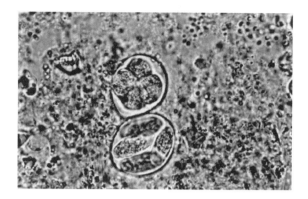

Figure 16. *Eimeria* sp. oocyst found in the posterior intestine of tilapia fry.

number of oocysts found per unit weight of feces. Signs of the disease caused by this parasite include emaciation, lethargy, and general poor health.

Trypanosoma sp. are the only flagellate protozoan endoparasite of tilapia. Wild fish are usually affected by small numbers of trypanosomes. These parasites seldom cause serious harm to their host unless there is a very heavy infestation. They have a direct damaging effect on the blood cells, and, in severe infestations, affected fish may show signs of anemia. Baker (1960) reviewed the trypanosomes of African freshwater fishes and reported *Trypanosoma mukasai* in different species of tilapias (*O. niloticus, O. exculenta, O. variubilis,* and *O. lencosticta*). *T. choudhuryi* was found in *O. mossambicus* collected from West Bengal, India (Mandal, 1977).

Monogenea

Monogenean trematodes are generally recognized as external parasites. However, *Enterogyrus cichlidarum*, one such monogenean parasite, has been recorded in the stomach of Nile tilapia in Africa (Paperna, 1963a) and during a parasitic survey in the Philippines (Bondad-Reantaso and Arthur, 1990).

Metazoa

Endoparasitic metazoans infesting the viscera muscle and eye include species of digenetic trematodes, cestodes, and nematodes. Digenes are predominantly endoparasitic. They possess unsegmented, oval or lanceolate bodies. They are equipped with two attachment organs, the oral sucker and the ventral sucker or acetabulum. Adult trematodes are often found in the intestine, gall bladder, or urinary bladder.

The life cycle of trematodes usually involves fishes as a primary or secondary intermediate host, but fish are also on occasion the final host of a digene. Eggs from the mature digenes are shed into the water, hatch, and seek the second intermediate host, where cercaria develops into a metacercaria. Adult digenes are considered either completely or virtually harmless in fish. Harmful effects occur during the invasion of cercariae that encyst to become metacercariae in fish tissues. Their pathogenicity depends on their numbers and on the paths they travel, as well as the organ where they are encysting. Effects can be severe when fry or fingerlings are infested. Among the internal parasites of tilapia, digenes are the predominant genera. Fish trematode can also post a public health problem. *Clinostomum complanatum*, when ingested with the fish by man, is capable of producing laryngopharyngitis.

The metacercariae of *Haplorchis pumilio* were first recorded in tilapias by Witenberg (1929) in Palestine and were later reported in various countries, including India, China, Taiwan, Japan, Philippines, Egypt, Tunisia, and Kenya (Sommerville, 1982), as well as in cultured hybrid tilapia

(O. aurea × O. niloticus) in Israel (Davidov, 1979; Farstey, 1986). Simultaneous invasion by large numbers of cercariae of *H. pumilio* into tilapia fry resulted in mortalities (Sommerville, 1982). The cercariae migrated through and eventually settled in the connective tissue associated with skeletal structures. Hunter and Hunter (1938) described a significant loss of weight in fish infected with metacercariae, and Meyer (1958) suggested that heavy infection of metacercariae greatly increased host mortality.

Clinostomum tilapiae has been reported in pond-cultured *O. mossambicus* in North Transvaal, South Africa (Britz et al., 1985). Large digenean parasites like *Clinostomum* sp. can cause bulging and distortion of the body profile.

Euclinostomum heterostomum was found in cultured *O. mossambicus* fingerlings in Malawi (Paperna, 1980). Mass mortalities due to heavy infections of this parasite in pond cultured *O. mossambicus* were observed in north Transvaal, South Africa (Britz et al., 1985).

Metacercariae of *Neascus* sp. have been reported in a wide range of tilapias from East and West Africa (Paperna and Thurston, 1968). The parasite has a limited effect on the fish host but if present in large numbers renders fish unmarketable.

Other endoparasites found in tilapias are cestodes, nematodes, and acanthocephalans. Cestodes have been reported in wild fish (Fryer and Iles, 1972). Several nematode species have also been reported (Goldstein, 1970; Fryer and Iles, 1972), but little is known of their significance. However, Paperna (1964a) and Scott (1977) reported pathological effects of nematode larval *Contracaecum* sp. infestations in tilapia muscle. Scott (1977) reported that up to seven parasites might be found in the pericardial cavity of *Sarotherodon alcalicus* in Kenya and that they had a significant effect on the growth of the fish. *Contracaecum* sp. is an unattractive parasite to the consumer, since it is a large encysted worm found throughout the muscle. *Bothriocephalus archeilognathi* has been found in the intestines of tilapia in Cuba (Williams and Williams, 1994). Adult cestodes inhabiting the alimentary canal may interfere with the absorption of food. Acanthocephalans may cause damage to the intestine due to their attachment organ. It has also been suggested that some *Acanthocephala* sp. produce toxins, secreted into the host tissues through pores present in the proboscis hooks (Kabata, 1985). However, there are no reports of severe disease in tilapia due to the above parasites.

BACTERIAL DISEASES

Intensive culture systems tend to disturb the natural equilibrium between fish and the environment. There are many ways in which the balance can be disturbed, for example, increased population densities, introduction of new fish species, rough handling of the fish, or increased organic load due to heavy feeding and fertilization. Many of these factors stress the fish, making them more susceptible to potentially pathogenic organisms in the environment (Snieszko, 1974). Tilapias cultured in ponds are generally kept under intensive conditions, which allows bacterial numbers to increase significantly. The bacterial infections are mostly opportunistic, occurring only in the presence of environmental stressors.

In the view of Roberts and Sommerville (1982), three clinical syndromes are associated with bacterial diseases of tilapias: a syndrome associated with pathogenic cytophaga-like bacteria (formerly known as myxobacteria) predominantly as skin lesions, hemorrhagic septicemia, and chronic granulomatosis.

Columnaris disease is one of the most common diseases of cultured tilapia. This disease is caused by *Flexibacter columnaris* (Figure 17). The affected fish shows opaque skin lesions with epithelial necrosis. It becomes very dark and slow moving and dies quickly. The gills may also be infected; this is associated with heavy mucous production. The incidence of this bacterial infection varies seasonally with water temperature and certain environmental conditions. Cases of columnaris disease increase when the fish are under environmental stress. In Taiwan, columnaris disease was common from July through December, with the peak of the infection in September and October

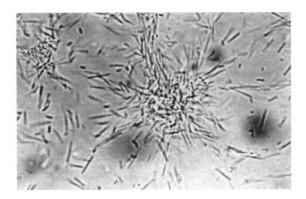

Figure 17. *Flexibacter columnaris* causes columnaris disease, which is one of the most severe bacterial diseases of cultured tilapia.

(Kuo et al., 1980). The pathogenicity of *F. columnaris* was studied by Hsu and Kou (1977), who observed that the disease evoked more rapidly by direct contact with infected fish than by introduction of the pathogen into the water.

Aeromonas hydrophila (Figure 18) is the most common opportunistic bacteria found in association with hemorrhagic septicemia in tropical fish including tilapias (Scott, 1977). Most epizootics of this motile aeromonad disease are stress related and is usually seasonal in occurrence. The organisms mostly enter the host through the digestive tract unless the fish have skin or gill abrasions, in which case the organism may enter by these routes.

Bacteria of the genus *Pseudomonas* are ubiquitous in natural freshwaters. Pseudomonad septicemia can occur in single fish or as epizootics. The etiological agents of pseudomonad septicemia are opportunist pathogens belonging to the genus *Pseudomonas*. *Pseudomonas fluorescens*, a nonspore-forming, Gram-negative, rod-shaped bacteria, is one example. Pseudomonad septicemia occurs primarily among cultured and aquarium fishes, but occasionally appears among wild fishes. Wu (1970) reported *Pseudomonas* sp. associated with disease in tilapia, causing a similar syndrome to aeromonad hemorrhagic septicemia.

Epizootics of *Edwardsiella* septicemia in channel catfish have been investigated since 1969. In some cases the pathogen appeared to be *Edwardsiella tarda*. External signs include lesions under the skin progressing to abscesses in the muscles, which may be associated with putrid gas. *Edwardsiella* sp. have been reported in tilapia by Wu (1970).

Streptococcus sp. Gram-positive cocci is the one disease that represented real danger to farmers engaged in warmwater aquaculture. Kitao et al. (1981) first reported *Streptococcus* sp. in tilapia,

Figure 18. *Aeromonas* septicemia disease caused by *Aeromonas hydrophila,* a common opportunistic bacteria.

both *Sarotherodon* and *Oreochromis* sp. in Japan, and later Miyazaki et al. (1984) reported an outbreak of streptococcosis, hemolytic strains, in Taiwan (Ming Chen et al., 1985). Streptococcosis in tilapia has been found mostly in chronic forms (Hubbert, 1989), characterized by erratic swimming in circles or nonmobility. Erythema is observed around the anus, and inflamed areas are found along the dorsal surfaces. In a later stage, exophthalmos is evident. The intestine is usually hemorrhagic and filled with pink mucous. There may also be inflammatory infiltration of the muscle tissue.

Fish mycobacteriosis is a chronic or subacute disease of many fishes. It is a well-known disease in many cichlids kept in aquaria (Nigrelli and Vogel, 1963). The causative agent of this disease, *Mycobacterium* sp., is a rod-shaped, Gram-positive, non-spore-forming, nonmotile, and acid fast bacterium. The most probable route of transmission of this bacteria is orally. Affected fish may develop unilateral or bilateral exophthalmia, and focal granulomata may occur in various internal organs, including the liver, spleen, and kidney.

Providencia rettgeri infection in *O. niloticus* in lower Egypt caused serious renal damage, resulting in high mortalities (Popp et al., 1988). Sakata and Koreeda (1986) made a numerical taxonomic study of the dominant bacteria isolated from tilapia intestine. They found two major bacteria, *Pleiseomonas shigelloides* and *A. hydrophila*, although there was no suggestion that these organisms were pathogenic.

MYCOTIC DISEASES

Saprolegniasis is a mycotic disease cause by fungi in the family Saprolegniaceae. This disease is most common when fish are held at low temperature. Most genera involved in fish disease are *Saprolegnia* sp., *Achlya* sp. (Figure 19), *Aphanomyces* sp. (Figure 20), and *Dictyuchus* sp. These fungi have long, branched, nonseptate hyphae. Disease signs are cotton-like white to gray-white or gray-brown organisms growing on the exterior surface of fish or dead eggs. These signs are seen much more easily when the fish are in the water than when they are removed from the water. *Achlya racemosa, Aphanomyces laevis, Dictyuchus sterile, Saprolegnia ferax, S. litoralis*, and *S. parasitica* were the species recorded from *Tilapia zilli* in ponds in Nigeria (Ogbonna and Alabi, 1991). Pillai and Freitas (1983) reported infection with *Fusarium* sp. in *O. mossambicus* in India, which caused mass mortalities in 1977.

Aspergillus sp. are terrestrial fungi often found in feed stored under damp, warm conditions. These fungi produce toxins (aflatoxins), which affect the fish when they eat the contaminated feed. The toxins from this fungus induce systemic disease in *O. niloticus* (Olufermi and Roberts, 1986). Infected fish become dark, inappetant, and lethargic and may show exophthalmia. It has been reported that these fungi are also capable of causing systemic disease in cultured tilapias.

Figure 19. *Achlya* sp. A fungus involved in mycotic disease of tilapia.

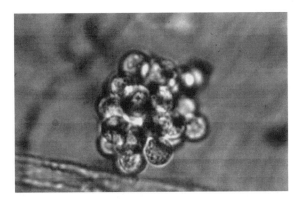

Figure 20. Zoospores of *Aphanomyces* sp., fungus involved in mycotic disease in tilapia.

VIRAL DISEASES

There are some reports of viral diseases in tilapias. Paperna (1974) made the first report of a viral disease (lymphocystis) in tilapia. Lymphocystis (Figure 21) disease is a chronic, slowly developing viral disease of connective tissue, especially in the dermal layer of the skin. The infected cells, usually fibrocytes, swell up to 100,000 times normal size and become visible to the naked eye. Other cells near or attached to affected cells remain normal. These infected cells may have the gross appearance of tumors, but they are not malignant. The disease is not usually fatal to infected fishes. The etiological agent of lymphocystis disease belongs to the Iridovirus family. This virus is a large, complex, naked virus with a deoxyribonucleic acid genome, which replicates in the cytoplasm of infected cells. There is no therapy for lymphocystis disease.

Birnavirus was found in *O. mossambicus* cultured in Taiwan (Chen et al., 1983). Hedrick et al. (1983) compared the four birnaviruses isolated from fishes in Taiwan with IPNV and found that one of the four viruses was closely related to the AB strain of IPNV. Ueno et al. (1984) has shown that infection with this virus, when injected intraperitoneally, can induce mortality among tilapia. A whirling viral disease of tilapia larvae in Israel was reported by Avtalion and Shlapobersky (1994). A viral particle of about 100 nm was observed in brain tissue. The whirling symptom was always preceded by a darkening of the tegument and anorexia.

An organism likely to be a member of the family Rickettsiaceae was detected for the first time from an outbreak of disease in pond-reared tilapia in Taiwan. Affected fish were dark in color, had a reduced appetite, and swam listlessly in circles at the surface or at the side of the pond.

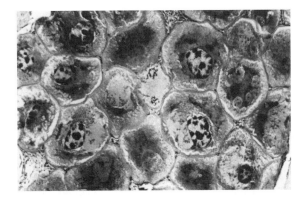

Figure 21. Lymphocystis: a viral disease. The infected cells swell up to 100,000 times normal size and become visible to the naked eye.

Exophthalmos, ulcerated skin, and swollen internal organs, especially an enlarged spleen, were also noted. Typical lesions in moribund fish included marked white nodules and microscopic granulomata formation all over the internal organs (Chen and Chao, 1994).

CHLAMYDIAL DISEASE

Epitheliocystis is a chronic and unique infection by a chlamydia-like organism that results in hypertrophied epithelial cells, typically of gills (Figure 22) and sometimes skin. It can affect a wide range of freshwater and marine fishes. Epitheliocystis cells alone are superficial and benign; however, their presence in gills can stimulate hyperplasia and fusion of adjacent lamellae. If the infection is extensive, respiration is impaired and death may result. Hyperinfection by epitheliocystis has caused mass mortalities among cultured fingerlings of many fish species (Paperna et al., 1981). Epitheliocystis was observed in gill epithelial cells of *O. niloticus* fry cultured in Thailand in ponds heavily loaded with organic matter. Infected fish showed no distinct clinical signs, and mortality was low (S. Chinabut, personal communication, 1994). There was also a report of a chlamydial organism in *O. mossambicus* from South Africa and hybrid tilapia (*O. aureus* × *O. niloticus*) from Israel (Paperna et al., 1981).

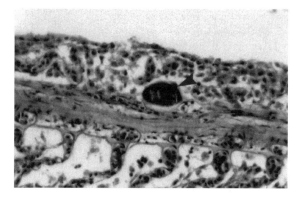

Figure 22.   Epitheliocystis on the gills of tilapia fingerling resulting in hypertrophy of the epithelial cells.

NEOPLASTIC DISEASE

Information on neoplasia of tilapias is very limited. Haller and Roberts (1980) reported both lymphoma on the flank and renal tubular adenocarcinoma in the kidney of *Sarotherodon spilurus* broodfish cultured in brackish groundwater in Kenya. The only clinical sign observed was a large swelling on the abdomen just above the anus. A kidney tumor, renal adeno carcinoma occurred in a Mozambique tilapia in Puerto Rico (Williams and Williams, 1994).

SOURCES OF INFECTION

Introduction of fish into new environments may also introduce parasites and diseases into that area. Wild fish species may act as major carriers for introduction of parasites into ponds. Many parasites are small, such as parasitic protozoa, and light infestation of these parasites has no adverse effects on the host. But when these infested fish get into the new environment, such as a pond, the parasites may quickly reproduce if the water quality and temperature are in the optimum range.

Parasitic protozoa of tilapia, such as *Trichodina* sp., *Ichthyobodo* sp., *Chilodonella* sp., and *Ichthyophthirius* sp., have free swimming larval stages. These larvae must find hosts within a short time. Therefore, in overcrowded conditions within a pond, the parasite has more opportunity to find hosts.

Inlet water can also be a major source of infestation. Okaeme (1987) reported that fish cultured in concrete tanks with a natural supply of pond or reservoir water often developed infestations with new organisms, mostly protozoa.

Aquatic plants may become a source of infestation. Intermediate stages of some external protozoa, such as *Ichthyophthirius* sp. or eggs of *Argulus*, may be attached to aquatic plants. Therefore, all plants being brought to the pond should be quarantined for some period of time and washed in mild disinfectant before introduction.

The life cycle of digenean trematodes involving fish consists of several succession of larval generations. It may use fish as intermediate host. Tilapia serves as intermediate host for several digenetic trematodes. Reduction or elimination of a link in the transmission cycle may be effective in reducing the number of flukes.

Lernaeids and argulids can be transmitted through the water supply or by the introduction of infested fishes. The wild fish infested with parasitic crustaceans can introduce large numbers of larvae into the pond. Amphibians moving from place to place may also be involved with transmission of argulids.

Flexibacter columnaris is ubiquitous to most freshwaters of the world. All freshwater fishes are susceptible to *F. columnaris* infection. Development of columnaris disease in susceptible fishes is related to temperature and stress. Once a fish is infected, the superficial lesion may release large numbers of bacterial cells into the water.

It is accepted that aeromonad infection is one of the stress-related diseases. Overcrowding in the pond causes reduction of oxygen, presence of large quantities of excretory products from the fishes, and general conditions conducive to invasion of opportunistic bacteria. Fishes on poor levels of nutrition are especially susceptible to secondary infections, as are fishes with injuries or damaged skin or gills.

The etiological agents of mycobacteriosis are considered to be *Mycobacterium marinum* in marine fishes and *M. fortuitum* in freshwater and brakishwater fishes. These organisms have been reported in a large number of fish species and other animals such as frogs, toads, salamanders, snakes, lizards, alligators, and turtles. The most probable route of transmission for these bacteria is orally. Feeding fish the viscera or other products from affected fish should be avoided. Another route of transmission is through injuries or abrasions of the skin. In case of tilapia, this probably is a more common route of infection. The source of infection is water contaminated with this bacteria.

Vibriosis of tilapia by *V. parahaemolyticus* has been found to be a serious pathogen of cultured tilapia in Israel (Hubbert, 1994). The disease is confined to the area where fish are cultured in brackish water and occurs as a direct result of handling. Externally the disease is manifested by red patches on the body surface and especially around the vent. The gills often appear cyanosed, having a purple coloration. The gut appears to be filled with gas and mucous. Internally, all the major organs and the blood are heavily infected with bacteria.

Source of infection of viral disease can be eggs, feces, or excretion. In case of eggs, transmission can be directly from gravid females to developing eggs. Infected fish will shed a tremendous number of viral particles with feces or excretion into the water. Many viruses are passed through water from infected to noninfected animals. Water discharged from infected ponds can carry the virus to noninfected ponds. Fish surviving from epizootics of viral diseases may become carriers and serve as a virus reservoir.

Fungi reproduce asexually, forming zoosporangia, and discharges zoospores into the water. These primary zoospores may form cysts, which release many secondary zoospores. Zoospores find a fertile area on susceptible fish, germinate, and begin to produce vegetative hyphae. Dead fish or eggs are a fertile medium for fungal growth and production of zoospores. The presence of dead fish in water increases the number of infective zoospores enormously, thus increasing the probability of other fish with only minor injuries to be infected.

DIAGNOSIS AND IDENTIFICATION OF THE PATHOGEN

External parasitic protozoa are found on skin, fins, and gills. To identify these parasites, skin scrapings and gill filament are mounted on a glass slide with a drop of water or normal saline and a cover glass. The organisms should be alive for accurate diagnosis. The stains commonly used are hematoxylin, carmine, Lugol's solution, acidified methol green, Noland's solution, Iodine-eosin, and Klein's silver impregnation technique.

Some parasites can be diagnosed from the clinical signs; for example, *Ichthyophthiriasis* produces easily seen white spots. Confirmation of *Ichthyophthirius* sp. infestation can be made by removing one or more of the white spots from skin or gill and examining them under the microscope.

Internal parasitic protozoa like *Eimeria* may be diagnosed from the fecal casts or intestinal smears. In case of severe infection, the oocysts can be easily observed. In light infection, oocysts can be concentrated by centrifugation.

Some monogeneans are relatively large and can be seen without magnification, but others are microscopic. Skin scraping or a piece of gill filament mounted with a few drops of water on a microscopic slide can be used to locate the parasites.

Diagnosis of digenetic trematode infestations is by removal of the development stage, examination for morphology, and the use of taxonomic keys for identification. Cysts containing the metacercariae are dissected and the metacercariae released.

Helminth parasites require relaxation prior to fixation; digeneans, cestodes and nematodes can be relaxed in Berland's fluid and fixed in 5% formalin or 70% ethyl alcohol. Monogeneans and nematodes can be identified without staining, but digeneans and cestodes specimens must be stained for taxonomic purposes.

Arthropod parasites, such as copepods and isopods, can be fixed in 5% formalin or 70% alcohol. Identification is based on morphology of the adult female. Lesions on skin fins and gills associated with contact by these organisms may be of assistance in diagnosis.

Bacterial diseases can be diagnosed from the appearance of the infected fish and through isolation and identification of bacterium. Definite diagnosis of motile aeromonad disease must be accompanied by assessment of biochemical activity of the organism and by observing growth and metabolic products on special media. Histopathological findings may also aid diagnosis.

Diagnosis of viral diseases depends on the history of the disease in the population, signs of disease, and histopathology of various tissue. Electron microscopy is also important for viral identification. Cell culture techniques are utilized in order to investigate the cytopathologal effect of the virus.

Mycotic disease can be diagnosed from clinical signs, cotton-like white to gray-white or gray-brown material growth on the exterior surface of fish or fish eggs. Identification can be done from hyphal structure and also sporangium at the tip of fertile hyphae.

TREATMENT

Prophylactic Treatment

Quite often the outbreak can be prevented by increasing water flow, reducing the biomass of fish, eliminating sources of skin or gill irritants, or improving the general health of the fish. On the other hand, the use of prophylactic treatment at regular intervals may reduce the chances of stressed fishes being infected by bacterial diseases. Maintaining good water quality, improving the environment, and good management practices are of great value for the success of aquaculture.

Chemotherapeutic Treatment

Chemotherapeutic agents can be applied to eliminate diseases. However, chemicals used in the pond must be cost effective, especially when used in a large pond.

Therapy for Parasites

Many chemicals have been used as treatment, particularly for external parasites. Some of them have been approved by the U.S. FDA in food fish. However, many of them have not yet been approved. Chemicals have specific effects. An advantage of a specific chemical use cannot be seen if chemicals are used indiscriminately. Again chemical used in the pond must be cost effective. In addition, some chemicals that are effective for one disease may not be effective for another. Some chemicals may have a wider effective range, enabling them to treat more than one disease. Selected chemotherapeutants for external protozoa in a pond condition with water temperature approximately 27 to 30°C are shown in Table 1.

Chemotherapy for the monogeneans is not usually satisfactory, especially in large ponds, unless the primary cause of the increased numbers of parasites is found and alleviated. There are several chemicals recommended for treatment of monogeneans in fish. They are shown in Table 2.

Ponds with high levels of phytoplankton should be closely observed after the application of formalin, since the chemical may kill the phytoplankton, which will result in increased dissolved oxygen.

Kabata (1985) has listed several chemicals ($KMNO_4$, salt, ammonium chloride, lysol, pyrethrum, DDT, lindane, neguvon, dipterex, and bromex) and many native remedies, including teaseed cake to castor oil plants, pine needles, and Chrysanthemum to combat crustacean parasites. To date, dipterex is likely to be recommended to treat parasitic crustaceans more than any other chemicals. The recommended doses are shown in Table 3.

Table 1   Selected Chemotherapeutic Agents for Parasitic Protozoa in the Pond

Target species	Chemical compound	Doses	Application	Remarks
Ichthyopthirius sp.	Formalin salt	25–50 ppm 1%	2–3 times at 3-d interval	
Tricodina and *Chilodonella*	Formalin	25–50 ppm	Indefinite	Repeat with the same dose if necessary
Epistylis	Formalin	25–50 ppm	2–3 times at 3-d interval	
Scyphidia	Salt	0.5–1%	Indefinite	
Ichthyobodo	Salt	0.5%	Indefinite	

Table 2   Chemotherapeutic Agents Recommended for Monogenes Treatment

Chemical compound	Dose	Application	Remarks
Formalin	30–50 ppm	Indefinite	
Dipterex (Trichlorofon)	0.25–0.5 ppm	Indefinite	
Salt	0.50–0.75 %	Indefinite	Recommended for freshwater monogens
	10–15 g/L	Dip 20 min	

Table 3   Recommended Doses of Dipterex to Treat Parasitic Crustaceans

Target species	Dose	Application	Remarks
Lernaea sp.	0.25–0.5 ppm	Indefinite	Spray over the pond surface; repeat 3–5 times at weekly intervals depending on the severity of infestation
Ergasilus	0.15–0.25 ppm	Indefinite	
Argulus	0.25–0.3 ppm	Indefinite	
Lamproglena	0.15–0.25 ppm	Indefinite	

Endoparasites are difficult to eliminate. It is best to use prophylactic measures, eliminating the first intermediate hosts, such as snails, and the definitive hosts that include small mammals or piscivorous birds. There are no economically viable specific therapeutic measures effective against these endoparasites. The only recommended chemical is di-n-butyl tin oxide, which is added to the diet at the rate of 3% of food weight. However, this chemical is not commonly available.

Therapy for Bacterial Diseases

Furanace is recommended for the treatment of *F. columnaris*, since all isolates are sensitive to this chemical. Furanace is not only absorbed rapidly but also discharged from the fish tissue rapidly. The dosage is 1 ppm added to the pond on a continuous basis for 3 d.

Many antibiotics can be used to treat aeromonad infection. They are applied by mixing them with food. Oxytetracycline is used in food at the rate of 50 mg/kg of fish for 10 d. Sulfamerazine can be used at 150 mg/kg of fish for 15 d. The external bacteriocides can also be used for superficial infections. Oxytetracycline is the recommend antibiotic for *Edwardsiella* septicemia, applied orally at the rate of 50 mg/kg of fish per day for 10 d.

There was a report that 25 strains of *Streptococcus* sp. in cultured freshwater tilapia, rainbow trout, and ayu in Japan were highly sensitive to 16 chemotherapeutic agents (Kitao et al., 1981). Several antibiotics have been suggested. Doxycycline, erythromycin, and spiramycin have been reported in the Japanese literature as having beneficial effects on the treatment of streptococcosis (Hubbert, 1994). Akoi et al. (1983) examined the effects of a range of antibacterial agents against 561 strains of nonhemolytic *Streptococcus* sp. isolated from yellowtail and found that there had been a doubling of the MIC value for the macrolide antibiotics over a period of 1 year since these antibiotics became popular for the control of *Streptococcus*, compared with their respective MIC values for strains isolated in the 6 years (1979 to 1980).

One of the most effective fungicides is zinc-free malachite green. Malachite green is an anilic dye product that has been found to be a teratogen and a mutagen compound; therefore, it should not be used on food fish. To increase its efficacy, malachite green should be used in combination with formalin at the ratio of 0.1:25 ppm. Salt may also be used to treat fungal infection by adding it into the pond as a permanent bath of 0.5 to 1%.

Treatment in Hatcheries and Nursery Ponds

Treatments in hatcheries are usually effective as they are performed under controlled conditions. Chemical concentrations can be accurately calculated. Observations can be made during the treatment, which can be altered if there is a problem.

At present, artificial incubation of tilapia eggs is a common practice. Fungal and bacterial infections frequently occur under such circumstances and may cause unacceptable losses in the hatching rate. Chemical disinfectants are widely used; however, the efficacy of these chemicals as disinfectants depends on various factors, including water quality, water temperature, active ingredient, and appropriate dosage. Many chemicals appear to be effective only at doses very close to the lethal level for the eggs. The efficacy of formalin, acriflavin, malachite green, and buffodine in the disinfection of tilapia eggs has been established by Subasinghe and Sommerville (1985). Their results showed that a dose rate of 100 ppm buffodine gave the highest hatching rate. Acriflavin and malachite green gave good results at higher doses.

The recommended doses for hatchery disinfectants and improving hatchability of tilapia eggs are shown in Table 4. Various chemicals are recommended, but economic return must be considered, as well as the efficacy when selecting a chemical treatment.

Table 4 Chemicals Recommended for Disinfecting Hatcheries

Disinfectant	Action	Doses	Application
Acriflavin	Bacteriocide for fish eggs	1:2000	Short bath (20 min)
Formalin	Protozoacide	1:4000	Short bath (1 h)
	Parasiticide (monogenetic skin and gill flukes)	25–50 ppm	Long bath (48 h)
Furanace	Bacteriocide	2 mg/L	Short bath (1 h)
Malachite green (zinc free)	Fungicide for treating eggs	2 ppm	Short bath (0.5–1 h)
	Parasiticide for protozoa by mixing with formalin	0.1 + 25 ppm	Continuous for 3 d
Masoten (Dipterex)	Parasiticide	0.25–0.5 ppm	Continuous for 3 d
Nitrofurans	Bacteriocide	1–2 ppm	Short bath (5–10 min)
		0.05–0.1 ppm	Long bath (3–5 d)
Oxolinic acid	Bacteriocide (Aeromonads)	10–30 mg/kg fish weight	10 d
Potassium permanganate	Bacteriocide	1–5 ppm	Short bath (1 h) repeating as necessary
Salt	Bacteriocide	3.5% (for fry)	Dip (1 min)
	Fungicide Parasiticide	1–1.5%	Short bath (20–30 min)

Treatment in Grow-Out Ponds

It is not only difficult but often uneconomical to undertake a chemical treatment in the large area of a grow-out pond. Farmers, therefore, should consider carefully before treating their grow-out ponds. The efficacy of various chemicals against parasitic infestations has been investigated. Recommended chemicals are shown in Table 5. Most treatment trials were conducted in tanks or aquaria; the farmer should use knowledge gained from previous experiences in order to treat ponds successfully. If formalin is applied to a pond, it should be used with care if there are phytoplankton blooms. Formalin will kill phytoplankton, causing rapid oxygen drop in the pond, which may asphyxiate the fish.

Antibiotics, at present, are widely used in aquaculture and residues in fish flesh are a cause of concern for consumers. It is recommended that the use of antibiotics be avoided in food fish, including tilapias. A much more practical approach is to improve water quality and pond bottom condition.

Table 5   Recommended Chemicals for Pond Treatment

Chemical compound	Target species	Dose	Application
Formalin	Parasiticide	30–50 mL/m^3	Prolonged treatment may be repeated 1–2 times at 3-d interval
Oxytetracycline	Bacteriocide	Mix with feed 50 mg/kg of fish/d	10 d
Masoten (Dipterex)	Parasiticide	0.25–0.5 g	At least 3 times at 3-d interval
Salt	Bacteriocide Fungicide and parasiticide	0.5–0.95%/m^3	Prolonged treatment

REFERENCES

Akoi, T., Takeshita, S., and Kitao, T., Antibacterial action of chemotherapeutic agents against non-haemolytic Streptococcus sp. isolated from cultured marine fish, yellowtail Seriola guinguerodiata., *Bull. Jpn. Soc, Sci. Fish.,* 49(11), 1673, 1983.

Amlacher, E., *Textbook of Fish Diseases,* T.F.H. Publications, Neptune City, NJ, 1970, 302.

Anonymous, Aquaculture of freshwater fishes in China, *Sci. Publ. Soc. China,* 598, 1973.

Avenant, A. and Van As, J. G., Observations on the seasonal occurrence of the fish ectoparasite *Dolops ranarum* (Stuhlmann, 1891) (Crustacea: Branchiura) in the Transvaal. S. Africa, *J. Wildl. Res.,* 16, 62, 1986.

Avtalion, R. R. and Shlapobersky, M., A whirling viral disease of tilapia larvae, *Bamidgeh,* 46, 102, 1994.

Baker J. R., Trypanosomes and dactylosomes from the blood of freshwater fish in East Africa, *Parasitology,* 50, 515, 1960.

Baker, J. R, Three new species of *Myxosoma* (Protozoa; Myxosporidia) from East Africa freshwater fish, *Parasitology,* 53, 289, 1963.

Balarin, J. D. and Hatton, J. P., Culture systems and methods of rearing tilapia, in *Tilapia, A Guide to Their Biology and Culture in Africa,* University of Stirling Press, 1979, 45.

Bondad-Reantaso, M. G., A survey of the parasites of Nile tilapia (*Oreochromisnilotica*) in the Philippines, Masters thesis, Department of Biology, De la Salle University, Manila, Philippines, 1989, 210.

Bondad-Reantaso, M. G. and Arthur, J. R., Trichodinids [Protozoa: Ciliophora: Peritrichida of Nile tilapia (Oreochromis niloticus)] in the Philippines, *Asian Fish. Sci.,* 3, 27, 1989.

Bondad-Reantaso, M. G. and Arthur, J. R., The parasites of Nile tilapia [*Oreochromis niloticus* (L.)] in the Philippines, including an analysis of changes in the parasite fauna of cultured tilapia from fry to marketable size, *Proc. of the Second Asian Fisheries Forum,* Hirano, R. and Harano, I., Eds., Asian Fish Society, Manila Philippines, Tokyo, Japan, 17–22 April 1989, 1990, 729.

Britz, J., Van As, J. G., and Saayman, J. E., Occurrences and distribution of *Clinostomum tilapia e* Ukoli, 1966 and *Euclinostomum heterostomum* (Rudolphi, 1809) metacercarial infections of freshwater fish in Venda and Lebowa, Southern Africa, *J. Fish Biol.,* 26, 21, 1985.

Brock, V. E. and Takata, M., Contributions to the problems of bait capture and mortality together with the experiments in the use of tilapia as live bait, *Rep. Ind. Res. Adv. Counc. Hawaii,* 49, 38, 1955.

Chen, R. S. and Chao, C. B., Outbreaks of a disease caused by rickettsia-like organism in cultured tilapias in Taiwan, *Fish Pathol.,* 29(2), 61, 1994.

Chen, S. N., Hedrick, R. P., Fryer, J. F., and Kou, G. H., Occurrence of viral infections of fishes in Taiwan, in *Proc. of Republic of China-Japan Cooperative Science Seminar on Fish Diseases,* Tonkang Marine Science Center, Tonkang, Taiwan, Republic of China, 15–17 November 1982, 1983.

Cone, D. K., Arthur, J. R., and Bondad-Reantaso, M. G., Description of two new species of *Gyrodactylus* von Nordmann, 1832, Monogenea from cultured Nile tilapia, *Tilapia nilotica* (Cichlidae), in the Philippines, *J. Helminthol. Soc. Wash.,* 62(1), 6, 1995.

Davidov, O. N., Growth, development and fecundity of *Bothriocephalus gowkongensis* (Yeh, 1955), a cyprinid parasite, *Hydrobiol. J.,* 14, 60, 1979.

Duncan, B. L., *Cichlidogyrus sclerosus* Paperna and Thurston from cultured *Tilapia mossambicus* Kalikasan, Philippines, *J. Biol.,* 2, 154, 1973.

Duncan, B. L., Urecolarid ciliates, including three new species, from cultured Philippines fishes, *Trans. Am. Microsc. Soc.,* 96, 76, 1977.

Farstey, V., *Centrocestus sp. (Heterophyidae)* and Other Trematode Infections of the Snail *Melanoides tuberculatus* (Muller, 1774) and Cichlid Fish in Lake Kinneret, Masters thesis, Hebrew University of Jerusalem, 1986, 118.

Fryer, G., The parasitic copepoda and Branchiura of Lake Victoria and the Victoria Nile, *Proc. Zool. Soc. Lond.,* 137, 41, 1961.

Fryer, G., The parasitic crustacea of African freshwater fishes: their biology and distribution, *J. Zool.,* 156, 45, 1968.

Fryer, G. and Iles, T. D., *The Cichlid Fishes of the Great Lakes of Africa,* Oliver and Boyd, Edinburgh, 1972, 641.

Goldstein, R. J. *Cichids.* TFH Publication, Inc. Ltd., Hong Kong. 1970.

Haller, R. D. and Roberts, R. J., Dual neoplasia in a specimen of *Sarotherodon spilurus* (Gunther) (*Tilapia spilurus*), *J. Fish Dis.,* 3(1), 63, 1980.

Hedrick, R. P, Fryer, J. L., Chen, S. N., and Kou, G. H., Characteristics of four birnaviruses isolated from fish in Taiwan, *Fish Pathol. Tokyo,* 18(2), 91, 1983.

Hsu, T. C. and Kou, G. H., Studies on the freshwater fish pathogenic myxobacterium, *Flexibacter columnaris,* *J. Fish Soc. Taiwan,* 52, 41, 1977.

Hubbert, R. M., Bacterial diseases in warmwater aquaculture, in Shilo, M. and Sarig, S., Eds., *Fish Culture in Warmwater Systems: Problems and Trends,* CRC Press, Boca Raton, FL, 1989, 179.

Hunter, G. W., III and Hunter, W. S., Studies on host reactions to larval parasites: I. The effect on weight, *J. Parasitol.,* 24, 477, 1938.

Kabata, Z., *Parasites and Diseases of Fish Cultured in the Tropics,* Taylor and Francis, London, 1985, 318.

Kaneko, J. J., Yamada, R., Brock, J. A., and Nakamura, R. M., Infection of tilapia, *Oreochromis mossambicus* (Trewavas), by a marine monogenean, *Neobenedenia melleni* (MacCallum, 1927) Yamaguti, 1963 in Kaneohe Bay, Hawaii, USA and its treatment, *J. Fish Dis.,* 11, 295, 1988.

Kitao, T., Aoki, T., and Sakoh, R., Epizootic caused by Haemolytic *Streptococcus* species in cultured freshwater fish, *Fish Pathol.,* 15(3/4), 301, 1981.

Kuo, S. C., Chung, H. Y., and Kou, G. H., Studies on identification and pathogenicity of the gliding bacteria in cultured fishes, *CAPD Fish. Ser.,* 3, 52, 1980.

Lansberg, J. H. and Paperna, I., Intestinal infection by *Eimeria s.i.vanasi n.* sp (Eimeridae, Apicomplexa, Protozoa) in cichlid fish, *Ann. Parasitol. Humaine Compare.,* 62, 283, 1987.

Mandal, A. K., *Trypanosoma choudhuryi* sp. nov. from *Tilapia mossambica* (Peters), *Acta Protozool.,* 16, 1, 1977.

Meyer, F. P., Helminths of fishes from Trumbull Lake, Clay Country, Iowa, *Proc. Iowa Acad. Sci.,* 65, 477, 1958.

Ming-Chen, T., Chen, S. C., and Tsai, S. S., General septicaemia of streptococcal infection in cage cultured tilapia. *Tilapia mossambica,* in southern Taiwan. CAO Fisheries Series No, 4, *Fish Disease Research (VII).,* 95, 1985.

Miyazaki, T., Kubota, S., Kaige, N., and Miyashita, T., A histopathological study of streptococcal disease in tilapia., *Fish Pathol.,* 19, 167, 1984.

Natividad, J. M., Bondad-Reantaso, M. G., and Arthur, J. R., Parasites of Nile tilapia (*Oreochromis niloticus*) in the Philippines, in Proceedings of the First Asian Fisheries Fourm, Maclean, J. L., Dizon, L. B., and Hosillos, L. V., Eds., 1986, 255.

Nguenga, D., A note on infestation of *Oreochromis niloticus* with *Trichodina* sp. and *Dactylogyrus* sp., in *Proceedings of the Second International Symposium on Tilapia in Aquaculture,* Pullin, R. S. V., Bhukaswan, T., Tonguthai, K., and Maclean, J. L., Eds., Manila, Philippines, 1988, 117.

Nigrelli, R. F. and Vogel, H., Spontaneous tuberculosis in fishes and in other cold-blooded vertebrates with special reference to *Mycobacterium fortuitum* (Cruz) from fish and human lesions, *Zoologica (N.Y.),* 48, 130, 1963.

Ogawa, K., Bondad-Reantaso, M. G., Fukudome, M., and Wakabayashi, H., *Neobenedenia girellae* (Hargis, 1955) Yamaguti, 1963 (Monogenia: Capsalidae) from cultured marine fishes of Japan, *J. Parasitol.,* 81(2), 223, 1995.

Ogbonna, C. I. C. and Alabi, R. O., Studies on species of fungi associated with mycotic infections of fish in a Nigerian freshwater fish pond, *Hydrobiology,* 220, 131, 1991.

Okaeme, A. N., Parasitic diseases in tilapia and carp production, *NAGA ICLARM Q.,* 10(3), 16, 1987.

Oldewage, W. H. and Van As, J. G., Parasites and winter mortalities of *Oreochromis mossambicus* South Africa, *J. Wildl. Res.,* 17(1), 7, 1987.

Olufemi, B. E and Roberts, R. J., Induction of clinical aspergillomycosis by feeding contaminated diet to tilapia, *Oreochromix niloticus* (L.), *J. Fish Dis.,* 9, 123, 1986.

Paperna, I., Studies on the monogenetic trematodes in Israel. 2. Monogenetic trematodes of cichlids, *Bamidgeh,* 12, 20, 1960.

Paperna, I., *Entergyrus cichlidarum* n. gen. n. sp. A monogenetic trematode parasite in the intestine of African freshwater fish, *Bull. Res. Counc. Israel (B. Zool.),* 11, 183, 1963a.

Paperna, I., Dynamics of *Dactylogyrus vastators* Nybelin (Monogenea) populations on the gills of carp fry in fish ponds, *Bamidgeh,* 15, 31, 1963b.

Paperna, I., Parasitic helminths of inland water fishes in Israel, *Israel J. Zool.,* 13, 1, 1964a.

Paperna, I., The metazoan parasite fauna of Israel inland water fishes, *Bamidgeh,* 16, 3, 1964b.

Paperna, I., Parasitic crustacea from fishes of the Volta basin and South Ghana, *Rev. Zool. Bot. Afr.,* 80, 208, 1969.

Paperna, I., Infection by *Ichthyophthirius multifiliis* of fish in Uganda, *Prog. Fish Cult.,* 34, 162, 1970.

Paperna, I., Lymphocystis in fish from East Africa Lakes, *J. Wild. Dis.,* 9(4), 331, 1974.

Paperna, I., Parasites and diseases of the grey mullet (Mugilidae) with special references to the seas of the Near East, *Aquaculture,* 5, 65, 1975.

Paperna, I., Parasites, infection and diseases of fish in Africa, *CIFA Tech. Pap.,* 7, 216, 1980a.

Paperna, I., *Amyloodinium ocellatum* (Brown, 1931) (Dinoflagellidae) infestations in cultured marine fish at Eilat, Red Sea: epizootiology and pathology, *J. Fish Dis.,* 3, 363, 1980b.

Paperna, I., Disease caused by parasites in aquaculture of warm water fish, in *Annual Review of Fish Diseases,* Faisal, M. and Hetrick, F. M., Eds., 1(1), 1991.

Paperna, I. and Lahav, M., New records and further data on fish parasites in Israel, *Bamidgeh,* 23, 43, 1971.

Paperna, I. and Overstreet, R. M., Parasites and diseases of mullets (Mugilidae), in *Aquaculture of Grey Mullets of Z. B. Oven,* Cambridge University Press, Cambridge, 1981, 411.

Paperna, I. and Thurston, J. P., Report on ectoparasitic infection of freshwater fish in Africa, in Third Symp. Comm. of the O.I.E. for the Study of Fish Disease, *Bull. Int. Epizoot,* 67(7/8), 1192, 1968.

Paperna, I. and Thurston, J. P., Monogenetic trematodes collected from cichlid fish in Uganda, including the description of five new species of *Cichlidogyrus, Rev. Zool. Bot. Afr.,* 79, 15, 1969.

Paperna, I. and Van As, J. G., Winter Diseases of Cultured Tilapia, *Fourth COPRAQ/IOE Session,* Cadiz, 1981.

Paperna, I., Diamant, A., and Overstreet, R. M., Monogenea infestations and mortality in wild and cultured Red Sea fishes, *Helgolander Meeresunters,* 37, 445, 1984.

Paperna, I., Sabnai, I., and Zachary, A., Ultrastructural studies in piscine Epitheliocystis: evidence for a pleomorphic developmental cycle, *J. Fish Dis.,* 4, 459, 1981.

Paperna, I., Van As, J. G., and Basson, L., A review of diseases affecting cultured cichlids, in *Proceedings of the International Symposium on Tilapia in Aquaculture,* Fishelson, L. and Yoron, Z., compilers, Nazareth, Israel, 1983, 174.

Pillai, C. T. and Freitas, Y. M., Fungal infection causing mass mortality of freshwater fish *Tilapia mossambica, Seafood Export J.,* 15, 15, 1983.

Popp, W., Faisal, M., and Refai, M., High mortality in the Nile tilapia *Oreochromis niloticus* caused by *Providencia rettgeri, Anim. Res. Dev.,* 29, 95, 1988.

Roberts, R. J. and Sommerville, C., Disease of tilapia, in *Proc. of the Int. Conf. on the Biology and Culture of Tilapias,* Pullin, R. S. V. and Lowe-McConnel, R. H., Eds., Bellagio, Italy, 1982, 247.

Sakata, T. and Koreeda, Y., A numerical taxonomic study of the dominant bacteria isolated from tilapia intestines, *Bull Jpn. Soc. Sci. Fish.,* 52(9), 1625, 1986.

Shariff, M. and Sommerville, C., Host parasite relationship of *Lernaea polymorpha* and *L. cyprinaces,* in *Parasitology — Quo Vadis Handbook* (VIICOPA Brisbane), Howell, M. J., Ed., Canberra, Australia, Abstract 599, 227, 1986.

Sarig, S., The prevention and treatment of diseases of warmwater fishes under subtropical conditions, with special emphasis on intensive fish farming, in *Diseases of Fishes,* Book 3, Snieszko, S. F. and Axelrod, H. R., Eds., T.F.H. Publications, Jersey City, 1971, 129.

Sarig, S., The status of information on fish diseases in Africa and possible means of their control, in *Symposium on Aquaculture in Africa,* Accra, Ghana CIFA/75/SRS, Food and Agriculture Organization of the United Nations, Rome, 1975, 6.

Schuberts, G., The injurious effects of *Costia necatrix, Bull. Int. Epizoot.,* 69, 1171, 1968.

Scott, P. W., Preliminary Studies on Disease in Intensively Farmed Tilapia in Kenya, M.S. thesis, University of Stirling, Stirling, Scotland, 1977, 159.

Sirikan, P., Ectoparasitic Digene on Mucous Along the Body of Nile Tilapia, Special Publication, Faculty of Fish., Kasetsart University, Thailand, 1983, 4.

Snieszko, S. F., The effects of environmental stress on outbreaks of infectious diseases of fishes, *J. Fish Biol.,* 6, 197, 1974.

Sommerville, C., The life history of *Haplorchis pumilio* (Looss, 1896) from cultured tilapias, *J. Fish Dis.,* 5(3), 233, 1982a.

Sommerville, C., The pathology of *Haplorchis pumilio* (Looss, 1896) infections in cultured tilapias, *J. Fish Dis.,* 5(3), 243, 1982b.

Subasinghe, R. and Sommerville, C., Disinfection of *Oreochromis mossambicus* (Peters) eggs against commonly occurring potentially pathogenic bacteria and fungi under artificial hatchery conditions, *Aquaculture Fish. Manage.,* 6(2), 121, 1985.

Ueno, Y., Chen, S. N., Kou, G. H., Hedrick, R. P., and Fryer, J. L., Characterization of a virus isolated from Japanese eel (*Anguilla japonica*) with nephroblastoma, *Bull. Inst. Zool. Acad. Sin.,* 1984.

Van As, J. G. and Basson, L., Checklist of freshwater fish parasites from Southern Africa, *S. Afr. J. Wildl. Res.,* 14, 49, 1984.

Vega, M. J. M., Who's working on tilapia and Carp diseases?, *NAGA,* 11(3), 18, 1988.

Williams, R. B. and Williams, E. H., Parasites of Puerto Rican Freshwater Sport Fishes, Department of Marine Sciences, Puerto Rico, 1994, 29.

Witenberg, G., Studies on the trematode family Heterophyidae, *Anim. Trop. Med. Parasitol.,* 23, 131, 1929.

Wu, S. Y., New bacterial disease of *Tilapia, FAO Fish Cult. Bull.,* 2, 4, 1970.

13 COMPUTER APPLICATIONS IN POND AQUACULTURE — MODELING AND DECISION SUPPORT SYSTEMS

Raul H. Piedrahita, Shree S. Nath, John Bolte, Steven D. Culberson,
Philip Giovannini, and Douglas H. Ernst

INTRODUCTION

Modeling and the development of decision support systems for pond aquaculture have received considerable effort and support under the Pond Dynamics/Aquaculture CRSP (PD/A CRSP). Models have been used as means for analyzing and organizing information and knowledge about aquaculture ponds. The models have served to test hypotheses of "how ponds work," and to design field experiments to test those assumptions. As the information base has improved, decision support systems have been designed for management purposes.

The current state of the art of pond modeling is very different from what it was at the initiation of the PD/A CRSP. At that time, Marjanovic and Orlob (1986) and Bernard (1986) conducted literature reviews on various aspects of pond modeling, especially as it referred to the types of ponds that were to be studied under the PD/A CRSP. In their reviews, they found that a great deal of related information existed but that there had been virtually no modeling of tropical aquaculture ponds. Related information found by Marjanovic and Orlob (1986) and by Bernard (1986) included models developed for lakes and reservoirs in temperate areas, as well as information about processes that are important in determining water quality in ponds.

General reviews of modeling of aquaculture systems have been conducted recently (Cuenco, 1989; Piedrahita, 1991), and their work will not be repeated here. The objective of this chapter is to review and highlight the contributions of the PD/A CRSP to the status of aquaculture pond modeling and to the development of decision support systems for pond aquaculture. The review will include examples of different types of models and of decision support systems developed under the PD/A CRSP. The examples will be preceded by a brief description of the data base established under the PD/A CRSP and of its significance to the development of aquaculture science.

CONCEPTUAL MODEL

One of the initial applications of modeling in the PD/A CRSP was the development and use of a conceptual model of the aquaculture pond ecosystem (PD/A CRSP, 1987). The conceptual model was used as the basis for the creation of a computer-based simulation model of the pond ecosystem. Subsequently, conceptual models have been used as aides in identifying research needs and in presenting paradigms of aquaculture pond ecosystem structure and function. Conceptual models have evolved as the CRSP work has progressed and the information base has improved, and as new research questions are posed. A recent version is shown in Figure 1 (PD/A CRSP, 1989; see also

0-56670-274-7/97/$0.00+$.50
© 1997 by CRC Press LLC

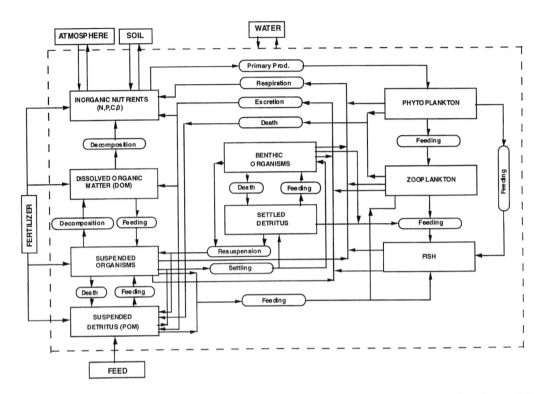

Figure 1. Conceptual model of the aquaculture pond ecosystem. The arrows connecting the model components represent the flow of mass. The rectangular blocks represent the main categories for mass accumulation in the system, while the rounded blocks represent the processes by which mass is transferred or transformed. The dashed line represents the boundary of the pond ecosystem, and lines crossing this boundary represent the flow of mass into or out of the pond.

Figure 6 in Chapter 2). In the context of the PD/A CRSP, a conceptual model of the pond has been useful for the coordination of research efforts of different scientists participating in the PD/A CRSP.

As depicted in Figure 1, the conceptual model shows the mass flow between the main biomass components in a typical aquaculture pond. These components include fish, phytoplankton, zooplankton, and suspended and benthic microfauna. In addition, detritus is considered as a separate term that may include living (bacteria) and nonliving (residues, fertilizers) components. Biological processes of production, respiration, grazing or feeding, and death are shown in the conceptual model. Physical processes such as resuspension and settling are also included in the model. The model also includes possible avenues for material entering or leaving the pond by fertilization or feeding, by gas exchange with the atmosphere, by dissolution or precipitation to the soils, and through water exchange.

AQUACULTURE CRSP DATA BASE

In addition to the many publications in the scientific literature originating from the PD/A CRSP, the PD/A CRSP data base constitutes one of the primary project contributions to the science of aquaculture. The PD/A CRSP endeavored to develop a set of standardized experimental protocols that included specific instructions on which variables were to be measured and on the procedures that would be followed in taking the measurements. The protocols were developed such that the resulting data would be suitable for inclusion in a data base that could be used for extensive computerized analysis. Furthermore, it was envisioned that the data base should be accessible to

PD/A CRSP scientists and to other interested aquaculture scientists. A description of data base attributes, including variables included and structure and access to the data base, are presented in this section.

Variables and Sample Frequency

There has been a continuous evolution in the choice of variables measured and in the frequency at which data have been taken. Large amounts of data were collected during the initial stages of the PD/A CRSP — many variables were measured at relatively high frequencies. As the results of the initial measurements were analyzed, it was realized that adjustments needed to be made in terms of variables measured and frequency of those measurements. The adjustments resulted in a reduction in the number and frequency with which soil and water quality parameters were monitored by providing a more focused research protocol. The variables monitored and frequency of measurements are shown in Tables 1 and 2 for the Second and Seventh Experimental Protocols (PD/A CRSP, 1984 and 1993) and illustrate the changes described above. In addition to the data on water quality, weather, and fish or shrimp biomass, a series of measurements were taken on lime, organic and inorganic fertilizers, and pond soils at the beginning and end of each experiment. Acid neutralization value was determined for lime. Each batch of fertilizer was analyzed at the time of delivery and at various other times to determine percent dry matter and nitrogen, phosphorus, and chemical oxygen demand (COD) content. The soils parameters monitored under the most recent work plans were pH, phosphorus, organic matter, total nitrogen, cation exchange capacity, metals (aluminum, iron, zinc), lime requirement, exchangeable hydrogen, and base saturation (PD/A CRSP, 1993).

Standardization of sampling procedures included not only the selection and frequency of variables, but also the analytical methods to be used, and the tools used to report the data to a central depository. Analytical methods have been largely based on Standard Methods for the Examination of Water and Wastewater (APHA et al., 1992), and have been summarized in a handbook of methods which is distributed to each of the PD/A CRSP research teams (PD/A CRSP, 1992).

Methods of Reporting

Data collected at the PD/A CRSP research sites are compiled and entered into standardized spreadsheets that are then submitted to the central data base manager for incorporation into the overall project data base. The spreadsheet templates used have undergone substantial evolution over the duration of the project as computer hardware and software have progressed. When the project was initiated, the recommended computers to be used for data entry and submission were Apple IIe®. Soon after, the Apple IIs were replaced by Apple Macintosh® computers, and PC-compatibles were introduced. Currently, both Macintosh and PC-type computers are supported, and there is a great deal of flexibility in the choice of hardware and software that can be used for data compilation and submission. Compatibility among systems is not the problem that it once was. Data are grouped in spreadsheet templates according to the frequency and nature of the measurements. Some of the templates used include *Weather* (daily weather measurements), *Diel* (diel cycle measurements), *Fish* (data on stocking, harvesting, and sampling of fish or shrimp), *plankton* (primary production and respiration rates), *Nutrient* (composition of liming materials and fertilizers), and *Inputs* (input addition rates to ponds).

Data Management

Data submitted by the field research teams are entered into a centralized data base. The data base is maintained on personal computers, and various software packages have been used in the past to access and maintain the data base: General Manager® for the Apple IIe was used initially,

Table 1 Water Quality and Related Parameters Monitored and Frequency of Measurements for the Second and Seventh Work Plans

Second Experimental Cycle

Daily[a]	Biweekly/weekly[b]	Monthly[d]	Occasional[e]
Solar radiation	Dissolved oxygen[c]	Alkalinity[c]	Pond soil, including pH, extractable bases, organic matter, total-N, nitrate-N, and ten additional parameters
Wind speed	Pond temperature extremes	Total hardness[c]	Pond morphology
Rainfall	Pond temperature[c]	Ammonia[c]	Hydrology, including surface inflow, outflow, and evaporation
Air temperature	pH[c]	Nitrate[c]	Water quality, including pH, ammonia, nitrate, chloride, sulfate, boron, and 12 additional parameters
Pond depth	Total kjeldahl N[c]	Total phosphorus[c]	
	Secchi disk depth[c]	Dissolved ortho-P[c]	
	Chlorophyll a[c]	Primary production	
		Phytoplankton comp.	
		Zooplankton comp.	
		Benthos composition	

Seventh Experimental Work Plan

Daily	Three times[f]	Diel[g]	Occasional
Solar radiation	Total kjeldahl N	Dissolved oxygen	Pond soil, including pH, organic matter, total-N, and eight additional parameters
Wind speed	Ammonia nitrogen	Temperature	Pond morphology
Rainfall	Total phosphorus	pH	Composition of lime and fertilizers: dry matter, N, P, COD, lime neutralization value
Evaporation	Secchi disk visibility	Wind (cumulative between sampling times)	
Mortalities	Chlorophyll a	Solar radiation (cumulative between sampling times)	
Pond depth	Dark bottle respiration		
Water inflow and overflow	Total suspended solids		
	Total volatile solids		
	Total alkalinity (3 depths: top, middle, bottom)		
	Primary productivity		

a Measurements are taken daily.

b Measurements are biweekly and weekly.

c Indicates parameters measured as part of monthly diel studies.

d Measurements are taken monthly.

e Occasional measurements are taken at the start and end of the experiments, and as needed.

f Measurements are taken at least three times in a given experiment, during intensive sampling sessions that include diel sampling. Sampling is carried out during the second week of an experiment, approximately midway through the experiment, and during the final week.

g Diel studies are conducted at least three times in a given experiment, and must coincide with the intensive sampling periods above. Samples for diel studies are collected at dawn, 1000, 1400, 1600, 1800, 2300, and at dawn the following day, or at higher frequencies if automated data logging equipment is available.

Data from PD/A CRSP, 1984 and PD/CRSP, 1993.

Table 2 Data Collected on Fish or Shrimp Populations

Stocking	Monthly sampling	Harvest
Total number	Total number in sample	Total number of stocked fish remaining
Total biomass	Total biomass in sample	Total biomass of stocked fish
Individual weights (of 10% sample)	Individual weights	Individual weights (10% sample of stocked fish)
Individual lengths (of 10% sample)	Individual lengths	Individual lengths (10% sample of stocked fish)
	Reproduction weight	Total number of recruits
		Total biomass of recruits

followed by various data base management packages on either PC-compatible or Apple Macintosh systems: RBASE5000®, RBASE®, and currently FoxPro®. Data received in the form of spreadsheets are entered into the data base by making each line (row) in a spreadsheet into a record in the data base. Each of the cells on a row in the spreadsheet format is transformed into a field in the data base record. Each record in the data base contains information identifying the date, location, and experiment number. By using this format for data base management, it is possible to extract information using more efficient search procedures than those that would be possible with a spreadsheet. This is especially important when data from more than one experiment or more than one site are to be analyzed.

Access to the Data Base

Access to the data base is open to PD/A CRSP and to other interested researchers. Data are normally requested from the data base manager, who extracts the data from the data base. Criteria for searches and extractions from the data base are dictated by the interests of the data user. Extractions may be based on a researcher's interest in analyzing or studying a group of variables at various sites and for several experiments or a complete data set for a particular site or experiment. Once the data are extracted from the data base, they may be formatted in various forms depending on the computer systems (Macintosh or PC-compatible) and software (various spreadsheet and data base formats) used by the individual requesting the data. Currently data are sent out in high density diskettes and more recently in the new removable hard disks. Distribution of the data on compact disk (CD-ROM) is being investigated, and a world wide web page and an ftp site are presently under construction.

Significance of the Data Base

The data base created under the PD/A CRSP constitutes a unique source of information to researchers around the world given the standardization of procedures for the design and conduct of field experiments, as well as for the collection and submission of data. In addition, the magnitude of the effort, with research sites in different geographic and climatic areas and multi-year data collection, makes it possible to study processes and phenomena that could not be studied using more limited and conventional data sets. The usefulness of the PD/A CRSP data base will continue long after the field experiments are completed and will probably expand as more researchers learn of its existence and gain access to the information included in it.

COMPUTER MODELS OF AQUACULTURE POND PROCESSES

There are many different types of computer models of aquaculture pond processes; the general characteristics of some of them will be reviewed in this section. The review will be limited to the types of models that have been developed under the PD/A CRSP. A general review of computer models used in pond aquaculture is beyond the scope of this chapter, and the reader is referred to other recent publications (Cuenco, 1989; Piedrahita, 1991).

Types of Models

Models of aquaculture pond processes may be classified in many different ways. Models are often classified by the degree of empiricism included in the development of equations and relationships. The two general categories recognized are empirical and mechanistic. Empirical models are based on statistical analysis of data. These models are highly site-specific and tend to be good predictive tools if applied in situations that closely resemble conditions found in the system from which the original data were collected. On the other hand and as a general rule, empirical models do not provide a great deal of insight into the functioning of the system being modeled.

Mechanistic models, also called analytical models, are constructed on the basis of a set of paradigms about how a system or process works. In a mechanistic model, processes affecting a particular state variable are identified, and variables that affect or determine process rates are also noted. Relationships or equations are then proposed for the various processes based on an understanding of basic physical, biological, chemical, and mathematical principles. Mechanistic models tend be more generally applicable than empirical models because of their foundation on fundamental principles. They also tend to be less accurate predictors and computationally more complex than empirical models. The major asset of mechanistic models is their usefulness in gaining some insights on how a particular system or process works, and in being able to identify how a system or process might perform under conditions different from those for which data are available.

Empirical models developed under the PD/A CRSP include models of fish growth and primary productivity. Other empirical models have been developed as components of the computerized decision support system for pond management, and those will be described in a later section. Mechanistic models have been developed to predict dissolved oxygen concentration and temperature in mixed and stratified ponds. In addition, an ecosystem model developed previously (Piedrahita, 1984) has been revised, calibrated, and validated using PD/A CRSP data (Piedrahita, 1990).

Models may also be categorized as deterministic or stochastic, depending to a certain extent on how input data are used in the model and on how a model is executed. A deterministic model yields a unique output for a given input data set. A stochastic model, on the other hand, has a certain degree of randomness associated with input data or with process rates. Because of this randomness or stochasticity, a stochastic model does not result in a unique result for a given set of input parameters. Results obtained with a stochastic model will vary due to the unpredictable variations in input data or in process rates. Stochastic models are normally run many times for a given set of input parameters, and the results are used to generate probability distributions for the output variables.

With very few exceptions, models used in aquaculture (including within the PD/A CRSP) have been deterministic. It is only recently that stochastic models have begun to be developed within the PD/A CRSP. This work is still in progress and will not be discussed here.

Data Requirements

Data requirements for modeling vary greatly, depending on the type of model being developed and on the overall complexity of the system being modeled. In general, data are required for model development, model calibration, model validation, and model execution and use. Data used for these purposes should be obtained in ways or from sources such that the data sets are distinct and independent from each other. The PD/A CRSP data base provides a wealth of data that can be used for all stages of model development. Given the magnitude of the data base, the standardization of analytical techniques, and experimental designs, it is possible to use data from different experiments or different sites for the various stages of model development. The magnitude of the data base also makes it ideal for the development of empirical models that may have more general applications than those developed on the basis of limited data sets for a single site or from a single experiment. Assets of the data base for mechanistic model development and testing are based on the scope of variables measured and on the frequency with which they have been measured.

Practical Utility and Limitations

Uses of models that have been developed under the PD/A CRSP have ranged from theoretical research studies of primary production and stratification to the generation of day-to-day pond management guidelines. The intended uses of the models are important considerations in their initial formulation and development.

EXAMPLES OF PD/A CRSP MODELS OF AQUACULTURE POND PROCESSES

A variety of models have been developed under the PD/A CRSP. Abbreviated examples of some of those models are presented below. Space limitation precludes the presentation of model listings and detailed analyses of the models.

Fish Growth

Fish growth models are useful both for analyzing factors that affect fish production and as management tools. As research tools, growth models provide a useful framework for analyzing and understanding the effects of different variables on fish performance.

Variables that significantly influence fish growth are size (or weight), food availability, photoperiod, temperature, dissolved oxygen, and un-ionized ammonia concentrations (Pütter, 1920; Fry, 1947; Winberg, 1960; Ursin, 1967; Warren and Davis, 1967; Brett et al., 1969; Stauffer, 1973; Huisman, 1976; Cuenco et al., 1985). These variables appear to affect fish growth via their impacts on food intake (Brett, 1979). Further, activities associated with the development and maturation of reproductive structures reduce growth because energy that might otherwise have been used for tissue build-up is diverted to these structures (Brody, 1945; Brett and Groves, 1979).

Consumed food is used to meet energy losses associated with bioenergetic processes, such as fecal and metabolite excretion, standard (maintenance) metabolism, stress response, heat increment (specific dynamic action), active metabolism (swimming), and gametogenesis (Winberg, 1960; Brett and Groves, 1979). Energy in excess of these losses is reflected in fish growth.

The fate of consumed food has been modeled for different fish species by the use of bioenergetic models (e.g., Machiels and Henken, 1986; Cacho, 1990). Bioenergetic models provide the most comprehensive and fundamentally sound basis for modeling fish performance, but they tend to be extremely detailed and data intensive. Simplified versions of bioenergetic models can be used in a practical context to analyze factors that affect fish growth (e.g., Ursin, 1967; Cuenco et al., 1985). Thus, following the classification scheme of Pütter (1920) and von Bertalanffy (1938), the bioenergetic processes listed earlier can be grouped into two categories, namely tissue synthesis or anabolism (which includes food intake) and tissue breakdown or catabolism (Ursin, 1967). The difference between anabolism and catabolism is realized as fish growth. Based on these concepts, Ursin (1967) developed a model (hereafter referred to as the Ursin model), modified versions of which have been used to describe the growth of Nile tilapia (*Oreochromis niloticus*) in warmwater ponds. These modified versions include one that has been used to describe Nile tilapia growth in fertilized ponds at Ayutthaya, Thailand (Liu and Chang, 1992). The latter model has since been expanded into a more generalized version called the BE model, which is discussed below.

BE Model Development

The BE model has been developed primarily as a tool that can be used to forecast fish growth and yields under different management and environmental conditions. Therefore, the use of traditional parameter estimation procedures that require actual growth data (e.g., nonlinear regression, as in Liu and Chang, 1992) has been minimized. The BE model has been parameterized for Nile tilapia (as an example species) using information from bioenergetic studies and by calibration.

A simplified version of the BE model, which assumed that fish growth is primarily a function of their stocking density and the pond water temperature was implemented in a decision support system (PONDCLASS© Version 1.2). A comprehensive version of the model has since been implemented in another decision support system (POND© Version 2.0). Both decision support systems are described elsewhere in this chapter.

Model Structure

The BE model assumes that energy losses due to gametogenesis are negligible (e.g., as might be the case in a monosex culture of Nile tilapia). In addition to the effects of size and food availability considered by Liu and Chang (1992), the BE model includes the effects of variables such as photoperiod, temperature, dissolved oxygen, and un-ionized ammonia concentrations on fish growth (Tables 3 and 4; see discussion below). The BE model assumes that such effects occur via the influence of these variables on food consumption, and that they are of an interactive nature (Cuenco et al., 1985).

Effects of Size

It is generally accepted that the growth rate of fish increases at a declining rate with size or weight (Pütter, 1920; Winberg, 1960). However, the Ursin model assumes that anabolism and catabolism may be paced at different rates in relation to fish weight, with subsequent effects on fish growth. This is achieved by use of the exponents m and n (Equation 1, Table 3). Because m cannot be easily estimated, it is assumed to equal 0.67 in the BE model (Hepher, 1988; Liu and Chang, 1992; Table 5). However, n can be estimated if oxygen consumption data are available for starving fish (Ursin, 1967). Such data on Nile tilapia (Farmer and Beamish, 1969) suggest that $n = 0.81$, which is similar to Ursin's (1967) estimate of $n = 0.83$ (from data on 81 fish species). Cuenco et al. (1985) also estimated similar values ($m = 0.64$ and $n = 0.85$) for *O. mossambicus* from the data of Suffern et al. (1978).

Effects of Photoperiod

Many cultured fish including tilapias tend to feed only during daylight hours (Caulton, 1982). Estimates of daily photoperiod for different sites can be obtained from sunrise and sunset hour angle calculations (Hsieh, 1986). The BE model uses a daylight scaler p based on such calculations to adjust daily food intake (Equation 3, Table 3). For example, a photoperiod of 12 h would result in $P = 0.5$.

Effects of Temperature

Food consumption tends to increase with temperature from a lower limit below which fish will not feed (T_{min}) until the optimum temperature (T_{opt}) for the given fish species is reached. Many fish species including tilapias (Caulton, 1978) tend to have a maximum food consumption rate within a temperature range rather than at a single optimum temperature. Because adequate data are not available to estimate this range for Nile tilapia, a function like the one used by Svirezhev et al. (1984), which is more or less flat around a known optimal temperature, may be appropriate to describe the effects of temperature on food consumption (Equation 5, Table 3).

The above temperature function appears to be appropriate for the effects of temperature on food consumption and therefore anabolism. However, catabolism increases exponentially with temperature within the tolerance limits for a given species (Ursin, 1967). In the BE model, the exponential function of the Ursin model has been modified to include the lower temperature tolerance limit for a given species (Equation 6, Table 3), which is assumed to be equivalent to T_{min}.

Available literature for Nile tilapia suggests values for T_{min}, T_{max}, and T_{opt} as listed in Table 5. The parameters k_{min} and s have been estimated by model calibration (Table 5). Using values for k_{min}, s, and T_{min} from Table 5, and assuming $T = T_{opt}$, Equation 6 predicts $k = 0.0327$. This predicted value is within the range of $k = 0.0319$ to 0.0468 estimated by Liu and Chang (1992).

Table 3   Listing of Equations for the BE Model

$$\frac{dW}{dt} = HW^m - kW^n \tag{1}$$

$$\frac{dW}{dt} = b(1-a)\frac{dR}{dt} - kW^n \tag{2}$$

$$\frac{dR}{dt} = h\,p\,f\,\tau\,\delta\,\upsilon\,W^m \tag{3}$$

$$\frac{dR_{max}}{dt} = h\,p\,\tau\,\delta\,\upsilon\,W^m \tag{4}$$

$$\tau = \exp\left\{-4.6\left[\left(T_{opt} - T\right)/\left(T_{opt} - T_{min}\right)\right]^4\right\}, \qquad \text{if } T < T_{opt}$$

$$\exp\left\{-4.6\left[\left(T - T_{opt}\right)/\left(T_{max} - T_{opt}\right)\right]^4\right\}, \qquad \text{if } T \geq T_{opt} \tag{5}$$

$$k = k_{min}\,\exp\left[s\left(T - T_{min}\right)\right] \tag{6}$$

$$\delta = 1.0, \qquad\qquad\qquad\qquad\qquad \text{if } DO > DO_{crit}$$
$$(DO - DO_{min})/(DO_{crit} - DO_{min}), \qquad \text{if } DO_{min} \leq DO \leq DO_{crit} \tag{7}$$
$$0.0, \qquad\qquad\qquad\qquad\qquad\qquad \text{if } DO < DO_{min}$$

$$\upsilon = 1.0, \qquad\qquad\qquad\qquad\qquad\qquad \text{if } UIA < UIA_{crit}$$
$$(UIA_{max} - UIA)/(UIA_{max} - UIA_{crit}), \qquad \text{if } UIA_{crit} \leq UIA \leq UIA_{max} \tag{8}$$
$$0.0, \qquad\qquad\qquad\qquad\qquad\qquad\qquad \text{if } UIA > UIA_{max}$$

$$f = 1.0, \qquad\qquad\qquad \text{if } SC < CSC \tag{9.1}$$
$$CSC/SC, \qquad\qquad\qquad \text{if } SC \geq CSC$$

$$S_{max} = \frac{dR_{max}}{dt}\,W^{-1} \tag{9.2.1}$$

$$S = \sum_{i=1}^{N} S_{max}\left(\frac{\dfrac{F_i}{c_i}}{1 + \displaystyle\sum_{j=1}^{N}\dfrac{F_j}{c_j}}\right) \tag{9.2.2}$$

$$f = \frac{dR/dt}{dR_{max}/dt} = \frac{SW}{dR_{max}/dt} \tag{9.2.3}$$

Note: Symbols are defined in Table 4.

Table 4   Definition of Symbols Used in the BE Model Equations Listed in Table 3

Symbol	Description	Units
a	Fraction of food assimilated used for feeding catabolism	None
b	Efficiency of food assimilation	None
c_i	Half-saturation constant for the feeding of fish on the ith natural food resource (e.g., phytoplankton)	[a]
CSC	Critical standing crop	kg/ha
DO	Dissolved oxygen level	g O_2/m^3
DO_{crit}	Critical DO level above which food intake is not affected	g O_2/m^3
DO_{min}	Minimum DO level below which fish will not feed	g O_2/m^3
f	Relative feeding level	None
F	Concentration of a natural food resource (e.g., zooplankton)	[a]
h	Coefficient of food consumption	g^{1-m}/d
H	Coefficient of anabolism	g^{1-m}/d
k	Coefficient of fasting catabolism	g^{1-n}/d
k_{min}	Coefficient of fasting catabolism at T_{min}	g^{1-n}/d
m	Exponent of anabolism	None
n	Exponent of catabolism	None
N	Total number of natural food resources	None
p	Daylight scaler	None
dR/dt	Food intake rate or daily ration	g/d
dR_{max}/dt	Maximum food intake rate	g/d
s	Constant to describe temperature effects on catabolism	$°C^{-1}$
S	Specific food intake rate	d^{-1}
S_{max}	Maximum specific food intake rate	d^{-1}
t	Time	d
T	Water temperature	°C
T_{min}	Minimum water temperature below which fish will not feed	°C
T_{max}	Maximum water temperature above which fish will not feed	°C
T_{opt}	Optimum water temperature for fish feeding	°C
UIA	Un-ionized ammonia level	gNH_3-N/m^3
UIA_{crit}	Critical UIA level below which food intake is not affected	gNH_3-N/m^3
UIA_{max}	Maximum UIA level above which fish will not feed	gNH_3-N/m^3
W	Fish weight	g
δ	Function to describe the effects of DO on food intake	None
υ	Function to describe the effects of UIA on food intake	None
τ	Function to describe the effects of temperature on food intake	None

[a] Depends on the particular resource. For example, units for phytoplankton and zooplankton are g/C/m^3 and g/m^3, respectively.

Table 5   Parameter Values Used for Simulations and Data Sources from Which They Were Estimated

Symbol	Value used in simulations	Data source
a	0.256	Brett, 1979
b	0.625 (0.53–0.70)	Meyer-Burgdorff et al., 1989
c_1 (phytoplankton A)	2	Calibration
c_2 (phytoplankton B)	10	Calibration
c_3 (zooplankton)	1	Calibration
h	0.8	Calibration
k_{min}	0.025	Calibration
m	0.67 (0.50–0.83)	Hepher, 1988
n	0.81	Farmer and Beamish, 1969
T_{min}	15	Gannam and Phillips, 1992
T_{max}	41	Denzer, 1967
T_{opt}	33 (30–36)	Caulton, 1982
s	0.015	Calibration

Note: Values in parentheses denote the range of the estimates. Some parameter values were estimated by model calibration.

Effects of Dissolved Oxygen and Un-ionized Ammonia

Dissolved oxygen typically does not affect food consumption if its concentration is above a critical limit that is species dependent, but a decrease in dissolved oxygen below this critical value causes a reduction in food consumption until lethal concentrations of dissolved oxygen are reached (Cuenco et al., 1985; Equation 7, Table 3). The effects of un-ionized ammonia are similar, with the exception that food consumption is affected if un-ionized ammonia levels exceed a certain critical concentration (Cuenco et al., 1985; Equation 8, Table 3). The effects of un-ionized ammonia and dissolved oxygen are assumed to be multiplicative in the BE model (Equation 3, Table 3). Dissolved oxygen and unionized ammonia parameters for tilapia are as follows: DO_{min} = 0.5 to 1.0 mg/L, DO_{crit} = 3 mg/L (Boyd, 1990; Stickney, 1994), UIA_{max} = 1.40 mg/L, UIA_{crit} = 0.06 mg/L (Abdalla, 1989).

Effects of Food Availability

The Ursin model includes a relative feeding level parameter f (0 to 1), which is the ratio of the actual food intake rate (dR/dt; Equation 3, Table 3) to the maximum possible intake rate or food intake rate at satiation (dR_{max}/dt; Equation 4, Table 3). During the initial phase of fish production, adequate natural food is produced within the pond to sustain fish at their maximum feeding level (i.e., f = 1), but once the standing crop of fish in the pond exceeds the "critical standing crop," natural food availability declines with fish biomass until the carrying capacity of the pond is reached (Hepher, 1978; Hepher et al., 1983; Hepher, 1988). A simple expression to approximate the relationship between food availability and fish standing crop is used in the BE model (Equation 9.1, Table 3).

Growth data from PD/A CRSP experiments suggest that critical standing crops vary substantially according to climate, water, and soil characteristics of a site, as well as management practices. Large differences in critical standing crop estimates have also been noticed among ponds that were treated identically. As a result of such variability, prediction of critical standing crops for ponds has not been very successful. In addition to this problem, the critical standing crop–based function for f (Equation 9.1, Table 3) does not account directly for the effects of pond fertilization on natural food availability.

A more involved approach of predicting f would be to actually simulate the dynamics of individual natural food resources in ponds and use an alternate feeding function such as the *resource substitution* model (Tilman, 1982; O'Neill et al., 1989). This function is essentially an extension of Monod uptake kinetics to multiple resources and is advantageous in that an unrestricted number of resources can be considered, and food preferences are inherently captured in the half-saturation constants used.

The resource substitution model (Equations 9.2.1 to 9.2.3, Table 3) is currently being tested as an alternative to the critical standing crop-based feeding function. The substitution model requires concentrations of different food resources (which vary over time) to be available, implying that dynamic simulation of such resources is also required. As of now, descriptions of only three different food resources (two pools of phytoplankton designated as A and B, and one of zooplankton) have been implemented in the systems model within POND©. Simulation of these food resources is based on models developed by Piedrahita and Giovannini (1991) and Svirezhev et al. (1984). Half-saturation constants to describe the feeding of fish on the two phytoplankton and one zooplankton pools have been estimated by model calibration (Table 5).

Bioenergetic Parameters

In addition to the above variables, the BE model includes three bioenergetic parameters, a, b, and h, from the Ursin model. The parameter a is the fraction of the food assimilated that is used for feeding catabolism, b is the efficiency of food assimilation, and h is the coefficient of food consumption (Table 4). Liu and Chang (1992) determined $h = 0.997$ by nonlinear regression. Calibration of the BE model suggests a value of $h = 0.80$ to be more appropriate. The term $b(1 - a)$ in Equation 2 (Table 3) represents energy that is available for growth and fasting catabolism. Although b actually decreases with food intake (Caulton, 1982; Meyer-Burgdorff et al., 1989), it is assumed to be constant in the BE model (Table 5). The value of 0.625 for b implies that excretory losses are about 37.5% of the gross energy intake.

An indirect estimate of a can be obtained by assuming 16% of the gross energy intake to be lost via heat increment and urinary wastes (Brett and Groves, 1979). Thus, overall energy losses associated with the processing of food (i.e., excretion, heat increment and urinary wastes) are about 53.5%. Alternately, the term $b(1 - a)$ is about 46.5% of the gross energy intake, which results in $a = 0.256$.

Results of Simulations with the BE Model

The BE model has been used to simulate the results of PD/A CRSP experiments with Nile tilapia at sites in Thailand (Knud-Hansen et al., 1990) and Honduras (Teichert-Coddington et al., 1989) (Figure 2). The simulations used the critical standing crop–based function (Equation 9.1, Table 3) to estimate the relative feeding level f. Critical standing crops were estimated from fish growth data reported for the above experiments. A daily time step was used. Daily water temperature was predicted by use of a model that assumes well-mixed conditions in the pond (Fritz et al., 1980). Solar radiation, daily minimum, and maximum air temperatures and wind speed data recorded in the PD/A CRSP data base for the above experiments were used as inputs to the water temperature model. The effects of dissolved oxygen and un-ionized ammonia were not considered in the simulations (i.e., $\delta = \upsilon = 1$) because inclusion of such effects is of limited use unless the dynamics of dissolved oxygen and un-ionized ammonia are modeled simultaneously.

Fish weight was consistently underpredicted by the BE model for the fertilization trial in Bang Sai, Thailand (Figure 2). This discrepancy may also have occurred because of errors in the estimation of the critical standing crop. However, model results for the fertilization experiment conducted in 1986 at the El Carao site were quite similar to observed values (Figure 2).

PD/A CRSP data indicate that fish growth varies substantially even within ponds that are treated identically. Some of this variability may be a function of prior pond history (Knud-Hansen, 1992). It is also possible that part of the variability is purely stochastic in nature. This implies that it may be beneficial to add a stochastic component to the BE model, in which case a distribution of model outcomes can be examined, as opposed to a single prediction.

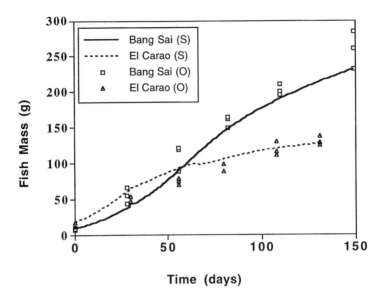

Figure 2.   Simulated (S) and observed (O) fish weights (g) for a 150 day experiment in Bang Sai, Thailand, and a 132 day experiment in El Carao, Honduras. The Bang Sai ponds were stocked at 2 fish/m² and received chicken manure at 75 kg ha/wk (dry matter basis), supplemented with adequate inorganic nitrogen and phosphorus inputs to provide a total of 5 kg ha/wk of nitrogen and 1.2 kg/ha/wk of phosphorus. The mean critical standing crop for the three ponds at this site was estimated to be 1975 kg/ha. The El Carao ponds were stocked at 1 fish/m² and fertilized with chicken manure at 500 kg/ha/wk (dry matter basis). The mean critical standing crop was estimated to be about 500 kg/ha.

The BE model in its current form is a useful tool for predicting fish growth under different management and climatic conditions with the caveat that recalibration may occasionally be necessary for some sites. Recalibration will also be required to describe the growth of fish species other than Nile tilapia. Further, although the critical standing crop–based function for the relative feeding level can be used to predict growth, there are two primary drawbacks that limit its use. First, critical standing crop estimates may vary substantially, even for identically treated ponds at the same site. Second, the effects of fertilization cannot be directly described by use of the critical standing crop–based function. The resource substitution function overcomes these drawbacks to some extent, but adds considerable complexity to the growth model and requires dynamic simulation of different natural food resources.

The BE model can also be used to address supplementary feeding requirements in ponds. For instance, if a target feeding level such as satiation feeding (i.e., feeding rate equal to dR_{max}/dt as estimated by Equation 4, Table 3) is specified, the model can be used to estimate the proportion of this requirement satisfied by natural food. The difference between the target requirement and the proportion met by natural food indicates the amount of supplemental feed to be added to the pond.

Another potential application of the BE model relates to scheduling tasks such as partial harvesting, where the goal may be to ensure that the standing crop is thinned such that fish are not food limited. Partial harvesting should therefore occur when the critical standing crop of the pond is reached. The culture period required to reach the critical standing crop can be predicted by the BE model. Such uses of the BE model are particularly relevant for situations where reasonably accurate estimates of critical standing crop are available.

Photosynthesis and Dissolved Oxygen

Much of the modeling work carried out under the PD/A CRSP has been related to dissolved oxygen. Recognizing that dissolved oxygen concentration in ponds is often the limiting water quality parameter determining fish stocking densities, water exchange rates, and feeding and fertilization rates, models have been developed to simulate dissolved oxygen concentration in ponds. Different types of models have been developed incorporating various degrees of empiricism and of temporal and spatial resolution. Some general characteristics of PD/A CRSP dissolved oxygen models will be reviewed in this section, including a discussion of water column respiration rates and of primary production optimization.

Primary Production and Respiration Rates

A multitude of models have been developed to simulate primary production rates by microalgae. A common model, and one that has been shown to be applicable in a variety of systems (Field and Effler, 1982), is that proposed by Steele (1962). Steele's equation has been used for aquaculture pond models with good results (e.g., Meyer, 1980; Piedrahita, 1984; Losordo, 1988; Culberson, 1993). Steele's equation expresses production rate in terms of three variables: light, phytoplankton light saturation, and maximum production rate, and can be written as

$$\text{Prod} = P_{max} \frac{I}{I_{sat}} \exp\left(1 - \frac{I}{I_{sat}}\right) \tag{1}$$

where Prod = photosynthetic production in an optically thin section ($mgO_2/m^3/h$), P_{max} = maximum photosynthetic production rate (mg $O_2/m^3/h$), I = photosynthetically active light radiation ($Einst/m^2/h$), and I_{sat} = optimum light intensity, saturation ($Einst/m^2/h$).

The gross production rate throughout the water column can be calculated by integrating Steele's equation (1962) over the depth of the pond, with light intensity at any depth given by Beer's law. Net production rate can be estimated by adding water column respiration to gross production rates calculated with an expression such as Equation 1. Net primary production rates can also be estimated using mass balance calculations based on measured values of dissolved oxygen concentration over diel cycles. The calculations have been based on mass balance equations where sources and sinks of dissolved oxygen are estimated based on information contained in the PD/A CRSP data base. As an example, rates of gas exchange with the atmosphere can be calculated using measured wind speeds and dissolved oxygen concentrations using the expressions proposed by Boyd and Teichert-Coddington (1992). The estimated rates of primary production obtained from the mass balance calculations have been used in conjunction with Steele's equation to investigate possible causes for differences in primary production rates per unit of incident light observed at different times of the day. A possible explanation for those changes is the apparent change in light sensitivity (as represented by the light saturation value in Steele's equation) of phytoplankton over a diel cycle (Figure 3).

Respiration by organisms in the water column (primarily phytoplankton) constitutes the largest dissolved oxygen sink in a typical pond oxygen mass balance, and relatively little is known about the factors affecting these respiration rates. In particular, there is little information on respiration rates during daylight hours. Normally respiration rates are estimated based on dark bottle tests or on analysis of nighttime whole-pond data. Both methods have been shown to provide unreliable estimates of respiration rates and therefore to result in unreliable estimates of gross primary production. Using equipment developed under the PD/A CRSP (Szyper et al., 1993), estimates of daytime respiration rates have been made by measuring dissolved oxygen concentration changes over periods of 15 to 30 min in darkened samples. Samples are collected automatically from the pond over a diel cycle, are held in a darkened chamber, and the rate of change in dissolved oxygen concentration is measured. Results obtained to date indicate that respiration rates are substantially higher during the daylight hours when compared to nighttime rates.

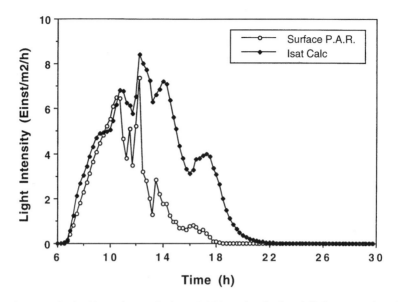

Figure 3. Photosynthetically active radiation (PAR) and calculated light saturation ($I_{satCalc}$) over a diel period.

Optimum Net Primary Production

Management strategies that reduce the zones of oxygen sink while increasing the zones of oxygen production increase the amount of oxygen available for fish respiration and organic matter oxidation and serve to decrease the dependence on mechanical aeration. Therefore, understanding the function of the pond's photosynthetic production system in relation to the light environment has immediate applications in production systems in terms of reducing costs of maintaining suitable dissolved oxygen levels and providing less variable growing conditions for cultured animals.

The complexity and difficulty in managing dissolved oxygen concentration in ponds increase as the intensity of the aquaculture operation increases, where intensity may be defined based on the level of fertilizer inputs, stocking density, water use, or feeding levels. In general terms, and for fertilized aquaculture ponds, as the productivity of the pond system and the standing crop of phytoplankton increase, so do the degree of fluctuation in diel dissolved oxygen concentration, the degree of vertical stratification of the water column, and the temporal fluctuations in water quality. High phytoplankton standing crops result in high turbidity, low light penetration, and low Secchi disk depths. In highly turbid pond systems the rapid extinction of incident light in the surface layers results in a large area close to the bottom of the pond acting only as a net oxygen sink. This zone of high oxygen consumption combined with the high-production zone in the surface layers during the day causes large fluctuations in diel dissolved oxygen concentration. As the turbidity decreases in a pond, so do the standing phytoplankton concentration, the magnitude of diel dissolved oxygen fluctuations, and the differences in dissolved oxygen concentration between surface and bottom waters.

An important challenge to a pond manager is to determine the combination of physical factors (pond depth, turbidity or light extinction coefficient, and light intensity), biological factors (primary production and respiration rates and sensitivity to light), and chemical factors (nutrient availability), that results in optimum net primary production rates. Recognizing the significance of this challenge, an analytical tool that could be used to study the interaction of factors mentioned above was developed under the PD/A CRSP (Giovannini and Piedrahita, 1994). The focus of the CRSP model was to study the interaction among physical factors of pond depth and light extinction coefficient, as these parameters may be controlled to some degree by a pond manager. Control of biological factors, on the other hand, is not feasible at this time, and the emphasis within the project was to develop reliable tools for measuring and predicting those factors. Nutrient saturation was assumed for the purposes of analysis of physical factor interactions.

The objective of the model was to define conditions for which optimum net oxygen production would be obtained. Net production, in turn, can be determined by subtracting respiration from gross production. If Steele's equation (1962) (Equation 1) is assumed to represent gross production, and if water column respiration is assumed to be a function of algal concentration, the net production equation becomes

$$NP_Z = \frac{\Phi_{max} I_{sat} (\varepsilon_{total} - \varepsilon_{other})}{\varepsilon_{total}} \left[\exp\left(\frac{-I_o \exp(-\varepsilon_{total} Z)}{I_{sat}} \right) - \exp\left(\frac{-I_o}{I_{sat}} \right) \right] - \frac{(\varepsilon_{total} - \varepsilon_{other}) R_Z}{K_c} \quad (2)$$

where NP_Z = total net primary production of water column (mg $O_2/m^2/h$), I_o = surface light intensity (Einst/m^2/h), z = pond depth (m), $p_{mac(c)} = \frac{\Phi_{max} K_c I_{sat}}{e} = \frac{P_{max}}{Chlor - a}$, $P_{max(c)}$ = maximum specific production rate (mg O_2/mg chlor-a/h), Φ_{max} = maximum quantum yield (mg O_2/Einst absorbed), K_c = light absorption coefficient of chlorophyll a (m^2/mg chlor-a), e = base of natural logarithm (exp(1) = 2.718...), ε_{total} = total light extinction coefficient (m^{-1}) = $\varepsilon_{chlor} + \varepsilon_{other}$, ε_{chlor} = light extinction coefficient due to chlorophyll a (m^{-1}), ε_{other} = sum of all sources of light attenuation other than chlorophyll, including water, and suspended and dissolved solids (m^{-1}), and R = specific respiration rate (mg O_2/mg chlor-a/h).

The process of deriving optimum values for either the depth or the light extinction coefficient from Equation 2 can be carried out by differentiating the net production function with respect to depth or turbidity, solving for the corresponding optimum value at the peak, and obtaining an expression for optimum depth or turbidity. The resulting equations are transcendental, but can be simplified to provide solvable expressions that yield approximate values for optimum depth and turbidity. The simplification is based on the assumption that the light reaching the bottom of the pond is effectively zero, a condition that is common in aquaculture ponds.

The simplified expression for Optimum Depth (z_{opt}) given a particular turbidity becomes (Giovannini and Piedrahita, 1994):

$$z_{opt} \cong -Ln\left(\frac{R}{P_{max(c)}} \frac{I_{sat}}{I_o} \right) \frac{1}{\varepsilon_{total}} \quad (3)$$

and the Optimum Total Turbidity ($\varepsilon_{T(opt)}$) for a particular pond depth is:

$$\varepsilon_{T(opt)} \cong \left(\frac{P_{max(c)} e \varepsilon_{other}}{Rz} \left(1 - \exp\left(\frac{-I_o}{I_{sat}} \right) \right) \right)^{\frac{1}{2}} \quad (4)$$

A sample graph of the net production rate per unit area (mg $O_2/m^2/h$) given by Equation 2, is shown in Figure 4 as a function of depth and turbidity. The overall maximum net production for this production equation is an asymptotic value as depth approaches zero and turbidity approaches infinity. At any particular depth, there is an optimum turbidity where net production is maximized. Similarly, at a given turbidity, there is an optimum depth that can be derived from Equation 3 as seen from Figure 4. Control of algal populations to achieve optimum turbidity may be achieved by a combination of physical (flushing) and biological methods (grazing), but none are considered to be very practical for on-farm implementation (Smith, 1987).

The expressions for optimum depth and turbidity (Equations 3 and 4) are written in terms of three physical and three biological parameters. The biological parameters are the respiration rate per unit chlorophyll, the maximum production rate per unit chlorophyll, and the saturation light

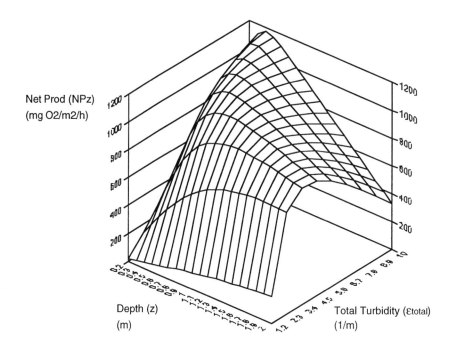

Figure 4. Sample relationship between net primary production, depth (z, meters) and total turbidity (ε_{total}, m^{-1}). Optimum net photosynthesis for a particular pond depth is obtained for the total turbidity that results in the maximum possible net production. Similarly for a particular total turbidity, there is a pond depth that maximized net production. (From Giovannini, P. and Piedrahita, R. H., *Aquaculture Eng.*, 13, 83, 1994. With permission.)

intensity of the phytoplankton. These represent the minimum information about a primary production system that is needed to characterize that system. As expected, the biological parameters are the most difficult to quantify and are subject to change over time as adaptation and successions occur in the phytoplankton population. This is characteristic of any biological process and points to the importance of developing efficient measurement techniques for routinely determining these parameters.

In aquaculture pond systems, it is important to understand the relationships between the physical and biological parameters of the primary production system. These models may be used to provide the baseline values for pond configuration, which can then be adjusted based upon the physical and biological requirements of the cultured animals. In general, pond gross photosynthetic production can be increased by reducing pond depth and/or total turbidity. The optimum value of pond depth or turbidity for net photosynthetic production can then be determined from Equations 3 or 4. This pond configuration can then be compared to what is feasible for the requirements of the culture system and adjusted accordingly. Special consideration needs to be given to the effects of particular depth-turbidity combinations on water temperature and pond stratification. In addition, the behavioral and physiological requirements of the species cultured will affect the ultimate value of depth-turbidity that is selected.

The equations presented for calculating optimum depth and turbidity are theoretical expressions based on a widely used and tested production equation. They have the advantage of using commonly available parameters that simplify their use. However the applicability of any production equation to a system under study must be determined by direct experiment. The techniques used in the derivation of these optimization equations can be applied to other production equations and thus expand the applicability of this type of analysis for pond management.

Stratification

Shallow aquaculture ponds such as those used for the culture of tilapia undergo diel cycles of stratification and destratification. Previous work carried out at the University of California at Davis (Losordo, 1988; Losordo and Piedrahita, 1991), resulted in models for the simulation of dissolved oxygen and temperature in stratified ponds. The models developed by Losordo (Losordo, 1988) required extensive data sets collected primarily with the aid of a data logger and resulted in high quality simulations for ponds at various locations and for different weather conditions. Under the PD/A CRSP, Losordo's original models (the *Losordo Models*) were revised and simplified to be executed with data sets extracted from the PD/A CRSP data base. Model revisions were governed by the reduction of data inputs required to execute the model and the tailoring of the data input requirements to more closely reflect those data commonly available (or easily collectible) to an aquaculturist. The two models (one for temperature and one for dissolved oxygen simulation) are described below in general terms.

Stratified Temperature Model

A detailed description of the stratified temperature model developed for execution with PD/A CRSP data and of the simulation results has been presented by Culberson (1993) and Culberson and Piedrahita (1992). Temperatures are calculated at three different depths in a simulated pond by doing an energy balance for layers of water centered around the three depths selected.

The three-layer model is taken to be representative of a water column in a generalized aquaculture pond and includes an additional sediment volume element.

The general form of the energy balance equation used in the stratified temperature model is based on an equation used by Losordo (1988)

$$\phi_{net} = \phi_{sn} + \phi_{at} - \phi_{ws} - \phi_e \pm \phi_c \pm \phi_{d,z} - \phi_{sn,z} \pm \phi_{sed} - \phi_{gw} \qquad (5)$$

where ϕ_{net} = net heat flux, ϕ_{sn} = penetrating short-wave solar irradiance, ϕ_{at} = net atmospheric radiation, ϕ_{ws} = water surface back radiation, ϕ_e = evaporative heat transfer, ϕ_c = sensible heat transfer, $\phi_{d,z}$ = effective diffusion of heat at volume element boundaries, $\phi_{sn,z}$ = solar irradiance at lower boundary of volume element, ϕ_{sed} = heat transfer between sediment and bottom water volume element, and ϕ_{gw} = heat loss from sediment volume element to groundwater.

This general form of the energy balance equation can be applied to different volume elements, such as surface, midwater, and bottom elements, by the elimination of terms that are inappropriate for the particular volume element. As an example, a midwater or bottom element would not include a term for evaporative heat transfer. Calculations of the different heat flux terms are based on a combination of theoretical (mechanistic) and empirical expressions. Detailed formulations for the various terms in the energy balance equation are beyond the scope of this publication and are available elsewhere (e.g., Culberson, 1993).

Validation of the revised model was conducted by comparing the predicted temperature simulations generated with measurements collected by Losordo (1988), and with simulation results obtained with the Losordo model. The detailed data set gathered by Losordo (1988) for Northern California pond sites was modified to resemble those data sets generated by PD/A CRSP experiment sites. That is, where Losordo's data base presents site weather data (solar irradiation, wind speed, wind direction, air temperature, relative humidity, wind vector) every 20 min (based upon 60-s time intervals averaged over the 20 min), the PD/A CRSP data base provides weather data only at those times where experimental protocols demand it — most commonly at 0600, 1000, 1400, 1600, 1800, 2300, and 0600 h the next day (local time), over a given diel cycle. Readings are generally cumulative readings over the time interval, from which average readings for the time interval are then calculated and reported. After Losordo's (1988) data base was reduced in this way to averages for the time intervals between 0600, 1000, 1400, 1600, 1800, 2300, and 0600 h the following day, the data were entered directly into the revised model.

After these validation runs were made, the revised model was initialized with data from PD/A CRSP experiment sites: Thailand, Honduras, and Rwanda. The model was initialized for each site using a pond selected at random to calibrate for wind vector and solar radiation estimates. The model was then run for different ponds from the same experimental site for the same day. Predicted temperature profiles for each pond were generated for the given diel cycle. Examples of the revised model's predicted temperatures compared to the measured temperature profiles are illustrated in Figure 5.

The temperature model has been tested with data from stratified as well as mixed ponds at several pond sites in different countries and climatic regions. The energy-balance approach used in these models has been shown to be appropriate when characterizing the overall heat content within ponds used for aquaculture. Data required for execution of the simplified temperature model developed under the PD/A CRSP can be collected with relatively simple instruments and for a

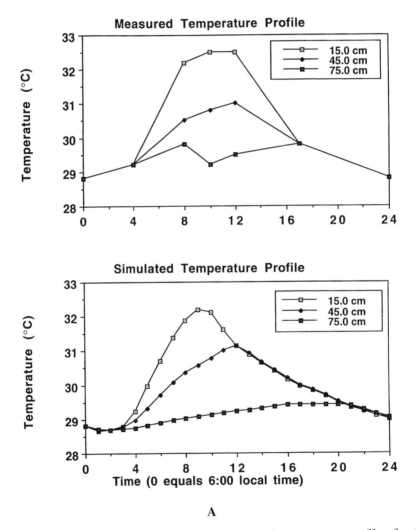

A

Figure 5. Sample comparison of measured and simulated temperature profiles for two PD/A CRSP ponds. The pond in Figure 5A undergoes stratification, while the pond in Figure 5B. remains approximately uniform (mixed) throughout the diel period. The mixed pond is shallow (approximately 45 cm deep), and only two layers are considered. (Part A from Culberson, S. D. and Piedrahita, R. H., *Ecologic Modelling*, 89, 240, 1996. With permission.)

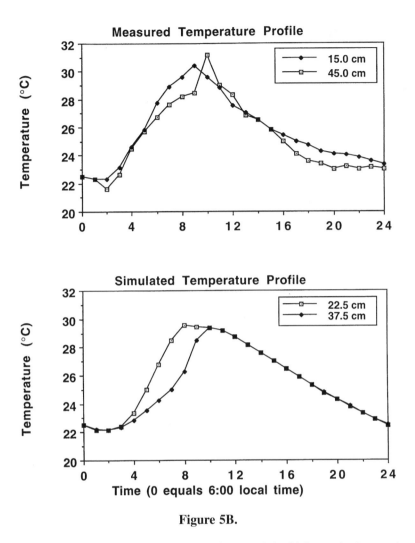

Figure 5B.

relatively small cost, as opposed to Losordo's original model which required extensive data sets. Adaptation of Losordo's original model has resulted in some loss of accuracy in the temperature predictions, but the onset of stratification, the degree of stratification from surface to pond bottom, the onset of mixing, and the simulation of completely mixed conditions remain remarkably accurate. Even with some resolution loss from Losordo's model, differences in instantaneous temperature predictions from measured temperature values across the pond depth rarely exceed 0.25°C.

Stratified Dissolved Oxygen Model

A detailed description of the stratified dissolved oxygen model developed under the PD/A CRSP has been presented elsewhere (Culberson, 1993; Culberson and Piedrahita, 1993). The dissolved oxygen model is coupled to the temperature model described above, and simultaneous calculations of temperature and dissolved oxygen are obtained as the model is executed. Just as temperature calculations were based on energy balances, dissolved oxygen estimates are based on mass balances. In the model, rates of production, consumption, and transfer of oxygen are quantified and used to calculate dissolved oxygen concentrations as functions of time. Diffusion of oxygen between water at different depths is estimated with diffusion coefficients similar to those used for the temperature model. A general mass balance equation for dissolved oxygen in a given pond volume element can be written as

$$MO_{2net} = MO_{2p} \pm MO_{2D} \pm MO_{2d,z} - MO_{2f} - MO_{2pr} - MO_{2wcr} - MO_{2sr} \qquad (6)$$

where MO_{2net} = net oxygen change for the volume element, MO_{2p} = oxygen production by phytoplankton (gross photosynthesis), MO_{2D} = oxygen diffusion to or from the atmosphere across the pond surface, $MO_{2d,z}$ = effective diffusion of oxygen at volume element boundaries, MO_{2f} = oxygen consumption by fish (respiration), MO_{2pr} = oxygen consumption by phytoplankton (respiration), MO_{2wcr} = oxygen consumption by organisms and processes in the water column (other than fish and phytoplankton), and MO_{2sr} = oxygen consumption in the sediment.

Whereas calculation of the various terms in the energy balance for the temperature model were based on considerations limited to physical processes, dissolved oxygen calculations are based on combinations of physical, biological, and chemical processes. The current level of understanding of many of the processes involved in the oxygen mass balance is much lower than that of energy-related processes. As a result, the expressions used for the dissolved oxygen mass balance components tend to be of a more empirical nature than for the energy balance calculations. There is, also, a greater uncertainty associated with the terms of the dissolved oxygen mass balance, and this is reflected in the relative accuracy of the simulations (Figure 6).

Perhaps the most notable change in the dissolved oxygen model when compared to Losordo's original model is that of the characterization of phytoplankton respiration. In the original Losordo model (Losordo, 1988), phytoplankton respiration at night was estimated as a percentage of the primary productivity value during the light period. Recent work on the characterization of diel fluctuations in phytoplankton productivity (e.g., Szyper et al., 1992) suggests that this straight percentage neglects a residual respiration component exhibited by the phytoplankton in the first 2 or 3 h following sundown. Inclusion of a time-lag element in the revised model for characterizing this residual respiration shows an improvement over the original (Losordo) characterization.

A second revision to the Losordo model includes recent information concerning the interpretation of dark-bottle respiration data as estimates of pond water column respiration rates. Teichert-Coddington and Green (1993) have outlined procedures for determining the best incubation schedule for obtaining mean daytime and nighttime water column respiration rates, and their results have been used to generate a regression equation which relates a single 4-h incubated dark-bottle measurement initiated at 8:00 a.m. (the usual format followed in PD/A CRSP research protocols) to an estimate of average overall nighttime pond water column respiration rate. Again, as with the above-mentioned revision to phytoplankton respiration, this modification has improved the dissolved oxygen concentration predictions over the previous Losordo model.

The dissolved oxygen model was tested and validated using Losordo's original data and following procedures similar to those used for the temperature model. Once the model had been tested with Losordo's original data, it was used to simulate dissolved oxygen concentrations in ponds from the currently active PD/A CRSP sites (Honduras, Rwanda, and Thailand). Sample results are shown in Figure 6. The revised model has been run for simulations describing ponds of depths from 45 cm to 120 cm, for ponds of surface areas ranging from 0.30 to 4.00 ha, for ponds in sheltered areas and in exposed ones, and for ponds in both the southern and northern hemispheres. Simulation results have generally remained consistently accurate for all these situations. Applied as either a research tool or used as an aide by extension agents or farmers themselves, the revised model can be of use when deciding where to place aquaculture ponds, how deep to build them, and what considerations might best be exploited locally to suit the needs of the particular animal to be cultured.

COMPUTERIZED DECISION SUPPORT FOR POND MANAGEMENT

As our collective knowledge of the fundamental processes controlling pond dynamics has improved, it has become apparent that computerized tools that can capture this knowledge and apply it to pond management issues are potentially valuable. Such tools, termed decision support

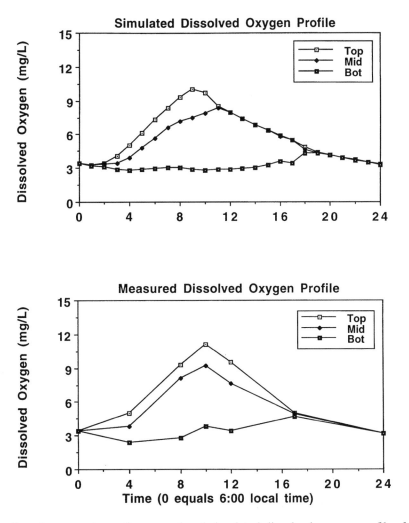

Figure 6. Sample comparison of measured and simulated dissolved oxygen profiles for a PD/A CRSP pond which undergoes substantial stratification, with a dissolved oxygen difference between surface and bottom as high as 8 mg/L. (From Culberson, S. D. and Piedrahita, R. H., *Ecol. Modelling*, 89, 247, 1996. With permission.)

systems, integrate knowledge in the form of mathematical models, rule-based (expert) systems, and/or data bases into user-friendly software systems focused on developing, analyzing, and optimizing management strategies. These tools address the problem of packaging a large domain of scientific and technical knowledge into a form that is of practical value to a diverse audience, including nonscientists (Lannan, 1993). The power of these systems results from their capability for representing and manipulating both quantitative and qualitative knowledge that describe objects in the domain of interest and their inter-relationships.

A key component of any decision support system is the knowledge base(s) upon which decisions are made. The question of *knowledge representation* becomes a critical one. Expertise exists in many forms, ranging from highly qualitative "rule of thumb" approaches useful for capturing subjective, historical experience, to data bases containing historical information available for "data mining," to more rigorous and quantitative mathematical algorithms that describe explicit relationships between components of the domain in question. Because knowledge can exist in many forms, decision support systems must be built using information technologies that can effectively represent, synthesize, and integrate this knowledge into a decision making process ultimately directed at answering questions posed by the user of the system.

Expert systems are an alternative to simulation models for describing relationships among components of a system in a decision-making framework. Typically, expert systems consist of three basic components (Figure 7): (1) a knowledge base, containing the factual and relational knowledge of the system, (2) an inference engine, providing the reasoning capabilities and control strategy for emulating the human reasoning process, and (3) a user interface to provide friendly user interaction with the system. Most systems to date have been *rule-based*, containing knowledge encoded as a series of if/then statements describing conditional relationships about the domain of interest. In such systems, the inference engine seeks to establish valid lines of reasoning from a problem statement to a specific solution (or solutions) via an interview process, asking the user questions corresponding to the conditions of the rules in an attempt to establish the validity of the rule. Depending on the control strategy used in the inference process, the inference engine will attempt to piece together chains of rules into a complete line of reasoning. Rule-based systems have several advantages over conventional approaches of encapsulating knowledge, including (1) capabilities for manipulation of both symbolic and numeric information, (2) separation of knowledge (in the knowledge base) and control (in the inference engine), allowing easier encoding and maintenance of the knowledge in the system, (3) use of natural language syntax in specifying the rule structure and content, allowing rapid prototyping and system development, and (4) capabilities for handling uncertain information ("fuzzy" inference) and explaining system behavior. A number of user-friendly expert system shells have been developed around the rule-based paradigm, which facilitates relatively easy implementation of these systems. Expert systems are particularly appropriate for knowledge that is poorly structured and readily formulated as "rules-of-thumb." Example applications areas where expert system approaches excel include diagnostics, classification, and identification.

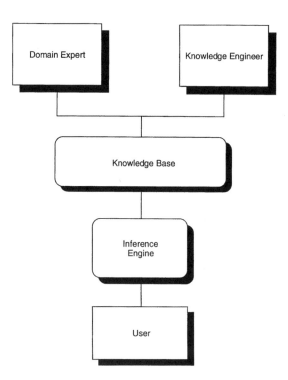

Figure 7. Components of an expert system.

To be effective, an aquacultural management decision support system should possess some or all of the following characteristics:

1. Be accessible to a range of potential end-users via a user-friendly interface
2. Be adaptable to a range of facility configurations
3. Address important issues effecting the viability of the pond facility from either an economic or biological perspective
4. Allow the user to ask "what-if" questions to explore different configurations, management strategies, or assumptions of the system
5. Present results in a manner that allows rapid and detailed understanding of the implications of particular scenarios on any given aspect of the facility or management plan under consideration

With the PD/A CRSP, two decision support systems have been developed. The first of these (PONDCLASS©) addresses issues relating to freshwater pond fertilization and determination of lime requirements. The second one (POND©) expands PONDCLASS© functionality to include consideration of fundamental processes controlling pond dynamics as well as enterprise-level analysis of production facilities. These systems are described below.

PONDCLASS©

Application rates of fertilizers to aquaculture ponds are typically arrived at by trial and error or adapted from strategies that appear to be optimal at one location. Experience suggests that the use of a fertilizer application rate appropriate for a particular site can result in substantially different fish yields when applied elsewhere. Such variability in fish yields may be due to differences in climatic, water quality, and soil characteristics among sites. PONDCLASS© Version 1.1 (Lannan, 1993) is a decision support system that represents perhaps the first attempt to consider these characteristics when estimating fertilizer requirements for freshwater ponds. The program also provides liming recommendations for ponds. A more recent release, PONDCLASS© Version 1.2 (Nath and Lannan, 1993), includes capabilities for simulating fish growth and pond water temperature (see earlier discussion on fish growth models).

Fertilizer Requirements of Ponds

It is generally accepted that fertilizer addition to ponds stimulates autotrophic and detrital feeding pathways leading to an increase in food availability, and therefore enhanced fish yields (Hickling, 1962; Boyd, 1979; Hepher, 1978; Schroeder et al., 1990). There is debate among aquacultural scientists as to whether autotrophic or detrital feeding pathways dominate in any particular system (see review by Colman and Edwards, 1987). The relative importance of these chains likely depends on a variety of factors including the type of fish species stocked, their food preferences, and the characteristics of the pond water and soil. In general, it appears that management of primary production in ponds is a key factor toward enhancing food resources for the stocked fish and therefore their yields (King and Garling, 1986; Yusoff and McNabb, 1989; McNabb et al., 1990).

Yusoff and McNabb (1989) and McNabb et al. (1990) suggested that effective fertilization procedures for ponds could be developed by identifying the nutrients (inorganic carbon, nitrogen, and/or phosphorus) most limiting to algal growth and calculating the quantity of fertilizers required to alleviate nutrient limitation. Such procedures implicitly assume that other factors (temperature and light) that influence algal growth are rarely limiting in most shallow, tropical ponds (McNabb et al., 1988).

The general approach of identifying nutrient limitation and recommending fertilizer amendments that is used in PONDCLASS© (Figure 8) is a synthesis of the concepts outlined by King and Garling (1986), Yusoff and McNabb (1989), and McNabb et al. (1990). The program uses standard limnological equations, which are described in detail elsewhere (Lannan, 1993), to compute nutrient requirements for ponds.

The starting point for fertilizer calculations in PONDCLASS© is an estimate of the maximum or potential net primary production (gC/m³/d) for a given site in the absence of nutrient limitation.

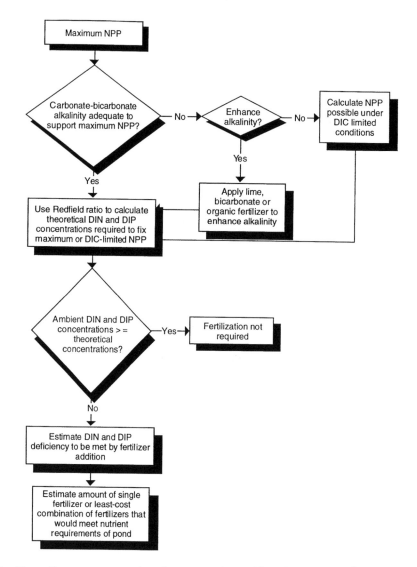

Figure 8. Flow diagram representing the approach used in PONDCLASS© to evaluate nutrient requirements of ponds, and generate fertilizer recommendations.

This is expected to be a function of climate characteristics at the site. The value for the maximum net primary production may be based on theoretical models (e.g., Piedrahita and Giovannini, 1991), productivity measurements (Vollenweider, 1974), or the user's experience. Carbon limitation is handled by examining whether the maximum net primary production can be supported by the dissolved inorganic carbon (g/m³) originating from the carbonate-bicarbonate alkalinity system of the pond water. If dissolved inorganic carbon concentrations are inadequate, alkalinity enhancement may be warranted, or the pond can be operated at a lower net primary production level. The program then computes fertilizer amounts that are expected to provide adequate dissolved inorganic nitrogen (g/m^3) and dissolved inorganic phosphorus (g/m^3) to meet the requirements of algae.

Fertilizer recommendations generated by PONDCLASS© are expected to vary over time, because dissolved nutrient requirements of a pond fluctuate in response to climatic conditions, carryover of nutrients from prior fertilizer applications, and other factors. Fertilizing ponds in accordance with PONDCLASS© guidance thus represents a departure from the traditional method of providing constant amounts of fertilizers at regular time intervals. Fertilizing at constant rates is likely to result in nutrient limitations at some times and wasting fertilizer by providing an excess at other times. In both cases the efficiency of fish production (in terms of cost per unit of fish produced) is compromised.

Fertilizer guidelines generated by PONDCLASS© are calculated on a volumetric basis rather than an areal basis, because ponds may have similar surface areas but substantially different volumes if the slopes of their levees or their operating depths are not the same. If an areal basis is used in fertilizer calculations for such ponds, large errors in the estimates of nutrient requirements may occur.

In pond aquaculture (as in agriculture), it may often be more economically efficient to use a mix of fertilizers rather than a single fertilizer to satisfy nutrient requirements of ponds. Therefore, PONDCLASS© includes a linear programming algorithm (the simplex technique; Lipschutz, 1966) which can be used to find the least cost combination of fertilizers that will satisfy nutrient requirements of the pond.

Apart from the assumptions described above, the fertilizer guidelines generated by PONDCLASS© also make the following assumptions about the pond environment: (1) earthen ponds with low rates of water exchange are used; (2) the pond water budget is balanced (i.e., a constant pond volume is maintained, seepage and overflow are negligible, and inflow replaces water loss due to evaporation); (3) the pond is completely mixed; (4) chemical fertilizers or manures are the only nutrient inputs to the pond and supplemental feeds are not used; and (5) appropriate liming practices have been used to neutralize pond soil acidity (e.g., Boyd, 1990; Bowman, 1992). This is because nutrient losses can be substantial and nutrient availability reduced if pond soils are acidic.

Lime Requirements for Ponds

The application of lime to ponds with acid muds or water of low alkalinity is a widely accepted aquacultural practice. Liming practices for ponds have been adapted from procedures developed for agricultural soils (Schaeperclaus, 1933; Boyd, 1990). The amount of lime expected to improve pond productivity depends on pond water and soil characteristics and is thus highly variable from pond to pond. To assist in estimating the amount of lime that would be beneficial to pond soils, PONDCLASS© includes a utility developed by Bowman (1992).

Lime requirement refers to the amount (kg/ha) of calcium carbonate, $CaCO_3$, required to neutralize the exchange acidity of a pond soil. The amount of exchange acidity to be neutralized depends on the difference between the initial and desired percent base saturation of the soil (Boyd, 1990). Lime requirement is generally computed as the product of the cation exchange capacity of the soil, the difference between the initial and desired percent base saturation, and a correction term that converts cation exchange capacity (meq per 100 g) to the amount of lime (kg $CaCO_3$) applied to a mass of soil (Adams and Evans, 1962; Peech, 1965). However, cation exchange capacity and initial percent base saturation measurements require analytical laboratory support which is not available in many situations.

An alternate approach was therefore developed by Bowman (1992). He examined agricultural data bases for cation exchange capacity, soil pH, and initial percent base saturation values corresponding to a wide range of soil types, and classified these soil types according to particle size and mineralogy. For each soil type in this classification, he determined the average cation exchange capacity and used curve fitting techniques to model the relationship between soil pH and percent base saturation. If the soil type and pH at a given site are known (both of which can be easily determined in the field), the initial percent base saturation can be estimated from the pH–percent base saturation relationship for the particular soil type. Bowman (1992) also modified the correction term in the original Adams-Evans lime requirement expression to enable consideration of soil density and the depth to which the liming reaction is expected to occur. In PONDCLASS©, the lime requirement for a given soil type is estimated by inserting the average cation exchange capacity and estimated initial percent base saturation into the Adams-Evans expression as modified by Bowman (1992).

PONDCLASS© Architecture and Implementation

Both versions of PONDCLASS© are menu-driven programs written in True BASIC®. Version 1.1 is available for computers that use either the DOS or Macintosh operating platforms, whereas Version 1.2 is available only for DOS machines. Both versions of PONDCLASS© provide separate activities (Planner, Manager, and Simulator) to address the needs of development planners, aquacultural producers, and aquacultural scientists, respectively (Lannan, 1993). An additional activity (File Manager) is used to manipulate data files for particular ponds.

Testing, Validation, and Discussion

The fertilization guidelines generated with PONDCLASS© have recently been tested at various PD/A CRSP sites. Preliminary analysis of the results suggest that fertilizer application rates recommended by running PONDCLASS© for a site in Thailand are lower than those traditionally used there, presumably because the software takes account of nutrients remaining in ponds from previous inputs (Szyper and Hopkins, in press). This translates into lower fertilization (and therefore fish production) costs and supports earlier results from an experiment that tested PONDCLASS© at a PD/A CRSP site in the Philippines (Hopkins et al., 1994). In both experiments, fish yields were within the range typically obtained from fertilized ponds. However, yields from Honduran ponds where PONDCLASS© was tested were lower than those typically obtained at the site. The low fish yields from ponds that were managed using PONDCLASS© appeared to be due to excessive un-ionized ammonia concentrations resulting from an overload of nitrogen in the fertilization regime suggested by the program (Teichert-Coddington, personal communication). High nutrient loading may be recommended by running PONDCLASS© if the initial net primary production value entered is very high. Indeed, arriving at the appropriate value for the maximum net primary production is somewhat difficult and may require some calibration and testing for individual sites. Alternately, primary productivity models (e.g., Piedrahita and Giovannini, 1991) may be of use in estimating the maximum net primary production for a site.

Enhancement of the carbonate-bicarbonate alkalinity may occasionally be warranted in ponds that are managed using PONDCLASS© if algal growth appears to be substantially limited by inadequate dissolved inorganic carbon or if declining alkalinities are observed in successive sampling intervals (Szyper and Hopkins, in press). This may be accomplished by the addition of soluble bicarbonate, lime, or organic fertilizers.

The lime requirement guidelines in PONDCLASS© have not been extensively tested as yet. Bowman (1992) suggested that they may be more appropriate for liming new ponds built on the soil types in his classification, because data published for agricultural soils are used to predict the lime requirement for a particular soil type. Such data may be inappropriate for older ponds because organic matter accumulation can affect the cation exchange capacity of the sediments and soil density may change over time. Bowman (1992) also recommended that measured values for variables like the cation exchange capacity and initial percent base saturation of pond soils be used whenever possible to increase the accuracy of lime requirement estimates.

POND© VERSION 2

PONDCLASS© has limited support for facility-level analysis (i.e., simultaneous analysis of an entire production facility), fish production simulation, and economic analyses capabilities. To address these limitations, an additional pond management decision support system, POND©, was developed under the auspices of the PD/A CRSP (Bolte et al., 1994). The POND© program was developed with several goals:

1. To create a software architectural framework that would provide more sophisticated decision support capabilities, incorporating both simulation and heuristic approaches to describing pond and facility processes and incorporating rule-of-thumb expertise into the decision-making process
2. To incorporate the functionality of the PONDCLASS© software in the areas of fertilizer and liming guideline development and fertilizer optimization into this framework
3. To develop and incorporate simple models describing facility and pond level dynamics with minimal user inputs that would allow rapid analyses of pond performance based on site considerations
4. To develop and incorporate more sophisticated models of fish growth and pond dynamics that would allow exploration of fundamental biological, physical, and chemical relationships controlling pond performance
5. To provide capabilities for conducting enterprise-level economic analyses
6. To provide a user-friendly interface for specifying inputs and interpreting results

A brief overview of the architecture and theoretical basis of POND© Version 2 is given below, and is shown schematically in Figure 9.

POND© Architecture

The POND© program is focused on providing a view of pond dynamics at both the individual pond level as well as at the facility level. This involves providing capabilities for simulating processes within a pond, as well as allowing the definition of multiple ponds and multiple fish lots (i.e., a population of fish stocked in a pond), each with its own characteristic data. A series of object classes are defined in POND© using an object-oriented programming paradigm. Each class represents expertise in an area such as aquatic biology, aquatic chemistry, fish biology, fish culture, aquacultural engineering, and economics. A typical POND© simulation might be conducted by creating a series of "experts" (e.g., an "aquatic chemist," an "aquatic biologist," an "aquacultural engineer," a "fish culturist," a "soil chemist," and an "economist") and a collection of facility entities (e.g., one or more fish pond instances, each representing a pond in the facility, and one or more fish lot instances, each representing a single population of fish of the same species). The simulation then proceeds by "asking" the "fish culturist" "expert" to manage the fish lot; it does so with the assistance of its associated "experts" according to the conditions present in each pond.

POND© simulations are dynamic, providing time series results for a range of variables. During a simulation, time series data for each variable are stored; these data may be viewed in plots or tables at the end of the simulation run. Because POND© includes an "economist" in its collection of supported object classes, economic analysis of the facility may be conducted at any time. Such analyses may use simulation data if available, as well as other fixed and variable costs and income streams as defined by the user. This approach to facility-level simulation has proven to be very

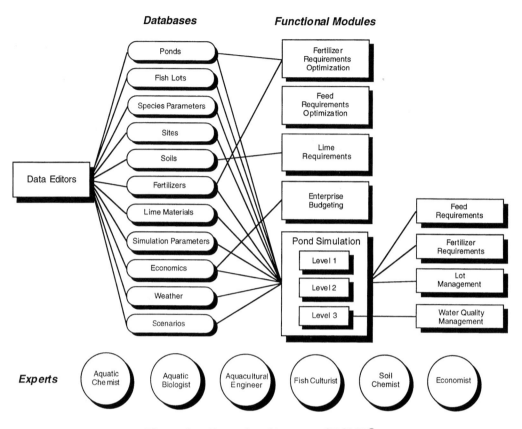

Figure 9.   General architecture of POND©.

flexible and effective for accomplishing POND©'s design goals, and the partitioning of expertise into well-defined components allows reuse of these components in other simulation contexts.

POND© contains a series of data bases, which are accessible to the various objects in the software. Data bases are maintained for each lot and pond in a facility, as well as for simulation scenarios, economic information, soils types, fertilizers and liming materials, site information, and weather characteristics.

Running POND© requires an IBM-PC compatible personal computer running Microsoft Windows® (version 3.1 or higher). It requires approximately 1.5 MB of available hard disk space and a minimum of 4 MB RAM. An 80386 CPU is required, and an 80486 or greater CPU is recommended.

POND© Simulation Models

The inclusion of various "experts" in POND© results in capabilities for simulating different aspects of production, such as fish performance (described elsewhere in this chapter), water temperature, water quality dynamics, and primary and secondary productivity. Models for simulating the various aspects of production are organized hierarchically into three levels, allowing users to perform different kinds of analyses based on data availability and output resolution requirements. Each of these levels is described below.

Level 1

Level 1 models require minimal data inputs and are intended for applied management and rapid analysis of pond facilities. At this level, the variables simulated are fish growth (based on a bioenergetics model) and water temperature. Consumption of natural food by fish is assumed to be a function of fish biomass and appetite. Fertilizer application rates are user specified, but the model optionally generates supplementary feeding schedules.

Level 2

Level 2 models provide a substantially more sophisticated view of pond dynamics than Level 1, allowing prediction of phytoplankton, zooplankton, and nutrient dynamics (carbon, nitrogen, and phosphorus) in addition to fish growth and water temperature. This modeling level is intended for detailed pond analysis, management optimization, and numerical experimentation. At this level, fish can feed from natural and/or artificial food pools; consumption of natural food (phytoplankton and zooplankton pools) by fish is predicted on the basis of a resource competition model and also depends on fish appetite; and a constant, user-specified concentration of pond nitrogen, phosphorus, and carbon is assumed. Mass balance accounting for each of these variables is maintained, allowing estimation of fertilizer requirements necessary to maintain steady-state levels. Both fertilization and feeding schedules are generated by the models.

Level 3

Level 3 models build on Level 2 models, and provide additional capabilities for simulating detrital pool dynamics, heterotrophic productivity, and nutrient dynamics. User-specified fertilization regimes, coupled to pond process-based nutrient mass balances, are used to estimate nutrient consumption and production. Level 3 models, while not currently fully supported in POND© Version 2, would be useful for exploring fundamental pond dynamics relationships, nutrient transformations, and dynamic limiting factors analyses.

Economics

Economic analyses of facilities in the form of enterprise budgets may be generated using POND©. Enterprise budgets allow for the accumulation of various types of costs and incomes, summarized and coupled with interest and depreciation expressions, to assess the overall economic viability of a particular production enterprise. Three cost categories are supported in POND©, (1) fixed, (2) depreciable, and (3) variable costs. Fixed costs are those costs that do not change over the course of facility operation (e.g., construction cost for a pond, a one-time cost that does not vary over time). Depreciable costs are related to fixed costs and typically are used for items that require up-front expenditures but that may have some back-end redeemable value after some period of time. Depreciation schedules describing the loss of value of the depreciable asset over time are incorporated in POND©. An example of a depreciable cost is a tractor, which has an initial cost as well as a resale value after some period in use at the facility. Variable costs are those costs that are not fixed or depreciable and typically vary according to the scale of production (e.g., labor costs, fertilizer and feed costs, and fuel and electricity costs).

To generate an enterprise budget, income sources are also required. In POND©, any number of income sources may be specified, based on either a per unit area, per unit of production, or per facility basis. Income sources relating to fish production are provided by the facility simulator. Additionally, interest rates used for calculating fixed and variable investment costs are required. After specifying each cost by an amount, a cost type (fixed, depreciable, or variable), a basis (per unit area, per unit of production, or per facility) and other related information, the economics module in POND© summarizes costs on an areal, per unit of production, or facility basis, balances those costs against income, and reports the results in a tabular form. By including and/or excluding particular costs/incomes, or adjusting cost/income details, one can quickly "experiment" to determine possibilities for the economic viability of various facility and management configurations.

Testing and Validation of POND©

The interaction between variables controlling pond dynamics may be considerable. This makes ecosystem-level validation of the models in POND© a difficult task. However, specific components of POND© (e.g., fish performance) have been validated with favorable results. These are discussed more fully elsewhere in this chapter. Future efforts on POND© will focus on more comprehensive validation in field situations as well as sensitivity analyses of specific model components to model parameters and operating criteria. Refinement of the model algorithms will occur as validation proceeds. This will allow new knowledge of fundamental pond processes to be captured and presented to the end user in a more generally applicable context.

CHALLENGES FOR THE FUTURE

Mathematical models and rule-based expert systems are two distinct strategies that can be followed for organizing, presenting, and using information on the dynamics of aquaculture ponds. These two strategies are sometimes combined in decision support systems such as PONDCLASS© and POND©. The process of developing models and expert systems of aquaculture pond dynamics provides an opportunity for systematically synthesizing knowledge from a wide variety of disciplines. As the information base related to aquaculture continues to grow and as computing power and ease of use develops, the level of sophistication and the usefulness of computer models and decision support systems will continue to grow. In many ways, computer models and decision support systems are in what could be termed "a first generation" of development and testing: data have been scarce in the past, much information and many techniques have been adapted from related disciplines, and the models and decision support systems have been difficult to distribute and test in the field. This is all changing rapidly, and the challenge to those of us developing models and decision support systems is to incorporate new information and techniques into our products and to tailor those products to users who might have a wide range of backgrounds, interests, and expectations for the new computerized tools.

REFERENCES

Adams, F. and Evans, C. E., A rapid method for measuring lime requirement of red-yellow podzolic soils, *Proc. Soil Sci. Soc. Am.,* 26, 355, 1962.

American Public Health Association, American Water Works Association, and Water Environment Federation, *Standard Methods for the Examination of Water and Wastewater,* 18th ed., American Public Health Association, Washington, D.C., 1992.

Bernard, D. R., A survey of the mathematical models pertinent to fish production and tropical pond aquaculture, in Lannan, J. E., Smitherman, R. O., and Tchobanoglous, G., Eds., *Principles and Practices of Pond Aquaculture,* Oregon State University Press, Corvallis, 1986, 231–244.

Bolte, J. P., Nath, S. S., and Ernst, D. E., *POND© Version 2 Users Guide,* Bioresource Engineering Department, Oregon State University, Corvallis, 1994, 36 pp.

Bowman, J. R., Classification and Management of Earthen Aquaculture Ponds with Emphasis on the Role of the Soil, Ph.D. dissertation, Oregon State University, Corvallis, 1992, 209 pp.

Boyd, C. E., 1979. *Water Quality in Warmwater Fish Ponds,* Agricultural Experiment Station, Auburn University, Auburn, AL, 1979, 359 pp.

Boyd, C. E., *Water Quality in Ponds for Aquaculture,* Birmingham Publishing, Birmingham, AL, 1990, 482 pp.

Boyd, C. E. and Teichert-Coddington, D., Relationship between wind speed and reaeration in small aquaculture ponds, *Aquacultural Eng.,* 11, 121–131, 1992.

Brett, J. R., Environmental factors and growth, in Hoar, W. S., Randall, D. J., and Brett, J. R., Eds., *Fish Physiology,* Vol. 8, Academic Press, New York, 1979, 599–675.

Brett, J. R. and Groves, T. D. D., Physiological energetics, in Hoar, W. S., Randall, D. J., and Brett, J. R., Eds., *Fish Physiology,* Vol. 8, Academic Press, New York, 1979, 277–352.

Brett, J. R., Shelbourn, J. E., and Shoop, C. T., Growth rate and body composition of fingerling sockeye salmon, *Onchorhynchus nerka,* in relation to temperature and body size, *J. Fish. Res. Bd. Can.,* 26, 2363–2394, 1969.

Brody, S., *Bioenergetics and Growth,* Reinholds Publ., New York, 1945, 1023 pp.

Cacho, O. J., Protein and fat dynamics in fish: a bioenergetic model applied to aquaculture, *Ecol. Modelling,* 50, 33–56, 1990.

Caulton, M. S., The effect of temperature and mass on routine metabolism in *Sarotherodon* (Tilapia) *mossambicus* (Peters), *J. Fish. Biol.,* 13, 195–201, 1978.

Caulton, M. S., Feeding, metabolism and growth of tilapias: some quantitative considerations, in Pullin, R. S. V. and Lowe-McConnell, R. H., Eds., *The Biology and Culture of Tilapias,* ICLARM Conf. Proc. 7, International Center for Living Aquatic Resources Management, Manila, Philippines, 1982, 157–180.

Colman, J. A. and Edwards, P., Feeding pathways and environmental constraints in waste-fed aquaculture: balance and optimization, in Moriarty, D. J. W. and Pullin, R. S. V., Eds., *Detritus and Microbial Ecology in Aquaculture,* ICLARM Conference Proceedings 14, International Center for Living Aquatic Resources Management, Manila, Philippines, 1987, 240–281.

Cuenco, M. L., Aquaculture Systems Modeling: An Introduction with Emphasis on Warmwater Aquaculture, *ICLARM Studies and Reviews 19,* International Center for Living Aquatic Resources Management, Manila, Philippines, 1989, 46 pp.

Cuenco, M. L., Stickney, R. R., and Grant, W. E., Fish bioenergetics and growth in aquaculture ponds: I. Individual fish model development, *Ecol. Modelling,* 27, 169–190, 1985.

Culberson, S. D., Simplified Model for Prediction of Temperature and Dissolved Oxygen in Aquaculture Ponds: Using Reduced Data Inputs, M.S. thesis, University of California, Davis, CA, 1993, 212 pp.

Culberson, S. D. and Piedrahita, R. H., Modification of Stratified Temperature Model to Accommodate Reduced Data Inputs: Identifying Critical Requirements, Presented at the May 1992 meeting AQUA-CULTURE 92, sponsored by WAS/AFS/NSA/ASAE, Paper No. AQUA-92-102, American Society of Agricultural Engineers, St. Joseph, MI, 1992.

Culberson, S. D. and Piedrahita, R. H., Model for predicting dissolved oxygen levels in stratified ponds using reduced data inputs, in *Techniques for Modern Aquaculture,* Proceedings American Society of Agricultural Engineers, June 1993, 1993, 543–552.

Culberson, S. D. and Piedrahita, R. H., Aquaculture pond ecosystem model: temperature and dissolved oxygen prediction — mechanism and application, *Ecol. Modelling,* 89 231–258, 1996.

Farmer, G. J. and Beamish, F. W. H., Oxygen consumption of tilapia nilotica in relation to swimming speed and salinity, *J. Fish. Res. Bd. Can.,* 26, 2807–2821, 1969.

Field, S. D. and Effler, S. W., Photosynthesis–light mathematical formulations, Proceedings of the American Society of Civil Engineers, *J. Environ. Eng. Div.,* 8, (EE1) 199–203, 1982.

Fritz, J. J., Meredith, D. D., and Middleton, A. C., Non-steady state bulk temperature determinations for stabilization ponds, *Water Res.,* 14, 413–420, 1980.

Henderson-Sellers, B., *Engineering Limnology,* Pitman Advanced Publishing Program, Boston, 1984, 356 pp.

Fry, C. E. J., Effects of the environment on animal activity, *Univ. Toronto Stud. Bio. Ser.,* 55, 1–62, 1947.

Giovannini, P. and Piedrahita, R. H., Modeling net primary production optimization in aquaculture ponds: depth and turbidity management, *Aquacultural Eng.,* 13, 83–100, 1994.

Hepher, B., Ecological aspects of warmwater fishpond management, in Gerking, S. D., Ed., *Ecology of Freshwater Fish Production,* Wiley Interscience, 1978, 447–468.

Hepher, B., *Nutrition of Pond Fishes,* Cambridge University Press, 1988, 388 pp.

Hepher, B., Liao, I. C., Cheng, S. H., and Hsieh, C. S., Food utilization by red tilapia — effects of diet composition, feeding level and temperature on utilization efficiencies for maintenance and growth, *Aquaculture,* 32, 255–275, 1983.

Hickling, C. F., *Fish Cultures,* Faber and Faber, London, 1962, 295 pp.

Hopkins, K. D., Lopez, E., and Szyper, J. P., Intensive Fertilization of Tilapia Ponds in the Philippines, *Eleventh Annual Administrative Report,* Pond Dynamics/Aquaculture CRSP, Corvallis, 1994, 16–20.

Hsieh, J. S., *Solar Energy Engineering,* Prentice-Hall, New Jersey, 1986, 553 pp.

Huisman, E. A., Food conversion efficiencies at maintenance and production levels for carp, *Cyprinus carpio* L., and rainbow trout, *Salmo gairdneri* Richardson, *Aquaculture,* 9, 259–273, 1976.

King, D. L. and Garling, D. L., A state of the art overview of aquatic fertility with specific reference to control exerted by chemical and physical factors, in Lannan, J. E., Smitherman, R. O., and Tchobanoglous, G., Eds., *Principles and Practices of Pond Aquaculture,* Oregon State University Press, Corvallis, 1986, 53–66.

Knud-Hansen, C. F., Pond history as a source of error in fish culture experiments: a quantitative assessment using covariate analysis, *Aquaculture,* 105, 21–36, 1992.

Knud-Hansen, C. F., Batterson, T. R., McNabb, C. D., and Jaiyen, K., Yield of Nile Tilapia (*Oreochromis niloticus*) in Fish Ponds in Thailand Using Chicken Manure Supplemented with Nitrogen and Phosphorus, *Eighth Annual Administrative Report,* Pond Dynamics/Aquaculture CRSP, Corvallis, 1990, 54–62.

Lannan, J. E., Users Guide to PONDCLASS©: Guidelines for Fertilizing Aquaculture Ponds, Pond Dynamics/ Aquaculture CRSP, Oregon State University, Corvallis, 1993, 60 pp.

Lipschutz, S., *Theory and Problems of Finite Mathematics,* Schaum Publishing, New York, 1966, 339 pp.

Liu, K. M. and Chang W. Y. B., Bioenergetic modelling of effects of fertilization, stocking density, and spawning on growth of Nile tilapia, *Oreochromis niloticus, Aquaculture Fish. Manage.,* 23, 291–301, 1992.

Losordo, T. M., The Characterization and Modeling of Thermal and Oxygen Stratification in Aquaculture Ponds, Ph.D. dissertation, University of California at Davis, CA, 1988, 416 pp.

Losordo, T. M. and Piedrahita, R. H., Modelling Temperature Variation and Thermal Stratification in Shallow Aquaculture Ponds, *Ecol. Modelling,* 54, 189–226, 1991.

Machiels, M. A. M. and Henken, A. M., A dynamic simulation model for growth of African catfish, *Clarias gariepinus* (Burchell 1822). I. Effect of feeding level on growth and energy metabolism, *Aquaculture,* 56, 29–52, 1986.

Marjanovic, N. and Orlob, G. T., Modeling the hydromechanical and water quality responses of aquaculture ponds — a literature review, in Lannan, J. E., Smitherman, R. O., and Tchobanoglous, G., Eds., *Principles and Practices of Pond Aquaculture,* Oregon State University Press, Corvallis, 1986, 207–230.

McNabb, C. D., Batterson, T. R., Premo, B. J., Eidman, H. M., and Sumantadinata, K., *Pond Dynamics/Aquaculture Collaborative Research Data Reports, Vol. 3: Indonesia,* Pond Dynamics/Aquaculture CRSP, Oregon State University, Corvallis, 1988, 63 pp.

McNabb, C. D., Batterson, T. R., Premo, B. J., Knud-Hansen, C. F., Eidman, H. M., Lin, C. K., Hanson, J. E., and Chuenpagdee, R., Managing fertilizers for fish yield in tropical ponds in Asia, in *Proceedings of the Second Asian Fisheries Forum,* Tokyo, Japan, Hirano, R. and Hanyu, I., Eds., The Asian Fisheries Society, Manila, Philippines, 1990, 169–172.

Meyer, D. I., Modeling Diel Oxygen Flux in a Simulated Catfish Pond, Masters thesis, University of California, Davis, CA, 1980, 123 pp.

Meyer-Burgdorff, K.-H., Osman, M. F., and Gunther, K. D., Energy metabolism in *Oreochromis niloticus, Aquaculture,* 79, 283–291, 1989.

Nath, S. S. and Lannan, J. E., *Revisions to PONDCLASS©: Guidelines for Fertilizing Aquaculture Ponds,* Pond Dynamics/Aquaculture CRSP, Oregon State University, Corvallis, 1993, 46 pp.

O'Neill, R. V., DeAngelis, D. L., Pastor, J. J., Jackson, B. J., and Post, W. M., Multiple nutrient limitations in ecological models, *Ecol. Modelling,* 46, 147–163, 1989.

Peech, M., Lime requirement, in Black, C. A., Ed., *Methods of Soil Analysis,* American Society of Agronomy, Inc., Madison, WI, 1965, 927–932.

Piedrahita, R. H., Development of a Computer Model of the Aquaculture Pond Ecosystem, Ph.D. dissertation, University of California, Davis, CA, 1984.

Piedrahita, R. H., Calibration and validation of TAP, an aquaculture pond water quality model, *Aquacultural Eng.,* 9, 75–96, 1990.

Piedrahita, R. H., Modeling water quality in aquaculture ecosystems, in Brune, D. E. and Tomasso, J. R., Eds., *Aquaculture and Water Quality,* World Aquaculture Society, Baton Rouge, LA, 1991, 322–362.

Piedrahita, R. H. and Giovannini, P., Fertilized non-fed pond systems, in Aquaculture Systems Engineering, ASAE Publication 02-91, 1991, 1–14.

Pond Dynamics/Aquaculture CRSP, *CRSP Work Plan: Second Experimental Cycle,* Oregon State University, Corvallis, 1984.

Pond Dynamics/Aquaculture CRSP, *Fifth Annual Administrative Report,* Oregon State University, Corvallis, 1987.

Pond Dynamics/Aquaculture CRSP, *Seventh Annual Administrative Report,* Oregon State University, Corvallis, 1989.

Pond Dynamics/Aquaculture CRSP, *Handbook of Analytical Methods,* Oregon State University, Corvallis, 1992.

Pond Dynamics/Aquaculture CRSP, *Seventh Work Plan,* September 1993 Revision, Oregon State University, Corvallis, 1993.

Pütter, A., Wachstumsähnlichkeiten, *Phleugers Arch. Gesamte Physiol. Menschen Tiere,* 180, 298–340, 1920.

Rabl, A. and Nielson, C. E., Solar ponds for space heating, *Solar Energy,* 17, 1–12, 1975.

Schaeperclaus, W., *Textbook of Pond Culture. Rearing and Keeping of Carp, Trout and Allied Fishes,* Paul Parey, Berlin, 1933, 261 pp.

Schroeder, G. L., Wohlfarth, G., Alkon, A., Halevy, A., and Krueger, H., The dominance of algal-based food webs in fish ponds receiving chemical fertilizers plus organic manures, *Aquaculture,* 86, 219–229, 1990.

Smith, D. W., Biological Control of Excessive Phytoplankton and Enhancement of Aquacultural Production, Ph.D. dissertation, University of California, Santa Barbara, CA, 1987.

Stauffer, G. D., A Growth Model for Salmonids Reared in Hatchery Environments, Ph.D. dissertation, University of Washington, Seattle, WA, 1973, 213 pp.

Steele, J. H., Environmental control of photosynthesis in the sea, *Limnol. Oceanogr.,* 7, 137–150, 1962.

Stickney, R. R., *Principles of Aquaculture,* John Wiley & Sons, New York, 1994, 502 pp.

Suffern, J. S., Adams, S. M., Blaylock, B. G., Coutant, C. C., and Guthrie, C. A., Growth of monosex hybrid tilapia in the laboratory and sewage oxidation ponds, in *Proceedings, Culture of Exotic Fishes Symposium,* Fish Culture Section, Smitherman, R. O., Shelton, W. L., and Grover, J. H., Eds., American Fisheries Society, Bethesda, MD, 1978, 65–81.

Svirezhev, Yu. M., Krysanova, V. P., and Voinov, A. A., Mathematical modelling of a fish pond ecosystem, *Ecol. Modelling,* 21, 315–337, 1984.

Szyper, J. P. and Hopkins, K. D., Management of carbon dioxide balance for stability of total alkalinity and phytoplankton stocks in fertilized fish ponds and PD/A CRSP Global Experiment — Thailand, *Twelfth Annual Administrative Report,* Pond Dynamics/Aquaculture CRSP, Corvallis, OR, in press.

Szyper, J. P., Rosenfeld, L. Z., Piedrahita, R. H., and Giovannini, P., Diel cycles of planktonic respiration rates in briefly-incubated water samples from a fertile earthen pond, *Limnol. Oceanogr.,* 37, 1193–1201, 1992.

Teichert-Coddington, D. R. and Green, B. W., Influence of daylight and incubation interval on dark-bottle respiration in tropical fish ponds, *Hydrobiologia,* 250, 159–165, 1993.

Teichert-Coddington, D. R., Green, B. W., and Rodriguez, M. I., Culture of Tilapia with combination of chicken litter and a commercial diet: water quality considerations, in *Seventh Annual Administrative Report,* Pond Dynamics/Aquaculture CRSP, Corvallis, 1989, 16–20.

Tilman, D., *Resource Competition and Community Structure,* Princeton University Press, Princeton, NJ, 1982, 296 pp.

Ursin, E., A mathematical model of some aspects of fish growth, respiration, and mortality, *J. Fish. Res. Bd. Can.,* 24, 2355–2453, 1967.

Vollenweider, R. A., *A Manual on Methods for Measuring Primary Production in Aquatic Environments,* IBP Handbook No. 12, Blackwell Scientific Publications, Oxford, 1974, 225 pp.

von Bertalanffy, L., A quantitative theory of organic growth, *Hum. Biol.,* 32, 217–231, 1938.

Warren, C. E. and Davis, G. E., Laboratory studies on the feeding, bioenergetics and growth of fish, in Gerking, S. D., Ed., *The Biological Basis of Freshwater Fish Production,* Blackwell Scientific Publications, Oxford, 1967, 175–214.

Winberg, G. G., *Rate of Metabolism and Food Requirements of Fishes,* Translation Series 194, Fish. Res. Bd. Can., Ottawa, Ontario, 1960, 201 pp.

Yusoff, F. and McNabb, C. D., Effects of nutrient availability on primary productivity and fish production in fertilized tropical ponds, *Aquaculture,* 78, 303–319, 1989.

14 EXPERIMENTAL DESIGN AND ANALYSIS IN AQUACULTURE

Christopher F. Knud-Hansen

INTRODUCTION

Although aquaculture as a farming practice dates back thousands of years, during the last three decades several simultaneous occurrences have stimulated scientific research of shellfish and finfish cultivation. First, per capita consumption of fish, long appreciated as an excellent source of dietary protein, is increasing across the globe. Second, in countries with rapidly expanding human populations, natural waters no longer meet the growing demand for fish due to overfishing and water quality degradation from poor watershed and waste disposal management (Edwards, 1991). And third, technological advances, such as hormonally induced spawning, sophisticated recirculating systems, and pelleted feeds, have moved the production of commercial species (e.g., the tiger prawn and channel catfish) into large-scale operations.

As aquaculture research rapidly expands in all directions, new species are constantly being considered for grow-out and market potential. Egg production and fry rearing strategies are improving with experimentation. Investigations into semi-intensive and integrated farming management often relate inputs (e.g., manures, cassava leaves, and urea) to water quality, primary production, and fish yield. Identification of nutritional requirements has helped develop more efficient formulated feeds for intensive fish culture systems. Progress has been made. But as the irony of science would have it, from each question answered springs forth more questions posed.

Identifying precisely what questions to ask is only slightly easier than devising the proper methodologies to answer them. A well-designed experiment will yield observations or measurements under controlled conditions and hopefully will expand existing knowledge of how the system works. With limited resources (human, material, time, and capital), it is essential for any investigation to determine not only what data to collect but what data not to collect. Objectives, hypotheses, treatments, sources of experimental error, and the types, methods, and frequency of measurements must all be clearly defined and understood in order to optimize research grants and to be in a position to reasonably request more. The limitation of available funds for research necessitates focused, cost-efficient experimental designs.

Knowledge of statistics, the foundation of experimental design, benefits the researcher in three very important ways. First, it gives the researcher the analytical skills to effectively test hypotheses, question assumptions, and tease apart relationships. Science is more than collecting data; it is knowing which data need to be gathered to effectively answer a specific problem. Too often piles of data are passed on to a statistician with the desperate plea "analyze this and tell me if anything is significant!" Sometimes the only "significant" observations are the sizes of piles and the anxiety over what to do with them, now that so much time, effort, and money had been spent in their creation. There must be thousands of diel temperature, dissolved oxygen, alkalinity and pH measurements just waiting, … and waiting.

0-56670-274-7/97/$0.00+$.50
© 1997 by CRC Press LLC

Second, statistics enable the researcher to have an objective measure of confidence in her or his results and interpretations. Some people believe something is true because they see it; others see it because they believe it is true. If you feel 99% sure that using a brand-named high-priced feed does not give any better fish yields than the bargain variety feed, it is easier to convince others of your conclusions when you let the numbers do the talking.

Third, statistical knowledge gives the scientist the power to critically review the literature, whether it be to discover design flaws and thereby alter stated conclusions, to enhance existing analyses, or to confirm in one's own mind the appropriateness of a given interpretation.

Unfortunately, most biologists' first (and often last) introduction to statistics took place in a large lecture hall with a professor at the lectern droning in uninspiring Greek. There was little joy memorizing formulas and grunting through calculations, which were difficult to relate to and never quite made much sense. Surviving such threats as pooled variances and mean square errors was the main objective. That the only mathematical talent one needs to gain sufficient statistical knowledge to do quality science is the ability to add, subtract, divide, and multiply is rarely made clear at the onset of introductory statistics courses.

This chapter attempts to further develop the design and analytical skills of aquaculturists who, on the average, tend to have more fear than background in mathematics/statistics. Major concepts of descriptive and inferential statistics are covered without plodding along the traditional well-worn "cookbook" paths found in most statistics books. Experimental design and data analysis are two sides of the same coin, and many aquaculture researchers focus on data analysis without ever appreciating the connection. Two analogies may help illustrate the importance of experimental design in research. First, the field of statistics can be thought of as a language. Methods of data analysis are like the words, but it is knowledge of experimental design that allows the scientist to use the words efficiently and to their fullest potential. Researchers who memorized words (i.e, statistical formulas) in statistics courses rarely learned how to put the words together. Scientific eloquence requires a knowledge of both data analysis (vocabulary) and experimental design (grammar). A second conceptual analogy is the toolbox. Data analyses are like tools, but experimental designs represent the knowledge of how, when, and where these tools should be used for maximum efficiency and utility. As there are already enough "cookbooks" and software packages to do the mechanics of data analyses, this chapter emphasizes the concept and philosophy of experimental designs. Except as foundational requirements, little emphasis is placed on mathematical computations of statistical theory. Where calculations are presented, emphasis will be on when, where, and why, and not how.

The practioner's approach to controlled experimentation used in this chapter is meant to complement the reader's own more theoretical statistical literature. Sections on "Basic Concepts" and "Scientific Approach to Aquaculture Research" provide basic statistical concepts and discuss hypothesis testing, respectively. Sections on "Treatments" through "System Characterization and Monitoring" show how to choose treatments, reduce experimental error through appropriate treatment allocation, and how to best characterize the experimental system. The section, "Data Analysis," discusses how data generated from the various designs are analyzed, while the section, "Quality Control," presents ways to improve overall quality control of data collection, processing, and evaluation. The "Conclusion" section briefly addresses issues regarding publication of research data. Standard calculations and equations presented here are not cited to any particular source, as they are found and more thoroughly discussed in most general statistics books. Among the basic reference books relied upon for this chapter are Heath (1970), Parker (1979), Steel and Torrie (1980), and Baily (1981). This chapter, however, does not present every possible analytical technique relevant to aquaculture research. Intelligent use of more complicated procedures, such as multivariate analysis, requires a solid foundation in basic research and experimental design theory. A critical examination of current aquaculture literature indicates that such a foundation is often lacking. This chapter tries to promote quality research and build that foundation by communicating with words and examples in a way fellow researchers can appreciate.

BASIC CONCEPTS

Although it is assumed that the reader already has had some exposure to statistics, a quick refresher of basic terminology may be useful. The field of statistics examines methods and procedures for collecting, classifying, processing, and analyzing data for the purpose of making inferences from those data. Statistics is an inductive process where attempts to understand the whole are based on examining representative parts through sampling and experimentation (Finney, 1988b).

The first task for the researcher, therefore, is to define what she means by the "whole." In statistical terms the "whole" is defined as the ***population***, and can be as broad or narrow as is scientifically reasonable. A population could be all the catfish raised in a particular pond, or all the catfish raised in all the earthen ponds of similar dimensions situated within tropical latitudes, or raised in concrete recirculating freshwater systems maintained within a specified temperature range. Characteristics of populations, such as mean fish weight, are called ***parameters*** (from the Greek words ***para*** and ***metron*** which mean "beyond measure"). Parameters are fixed values without variability. If there are 10,000 fish in a pond, there is only one true mean weight at harvest for those 10,000 fish.

For populations that are either too large, indefinite, or indeterminate (e.g., all tropical earthen ponds), it is generally both impractical and impossible to determine parameter measurements (hence the name parameter). Always remember that it is the population, however you have defined it, that you want to understand. In order to gain this understanding ***samples*** are taken from the population. Characteristics of the samples, such as sample mean fish weight, are ***statistics***. Unlike parameters, statistics can vary with each sample. Sample statistics estimate population parameters. For example, the sample mean estimates the true population mean, the sample variance estimates the true population variance, and so on. Since we do not know the exact values of the population parameters (if we did there would be no need to take samples), sample statistics should always be given with some indication of their variability and level of confidence of how well they represent corresponding parameter values.

The starting point for determining variability and levels of confidence is knowing the underlying population distribution. Frequencies of measurements or observations often follow a ***normal distribution***, with values distributed symmetrically on either side of the true mean (Figure 1). This normal distribution is described by two parameters, the true arithmetic mean (μ) and the true

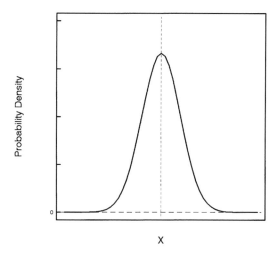

Figure 1. A typical normal distribution curve of X illustrating the characteristic "bell-shaped" curve, symmetrical on either side of the mean.

variance (σ^2). If n = the total number of observations and x_i = the ith observation, then μ = sum of all values divided by n, or

$$\mu = \frac{\sum\limits_{i=1}^{n}(x_i)}{n} \tag{1}$$

The **variance** measures the dispersion of individual values in relation to the mean, and equals the average square of the difference between each observation and the true mean.

$$\sigma^2 = \frac{\sum\limits_{i=1}^{n}(x_i - \mu)^2}{n} \tag{2}$$

As can be seen from Equation 2, the greater the dispersion (i.e., more observations farther away from μ) the greater the variance. If all the observations were the same, then the variance would be zero. The variance can be calculated using an alternative equation, which is both simpler to use and illustrates two major components of the variance, the "Sum of Squares" and the "Square of Sums." The variance then becomes the sum of squares minus the square of sums divided by n, or

$$\sigma^2 = \frac{\sum\limits_{i=1}^{n} x_i^2 - \dfrac{\left(\sum\limits_{i=1}^{n} x_i\right)^2}{n}}{n} \tag{3}$$

Since variances are in squared units, a more practical indicator of population variability is by taking the square root of the variance. This value, called the **standard deviation** (σ) of the population, now has the same units as the observations from which it was calculated.

When a sample is taken from a population, the parameter values of μ, σ, and σ^2 are estimated by the sample statistics ($\bar{x}$, s, and s^2), respectively. Calculations are similar, except $n - 1$ is used instead of n when calculating the s and s^2.

$$\bar{x} = \frac{\sum\limits_{i=1}^{n}(x_i)}{n} \tag{4}$$

$$s^2 = \frac{\sum\limits_{i=1}^{n} x_i^2 - \dfrac{\left(\sum\limits_{i=1}^{n} x_i\right)^2}{n}}{n - 1} \tag{5}$$

When using electronic calculators or computers to determine variances and standard deviations, you must determine whether they use n or $n - 1$ in these calculations. Most experiments are designed to represent larger (often theoretical) populations. For example, a simple experiment comparing shrimp growth using feed A vs. feed B may use three ponds for each treatment. If you are only interested in those specific ponds, then the variance is calculated using n (= 3). If, however, data are to be used for inferring possible results in other nontested ponds (which is nearly always the case), then the three ponds per treatment are really samples of the indeterminate number of shrimp ponds that may utilize either feed A or B in the future. Since the three ponds per treatment are now considered samples of a larger population of ponds, the variance of each treatment should be calculated using $n - 1$ (= 2) instead of n. When n is small, the resulting error in estimating true population variance can be great (see Hypothesis Testing in the section on Scientific Approach to Aquaculture Research). For larger sample sizes (i.e., $n > 35$), however, using $n - 1$ instead of n loses computational relevance.

Another measurement of variability is the ***standard error*** (SE). The standard error is the standard deviation of the distribution of a statistic for a random sample of size n. Let's say, for example, that you want to know the average weight of the 10,000 tilapia being raised in one of your ponds. You collect 25 fish, weigh them individually, and calculate the $\bar{x}$ and s^2. Then you collect 25 more fish and do the same thing (most likely getting different values for $\bar{x}$ and s^2), and repeat this process 20 more times. If you plot your means and variances on a graph, you will find they too have a normal distribution. The standard deviations of these distributions of means and variances represent their respective standard errors. In aquaculture we are often concerned primarily with the SE of the mean. And rather than taking a series of samples, the SE can be calculated dividing the standard deviation (s) by the square root of n.

$$\text{SE of } \bar{x} \ = \ \frac{s}{\sqrt{n}} \qquad\qquad (6)$$

Note the distinction between standard deviation and standard error. Standard deviation describes the variability of ***observations*** about a sample mean, while standard error describes the variability of ***means*** about a sample mean. Whenever the scientific interest concerns comparing sample means, therefore, report mean values with standard errors (e.g., $\bar{x} \pm 1$ SE), *not* with standard deviations. On the other hand, population variability should be reported with a standard deviation.

Also note some important characteristics of the standard error. It is proportional to the standard deviation, and inversely proportional to n. That is, as the sample size increases, the variability of sample means gets smaller. Notice too that the size of the population is totally unrelated to the standard error of the mean. As long as observations are taken randomly, how well our sample mean estimates the true mean depends *only* on the sample standard deviation and the number of observations taken for that calculation. Some field manuals erroneously recommend making sample sizes equal to 10% of total population size. But as indicated by Equation 6, population size is totally irrelevant with regard to estimating the true population mean and variance.

The sample mean estimates the true mean, and the standard error describes the variability of that estimation. This variability can be conveniently expressed in terms of probabilities by calculating ***confidence intervals*** (CI). For example, assume a sampling of 25 catfish (in the pond of 10,000) gives a mean weight of 350 g/fish and a standard deviation (using $n - 1$, Equation 5) of 75 g/fish. We know that the true mean is unlikely to be exactly 350 g/fish, but what is the range of means within which we can say with a certain level of confidence that the true mean falls? First we calculate the standard error (Equation 6), which gives the variability of the theoretical distribution of sample means. Here, the SE of the mean is 75 $(\sqrt{25})^{-1}$, or 15 g/fish. If we make our confidence interval ± 1 SE, we are 68% confident that the true population mean is between $350 + 15$ g/fish and $350 - 15$ g/fish. In other words, there is a 0.68 probability (P) that the true mean is somewhere between 335 and 365 g/fish. The upper and lower mean values are called the ***confidence limits*** (CL) of the mean.

In aquaculture, as with most fields of science, it is common to give sample means with 95% confidence ($P = 0.95$) intervals. This interval is calculated by simply multiplying the SE by a t value (Equation 7) taken from a Student's t Table located in the Appendix of virtually every statistics book on the market. On the left-hand side of the table is a column of *degrees of freedom* (*df*), which equals $n - 1$. The horizontal heading values are probabilities, often ranging from 0.5 to 0.001. If we make our desired level of confidence = α, then the probabilities in the Student's *t* Table equal $1 - \alpha$. Within the table matrix are *t* values. In our example we have $n - 1$ or 24 *df*, and the *t* value at $P = 0.05$ ($= 1 - 0.95$) is 2.064. So the 95% CI and CL equals:

$$\text{CI at } P = \alpha, \text{ or } \alpha(100)\% = \left(t_{(1-\alpha)}\right)(\text{SE}) \tag{7}$$

$$= (2.064)(15 \text{ g/fish}) = 31.0 \text{ g/fish}$$

$$\text{CL at } P = \alpha \qquad = \bar{x} \pm \left(t_{(1-\alpha)}\right)(\text{SE}) \tag{8}$$

$$= 350 \pm 31.0 \text{ g/fish} = 319.0 - 381.0 \text{ g/fish}$$

Therefore, we are 95% confident that the true mean of the 10,000 catfish is somewhere between 319.0 and 381.0 g/fish. In other words, there is a probability (*P*) of 0.05 that the true mean is less than 319.0 g/fish or greater than 381.0 g/fish. Since the normal distribution is symmetrical, $P = 0.025$ that $\mu < 319.0$ g/fish, and $P = 0.025$ that $\mu > 381.0$ g/fish.

Notice in the Student's *t* Table that *t* values increase as we demand greater and greater confidence. For instance, if we wanted to be 99.9% confident, our CI would be (3.745) (15 g/fish) or 56.2 g/fish. Notice also that *t* values decrease as the sample size increases. If we had sampled 250 fish instead of 25, assuming the same SE, the 95% CI would decrease to (1.960) (15 g/fish) or 29.4 g/fish. The mean with $n = 250$ is not necessarily more accurate (i.e., closer to the true mean) than the mean with $n = 25$, but the CI about the mean gets smaller. In actuality, increasing n should decrease the SE (see Equation 6) as well as the *t* value, thereby increasing the precision of estimating the true mean.

The last descriptive analysis discussed in this section is the *coefficient of variation* (CV). The CV is a relative, unitless measure of variation that describes as a percent the relationship between the sample standard deviation and the sample mean (Equation 9).

$$\text{CV} = \frac{s(100)}{\bar{x}} = \text{standard deviation as percent of the mean} \tag{9}$$

Researchers often report CVs, but there is very little analytically that can be done with them. Since their values are entirely relative, there is no such thing as an inherently good or bad CV. They can be used, however, to check consistency of data collection. For example, standard deviations of mean fish weights tend to increase proportionally with mean values, so CVs are often fairly consistent whether the fish mean weight is 50 g/fish or 500 g/fish. A relatively low or high CV may indicate a problem with the sampling technique.

SCIENTIFIC APPROACH TO AQUACULTURE RESEARCH

Introduction

The starting point for scientific research is the *hypothesis* (H), which is an educated guess consistent with observations from a "natural" system (i.e., natural history) and/or existing scientific literature. The hypothesis predicts how a system will respond when manipulated. Predictions may be anything, including nonquantitative effects or mathematical relationships.

To test hypotheses, experiments are conducted under controlled conditions. Experiments must be designed carefully so that the nature, timing, frequency, and accuracy of variable measurements account for all important sources of variation. Variation not accounted for in data analysis can be referred to interchangeably as *noise*, *experimental error*, or *residual error* (often represented by the Greek letter ε, or by R). If the noise is great enough, it may obscure actual significant relationships. After data collection and analysis, experimental results are compared to predicted hypothetical results. If the data analysis does not support the original hypothesis, further hypothesis modification may be necessary before designing more focused experiments and repeating the scientific process.

Hypothesis Testing

It is important to keep in mind that although experimental conclusions may have matched predicted results, we still cannot say that the experiment "proved" a causal relationship. It is usually more constructive to analytically test the *null hypothesis* (H_0) rather than the hypothesis itself. If the hypothesis predicts that a particular relationship exists, the corresponding null hypothesis predicts that this relationship does not exist. From data analysis a probability (P, where $P = 1.0$ is absolute certainty) is obtained, which equals the probability that the observed relationship or treatment differences did not just happen by chance. This probability forms the basis for accepting or rejecting null hypotheses.

For example, assume that a researcher believes (i.e., hypothesizes) that her homemade formulated pelleted feed will give greater tilapia yields than a high-priced commercial variety. The null hypothesis to be tested is that the two types of feed will not produce significantly different yields of tilapia. To test her hypothesis, the researcher conducts an outdoor tank experiment with four tanks for each type of feed. Fish growth measurements after 3 months gave mean weights (± 1 SE, $n = 4$) of 157.3 ± 10.1 g/fish and 189.6 ± 9.8 g/fish for the homemade and commercial feeds, respectively. Should the null hypothesis be accepted, or is the difference between the two means so great that differences could not have happened by chance and therefore the null hypothesis should be rejected?

The first step is to understand what is meant by significance. Significance can be viewed from two perspectives, *"true" significance* and *statistical significance*. True significance is to statistical significance what a parameter is to the corresponding statistic. Similar to a parameter, whether the difference between sample means is truly significant (i.e., $P = 1.0$) is not really known. Statistical significance, therefore, provides a level of confidence or probability < 1.0, by which we can infer that a truly significant difference/relationship does or does not exist. Statistical significance is based on probabilities determined from experimentation. In aquaculture, a $P < 0.05$ that the observed relationship ordifference between means, or whatever happened just by random chance is the most common level of statistical significance. That is, we are more than 95% confident ($P > 0.95$) that there is a true significant relationship. This 0.05 probability is the dividing line between statistical significance and statistical nonsignificance in many other scientific disciplines as well.

It is important to remember, however, that there is no magic probability for statistical significance, and it is the researchers themselves who determine that level. Statistical significance is actually a highly flexible concept. For example, scientists working on a cure for AIDS may consider $P < 0.40$ an adequate limit for accepting statistical significance (i.e., there is a $P > 0.60$ that a certain chemical treatment is effective against AIDS). On the other hand, scientists working on a cure for the common cold may consider a probability of 0.0001 the highest acceptable probability that the alleged cure does not cause lethal side effects. The scientist's real objective is to understand how the system works, not to find mindless and mechanistic statistical significance (see Yoccoz, 1991).

In our tilapia feed example above, the researcher has four possible outcomes based upon her decision whether to accept or reject the null hypothesis and what is actually the truth. These four outcomes are best understood by comparing the statistical vs. true significance of her experiment (Figure 2). She will be correct if she rejects the null hypothesis when the null hypothesis is false (saying the observed difference is statistically significant when it truly is), or if she accepts the null hypothesis when the null hypothesis is true (saying there is no statistically significant difference when there truly is no difference). With the two other outcomes she will be incorrect. ***Type I error*** occurs when, based on statistical significance, she rejects the null hypothesis and claims a truly significant difference when there is none. In contrast, if she says there is no truly significant difference when there actually is one (accepting the null hypothesis when it is true), this is called a ***Type II error***. The probabilities of committing these two errors are typically represented by the Greek letters α and β, respectively.

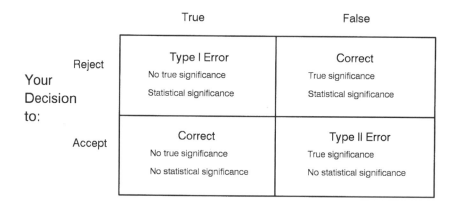

Figure 2.   A diagram showing a results' matrix for hypothesis testing based on the researcher's decision to reject or accept the null hypothesis when the null hypothesis is either true (i.e., no true significance) or false (i.e., true significance).

Since much of science involves determining probabilities whether or not observed hypothesized treatment effects, relationships, etc. are truly causal, we are never certain when making our conclusions that we are not committing a Type I or Type II error. If we want to reduce the possibility of committing a Type I error, we can increase our level of statistical significance from 95 to 99% confidence by reducing α from 0.05 to 0.01. In other words, we will not claim true significance for a relationship or difference between means unless there is a $P < 0.01$ that our observations happened by chance. By doing this, of course, we also increase the probability of committing a Type II error. Scientists generally prefer to decrease the possibility of committing a Type I error because it is more professionally prudent to miss a significant relationship when there was one than to claim significance when it was not there (e.g., claims of energy released from cold fusion).

Nevertheless, missing a significant relationship when there was one (i.e., committing a Type II error) should also be avoided. As the probability of committing a Type II error equals β, the probability of *not* committing a Type II error equals $1 - \beta$. This latter probability is know as the ***statistical power*** (Cohen, 1988). This is an important concept, particularly when analyzing nonsignificant results. Statistical power calculations involve treatment means, experimental variability, and the number of replicates per treatment. For example, statistical power analysis can determine if there were too few replicates to reveal any significant differences between treatment means given the closeness of the mean values, extent of experimental variability, and the desired power (Cohen,

1988). A probability of 0.80 (i.e., $\beta = 0.20$) has been proposed as the minimum acceptable statistical power, and may be higher depending on the financial consequences of the results (e.g., environmental impact assessments) (Searcy-Bernal, 1994; Cohen, 1988). Searcy-Bernal (1994) provides a clear and concise application of power analysis to aquaculture research, and thorough reading is highly recommended (see also Peterman, 1989).

Using a $P < 0.05$ (i.e., $\alpha = 0.05$) is generally an acceptable balance between committing Type I or Type II errors. But whether or not the researcher subjectively thinks a difference is significant is secondary to the actual probabilities generated by the analysis. Therefore, always report probabilities with statistical analyses. A relationship with a $P < 0.10$ may not be statistically significant, but it may be truly significant though not "seen" statistically because of too few replicates or too much noise in the experiment. With probabilities less than 0.01, it is often better to report probabilities to the nearest number other than zero (e.g., $P < 0.005$, $P < 0.001$, $P < 0.0008$, etc.). Reporting probabilities allows the reader to decide for him or herself just how much the data support or refute the tested null hypothesis and not to rely only on an author's claim of "highly significant" differences.

With the feeding experiment described above, the researcher conducted a t-test (see the section on Data Analysis) to determine whether the two means (157.3 ± 10.1 g/fish and 189.6 ± 9.8 g/fish for the homemade and commercial feeds, respectively) were significantly different from each other. The analysis gave a probability of $0.2 < P < 0.3$ that the commercial feed was truly no better than the homemade feed; or in other words, there is >70% chance that the commercial feed was truly better than the homemade feed. Although mean weights were not statistically significant, she may wish to review her experimental methodology (e.g., confirm all tanks were environmentally equal, all fish came from the same stock, etc.), calculate the statistical power, and perhaps rerun the experiment with more replicates before accepting the null hypothesis that the two feeds were not really different with respect to tilapia growth.

TREATMENTS

Introduction

After a general hypothesis has been formulated, the next step is to identify appropriate treatments necessary to test the resultant null hypothesis. Treatments are selected because the researcher hypothesizes they will (or will not) make a difference to a particular response variable(s). Remember, the main scientific objective is to identify and quantify sources of variability in response variables. In aquaculture, response variables are often related to aspects of growth, reproduction, productivity, and water quality. Understanding variability and how to manage it is at the heart of experimental science. Variability does not just happen, it happens for a reason, and the job of aquaculture researchers is to find out both how and why.

To understand the role of treatments, it is useful first to examine a system without treatments. For example, anyone who has worked with earthen ponds knows that if you have 16 tilapia ponds stocked and fertilized identically, you will have 16 different yields at harvest. This result can be expressed by the following model:

$$Y_i = \mu + \varepsilon_i \tag{10}$$

where Y_i = yield in the ith pond, μ = the overall mean of all ponds, and ε_i = the residual (or "experimental error") for the ith pond.

To better understand residuals, let us assume that for these 16 ponds the overall mean for the tilapia harvest was a net fish yield (NFY) of 23.5 kg/ha/d, and ponds 3 and 7 had yields of 19.9 and 25.1 kg/ha/d, respectively. As mentioned earlier, the residual represents the sum of all unidentified sources of variation. If the overall mean is thought of as the predicted value of the experiment, then the residual is simply the observed value minus the predicted value. The NFY residuals for ponds 3 and 7 would then be −3.6 and +1.6 kg/ha/d, respectively. Although residuals will be discussed throughout this chapter, the idea that the residual equals the observed value minus the predicted value will be particularly evident with regression analysis discussed in the section on Data Analysis.

At this point, however, the researcher's primary objective is to improve the predictability of μ by decreasing ε. Decreasing ε can be achieved by focusing research on *manageable* sources of variation, and evaluating factors that influence these sources. For example, Nile tilapia will grow proportionally (within limits) to the rate of natural food production (Knud-Hansen et al., 1993). The issue then becomes identifying manageable factors that influence the variability of natural food production, which in turn affects the variability of tilapia production.

More generally, treatments are selected for their hypothesized effect (or relationship) with the response variable. A far from exhaustive list of treatment possibilities for affecting tilapia yields includes stocking density, pond sediments, natural food availability, temperature, inorganic turbidity in the water column, and rates of nutrient input. Treatment effects on Y can be shown by expanding the above model (Equation 10) as follows:

$$Y_i = \mu + \tau_i + \varepsilon_i \qquad (11)$$

where τ_i = deviation due to treatment i. Now there are two sources of variation in the equation, the treatment and the residual. The hypothesis tested here is that τ is a significant source of variation of Y. The section on Data Analysis discusses how to quantitatively test this hypothesis.

Since one of the purposes of science is to improve upon existing knowledge or understanding of a particular system, the choice of treatment(s) to test a given hypothesis must be carefully thought out. The subsections below describe different aspects and guidelines for determining basic treatment selection, depending on the type of hypothesis the researcher wishes to test.

Unstructured, Structured, and Factorial Experiments

The first cross-road in choosing treatments is to decide whether the experiment will be *unstructured* or *structured*. Unstructured experiments are those in which the researcher wishes to compare, in a matter of speaking, apples with oranges. One treatment cannot be expressed as a function of another. In aquaculture, examples of such experiments include comparing different culture species (or genetic varieties of the same species) under identical conditions, or comparing the efficacy of different brands of shrimp feeds. Often the researcher's primary interest is to determine which of the lot performs the best and whether it is significantly better than the rest. Such treatments are unstructured because there is no quantitative way to rank or arrange them or to demonstrate relationships between treatments. For this reason results from unstructured experiments are presented as vertical rankings of data (in either ascending or descending order) and are not based on any logical order of treatments. It is worth mentioning here and will be repeated in the section on Data Analysis that multiple range tests should be used *only* with unstructured experiments.

There are two types of unstructured experiments, both based on the nature of the researcher's hypothesis. The first type is when he tests the null hypothesis that none of the treatments (e.g., feed types, species of fish) produces a response significantly different from any other treatment. Screening trials typify this type of experiment, where frequently the objective is to determine whether the best is significantly better than the rest. In the second type of unstructured experiment, the researcher wishes to compare unrelated treatments to some benchmark treatment. This benchmark could be, for example, the local strain of tilapia or the main commercial brand of feed used at the research station. In a sense this benchmark treatment could be thought of as the "control." But in actuality, the researcher is testing the null hypothesis that none of the other treatments produces significantly different results than the benchmark treatment. The distinction between the two types of unstructured experiments determines how the data are analyzed (see the section on Data Analysis).

Structured experiments are those in which there is a logical ordering of quantitatively or qualitatively defined treatments. Each treatment can be expressed as a function of another. Simply stated, if you can draw a line graph of the results with treatments aligned along the x-axis, then the experiment was structured. The primary objective with structured experiments is to determine *relationships* and/or *treatment interactions*, and *not* just to find which treatment gave the "best" results. For this reason, among many others discussed in the section on Data Analysis, multiple range tests should *never* (repeat *never*) be used to analyze structured experiments.

The two most common types of experiments used to evaluate relationships are those examining changes over time and those examining response changes with increasing levels/concentrations of a treatment variable. The latter is also known as a dose–response experiment. The purpose of such experiments is to estimate and identify trends. The hypotheses tested generally reflect a mathematical relationship (e.g., linear, quadratic, exponential, asymptotic, etc.) that the tested relationship hypothetically demonstrates. Regression analysis (described below in the section on Data Analysis) is normally used to test the null hypothesis that there is no relationship over time or with variable treatment dosages.

Structured experiments are among the most common in aquaculture. Examples of the "dose–response" variety include fertilization experiments (e.g., relationship between increasing nutrient input levels and yield), stocking density experiments (e.g., relationship between stocking density and yield), hatchery studies (e.g., relationship between flow rate and hatching success), and feeding trials (e.g., relationship between lipid content in feed and its digestibility). The key concept is relationship, and for that reason such investigations are particularly useful for model building. Note that each treatment level is a function of another (e.g., stocking densities of 1 m^{-2}, 2 m^{-2}, 3 m^{-2}, and 4 m^{-2} can be expressed as multiples of 1, 2, 3, and 4 times 1 m^{-2}, and therefore the experiment is structured.

When the experimental objectives include testing the hypothesis that two treatment variables interact with each other to produce a nonadditive response, a *factorial* design is chosen. To illustrate a positive interaction, consider a nutrient-poor pond. If you add only phosphorus (P), you may get a little algal growth. If you add only nitrogen (N), you may again get some minor algal response. But if you add P and N together, you may get an algal bloom, a response much greater than the individual N and P responses added together. Our model equation would now be as follows:

$$Y_{ij} = \mu + N_i + P_j + (NP)_{ij} + \varepsilon_{ij} \qquad (12)$$

where N_i = ith level of nitrogen treatment, P_j = jth level of phosphorus treatment, and NP_{ij} = deviation due to the interaction between N and P.

N and P in the above example are called *factors*. Within each factor there may be two or more *levels* of that factor. What characterizes a factorial experiment is that every factor is represented in all treatments, and the treatments represent every possible factor and level combination. A simple factorially designed experiment would have two factors (e.g., N and P input), and two levels of each factor (e.g., no input and input). This experiment is called a 2 by 2 factorially designed experiment because there are 2 levels (input amounts) of 2 factors (N and P). There would be a total of 2 × 2, or four different treatments in this experiment: (1) no N input, no P input, (2) N input, no P input, (3) no N input, P input, and (4) N input, P input (Figure 3). Although there are only four treatments, three different null hypotheses are tested:

1. There is no significant response with the addition of N.
2. There is no significant response with the addition of P.
3. There is no significant interaction between and N and P with regard to the response variable.

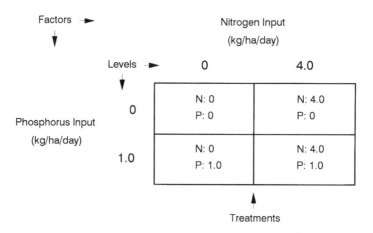

Figure 3. A diagram illustrating the four treatment combinations of a 2 by 2 factorially designed experiment with two levels (hypothetical input rates) for two factors (nitrogen and phosphorus input).

More complicated factorial experiments can reveal a tremendous amount of information because of the great number of hypotheses tested. For example, assume a researcher wants to test for any interaction between lipid and protein concentrations in shrimp feeds as reflected in the growth of two different varieties of shrimp. In the experiment she tests three levels (i.e., concentrations in the feed) of lipid and four levels of protein. As a factorial experiment, there are three factors (shrimp variety, lipid concentration in feed, and protein concentration in feed), with two levels of shrimp, three levels of lipid, and four levels of protein. The experiment is, therefore, a 2 by 3 by 4 factorially designed experiment, with a total of $2 \times 3 \times 4 = 24$ treatments (Figure 4).

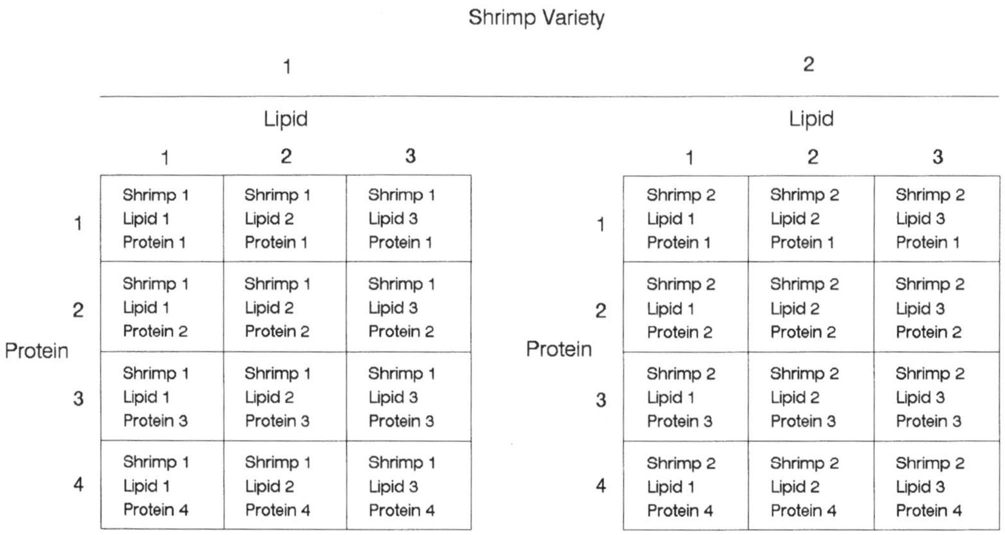

Shrimp Variety

	1				2		
	Lipid				Lipid		
Protein	1	2	3	Protein	1	2	3
1	Shrimp 1 Lipid 1 Protein 1	Shrimp 1 Lipid 2 Protein 1	Shrimp 1 Lipid 3 Protein 1	1	Shrimp 2 Lipid 1 Protein 1	Shrimp 2 Lipid 2 Protein 1	Shrimp 2 Lipid 3 Protein 1
2	Shrimp 1 Lipid 1 Protein 2	Shrimp 1 Lipid 2 Protein 2	Shrimp 1 Lipid 3 Protein 2	2	Shrimp 2 Lipid 1 Protein 2	Shrimp 2 Lipid 2 Protein 2	Shrimp 2 Lipid 3 Protein 2
3	Shrimp 1 Lipid 1 Protein 3	Shrimp 1 Lipid 2 Protein 3	Shrimp 1 Lipid 3 Protein 3	3	Shrimp 2 Lipid 1 Protein 3	Shrimp 2 Lipid 2 Protein 3	Shrimp 2 Lipid 3 Protein 3
4	Shrimp 1 Lipid 1 Protein 4	Shrimp 1 Lipid 2 Protein 4	Shrimp 1 Lipid 3 Protein 4	4	Shrimp 2 Lipid 1 Protein 4	Shrimp 2 Lipid 2 Protein 4	Shrimp 2 Lipid 3 Protein 4

Figure 4. A diagram illustrating the 24 treatment combinations of a 2 by 3 by 4 factorially designed, hypothetical shrimp feed experiment. There are three factors of shrimp variety, and feed concentrations of lipid and protein, with two, three, and four levels for the three factors, respectively.

This shrimp feed experiment tests the following null hypotheses for *each* response variable (e.g., shrimp growth, productivity, etc.):

1. Null hypotheses based on factors:
   a. No difference in response between the two shrimp types (results pooled over all protein-lipid combinations)
   b. No difference in response between the three lipid levels (results pooled over all shrimp-protein combinations)
   c. No difference in response between the four protein levels (results pooled over all shrimp-lipid combinations)
2. Null hypotheses based on nested designs within the factorial experiment at different levels of each factor:
   a. No difference in response between the two shrimp varieties at
      1. Lipid level 1 and protein level 1
      2. Lipid level 1 and protein level 2
      3. Lipid level 1 and protein level 3
      4. Lipid level 1 and protein level 4
      5. Lipid level 2 and protein level 1
      6. Lipid level 2 and protein level 2
      7. Lipid level 2 and protein level 3
      8. Lipid level 2 and protein level 4
      9. Lipid level 3 and protein level 1
      10. Lipid level 3 and protein level 2
      11. Lipid level 3 and protein level 3
      12. Lipid level 3 and protein level 4
   b. With each shrimp variety, there is no relationship between the response variable and
      1. Increasing lipid concentration when protein is kept at level 1
      2. Increasing lipid concentration when protein is kept at level 2
      3. Increasing lipid concentration when protein is kept at level 3
      4. Increasing lipid concentration when protein is kept at level 4
      5. Increasing protein concentration when lipid is kept at level 1
      6. Increasing protein concentration when lipid is kept at level 2
      7. Increasing protein concentration when lipid is kept at level 3
3. There is no two-way interaction as reflected in the response variable between
   a. Lipid and protein concentrations (pooled over both shrimp varieties)
   b. Shrimp variety and lipid concentration (pooled over all protein levels)
   c. Shrimp variety and protein concentration (pooled over all lipid levels)
4. There is no three-way interaction between shrimp variety, lipid concentration, and protein concentration

Schematic representations of treatments (e.g., Figure 4) greatly facilitate identifying testable hypothesis. An advantage of factorially designed experiments is that unusual treatment combinations may reveal relationships or interactions not previously considered. On the other hand, some interactions may not make any biological sense. A statistically significant three-way interaction between shrimp variety, lipid, and protein concentration would probably defy biological understanding. In reality, however, only a few of the above possible hypotheses may have real importance to the researcher. By identifying testable hypotheses *prior* to conducting the experiment, the researcher can adjust treatment levels to ensure the primary hypotheses will be adequately tested.

Note that the shrimp feed experiment has both unstructured and structured components when levels of each factor are analyzed. The factor of shrimp variety is unstructured since the two varieties are distinctly different with no definable relationship (like apples and oranges). Of the lipid and protein factors, however, each has different levels related to each other; i.e., each level is a fraction of another. Both protein and lipid treatments are of the dose–response kind; therefore, that part of the experiment (and analysis) is considered structured.

Factorially designed experiments are probably the most complicated experiments aquaculture scientists will encounter. A review of the literature shows that most researchers employing factorial experiments have grossly underutilized the analytical potential of their research. A great deal more is yet to be learned from existing data sets, and hopefully this brief discussion will induce interested scientists to review their favorite statistics book with a renewed purpose.

Scope of Experiments

Existing scientific knowledge together with the researcher's ambitions (as well as available funding) usually determine the scope of research. The two most common problems are trying to "prove" too much and not developing an overall game plan. Not surprisingly, these problems tend to go hand-in-hand.

The researcher who tries to establish a major scientific principle with a single experiment or analysis almost always ignores the myriad of other relevant factors and influences, which eventually show up as unidentified sources of variation (i.e., noise). Instead of testing an hypothesis, the researcher attempts to "prove" a particular personal belief. This is both dangerous and unethical. Ironically, it is the narrow mind that produces an overbroad experimental design and an open mind that produces a narrowly focused design. The latter is by far preferred, while the former is a waste of time, resources, and money.

The second problem of lack of game plan can be avoided by having a coherent, systematic approach to the problem. This is the heart of a good research proposal, and it is at that stage where designs of *all* experiments necessary to attain stated objectives must be made. To revive the cooking analogy, in making a cake there is a time to add the dry ingredients, a time to add the wet ingredients, a time to mix, a time to bake, etc. In order to produce the desired objective (i.e., an edible cake), the whole process must be visualized and then followed according to plan (i.e., recipe).

Good scientific research is conceptually no different from baking a cake, except that the results are published instead of eaten. The objectives must be clearly stated up front. Most objectives are to gain scientific understanding through testing specific hypotheses. Too often, however, objectives are stated as "to measure . . . , to monitor . . . ," etc. These are not objectives, but means of attaining objectives. The difference is significant. The objective is not to measure flour or heat a mixture of ingredients at 350°F, but to produce a cake.

And like good cooking, good research is rarely conducted *ad hoc*. The first ingredient is a thorough understanding of existing scientific knowledge in order to know what assumptions can be made or should first be tested. The importance of identifying potential sources of error right from the start cannot be overemphasized. These assumptions include all aspects of research from equipment (Are all pumps really operating the same? Do all ponds have the same size and characteristics?), to methodology (Is my sampling truly random or representative? Should I trust my DO meter? should I believe my primary productivity or chlorophyll measurements?), to design (Is this variable truly not important and therefore should not be measured?). It is difficult to repair the damage after discovering that assumptions were falsely made and probably not even considered. It is particularly embarrassing when a reviewer of the subsequent manuscript points out these unconsidered (and possibly fatal) assumptions. Not all assumptions should necessarily be tested, but the competent researcher will carefully consider all possible sources of error (i.e., sources of variation) and make every reasonable attempt to minimize them before proceeding.

Identifying and evaluating necessary assumptions is a critical step in any comprehensive research design. If more than one experiment is required to satisfy research objectives, then these experiments must be planned and coordinated logically and efficiently. Two common schemes are the "wagon-wheel" approach and the "funnel" approach. In the wagon-wheel plan, experiments are like the wheel's spokes, which address a central objective (the wheel's hub) from different angles. An example would be looking at the role of chicken manure fertilization in the production of Nile tilapia (Knud-Hansen et al., 1993). In that study, chicken manure was examined as source of N, P, and inorganic carbon for phytoplankton production, particulate carbon as a direct food source, impacts on pond ecology, and economic feasibility. Another example is systems modeling, where each experiment may evaluate an identified relationship in the system. The hub of the wheel is like a jigsaw puzzle, and each experiment provides missing piece(s).

The funnel approach is appropriate where the objectives are more focused and background scientific knowledge is less understood. Experiments are more linearly planned, beginning with broad ones and progressing to those more narrowly defined. Assume, for example, that the research objective was to determine the optimal input rate of the best available local plant to feed a culture of native herbivorous fish. The first experiment would be to test all possible candidates (i.e., local plants) in an unstructured experiment. The clear winner(s) (based on fish yield, plant availability, convenience, etc.) would then be examined in a dose–response experiment to determine optimal input rates based on predetermined criteria.

The wagon-wheel and funnel approaches are not mutually exclusive, and both can be easily incorporated in a comprehensive research scheme. Regardless of the approach taken, however, the researcher must be clear from the start about how results from each experiments fit into the big picture. The overall plan must be flexible where necessary, but potential contingencies should be already outlined in the research proposal. One research direction may be chosen over another depending on whether or not the preliminary experiment showed any significant relationships or differences.

EXPERIMENTAL UNITS

At this point the researcher has identified the hypothesis(es) to be tested, and has determined the most appropriate designs (e.g., unstructured, dose–response, factorial) to meet specified research objectives. Part of this preparatory process also entails defining what the experimental units for each experiment will be.

Types

Experimental units are individual entities, representing a population of similar entities, each of which receives an individual treatment. In aquaculture the most common experimental units are aquaria, tanks, hapas (net enclosures), and ponds. For example, in a fertilization study conducted in ponds, each pond is an experimental unit because treatments (i.e., different fertilization strategies) are applied on a per pond basis. Similarly, with a fry density experiment conducted in 30 hapas with 10 hapas in each of three ponds, the hapas would be the experimental units if fry density (i.e., treatment) varied on a per hapa basis.

Recall the discussion in the section on Basic Concepts regarding samples and populations. Experimental units can be thought of as a subset of a population of like units (aquaria, ponds, etc.) similarly situated; or, the population could also be the same aquaria but at different times. For example, results in your experimental ponds will be used to predict what will happen in other similarly situated ponds receiving the same treatments or in the same ponds at a latter time.

Individual organisms can also be experimental units as long as there is some way to mark, tag, or otherwise distinguish one individual from another. Types of data collected include observed changes within each organism over time. For example, assume eight common carp (tagged for identification) were treated with a test solution to kill known parasites. The null hypothesis is that the solution has no effect on parasite infestation. A parasite count is performed on each fish before and after treatment. The change in parasite number for each fish is determined, and the eight values are analyzed using a paired sample *t*-test or Wilcoxon's Signed Rank test (See the section on Data Analysis) to see if the treatment had a significant effect on parasite infestation. In this experiment each individual fish is an experimental unit.

Replications

To appreciate the importance of treatment replication in designing experiments, the researcher must first understand two points, first, why replication is necessary and, second, what needs to be replicated. First, replication of treatments is necessary because of noise. Remember that there are generically two sources of variation, known and unknown. Whatever variation we cannot account for by treatments and other measured variables is called the residual experimental error, or noise. Through identical replicated treatments, the residual error can be partitioned from total experimental variability (as indicated, for example, in Equations 10 through 12). As will be discussed more fully in the section on Data Analysis, the ratio of known to unknown variation is the foundation of analysis of variance (ANOVA). With regard to the second point, experimental units (aquaria, ponds, etc.) must be replicated to determine within-treatment variability.

Treatments can be replicated spatially or temporally. With spatial replication the same treatment is repeated in two or more experimental units in the same experiment. This is by far the more common approach and is generally preferred over temporal replication, in which a treatment is replicated in sequential experiments over time. Temporal replication may be used where experimental and environmental conditions are highly controlled, such as with some hatchery experiments. For outdoor investigations, however, variable climatic conditions add too much experimental error to make temporal replication a reasonable option.

Now comes the important question: how many replicates are enough? Simplistically, the greater the anticipated experimental error (noise) and/or desired precision of analytical estimates, the more replicates you need. Practically, however, the type of experimental units (e.g., ponds vs. aquaria) and the scope of the experiment often influence the chosen number of replicates per treatment. For example, inherent variability would be expected to be relatively high in earthen ponds as compared to tanks or aquaria. Similarly, flow-through systems with constant water quality may be less variable than static water cultures, and incubation under controlled indoor conditions would likely be less variable than outdoor culture systems. A limitation of the number of available experimental units can also affect the number of different treatments or testable hypotheses for a given experiment. Remember from Equation 6, however, that decreasing the number of replicates per treatment generally increases the standard error of the treatment mean and therefore decreases the precision of how well this sample mean estimates the true mean. So trying to squeeze in too many treatments at the expense of too few replicates only increases the possibility of erroneous conclusions. For unstructured aquaculture experiments, where the hypothesis involves comparing one treatment mean with another, three should be the minimum number of replicates per treatment; four replicates per treatment is generally better. In fact, statistical power analysis (Searcy-Bernal, 1994; Cohen, 1988; and discussed briefly in the section on Scientific Approach to Aquaculture Research) may indicate the need for even more replicates to reduce the probability of missing significant treatment differences (i.e., committing Type II errors).

Replication in dose–response type experiments can be viewed differently from unstructured experiments. There is often only one treatment, with several levels of that treatment. Stocking density may be the treatment, and stocking densities of 1 m^{-2}, 2 m^{-2}, 3 m^{-2}, and 4 m^{-2} are different levels of that treatment. In contrast to unstructured experiments, where the null hypothesis involves comparing two or more treatment means, with dose–response experiments the null hypothesis involves comparing the observed relationship with a predicted model. The residual (experimental error) is measured by summing the absolute differences between observed and predicted values based on the model. Since there is really only one treatment with several levels, there is no reason to replicate each level (i.e., stocking density) since the residual is *not* based on within-level variation.

For example, assume a researcher wants to know the relationship between the rate of urea input and the growth rate of a particular herbivorous fish. She has twelve ponds in which to conduct the experiment. Since the experimental objective is to determine a ***relationship*** (and not differences between means), it is better to have twelve different levels of urea input with no replicate levels (Figure 5) than to have, say, four levels with three replicates of each level (Figure 6a). Rather than testing a relationship, the latter design tests whether one treatment mean is significantly different from another (Figures 6b, 6c). Remember that different urea input rates just represent different levels of a single treatment, namely urea input. In a sense both designs, therefore, have twelve replications of a single treatment.

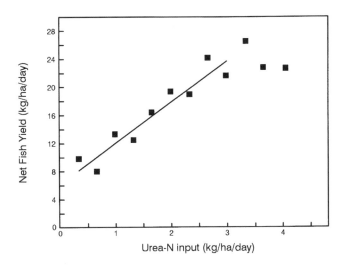

Figure 5. Results from a hypothetical dose-response experiment using 12 different input levels to examine the relationship between urea-N fertilization and net fish yield. The regression line shows a linear relationship ($r^2 = 0.91$, $P < 0.001$) up to about 3 kg urea-N/ha/d.

There are at least two reasons why it is better not to replicate levels in a single treatment, dose–response experiment. First, the greater the number of levels (i.e., urea input rates), the more likely any true relationship between urea input and growth will be detected (compare Figure 5 with Figure 6). Second, any relationship between urea input and the individual residuals (observed fish growth minus predicted fish growth) which constitute experimental error can be identified. A significant relationship here may reveal the need to transform and reanalyze the data, and any relatively unusual or extreme data can be readily identified and reexamined if necessary. These aspects of residual analyses are discussed more fully in the section on Data Analysis.

Unfortunately, nearly all dose–response experiments in aquaculture have replicated treatment levels to test differences between treatment means and then add regression lines to show relationships. It is not that these studies are wrong; it is just that researchers could have attained more valuable information with the same amount of effort. Since researchers have taken this hybrid approach, the reported relationships have limited interpretive value because either there were sizeable gaps between treatment levels or the range of levels was too narrow to infer a broader, meaningful relationship. Analysis of treatment means is only meaningful for those few, somewhat arbitrarily selected treatment levels. As an example, if the underlying objective in the urea input study mentioned above were to find which input rate gives the best fish yield, increasing the number of treatment levels clearly provides more useful results (again, compare Figure 5 with Figure 6).

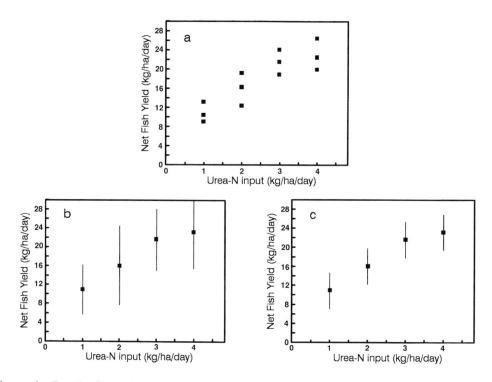

Figure 6. Results from a hypothetical treatment experiment using four different levels of urea-N input as four treatments with three replicates per treatment. Graph (a) shows individual treatments, graph (b) gives treatment means with 95% confidence limits based on individual treatment standard errors (i.e., t value based on 2 degrees of freedom), and graph (c) gives treatment means with 95% confidence limits based on the pooled standard error for all treatments with t value based on degrees of freedom for residual error in an ANOVA (i.e., 8 degrees of freedom, see Tables 1 and 2).

The threshold question in choosing one design over the other is whether the ultimate objective is to compare treatment means or to determine relationships.

However, there are situations in which replicates in dose–response experiments are justified, such as when relationships and treatment mean comparisons are both important experimental objectives. Furthermore, factorially designed experiments with a dose–response component (e.g., the lipid-protein shrimp feed experiment described above) also represent a hybrid approach in which replicates must be utilized in order to properly analyze possible interactions between factors (see the section on Data Analysis).

Allocation of Treatments to Experimental Units

Estimating experimental error is an essential element of data analysis. As was just discussed, whether residual determinations are based on within-treatment variation or on observed variation about a hypothesized model depends upon the chosen experimental design. The greater the experimental error (residual variation) as a percentage of total variation, the more difficult it becomes to find a statistically significant treatment effect, even if there truly is one. There is also a greater likelihood of committing a Type II error. A primary objective in designing experiments, therefore, is to identify and account for other possible sources of variation.

Perhaps the most common source of experimental error comes from the allocation of treatments to experimental units. By better understanding any nonuniformity of experimental units, variation due to suspected nonuniformity can be partitioned from other sources of variation, including experimental error. For example, are all the ponds identical? Are some ponds leakier than others? Are they really all the same size? In a tank experiment, are some tanks shaded more than others? Do aquaria on the top shelf have the same water temperatures as those on the bottom shelf? Such examples of potential sources of error are endless. Depending on the which experimental units are different, treatment effects could be artificially enhanced or diminished.

Completely Randomized Design (CRD)

If the researcher believes that all experimental units are identical, then treatments should be allocated to experimental units in a completely random fashion. This type of treatment allocation is called a *completely randomized design* (CRD). With this design, experimental units are either presumed to be identical to each other or any actual differences are nonsystematic. That is, there is no way to characterize any differences. The representative model is the same given in Equation 11 above:

$$Y_i = \mu + \tau_i + \varepsilon_i \tag{13}$$

where μ = overall mean, τ = deviation due to treatment, and ε = residual, or deviation due to experimental error.

The simplest way to randomize treatment allocation is by using a random numbers table. Such tables are located in the appendices of most statistics books or can be computer-generated using statistical software. First, make a vertical column listing all treatments. Then identify each experimental unit with a number from 1 to however many are needed. For example, a feeding trial conducted in tanks that tests four treatments with four replicates per treatment requires 16 tanks. Number the tanks 1 through 16. Then randomly pick a point in the random number table and read the last two digits in the column. If that number is not between 01 and 16, then continue reading down. The first suitable number, say 08, will be the tank number for the first replicate of the first treatment as listed in the column of treatments. Then, continue down the random number column until another number between 01 and 16 (but not 08) appears, and that will be the tank number for the second replicate of the first treatment listed. Continue this process until every treatment and replicate has a different tank number. You have now completely randomized treatment allocation to your experimental tanks.

Some advantages and disadvantages of the CRD design, as well as other designs to follow, are discussed in the Data Analysis section using example analysis of variance (ANOVA) tables for each treatment allocation scheme. Such analytical considerations are essential at the experimental design stage, and the reader is strongly encouraged to familiarize him/herself with the section on Data Analysis prior to finalizing any research plans.

Randomized Complete Block Design (RCBD)

Complete randomization is ideal when appropriate but will result in unnecessary experimental error if there are systematic differences between experimental units. If experimental errors are not random but instead relate to some definable differences between experimental units, then this variation can be separated out.

The simplest case occurs when variability among experimental units can be blocked out. For example, you find that water in aquaria on the top shelf is warmer than in those on the bottom shelf; or you have 16 cement tanks in a 4 by 4 square, and a dirt road passes along one side (noise/cars make catfish very nervous, and dust has a lot of phosphorus in it); or a water source canal passes by some ponds but not others; or you are doing a fry nursing experiment in hapas (i.e., experimental units) with 15 hapas in each of three ponds, but you know the ponds are not identical. In each of these examples, variability among experimental units can be blocked out by allocating treatments according to a *randomized complete block design* (*RCBD*).

The main idea of a RCBD design is to *maximize* experimental unit variation *between* blocks, and *minimize* variation *within* blocks. So with the above examples, each shelf of aquaria would be a block; each row of tanks parallel to the road would be a block; ponds adjacent to the canal would be one block, while those away would be another block; and each pond (with 15 hapas in each pond) would be a block. The model for each of these designs would be the following:

$$Y_{ij} = \mu + \tau_i + \beta_j + \varepsilon_{ij} \tag{14}$$

where β = variation due to blocks.

The RCBD restricts randomization of treatment allocation in order to better isolate treatment variation while minimizing residual error. There are three requirements in setting up a RCBD. First, there must be the same number of blocks as there are replicates per treatment. Second, each treatment must be represented in each block. And third, treatment allocation within each block should be done randomly. In the hapa experiment, for example, there would be as many replicates per treatment as there are ponds. Each treatment would be represented once in each pond, and randomly allocated to a hapa within each pond.

An added benefit to using a RCBD is the extra hypothesis being tested with no additional effort. The RCBD tests the null hypothesis that the blocking (or the reason for blocking) has no significant effect on the response variable. Using the above examples, null hypotheses may include (1) that there is no road effect on catfish growth in tanks adjacent to a road or (2) that there is no difference in shrimp growth due to the horizontal placement of aquaria on shelves. In the former example, three blocks may be statistically the same, with only the block adjacent to the road causing a significant block effect. In the latter experiment, a significant block effect may reflect a horizontal gradient in water temperatures. For example, with four shelves and accurate temperature measurements, a dose–response experiment revealing a growth relationship with water temperature can be incorporated solely by allocating treatments to aquaria using a RCBD. The section on Data Analysis discusses how to analyze data, test hypotheses, and make all relevant comparisons.

To most effectively use the power of a RCBD, blocks must be selected carefully. Remember that the goal of blocking is to maximize differences between blocks while keeping experimental units within blocks as similar as possible. And always keep in mind that the reason for restricting randomization of treatment allocation is to identify and quantify sources of variation.

Latin Square Design

Whereas the RCBD identifies differences between experimental units across one dimension (e.g., between ponds in the hapa experiment or between shelves in the aquaria experiment), the *Latin Square design* attempts to quantify differences in experimental units when two sources of variation are suspected. The representative model looks like the following:

$$Y_{ijk} = \mu + \tau_i + r_j + c_k + \varepsilon_{ijk} \tag{15}$$

where r = variation due to rows, and c = variation due to columns.

The Latin Square design severely restricts randomization and therefore has fairly rigid design requirements. Typically, experimental units are laid out in the shape of a square (Figure 7), although variations are possible. The design requires that the number of replicates per treatment equals the number of rows and columns in the square. It is also essential that a treatment be represented only once in each row and column (see, for example, Figure 7). For analytical reasons (see the section

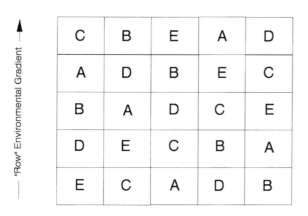

Figure 7. A schematic diagram of an experiment where five replicates of each of five treatments (A through E) were allocated to experimental units using a Latin Square design to account for suspected environmental gradients in two directions. Notice that each column and row is represented by exactly one replicate of each and every treatment.

on Data Analysis) a 4 by 4 Latin Square is the smallest size that should be considered. For practical reasons, a 6 by 6 (with 36 experimental units) is probably the largest Latin Square manageable.

Thankfully, the Latin Square design has limited utility in aquaculture. One possible situation may be in an experiment using aquaria as experimental units. If we assume that the aquaria are in a poorly air-conditioned room with large windows on a wall perpendicular to the aquaria, there could be a horizontal temperature gradient (i.e., row variation) and a vertical light gradient (i.e., column variation). A Latin Square design could identify variations caused by these gradients, thereby reducing experimental error. It might be easier, however, to move the aquaria to the wall opposite the windows, fix the air conditioner, and use a CRD.

The RCBD and Latin Square designs are complete block designs, where every treatment is represented in each block. There are also more complicated ways to block out suspected variation among experimental units utilizing incomplete block designs such as the ***split-plot design***, ***balanced incomplete blocks***, and ***partially balanced lattices*** (Steel and Torrie, 1980). Although discussion of these designs is beyond the scope of this chapter and best left for statistics textbooks, it is worth noting that the best designs are often the simplest, and increasing design complexity without increasing precision makes no sense.

Analysis of Covariance (ANCOVA)

The last design consideration applies when you suspect differences in your experimental units, but there is no way to systematically block out those differences. This is less of an issue when aquaria or cement tanks are used as experimental units but can be a serious concern when earthen ponds are used. The way to account for suspected nonuniform sources of variation is through an analytical technique called ***Analysis of Covariance*** (ANCOVA). The section on Data Analysis presents further aspects regarding ANCOVAs, but it is important to understand some basic concepts when designing experiments.

As an example, this technique was employed when effects of previous pond fertilizations (i.e., pond history) were shown to significantly affect tilapia production in a subsequent grow-out experiment (Knud-Hansen, 1992a). Since the ponds had received variable inputs (mainly chicken manure, urea, and TSP) during prior experiments, each pond had a unique fertilization history.

Analysis of variance of the CRD grow-out experiment revealed that treatments accounted for only 55% of the observed variation, while the residual accounted for about 45% of total variation. ANCOVA showed that over 90% of the residual variation, however, was due to variable chicken manure, urea, and TSP fertilizations from prior experiments within the previous two years. The greater the amount of previous fertilization, the greater the positive influence on algal productivity, and therefore tilapia productivity (see also the section on Data Analysis under residual analysis).

Based on the pond history example, it may be argued that it would be better to scrape the pond bottoms prior to each experiment in order to make the ponds more uniform. But this would be wrong for many reasons, not least of which would be the high cost and the practical impossibility of scraping pond sides and bottoms uniformly. In fact, anything but precision scraping would likely cause more harm by reducing the ability to identify and partition out sources of between-pond variability. Experimental units do *not* have to be uniform as long as the reasons for their lack of uniformity can be identified and tested. The message, therefore, is to identify possible differences among your experimental units *before* beginning the experiment and make the necessary measurements to be later used in ANCOVA.

SYSTEM CHARACTERIZATION AND MONITORING

Before beginning this section on system characterization and monitoring, it is useful to see where we are with respect to the experimental design process. So far we have conceptually:

1. Identified a research topic
2. Identified primary objectives of the research
3. Stated these objectives in the form of null hypotheses
4. Determined which type(s) of experiment(s) (e.g., unstructured, dose–response, etc.) are best suited to test stated hypotheses
5. For each experiment determined the scope, treatments, treatment levels, and number of replicates per treatment most appropriate for testing stated hypotheses
6. If more than one experiment is necessary, arranged in logical systematic order the appropriate sequence for conducting several experiments
7. Identified the most appropriate allocation of treatments to experimental units (e.g., CRD, RCBD, and Latin Square)
8. Identified any differences among experimental units that may affect experimental results and that can be measured before beginning the experiment

The last remaining task is to identify which variables (i.e., system characteristics) should be measured during the experiment and to determine how to go about measuring these variables.

Identification of Variables

Everyone agrees that you cannot measure everything, so there must be some sort of limitation on both what you measure and how frequently you measure it. Factors that limit the scope of variables include time, money, effort, and utility of data collected. As a practical matter, utility of data affects more the decision of which variables should be measured, while time, money, and effort affect more the determination of measurement frequency.

There are three types of variables generally measured in an experiment. The first type represents those variables that are directly part of your null hypothesis. If the null hypothesis states that brand X feed has no greater beneficial effect on fish growth than brand Y, then you must measure fish growth in order to test the hypothesis. All variables essential for testing stated null hypotheses must be measured. These will be the response variables in unstructured experiments, and the relationship variables in dose–response experiments. For example, in a fertilization experiment that also includes the null hypothesis that algal productivity has no effect on fish yield, both algal productivity and fish yield must be measured.

The second type includes those variables that are not essential for testing null hypotheses but may reflect other sources of variation affecting response variables. Water quality measurements most frequently fall into this category. Which water quality variables should be measured depends on the nature of experimental units and how the researcher intends to analyze resulting data. With outdoor tank and pond studies, for example, variables that affect both the culture species and algal growth should be included. Culture species are generally affected by temperature, dissolved oxygen (DO) concentrations, un-ionized ammonia concentrations, and turbidity (Meade, 1985). Algal productivity is affected by temperature, light availability, and availability of inorganic N, P, and C (Knud-Hansen and Batterson, 1994). Monitoring the following variables would provide information used to evaluate possible effects on response variables: water temperature, DO, total ammonia-N, nitrate-nitrite-N, soluble reactive P, pH, Secchi depth, total and ash-free suspended solids, and total alkalinity. Chlorophyll *a* (i.e., primary production) would not necessarily be included because its relationship with the *rate* of algal growth (i.e., primary productivity) is extremely tenuous. What affects the *rate* of growth for detrivorous species like Nile tilapia is primary productivity, not primary production (Knud-Hansen et al., 1993).

Variables that are used for broader purposes beyond the immediate experiment represent the third type. Such data provide baseline information useful for comparing separate experiments. For example, weather measurements may not provide any insight into variations observed in one experiment but can be extremely useful when making comparisons between seasons or climatic regions.

It is essential to understand how you intend to use and analyze the data being collected. Measuring Kjeldahl N simply because you have the equipment available is a waste of your time, money, and energy if all you intend to do with the data is create summary tables. Carefully consider all reasonable variables and eliminate those that do not help test your hypotheses and meet your experimental objectives. If the Kjeldahl digester needs to be used, then design an experiment where those data have importance.

Selection of Methods

Once you have identified which variables should be measured to test your hypotheses, the next issue is how should they be measured. There are a number of method manuals and books offering a variety of ways to get the same information. But the fact that these methods are in print does not necessarily make them the best or even necessarily adequate. Standard Methods for the Analysis of Water and Wastewater (APHA, 1985) is one of the few that present methods that have been actually comparatively tested, but they too might not be suitable for your particular needs.

It is worth the effort for an aquaculture research facility or program to select methods based on rational comparisons. There are five factors that should influence the choice of one method over another: (1) precision or sensitivity, (2) accuracy, (3) reproducibility, (4) ease of operation, and (5) cost. For field and laboratory instruments, decisions often can be made by comparing hardware specifications. An instrument's reputation regarding ease of operation and repair history can also be quite useful. Inherent in such decisions is an understanding of how the data can foreseeably be used, because accuracy and precision must often be balanced against cost. For example, is a digital pH meter accurate to 0.001 pH unit really cost effective? Is that level of sensitivity really important?

With water chemistry such comparisons are essential, not only to understand the limits of a particular method but to ensure that the most rational method is being used. As an example of how the above decision criteria can be applied, several years ago the aquaculture laboratory at the Asian Institute of Technology compared the nitrate-nitrate analysis using the cadmium reduction column

(APHA, 1985) with another method using hydrazine reduction (Kamphake et al., 1967). To determine precision, the average slope of five standard curves (linear relationship between nitrate concentration and spectrophotometric absorption; APHA, 1985) for each method was compared. Hydrazine was slightly more precise (i.e., had a higher slope), but both methods were adequate for aquaculture investigations. For accuracy, five internal standards per method were used. Internal standards gave the percent recovery of known nitrate and nitrite spikes added to pond water (Knud-Hansen, 1992b). Cadmium reduction gave slightly higher percent recoveries than hydrazine reduction. Reproducibility was based on standard deviations of five replicate samples, and there was little difference between methods. Ease of operation favored hydrazine reduction. A liquid reagent was much easier to handle than dealing with the vagaries of the cadmium reduction column. Differences in cost were not determinative. What tipped the scales in favor of the cadmium column method, however, was the current laboratory expertise in the method, the fact that the laboratory had been using the cadmium column method for years, and the lack of significant analytical improvement by switching to the hydrazine method. The important point here is that a rational decision was made using specific comparable criteria.

Aquaculture laboratories should also be on the look-out for new methods that offer significant improvement in one or more of the above decision criteria. A good example is with chlorophyll *a* measurements. A review of the aquaculture literature indicates continued widespread use of the acetone extraction method (Strickland and Parsons, 1968), whereas a review of the limnological and oceanographic literature suggests that most laboratories have rejected that method for others that are much easier and less prone to experimental error (e.g., Holm-Hansen and Riemann, 1978).

Finally, field methods should be tested before blind adoption. For example, primary productivity is commonly measured by *in situ* incubation of pond water suspended in light and dark bottles at multiple depths (Boyd, 1990). This method may be appropriate in oceans and lakes with photic zones of several meters or greater, but it also has been commonly used in aquaculture studies in ponds with Secchi depths of < 15 cm. Data are reported as if resulting numbers actually represented pond algal productivity (severe vertical variations and unmeasured floating algal mats notwithstanding). There may be no practical way to measure true algal productivity in shallow, highly productive aquaculture ponds. Nevertheless, it is absolutely essential that researchers understand the analytical limitations of all methodologies employed in their research.

Representativeness of Samples

Both sampling and field measurements are founded on the implicit assumption that they represent larger populations. Fish samples characterize the pond's population of fish in the same way water and sediment samples characterize the pond itself. In order to obtain representative samples, however, the researcher must appreciate the spatial and temporal heterogeneity of aquaculture systems.

Studies conducted in aquaria and recirculating systems present the fewest problems regarding collecting representative samples. Water in aquaria can be easily mixed, and recirculating systems by their nature are generally more homogenous.

Perhaps the greatest concern is in earthen ponds, where so many dynamic processes occur at once. There are diel temperature and oxygen changes, daily thermal stratifications, and variable weather-induced effects just to name a few. The following discussion does not pretend to be exhaustive but only alerts researchers to the more common considerations regarding biological, water, and sediment sampling.

The main problems dealing with representative biological sampling concern collecting fish, zooplankton, and phytoplankton samples. Fish sampling can be problematic, depending on the method of sampling. Seining can give a biased representation if net avoidance is related to size or species. One way to test your method is to sample each pond the day before draining and completely harvesting the ponds. A comparison of sample means and distributions with each pond's population mean and distribution should indicate the adequacy of samples (see Hopkins and Yakupitiyage, 1991). Collecting representative zooplankton samples is particularly difficult because of their diel vertical migrations, the inherent patchiness of their distributions, and their ability to avoid nets and traps (Wetzel and Likens, 1979). The primary difficulty with representative phytoplankton sampling is how to quantify surface mats of blue-green algae. Unless the experimental hypothesis focuses on zooplankton and/or phytoplankton population dynamics, sampling (and subsequent analyses) should probably be restricted to a qualitative rather than quantitative level.

Water samples collected for chemical analyses should also be checked for representativeness. A 200-m^2 pond 1-m deep contains 2×10^5 L of water. How well does a one liter sample represent that pond (see Likens, 1985)? Spatial heterogeneity occurs both areally and vertically with depth. Afternoon soluble ammonia and phosphorus concentrations can vary several mg L^{-1} in highly productive fish ponds. Thermal destratification (i.e., complete pond mixing) can occur nightly or after an afternoon thunderstorm. Based on these and other similar considerations, researchers must select water sampling methods and times of collection carefully to provide the necessary representative data.

Water samplers that collect an integrated vertical sample of the entire water column are useful for whole pond nutrient budgets but are entirely inadequate for studying dynamics of thermally stratified ponds. Collecting and mixing water samples from different parts of the pond may help reduce problems associated with areal heterogeneity, but it would be extremely useful to determine the extent of variability through at least one synoptic sample of a pond (i.e., sampling many locations of the pond "simultaneously"). Timing of sampling should be based on pond dynamics, not convenience. If anoxia is a concern, DO samples should be taken at pre-dawn and not some fixed hour regardless of sunrise. If un-ionized ammonia is a concern, temperature, pH, and total ammonia should be measured several times during the day and particularly at mid-afternoon when the pH is the highest.

Trying to collect representative sediment samples may be the most difficult of all. Spatial heterogeneity over the pond bottom must be determined before any meaningful analysis can be made. Just collecting ten or so samples and mixing them without knowing the variability of these samples can readily lead to false conclusions. Synoptic sampling may be required in three or four ponds, but the effort is necessary to confirm that mixing a given number of scattered samples does in fact represent well the pond bottom. An additional problem is determining how deep the samples should be. Decisions regarding both the number of samples per pond and the depth of the samples should be rationally based after testing necessary assumptions. Sediment analyses tend to be time-consuming and costly, so it is important that the samples truly represent what the researcher believes they represent.

One last point on representativeness regards the concept of samples vs. subsamples. Individual samples are used to infer characteristics of a population. A water sample may represent the entire pond volume or just the upper photic zone, depending on how the researcher defines the population. Collecting and analyzing many samples acquired at the same time characterizes that population, whether it be the pond's water or sediments. These are replicate samples taken from a single population. But taking one sample and splitting it into several parts is called taking subsamples. Subsamples are good for evaluating variability of analyses (see discussion above on selection of methods) and as a backup reserve if/when rerunning an analysis becomes necessary. Subsamples do not, however, provide any indication of pond water or sediment variability.

Sample Size

The question of how large a sample size must be collected to adequately represent a population comes up often in pond studies. In particular, these concerns arise mainly with respect to sampling culture organisms and pond water. Sampling culture organisms in studies conducted in aquaria or tanks usually involves collecting entire populations (of each tank, etc.) for routine monitoring.

When sampling larger populations in pond and large tank studies, however, an adequate sample size must be determined.

One fallacy often perpetuated in aquaculture is that the sample size of culture organisms should be 10% of the total population. Recall the discussion in the section on Basic Concepts regarding the standard error of the mean (Equation 6) and confidence intervals. The standard error of the mean is based on the standard deviation of sample observations and the number of observations made. Confidence intervals about the mean are based on the standard error of the mean, the sample size, and the level of confidence (e.g., 95%) desired by the researcher. Nowhere does *population* size play a role in determining these statistics. Sample size simply does *not* depend on population size.

The sample size does depend on how precise (i.e., how small a confidence interval about the sample mean) you want or, in other words, the maximum acceptable difference at a given probability between the sample mean and the true mean. To determine the appropriate sample size, first take a preliminary sample, and use the information in the following standard equation:

$$t_\infty = \frac{(\bar{x} - \mu)}{s / \sqrt{n}} \tag{16}$$

where: t_∞ = tabular t value at ∞ degrees of freedom at the desired probability (level of confidence)

$\bar{x} - \mu$ = maximum acceptable difference between a sample mean and true mean at a desired probability

s = standard deviation of preliminary sample

n = number of observations in preliminary sample

$s / \sqrt{n}$ = standard error of preliminary sample mean

As an example, assume a farmer needed to know the mean weight of about 20,000 fish raised in a pond. He also wants to be 95% confident that the sample mean weight is no more than 10 g away from the true mean. A preliminary random sample of 30 fish gave a standard deviation of 42.0 g. By restructuring Equation 16 and noting from the Student's t Table in his handy statistics book that t_∞ at $P = 0.05$ equals 1.96, the farmer does the following calculation:

$$n = \left[\frac{ts}{(\bar{x} - \mu)}\right]^2 = \left[\frac{(1.96)(42)}{10}\right]^2 = 67.8 \approx 68 \text{ fish}$$

Therefore, the farmer must randomly sample 38 more fish to attain a sample mean that he can be 95% confident is within 10 g of the true mean of all 20,000 fish. Again note that population size, as well as the sample mean value, was unimportant in the sample size determination.

In practice, however, most researchers give little thought to how well the sample mean should represent the true mean. It is obvious that the appropriate sample size will differ depending upon population variability. But as a general rule, good random samples greater than 30 to 40 organisms per experimental unit are rarely cost effective. To be sure, use Equation 16.

Water samples collected for chemical analyses should be as large as manageable for two reasons. First, the greater the sample volume, the more likely it will reasonably represent the entire volume. Multiple subsamples combined in a mixing vessel helps reach this first objective. Second, the greater the sample size, the less relative effect any contamination (e.g., from handling, dirty glassware, etc.) may have on the sample. These two general principles help combat the twin problems of system heterogeneity and sample contamination.

Frequency of Sampling

The first criterion for determining sampling frequency is to identify the purpose of sampling in the first place. You want to strike a balance between sampling frequently enough to ensure the utility of the data, and yet not so frequently as to become a relative waste of time and money. With culture organisms, increased handling and labor costs often define the maximum reasonable frequency. With water sampling, the decision becomes more difficult.

For example, assume you are conducting a 5-month grow-out experiment and monitor pond water chemistry to help understand dynamic relationships that hypothetically affect yields. Monthly sampling may be sufficient, but changes in weather, etc. may cause short-term effects, thus increasing overall variability. Because of such variability, any real trends over time may or may not be seen. The conscientious scientist would find any conclusion based only on monthly samples unconvincing. At the other extreme are data loggers, which can amass mountains of data at the flip of a switch. They should be used only where cost effective, that is, where all the information they provide has planned utility. Data loggers that make many measurements, for example, every half hour, are best used for short-term (< 2 weeks) experiments (e.g., Green and Teichert-Coddington, 1991). Unless there are specific analytical objectives for mountains of data, they are most likely wasted when used for monitoring long-term experiments. For a 5-month grow-out experiment, sampling every 1 to 2 weeks would probably be adequate for the purposes of identifying other sources of variability. Or, as an alternative, sample daily for 1 week each month.

Again, the real issue for the researcher is knowing the reasons for monitoring chosen variables. Make sure that you sample enough to allow the data to achieve analytical objectives. Most researchers have a sense of how often they need to sample. But it is well worth the time and effort to think through each variable while designing the experimental protocol. When analyzing data from one experiment, researchers should also ask themselves whether or not sampling frequencies were adequate or sufficient. Addressing such questions will only improve future research.

DATA ANALYSIS

If by chance you jumped to this section in the hope of learning how to analyze your recently collected data set, sorry. Shortcuts do not work. To benefit from this section, you must get here by starting at the beginning of the chapter. Those who started at the Introduction section should now well appreciate that data analysis begins with the first hypothesis formulation. By the time you have completed the experimental design, you already know why, what, and conceptually how to analyze your data.

From your selection of treatments, experimental design, and treatment allocation to experimental units you already understand the basics of how you will analyze your results. This section focuses more on post-experiment considerations, although example calculations will be kept to a bare minimum. As indicated in the Introduction, this chapter minimizes recipes common to most statistics books. Those who wish to do ANOVAs and regressions by hand can choose one of those books for guidance. It is assumed here that the majority of researchers rely on computers to run their statistical calculations. A primary objective of this section is to help the researcher know what she is feeding her computer, and to better understand what the computer gives back in return.

Data Reduction

Conceptually, there are two types of data reduction, distinguished by their different purposes. The first purpose is to summarize data descriptively using means, standard errors (see the section on Basic Concepts) and sometimes *ranges*. Ranges are just the differences between minimum and maximum observed values. Ranges have limited analytical utility since they represent extreme values, and tend to increase with increasing number of observations (Steel and Torrie, 1980).

Another way of summarizing data is through *frequency tables* and *histograms*. Frequency tables are commonly used to summarize size classes of culture organisms or any other types of data which can be expressed in discrete form (i.e., discrete counts or continuous data (e.g., lengths, weights, etc.) expressed in discrete form). Histograms are bar-graphs used to illustrate frequency distributions (e.g., Figure 8). To adequately present a frequency distribution, you need enough classes to show differences in frequencies, but not too many or the histogram will appear flat. Usually about 10 to 15 classes adequately represent underlying distributions (compare Figure 8b with Figures 8a and 8c).

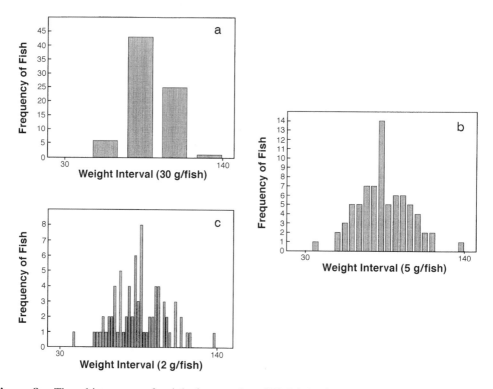

Figure 8.   Three histograms of weight frequencies of 75 fish (weights ranged from 34 to 138 g/fish) using weight intervals of 30 g/fish (a), 5 g/fish (b), and 2 g/fish (c).

The second purpose of data reduction, particularly for relatively large data sets, is to facilitate data analysis. Data may require some form of data reduction or consolidation to become more analytically manageable. As a general rule, reduce data *within* an experimental unit and not between experimental units. For example, imagine that you conducted a five-treatment grow-out experiment in tanks with three replicates per treatment. The experiment lasted 10 weeks, and water chemistry was monitored weekly. Assuming alkalinity did not show any trend over time but did vary with treatments, you could then average the ten measurements to give one mean value per tank (i.e., one value per experimental unit). Treatment means would then be based on the three tank values per treatment (i.e., $n = 3$). Do not average the 30 alkalinity values per treatment in order to get a treatment mean; otherwise, you will lose all information regarding within-treatment variability (i.e., variability among experimental units with the same treatment).

Some data reduction may be planned in the experimental design, but always make sure that no important information or insightful data analysis is lost in the process. Reduce data for a purpose, and understand that purpose and consequential implications before reducing.

Data Transformations

An essential requirement for employing *t*-tests, ANOVAs, regressions, correlations, and any other parametric analysis (e.g., using the statistics to estimate the parameters of true mean and variance, see Basic Concepts section) is that observations must have distributions not significantly different from normality. There are statistical tests, such as the Chi-Square test, which test the null hypothesis that your data set is not significantly different from normality. If the null hypothesis is rejected, non-normal data sets can often be normalized through transformations.

A few basic transformations are commonly used for certain types of data that are already known or assumed not to be normal. The *square root transformation* ($\sqrt{x}$) normalizes counting data, which typically exhibit a Poisson distribution. If counts are low and include 0, $\sqrt{(x + 0.5)}$ can be used. If counts are all much greater than 100, then transformation is probably not necessary before parametric statistical analysis.

The square root transformation also normalizes binomial percentage data that are generally <30% or >70%. Percent mortalities and survivals often fall in this range and therefore should be transformed before running ANOVAs. Note that the entire data set would be transformed, not just the ones falling within the specified range. If percentage data range widely from near 0% up to around 100%, the *arcsine of the square root* (arcsine $\sqrt{x}$) is used. On the other hand, if percentages range between around 30 to 70%, data transformations are generally unnecessary.

The *logarithmic transformation* is used when the standard deviation (or variance) is proportional to the mean. A classic example are fish grow-out trials with widely differing results. A treatment mean of 80 g/fish may have a standard deviation of 15 g, whereas a treatment mean of 500 g/fish may have a standard deviation of 100 g. As the mean gets bigger, so does the standard deviation. The Bartlett's test can be used to test the null hypothesis that treatment variances are not significantly different from each other (Steel and Torrie, 1980). The logarithmic transformation should be used with positive numbers only. If there are numbers less than 10, data transformation should be with $\log(x + 1)$ instead of $\log(x)$. Logarithms can be at any base (e.g., base 10 or the natural log).

It is a wise practice always to check data for normality before examining treatment comparisons and relationships. Virtually all statistical software packages have this capability, and with a computer it only takes a few seconds. If a particular data set is not normally distributed, try using the square root and logarithmic transformations, and retest the data for normality.

When data transformation is necessary, you must use the transformed data for making treatment comparisons, confidence intervals, etc., because all these analyses assume that the data have normal distributions. Therefore, do not change variances and standard deviations back to the original scale. Only treatment means and ranges should be given/reported in the original scale.

Comparisons of Means and ANOVA

This relatively small subsection covers what many researchers think of when they hear the words, "data analysis." Although an oversimplification, comparisons and relationships are often at the heart of hypothesis testing. All the analyses presented here are parametric and assume normality of the data. Other important assumptions are best left to the statisticians to describe, and consultation of your favorite statistics book is highly recommended.

The basic idea of treatment comparisons is to see whether treatment means are significantly different from each other. When the experiment consists of only two treatments, the *t*-test is used to test the hypothesis that one treatment mean is not significantly greater than the other. The *t* statistic is calculated by the following equation:

$$t = \frac{\text{difference between means}}{\text{standard error of difference between means}} \qquad (17)$$

The standard error of the difference between means is based on the weighted mean of the two treatment variances, also called the **pooled variance** (s_p^2).

$$s_p^2 = \left(\frac{s_1^2}{n_1} + \frac{s_2^2}{n_2} \right)^{1/2} \tag{18}$$

The pooled variance allows for comparisons of treatments with different sample sizes (n_1 and n_2). Conceptually, consider two ponds (numbered 1 and 2) each containing thousands of fish. You sample both ponds and get a difference between mean fish weight between the ponds. You sample both ponds several more times and get more differences between mean fish weights. The t-test calculates the standard error of the difference between means using the sample variances for each pond. In practice, however, each pond is usually sampled just once. The probability that the true means for ponds 1 and 2 are significantly different from each other is based on the ratio of the observed difference between sample means from the two ponds and the standard error of the difference between means (Equation 17). In other words, the difference between mean fish weights in ponds 1 and 2 is more likely to be significant as the variability of fish weights within each pond gets smaller.

The computed t statistic is compared to t values in the t-Table at $n_1 + n_2 - 2$ degrees of freedom. Probabilities are given horizontally in the Table. The probability range (e.g. $0.01 < P < 0.05$) for a given calculated t value is found by moving horizontally in the table at the appropriate degree of freedom. These probabilities, as well as those given by statistical software, represent the probability that there was no true difference between the two means, i.e., accepting the null hypothesis.

Experiments with more than two treatment means use **analysis of variance (ANOVA)** to test the null hypothesis that no two means are significantly different from each other. Simply stated, this hypothesis is tested by comparing the variation *between* treatments with the variation *among* treatments. A proportionally large between-treatment variation in relation to within-treatment variation (i.e., experimental error) results in a small probability that observed treatment differences happened by chance. The following discussion will examine ANOVAs for the experimental designs described in the section, Experimental Units.

Completely Randomized Design (CRD)

Table 1 provides the basic ANOVA table for a CRD experiment. There are only two sources of error, from treatments (between-treatment variation) and from experimental error (within-treatment variation). The sum of squares represents total variation for each source, and the mean square represents the amount of variation per degree of freedom. The F statistic is the ratio between mean square treatment (MST) and mean square error (MSE, which also equals the pooled variance, s_p^2).

Table 1   Example ANOVA Table of a Completely Randomized Designed (CRD) Experiment

Sources of variation	Sum of squares	Degrees of freedom	Mean squares	F
Treatments	SST	k − 1	$\dfrac{SST}{k-1} = MST$	$\dfrac{MST}{MSE}$
Residual error	SSE	k(n − 1)	$\dfrac{SSE}{k(n-1)} = MSE = s_p^2$	
Total	SS	kn − 1		

Note: k = number of treatments; n = number of replicates per treatment.

This calculated F statistic is compared to a table of F values at (k–1) degrees of freedom for the numerator and k (n–1) for the denominator. At given degrees of freedom, the greater the F value (i.e., ratio between MST and MSE), the smaller the probability that the difference between at least two of the treatment means happened by chance.

Table 2 gives an abbreviated ANOVA for a hypothetical CRD experiment with five treatments and five replicates per treatment. Note that 20 out of 24 degrees of freedom are for the residual. As indicated by Table 1, the more replicates per treatment, the more degrees of freedom for error. If there were only three replicates per treatment, 10 out of the available 14 degrees of freedom would be for the residual; with the same amount of experimental variation, MSE would be twice as large (and F half the value) than with five replicates and 20 degrees of freedom for residual. The more degrees of freedom for error, the smaller the MSE and the larger the MST/MSE ratio (i.e., F statistic). This observation highlights the importance of both reducing experimental error and maximizing the number of degrees of freedom for error. The high proportion of degrees of freedom for the residual is an advantage of the CRD, and so the CRD is better suited for smaller experiments with fewer experimental units. The main disadvantage of the CRD occurs when experimental units are not homogenous, thereby increasing experimental error.

Table 2   Example ANOVA Table of a Completely Randomized Designed (CRD) Experiment with Five Treatments and Five Replicates Per Treatment

Sources of variation	Sum of squares	Degrees of freedom	Mean squares	F
Treatments	SST	5 – 1 = 4	$\dfrac{SST}{4} = MST$	$\dfrac{MST}{MSE}$
Residual error	SSE	5(5 – 1) = 20	$\dfrac{SSE}{20} = MSE = s_p{}^2$	
Total	SS	(5)(5) – 1 = 24		

Randomized Complete Block Design (RCBD)

As discussed in the section Experimental Units, when the researcher suspects systematic differences in experimental units, a RCBD design is often chosen to account for those differences. Table 3 provides the generalized ANOVA formulation, while Table 4 illustrates an example ANOVA with five treatments and five replicates per treatment, with each replication representing an individual block. Notice that the example ANOVA for the RCBD has the same total number and treatment degrees of freedom as the CRD. The only difference is that four degrees of freedom have been moved from the residual in order to test a second null hypothesis that there are no significant differences between blocks. The F value generated by the block analysis (MSB/MSE) is analyzed in the same way as that for treatments, and compared to tabular F values (and associated probabilities) at (b–1) degrees of freedom for the numerator and (k–1)(b–1) for the denominator (Table 3).

Although the RCBD experiment restricts randomization and reduces the number of degrees of freedom for residual error (e.g., from 20 to 16, compare Tables 2 and 4), the variation due to blocks is removed from residual variation (SSE). A large block variation will reduce the SSE so that even with fewer degrees of freedom for error, the MSE is smaller than it would have been without the block design. Not only has the RCBD experiment tested a second hypothesis without any additional effort, but the smaller MSE results in a larger F value for treatment making that analysis more sensitive.

Table 3   Example ANOVA Table of a Randomized Complete Block Design (RCBD) Experiment

Sources of variation	Sum of squares	Degrees of freedom	Mean squares	F
Treatments	SST	k − 1	$\dfrac{SST}{k-1} = MST$	$\dfrac{MST}{MSE}$
Blocks	SSB	b − 1	$\dfrac{SSB}{b-1} = MSB$	$\dfrac{MSB}{MSE}$
Residual error	SSE	(b − 1)(k − 1)	$\dfrac{SSE}{(b-1)(k-1)} = MSE = s_p^{\,2}$	
Total	SS	bk − 1		

Note: k = number of treatments and b = number of blocks.

If you do not find a significant block effect, reanalyze the data as if it were a CRD experiment (in effect adding the block variation (SSB) and degrees of freedom to the residual variation (SSE) and degrees of freedom). Report, however, that the original design was a RCBD and there was no significant block effect with that particular response variable (i.e., you accepted the null hypothesis for blocks). If you had blocked against a measurable variable such as temperature or light, the lack of a block effect could nevertheless be scientifically significant. If you had blocked based on the physical location of tanks or ponds, a lack of a block effect would provide further insight into the nature of your research system. Remember also that your blocks may affect one response variable and not another, so always first analyze your data as a RCBD.

Latin Square Design

Whereas the RCBD tests one block hypothesis, the Latin Square design tests two block hypotheses. As discussed in the section Experimental Units, this design severely restricts randomization and has limited application in aquaculture. Tables 5 and 6 provide the generalized ANOVA and an example ANOVA with five treatments and five replicates, respectively. With the 5 by 5 Latin Square, only 12 of the 24 degrees of freedom are for residual error (Table 5). A 4 by 4 and 3 by 3

Table 4   Example ANOVA Table of a Randomized Complete Block Design (RCBD) Experiment with Five Treatments and Five Blocks (i.e., Replicates per Treatment)

Sources of variation	Sum of squares	Degrees of freedom	Mean squares	F
Treatments	SST	5 − 1 = 4	$\dfrac{SST}{4} = MST$	$\dfrac{MST}{MSE}$
Blocks	SSB	5 − 1 = 4	$\dfrac{SSB}{4} = MSB$	$\dfrac{MSB}{MSE}$
Residual error	SSE	(5 − 1)(5 − 1) = 16	$\dfrac{SSE}{16} = MSE = s_p^{\,2}$	
Total	SS	(5)(5) − 1 = 24		

Table 5   Example ANOVA Table of a Latin Square Designed Experiment

Sources of variation	Sum of squares	Degrees of freedom	Mean squares	F
Treatments	SST	$k - 1$	$\dfrac{SST}{k-1} = MST$	$\dfrac{MST}{MSE}$
Rows	SSR	$r - 1$	$\dfrac{SSR}{r-1} = MSR$	$\dfrac{MSR}{MSE}$
Columns	SSC	$c - 1$	$\dfrac{SSC}{c-1} = MSC$	$\dfrac{MSC}{MSE}$
Residual error	SSE	$(k-1)(k-2)$	$\dfrac{SSE}{(k-1)(k-2)} = MSE = s_p^{\,2}$	
Total	SS	$k^2 - 1$		

Note: k = number of treatments, r = number of rows, and c = number of columns.

Latin Square has only six and two degrees of freedom for residual error, respectively. Both latter totals are too few to get a meaningful MSE, and so a 5 by 5 Latin Square is probably the smallest size statistically manageable. On the other hand, a 4 by 4 Latin Square could be finessed by analyzing it as a RCBD first using rows as the block and then reanalyzing the data using columns as the block. This technique would increase degrees of freedom for experimental error; and if the block effects from rows or columns are not significant, the data can be analyzed as either a RCBD or even as a CRD (when neither rows nor columns show significant block effects).

Factorial Experiments

If we go back to the CRD ANOVA (Table 1), there are only two sources of variation, treatments and residual error. The main purpose of RCBD and Latin Square designs is to identify and partition

Table 6   Example ANOVA Table of a 5 by 5 Latin Square Designed Experiment with Five Treatments, Five Rows and Five Columns (i.e., Five Replicates Per Treatment

Sources of variation	Sum of squares	Degrees of freedom	Mean squares	F
Treatments	SST	$5 - 1 = 4$	$\dfrac{SSE}{4} = MST$	$\dfrac{MST}{MSE}$
Rows	SSR	$5 - 1 = 4$	$\dfrac{SSR}{4} = MSR$	$\dfrac{MSR}{MSE}$
Columns	SSC	$5 - 1 = 4$	$\dfrac{SSC}{4} = MSC$	$\dfrac{MSC}{MSE}$
Residual error	SSE	$(5-1)(5-2) = 12$	$\dfrac{SSE}{12} = MSE = s_p^{\,2}$	
Total	SS	$(5)(5) - 1 = 24$		

out sources of variation from the residual error. The main purpose of factorial experiments, however, is to identify and partition out sources of variation (additive and interactions) from two or more factors. The former designs concern location of experimental units, while factorial experiments describe treatment combinations. So depending on the layout of experimental units, the researcher may allocate factorially determined treatments in either a CRD or a RCBD. (Although theoretically possible, a Latin Square design would be extremely impractical.)

The simplest factorial is the 2 by 2 in a CRD (Figure 3). There are a total of four treatment combinations (2 factors with 2 levels for each factor). The four treatments use three degrees of freedom, which are partitioned out one for each factor, and one for their interaction (Table 7). The example given in Figure 3 illustrates two factors (nitrogen (N) and phosphorus (P) input) and two levels of each factor. The result is a four-treatment experiment examining individual effects of N and P individually and together.

Table 7   Example ANOVA Table of a Two Factor Factorially Designed Experiment in a CRD

Sources of variation	Sum of squares	Degrees of freedom	Mean squares	F
Treatments				
A	SS(A)	$a - 1$	$\dfrac{SS(A)}{a-1} = MS(A)$	$\dfrac{MS(AC}{MSE}$
B	SS(B)	$b - 1$	$\dfrac{SS(B)}{b-1} = MS(B)$	$\dfrac{MS(B)}{MSE}$
AB	SS(AB)	$(a-1)(b-1)$	$\dfrac{SSB}{b-1} = MS(AB)$	$\dfrac{MS(AB)}{MSE}$
Residual error	SSE	$ab(n-1)$	$\dfrac{SSE}{ab(n-1)} = MSE = s_p^2$	
Total	SS	$abn - 1$		

Note: a = number of levels of treatment A, b = number of levels of treatment B, and n = number of replicates per treatment combination.

Factorial experiments become progressively more complicated as the researcher adds more factors and levels to an experimental design. For example, Figure 4 illustrates schematically the shrimp feed experiment first discussed in the section, Treatments. In that experiment there were three factors (shrimp variety, and protein and lipid concentrations in shrimp feed), with two varieties of shrimp, three concentrations of lipid, and four concentrations of protein, resulting in a 2 by 3 by 4 factorial design with a total of 24 treatments. Table 8 gives the generalized ANOVA for this experiment. Assuming three replicates per treatment with each replicate in a separate block, Table 9 gives the ANOVA for the resulting RCBD factorially designed experiment. Notice that the ANOVA analyzes interactions between all factor combinations. Remember also that this ANOVA only represents treatment analysis and does not include individual relationship analyses (e.g., the relationship of protein concentrations in feed and shrimp yield at one level of lipid concentration) as previously listed. The following subsection discusses how to proceed with relationship analyses.

Table 8   Example ANOVA Table of a Three Factor Factorially Designed Experiment in a RCBD

Sources of variation	Sum of squares	Degrees of freedom	Mean squares	F
Treatments				
A	SS(A)	$a - 1$	$\dfrac{SS(A)}{a - 1} = MS(A)$	$\dfrac{MS(AC)}{MSE}$
B	SS(B)	$b - 1$	$\dfrac{SS(B)}{b - 1} = MS(B)$	$\dfrac{MS(B)}{MSE}$
C	SS(C)	$c - 1$	$\dfrac{SS(C)}{c - 1} = MS(C)$	$\dfrac{MS(C)}{MSE}$
AB	SS(AB)	$(a - 1)(b - 1)$	$\dfrac{SS(AB)}{(a - 1)(b - 1)} = MS(AB)$	$\dfrac{MS(AB)}{MSE}$
AC	SS(AC)	$(a - 1)(c - 1)$	$\dfrac{SS(AC)}{(a - 1)(c - 1)} = MS(AC)$	$\dfrac{MS(AC)}{MSE}$
BC	SS(BC)	$(b - 1)(c - 1)$	$\dfrac{SS(BC)}{(b - 1)(c - 1)} = MS(BC)$	$\dfrac{MS(BC)}{MSE}$
ABC	SS(ABC)	$(a - 1)(b - 1)$	$\dfrac{SSB}{b - 1} = MS(AB)$	$\dfrac{MS(AB)}{MSE}$
Blocks	SSB	$n - 1$	$\dfrac{SSB}{n - 1} = MSB$	$\dfrac{MSB}{MSE}$
Residual error	SSE	$(abc - 1)(n - 1)$	$\dfrac{SSE}{(abc - 1)(n - 1)} = MSE = s_p{}^2$	
Total	SS	$abcn - 1$		

Note: a = number of levels of treatment A, b = number of levels of treatment B, c = number of levels of treatment C, and n = number of blocks (replicates per treatment combination).

Relationships

Evaluating relationships between variables represents a key component of most data analyses and the primary objective of structured experiments. Similar to many other topics presented in this chapter, however, one does not have to look too closely in the aquaculture literature to find basic conceptual and analytical errors regarding such analyses. But because of the vast complexity of such analyses, the following discussion is necessarily limited in scope and aims primarily to equip the reader with a more solid foundation of elementary regression and correlation analyses. This foundation will hopefully enable the reader to explore more knowledgeably (i.e., through better understanding of purposes, assumptions, and limitations) advanced analytical techniques (e.g., path analysis, canonical analysis, etc.) found in some standard and specialty statistics books.

Table 9   Example ANOVA Table of a Three Factor Factorially Designed Experiment in a RCBD

Sources of variation	Sum of squares	Degrees of freedom	Mean squares	F
Treatments				
Shrimp	SS(A)	$2 - 1 = 1$	$\dfrac{SS(A)}{1} = MS(A)$	$\dfrac{MS(AC}{MSE}$
Lipid	SS(B)	$3 - 1 = 2$	$\dfrac{SS(B)}{2} = MS(B)$	$\dfrac{MS(B)}{MSE}$
Protein	SS(C)	$4 - 1 = 3$	$\dfrac{SS(C)}{3} = MS(C)$	$\dfrac{MS(C)}{MSE}$
Shrimp × lipid	SS(AB)	$(2 - 1)(3 - 1) = 2$	$\dfrac{SS(AB)}{2} = MS(AB)$	$\dfrac{MS(AB)}{MSE}$
Shrimp × protein	SS(AC)	$(2 - 1)(4 - 1) = 3$	$\dfrac{SS(AC)}{3} = MS(AC)$	$\dfrac{MS(AC)}{MSE}$
Lipid × protein	SS(BC)	$(3 - 1)(4 - 1) = 6$	$\dfrac{SS(BC)}{6} = MS(BC)$	$\dfrac{MS(BC)}{MSE}$
Shrimp × lipid × protein	SS(ABC)	$(2 - 1)(3 - 1)$ $(4 - 1) = 6$	$\dfrac{SSB}{6} = MS(AB)$	$\dfrac{MS(AB)}{MSE}$
Blocks	SSB	$3 - 1 = 2$	$\dfrac{SSB}{2} = MSB$	$\dfrac{MSB}{MSE}$
Residual error	SSE	$(2 \cdot 3 \cdot 4 - 1)(3 - 1) = 46$	$\dfrac{SSE}{46} = MSE = s_p{}^2$	
Total	SS	$2 \cdot 3 \cdot 4 \cdot 3 - 1 = 71$		

Note: There are two levels of factor A (shrimp variety), three levels of factor B (lipid concentration in feed), four levels of factor C (protein concentration in feed), and three blocks (replicates per treatment combination).

Regression Analysis

Regression analysis determines the presence or lack of relationships between a dependent variable (y), and an independent variable (x). The independent variable is assumed to be constant, not subject to sampling or measuring error. Examples would include the number of eggs produced (dependent) vs. time after injection (independent), fish yield (dependent) vs. nitrogen input (independent), shrimp growth rate (dependent) vs. water temperature (independent), and standard curves in water chemistry (absorption (dependent) vs. concentration (independent)). Two other assumptions for regression analysis are that for any x value the distribution of y is normal, and the variance of the distribution of y is the same for any value of x (Draper and Smith, 1981; Steel and Torrie, 1980).

The simplest regression assumes a linear relationship between x and y. Equation 19 below represents total experimental variability assuming a linear relationship.

$$y = a + bx + \varepsilon \tag{19}$$

where a = y-intercept, b = slope of line (regression coefficient), and ε = sum of residuals (sum of observed minus expected values predicted from equation). Computers determine the linear equation that best fits the observed points (Figure 9) through a process known as analysis of Least Squares. Least Squares refers to the sum of the square of each individual residual (i.e., Σ (observed y_i – predicted y_i)^2). The regression line determined through Least Squares analysis gives the smallest sum of residuals squared of all possible straight lines. In another words, any other line would give a greater total of the sum of individual residuals squared.

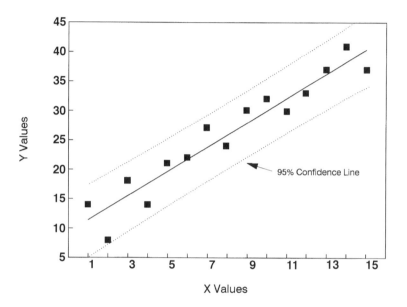

Figure 9. Linear relationship between X and Y, with both the regression line and 95% confidence limits indicated.

The linear equation provided through regression analysis is $y = a + bx$, and does not include the residual term. Implicit in this equation are several hypotheses that can be easily tested using standard information provided by most software packages when conducting a regression analysis. The ***standard error of the y estimate*** (which describes the variability of y at a given x value) can be used to test the null hypothesis that a is not significantly different from a value of 0. If $a > (t)$ (standard error of y estimate), where n = number of pairs of observations and t is the tabular value at $n - 2$ degrees of freedom at $P = 0.05$, then a is not significantly different from 0. For example, an a value significantly greater than 0 in the regression between nitrogen input and fish yield may indicate that the fish had an additional source of food not necessarily related to experimental inputs.

A second hypothesis utilizes the ***standard error of b*** (i.e., slope of regression line), and tests whether b is significantly different from 0. If it is not, then there is not a statistically significant linear relationship between x and y. To test the null hypothesis that b is not significantly different from 0, calculate a t value based on the following equation:

$$t = \frac{b}{\text{standard error of } b} \tag{20}$$

Compare the computed t value with tabular values at $n - 2$ degrees of freedom to determine the probability (P) that the observed hypothesized linear relationship happened by chance.

To test the null hypothesis that one linear relationship is significantly different from another (i.e., b_1 is not significantly different from b_2), calculate the 95% confidence intervals for both bs using the formula $b \pm (t_{0.05})$ (standard error of b) using the t value at $n - 2$ degrees of freedom. If 95% confidence limits for b_1 and b_2 do not overlap, then the two linear relationships have significantly different slopes.

An important analytical question is how well does the predicted linear equation fit the observed data points. The standard error of b gives some idea with respect to the actual data. If the standard error of $b = 0$, then the predicted values equal the observed values (i.e., sum of residuals squared $= 0$). If the standard error of b is relatively large in relation to b, then the hypothesized linear fit is not statistically significant. The ***correlation coefficient (r)*** standardizes the variability of b between -1 and $+1$, where $r = 0$ indicates no relationship at all. When $r = -1$ or $+1$, the standard error of $b = 0$, and b (slope of regression line) is either negative or positive, respectively.

Conveniently, r^2 ($= $ ***coefficient of determination***) equals the observed variability explained by b, whereas $1 - r^2$ represents the remaining residual variability (i.e., ε in Equation 19). Values of r^2 vary from 0 to 1.0, with 1.0 a perfect relationship without any residual variation.

It is extremely important to understand that r^2 only gives a relative measure of the variability of y "explained" by the relationship with x, and alone does not impute statistical significance. A high r^2 does not necessarily mean that there is a significant linear relationship. As shown above, that hypothesis is based on both the standard error of b and the number ($n - 2$) of degrees of freedom. Similarly, statistical significance (i.e., P) is also based on r (or r^2) at $n - 2$ degrees of freedom. For example, a linear relationship is not significant ($P > 0.05$) when $r = 0.87$ ($r^2 = 0.76$) with 3 degrees of freedom (i.e., $n = 5$), but is statistically significant ($P < 0.05$) when $r = 0.28$ ($r^2 = 0.08$) with 50 degrees of freedom. In the former case 76% of the observed variation was "explained" by b, but there were too few observations to give sufficient confidence to the relationship. In the latter case, b "explained" only 8% of total variation, but the large number of observations rendered the hypothesized linear relationship statistically significant. Since the foundational experimental objective is to understand sources of variation in a particular system, the former result (although not statistically significant) gave strong support for further experimentation; whereas the latter result (although statistically significant) indicated that x really had very little impact on the variation of y.

Multiple linear regression examines the relative impacts on the variability of dependent variable y with several independent variables, as indicated by Equation 21.

$$y = a + b_1 x_1 + b_2 x_2 + b_3 x_3 + \ldots + \varepsilon \qquad (21)$$

where b_1, b_2, and b_3 are regression coefficients for independent variables x_1, x_2, and x_3, respectively. For example, such an analysis could examine the hypothesized relationship that fish yield is a function of nitrogen input, stocking density, and survival (see Van Dam, 1990 for an example using rice-fish culture data). There are several types of multiple linear regressions available with many statistical software packages, and the reader is highly encouraged to consult proper authorities (e.g., Draper and Smith, 1981) regarding assumptions, limitations, etc. before blindly punching in numbers.

Many biological relationships are nonlinear and may be better described by ***curvilinear regression*** equations. Figure 10 illustrates representative graphs and equations for four common nonlinear relationships of quadratic, polynomial, exponential, and logarithmic. Linear regressions also include those nonlinear relationships in which (1) the model parameters are additive and multiple regression techniques are available or (2) when a transformation will make the relationship linear (Draper and Smith, 1981; Steel and Torrie, 1980). Quadratic and higher order polynomials fall into the first category, while exponential and logarithmic relationships (which can be linearized through log transformation) are examples of the second category. Nonlinear models that cannot satisfy either (1) or (2) above (e.g., logistic or Gompertz) can be analyzed by nonlinear regressions, discussion of which is beyond the scope of this chapter.

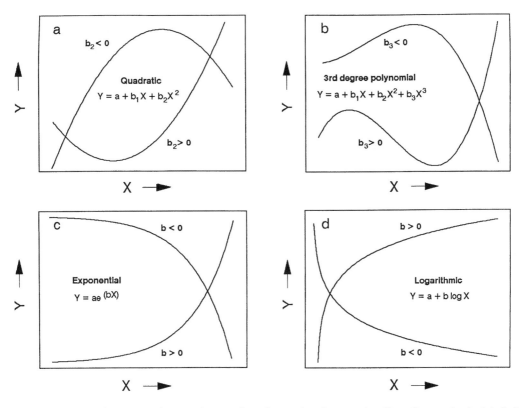

Figure 10. Nonlinear equations and examples of associated regression lines for quadratic (a), 3rd degree polynomial (b), exponential (c), and logarithmic (d) functions.

Nonlinear equations should be used *only* when sound biological reasoning supports such usage. However, because of the unfortunate and common misconception that r (or r^2) equals significance, the quadratic model (Equation 22) as well as third and fourth order polynomial equations have been reported only because the researcher found marginally higher r values by adding terms.

$$y = a + b_1 x + b_2 x^2 + \varepsilon \tag{22}$$

Two points require emphasizing. First, even though r may increase slightly, P could decrease slightly because of one less degree of freedom for each additional term. Second, and more importantly, P values must be determined for *each* b in the equation. For example in Equation 22, *both* b_1 and b_2 must be significantly different from 0. If b_2 is not significantly different from 0, the relationship cannot be said to be quadratic regardless of any increase in r by adding the quadratic term.

There are several other points worth mentioning before beginning any regression analysis. First, *always* plot your data points to visually see what sort of relationship you have before assuming linearity. A residual plot of independent variable x vs. residuals of y will provide a visual tool to help evaluate linearity, as well as indicate any need for data transformation (e.g., Figure 11 is a residual plot of the linear relationship in Figure 9). Second, never plot your regression line beyond your data points. The range of your data limits the predictive powers of your regression equation (e.g., Figure 9). This is one of the primary abuses of regression analysis by economic forecasters. Third, be aware that 95% confidence limits for y are not parallel to the regression line but increases (i.e., widens) toward each end of the line, giving a slightly fluted appearance (Figure 9).

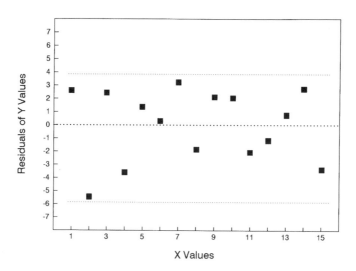

Figure 11.   Plot of the residuals of Y values from Figure 9 against values of X. This residual plot illustrates no relationship or systematic pattern between X and experimental errors from predicting Y values.

Linear Correlation

Linear correlation is conceptually similar to linear regression, except there are no clear dependent and independent variables. All variables (there are more than two with multiple correlation) must be continuous, each one normally distributed, and collected from random samples (Steel and Torrie, 1980). This is different from regression analysis, where the independent variable was fixed with no variation. Like regression analysis, the correlation coefficient r can range from -1.0 to $+1.0$, with 0.0 indicating no linear relationship whatsoever.

Although correlation analysis can be quite revealing and frequently an important component of experimental data analyses, interpretation of these results should be done very carefully. Always remember that resulting probabilities and correlation coefficients only provide the degree of association; a causal relationship between variables is *never implied*. A serious danger arises when a superficially plausible explanation for an observed association is accepted without further inquiry. For example, suppose that a researcher found a nice inverse relationship between tilapia yield and ammonia concentration in ponds fertilized with TSP and urea. Without supporting evidence, it may appear that high ammonia concentrations reduced fish yield (e.g., Meade, 1985). Knud-Hansen and Batterson (1994) found such a situation, but additionally found that high ammonia concentrations were also related to low algal productivity. Ammonia accumulated in the water when algal productivity was reduced by limitations of light, phosphorus, and/or inorganic carbon. The causal relationship with ammonia concentrations was with phytoplankton productivity and not tilapia yield. Tilapia yield, on the other hand, was strongly related to algal productivity. Further analysis of ammonia concentrations showed that the toxic un-ionized component was well below toxic levels. The observed inverse association between ammonia concentrations and tilapia yield, therefore, was without any directly causal links. For another example, see Bird and Duarte (1989) regarding possible spurious correlations between bacteria and organic matter in sediments.

To guard against jumping to false conclusions of causality from observed associations, treat results from correlation analyses as a single piece to a larger puzzle. A scientific conclusion should not be based solely on a causal interpretation from a single correlation analysis. Use whatever available data to provide other pieces of the puzzle and to help understand the basis for the observed association. If no other relevant information was collected or exists, then treat the observed correlation as the foundation for a hypothesis and design further research to test it. To repeat, a significant correlation only indicates a statistically significant association and not a causal relationship between variables. Maintaining this proper perspective promotes scientific objectivity and reduces the possibility of committing Type I errors.

Multiple Comparisons of Means and Range Tests

A review of the literature would suggest that, regardless of the experimental design employed, most aquaculture scientists, primary (if not only) objective is to compare treatment means. Such an experimental objective is appropriate for unstructured experiments, but too often scientists use multiple range tests to compare treatment means for structured experiments as well, completely ignoring hypothesized relationships and interactions incorporated into the experimental design. It is clear these scientists did not appreciate the inherent power of their designs. It is also clear that the virulent nature of multiple range tests has gotten out of control via a positive feedback loop; the more scientists see multiple range tests used, the more they feel encouraged to use them (whether or not they understand implicit assumptions and limitations), and so on. Other fields of science have also been inflicted and dealt with the multiple range test plague (e.g., ecology; Day and Quinn, 1989). This subsection will hopefully be a statistical antibiotic for aquaculture scientists, providing some guidance on when and how to compare means following ANOVA.

After conducting an ANOVA, most software packages provide a table of means by treatments, blocks, etc. In some cases both the *actual* individual standard errors as well as standard errors **based on the pooled variance** (s_p^2 or MSE from ANOVA; e.g., Table 1) are given. The general formula for the latter is:

$$\text{Pooled standard error of mean} = \sqrt{\left(s_p^2\left(1/n_1 + 1/n_2\right)\right)} \tag{23}$$

where s_p^2 = pooled variance = MSE and n_1 and n_2 = number of replicates means of treatments 1 and 2, respectively. The pooled standard error is then used to calculate confidence limits for individual treatments or blocks (see Figures 6b and 6c). If individual treatment variances are more or less similar, then using the pooled standard error to determine confidence limits for all treatment means makes sense. The difficulty arises if there is great variability among treatment variances, in which case the pooled standard error of the mean hides and possibly misrepresents observed experimental variability.

Unstructured Experiments

As discussed earlier, comparing means (with each other or with a control) is often the primary objective of unstructured experiments. The general analytical procedure is to rank the means and determine which ones are significantly different (at a given P value) from each other or the control. There are a great number of multiple comparison and range tests (e.g., the **Least Significant Difference (lsd) test**, **Duncan's Multiple Range test**, and **Tukey's w Procedure**, to name a few) with which to carry out the analysis. Interested researchers are strongly encouraged to consult Day and Quinn (1989) before choosing one method over another. The wrong way to make multiple comparisons, however, is to conduct a series of individual t-tests.

To illustrate both the inefficiency of individual t-tests and general characteristics of multiple comparison tests, the *lsd* test will be briefly described (for a humorous explanation and defense of the lsd test, see Carmer and Walker, 1982). Assume an aquaculturist examined six unrelated types/sources of shrimp feed for their ability to increase shrimp yield. There were three replicates per treatment, and an ANOVA revealed that at least two feeds gave significantly different results. In doing individual t-tests, the t value (at a given P value) would be at $n_1 + n_2 - 2$ degrees of freedom, or $3 + 3 - 2 = 4$ degrees of freedom (see above discussion of t-tests). In contrast, the lsd test uses the s_p^2 to calculate the standard error of the difference between the means (Equation 18) and the degrees of freedom for SSE. In this case, the degrees of freedom would equal $k(n - 1) = 6(3 - 1) = 12$. The lower t value with the higher degrees of freedom (2.179 vs. 2.776 at $P = 0.05$) makes the lsd much more sensitive (i.e., less likely to commit a Type II error) than a series of individual t-tests.

The lsd test calculates the largest difference between means that is not significant at a specified P value, hence the name Least Significant Difference test. For example, assuming $s_p^2 = 12.1$ g, the lsd value at $P = 0.05$ between any two treatments in the feeding trial would be

$$\text{lsd}_{0.05} = t_{0.05}\left(\sqrt{\left[s_p^2\left(1/n_1 + 1/n_2 \right) \right]} \right) \qquad \text{and, } df = k(n-1)$$

$$= 2.179\left(\sqrt{\left[12.1\left(1/3 + 1/3 \right) \right]} \right) \tag{24}$$

$$= 6.19 \text{ g}$$

where $\sqrt{(s_p^2\,(1/n_1 + 1/n_2)}$ = the standard error of the difference between means, and s_p^2 = pooled variance = MSE. For presentation purposes, shrimp yields (or any other response variable) would be ranked either in ascending or descending order. If descending, next to the highest yield would be placed the letter "a." The same letter would be placed next to all lesser yields where the difference between means with the highest yield was less than 6.19 g (the lsd value). If all yields have the letter "a", then there is no significant difference between any means at the specified P value. Otherwise, the first yield in descending order with a difference greater than 6.19 g when compared to the highest yield would then be designated "b." All other yields (both ascending and descending) with means within 6.19 g will also be designated with the letter "b." The process continues until all means are designated with a letter(s). The end result will be that all means designated with the same alphabet notation are not significantly different from each other. This provides an easy visual description of experimental results.

It is important to note that multiple range tests should be conducted only *after* an ANOVA has revealed significant differences between two or more treatments. Remember also that the lsd test, as well as other multiple range tests, provides only the most elementary form of analysis and does not provide any insight into reasons for observed differences. So even if the experiment were unstructured and a multiple range test were appropriate, more analyses may be warranted. In the feeding trial, for example, although the feeds came from independent sources they may have some identical ingredients. Was there a relationship between shrimp yield and protein or lipid concentrations or perhaps digestibility? Did any feeds have any unusual ingredients? These and many other questions may help explain your results or at least develop new research hypotheses.

Structured Experiments

The objective of structured experiments should be to evaluate system relationships. There are times in structured experiments, however, where comparing means represents an important part of data analysis. When this need arises, *never* use a multiple range test. Although the statistical community is quite clear on this point (see Chew, 1976; Peterson, 1977), the aquaculture community (among other scientific disciplines) has yet to appreciate the inappropriateness and gross inefficiency of multiple range test usage in analyzing structured experiments.

The way to compare two means in a structured experiment is with the t-test (Equation 17) using the pooled variance (s_p^2) to calculate the standard error of the difference between two means (Equation 18). The following is the standard equation:

$$t = \frac{\text{mean } 1 - \text{mean } 2}{\sqrt{\left[s_p{}^2 \left(1/n_1 + 1/n_2 \right) \right]}} \qquad (25)$$

where n_1 = number of observations for mean 1, n_2 = number of observations for mean 2, and degrees of freedom = total for SSE in the ANOVA. This formula can be used with any design and with any pairwise comparison. Since n_1 and n_2 can be different values, the researcher can aggregate treatment or block results for more meaningful comparisons. For example, in a sixteen tank/four treatment RCBD catfish grow-out experiment with four tanks next to a dirt road as one of the four blocks (as described in the RCBD discussion in the Treatments section), a *pairwise comparison* of the four "road" tanks (mean 1, n_1 = 4) vs. the "non-road" tanks (mean 2, n_2 = 12) tests the null hypothesis that the road had no effect on experiment response variables (e.g., fish growth). Similar to Equation 17, Equation 25 also allows comparison of means with an unequal number of replicates, which could happen if something went wrong with an experimental unit (e.g., an aquarium accidently breaks during mid-experiment, or the wrong pond gets fertilized with a half a ton of chicken manure).

Residual Plots

Residual plots are graphs that illustrate the relationship between experimental error (i.e., unidentified variation) and some variable. The variable can be a treatment, or it can be any other measured variable such as dissolved oxygen or net fish yield. Remember that residuals are just the differences between observed and expected values. A negative residual means that the observed value for that experimental unit was less than expected. Large residuals (either positive or negative) indicate large experimental errors for those experimental units. A random residual pattern suggests that experimental errors were also random with respect to the variable that residuals were plotted against (Figure 11).

A residual plot with a discernable nonrandom pattern provides valuable insight into the nature of experimental relationships. For example, a residual plot that looks like a funnel on its side (Figure 12b) indicates increasing residual variability with increasing variable values (Figure 12a). Logarithmic transformation of the data may reduce experimental variability in such situations. Similarly, a convex- or concave-shaped residual plot (Figure 13b) may reveal a linear model incorrectly applied to a quadratic relationship (Figure 13a).

Residual analysis can be a powerful tool to identify a significant linear effect of a nontreatment variable. The relationship between experimental errors and pre-existing experimental conditions is the conceptual heart of analysis of covariance (ANCOVA). As discussed earlier, Knud-Hansen (1992a) used residual analysis to demonstrate a significant positive relationship between the amount of previous fertilization inputs to individual ponds and residual net fish yields (i.e., observed minus expected yields) for each pond in a subsequent fertilization experiment (Figure 14). Most of the experimental error could be attributed to nutrient carryover effects from previous fertilizations. Residual analysis identified previous fertilizations as a significant source of net fish yield variability and, therefore, removed this variability from experimental error.

Used in this fashion, residual analysis can (and should) become an important means for testing further hypotheses. For example, algal productivity and tilapia yields often show a strong linear relationship (e.g., Knud-Hansen et al., 1993). Tilapia yield variability not explained by a regression equation (i.e., residuals) can be plotted against hypothetical sources of variation to determine the significance of that variable. For example, tilapia yield residuals could be plotted against mean (for each experimental unit) dissolved oxygen concentrations at dawn or un-ionized ammonia concentrations to see whether any sublethal growth inhibition occurred during the experiment. Care must be taken not to test variables that also correlate with the primary independent variable. For example, where algal productivity is the independent variable, plotting the tilapia yield residuals against

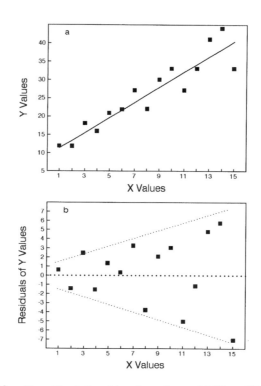

Figure 12.   Example of an X vs. Y relationship where the variability of Y increases with X values.
This becomes more evident when a linear regression line is made (a), and residuals
of Y plotted against X show a "funnel on its side" shape (b).

suspended solid concentrations could provide confusing results, since turbidity (i.e., light
availability) can affect algal productivity.

Thanks to computers, making residual plots, like testing for normality, should become a routine
part of data analysis. Again, most statistical software packages will make residual plots of your
results at the push of a button (or depression of a key). This is a quick way for determining if
experimental errors are random or if other identifiable sources of variation are present.

Nonparametric Tests

All discussions so far have concerned parametric statistics, where sample means and variances
were used to estimate corresponding parameters of normally distributed populations. In contrast,
nonparametric tests make no assumptions about population distributions and have their greatest
application in situations where data can be ranked rather than averaged. Surveys, which often collect
qualitative data (e.g., preference rankings), frequently employ nonparametric statistics. Nonparametric statistics can also be used where the underlying distributions are not normal or when sample
data have heterogenous variances, and subsequent transformations fail to adequately normalize the
data. Some researchers prefer to use nonparametric tests because they are quick and easy to do,
and require minimal mathematical skills.

The decision to use nonparametric statistics, however, should be made carefully and reluctantly.
To begin with, they are only tests. They do not give any information regarding variances, standard
errors, or confidence limits. Generally, nonparametric tests are less sensitive than corresponding
parametric tests, and the researcher is more likely to commit a Type II error (i.e., accept the null
hypothesis when it is false). In short, nonparametric statistics are useful only in situations when

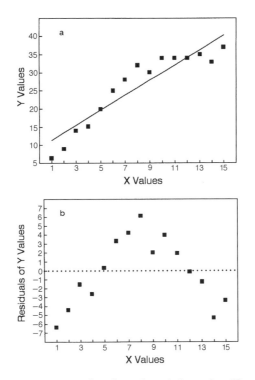

Figure 13.   Example of a linear regression line plotted through a X vs. Y relationship (a). The relationship reflects more of a quadratic or asymptotic function, which is reflected in the convex shaped residual plot (b).

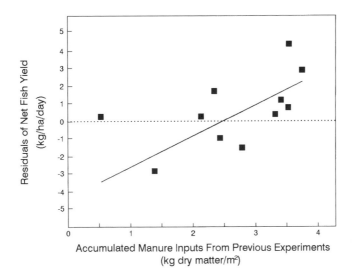

Figure 14.   Relationship between the residuals of net fish yield calculated from a tilapia grow-out experiment with different fertilization inputs and total chicken manure inputs accumulated in ponds from fertilizations from previous experiments. The graph shows an overall positive effect ($P < 0.01$) on net fish yields in ponds with historically greater manure inputs. (Adapted from Knud-Hansen, C. F., 1992a.)

parametric statistics cannot be used. If parametric statistics can be used, then that should be the preferred choice.

Most statistical software packages are capable of doing a variety of nonparametric tests. The following is a representative list and description of several tests, with the analogous parametric analysis given in parentheses. The **Kolmogorov-Smirnov One-Sample Test** is a goodness-of-fit test that examines the null hypothesis that the distribution of your data is not significantly different from a specified parametric distribution (e.g., normal, Poisson, binomial, etc.). The **Kolmogorov-Smirnov Two-Sample Test** tests the null hypothesis that two independent samples have the same underlying population distribution. **Wilcoxon's Signed Rank Test** (paired sample t-test) is used for detecting differences with paired treatments (i.e., changes due to treatment within experimental units). The **Mann-Whitney Test** (t-test) compares two sample means based on ranks. The **Kruskel-Wallis Test** (ANOVA) compares three or more sample means based on ranks. **Spearman's Rank Correlation** (correlation) measures the association of two variables based on ranked observations of each variable. The calculated Spearman's rank correlation coefficient (r_s) only gives a probability of association and says nothing about the nature of the tested association.

Perhaps the most common nonparametric test used in aquaculture investigations is the **Chi-square Test**. Chi-square analysis deals with frequency data, such as fish size distributions, and tests the null hypothesis that observed frequencies are not significantly different from expected. The greater the difference between observed and expected frequencies, the higher the Chi-square value and the less likely such differences happened by chance. Expected frequencies can be theoretical (e.g., probabilities based on genetics), or based on a given distributional frequency (e.g., normal and Poisson distributions) similar to the Kolmogorov-Smirnov One-Sample goodness-of-fit test. As with other nonparametric tests, the primary sampling requirement is that observations are made randomly and independent of each other.

Chi-square analysis can also test the null hypothesis that there is no association between two mutually exclusive classifications. For example, Chi-square analysis could be used to see if sex reversal of tilapia fry has any effect on their size distribution when they reach maturity. Frequency data are put in a contingency table with rows (r) representing classes of one variable (e.g., sex reversed or not), and columns (c) representing classes of the other variable (e.g., intervals of fish lengths). Expected frequencies are calculated from observed frequencies.

Although your actual contingency table may have classes with only a few or even no observations, it is important that the *expected* frequencies follow several general rules of thumb (Steel and Torrie, 1980). First, there should be no expected frequency < 1. Second, not more than 20% of the expected frequencies should be < 5. For example, if you make eight size classes each of sex-reversed and non-sex-reversed tilapia (i.e., a total of 16 frequencies) and find either rules of thumb violated, you have two options. You can either make more observations (i.e., measure more fish), or you can reduce the number of size classes, thus increasing the average number of fish per frequency. Since calculated Chi-square values are compared with tabular values at $(r - 1)(c - 1)$ degrees of freedom, reducing the number of classes will also reduce the test's sensitivity. Optimally, therefore, you want to make as many observations as practically feasible, and make as many size classes (e.g., class intervals every 1 cm vs. 2 cm) as possible within the above constraints.

QUALITY CONTROL

A potentially major source of experimental error comes from inadequate quality control of data collection, processing, and examination. Although most preventive measures are just plain common sense, they still merit discussion. Some useful information may be salvaged from a poorly designed experiment, but even the best designed investigation can be rendered completely worthless by carelessness and poor quality control.

Data Collection

The key here is to collect data systematically. Use the same instruments, methods, people, and times of day throughout the study. Make as few changes as possible to minimize any additional sources of variation. If a technical staff is used to collect field or laboratory data, make sure they understand the importance of both the data they are collecting and their roles in the research. Good research is frequently a team effort, and participants perform better when they feel like they are part the research team.

Essential to systematic data collection are comprehensive data sheets for all field/laboratory measurements and routine calculations. Each data sheet should provide (1) name of the project, (2) name of experiment within the project, (3) name of person(s) collecting the data, (4) date and time of day samples were collected or field measurements made, (5) date samples were analyzed/processed, (6) table headings with appropriate units, and (7) a place to add any miscellaneous notes (e.g., unusual weather, instrument malfunctions, and notable observations).

Data recording should be made with an indelible pen and should be sufficiently legible so that anyone could understand what was written (e.g., avoid confusion with the numbers 1 and 7). Unusual abbreviations should be defined on the data sheet. If a data entry needs correcting, *do not* erase or "white-out" the alleged incorrect number. Simply draw a single line through it and write the "correct" value next to it. Too often the suspected incorrect value turns out to be the correct one. And even if the correction is legitimate, the previous value could reveal a problem with the instrument or method of measurement.

Finally, file all raw data sheets systematically for future easy access. Raw data sheets represent the foundation for all subsequent data analyses. Original data sheets allow for easy detection of copying errors and discovery of systematic variations (e.g., a particular technician was later found to be routinely careless, or a reagent used during part of the experiment was later found to be incorrectly made). When appropriate, discovered errors can either be corrected or removed from further analysis. Additionally, written information that may be useful for future analyses might never enter the computer spreadsheet (e.g., miscellaneous notes, or individual fish weights and lengths when only sample means and standard deviations are analyzed).

Data Processing

Nowadays nearly all data end up on a computer spreadsheet. The *only* time data should be hand copied is when transferring data from raw data sheets to the spreadsheet. It can be quite helpful if both data sheets and spreadsheets list data column headings, treatment names, and experimental unit classifications in the same order. This foresight greatly facilitates data entry and minimizes copying errors.

No matter how careful the person, however, copying errors always occur. At a minimum, the person entering data should understand what the numbers mean. A data processor may not recognize that field data entries for ponds A1 and B1 were inadvertently reversed, or that pH values were accidently written in the DO column and vice versa, or the incongruity when on one sampling date all soluble reactive P measurements were greater than total P values for the same water samples. On the other hand, entering data is a good way for the researcher to get the feel of the numbers, sense trends, and discover errors, and thus can provide an important check on data quality control. Although time-consuming, thoughtful and careful data entry is often a cost-effective way to reduce experimental error and improve research quality.

Two further data processing considerations are the concepts of ***significant numbers*** and ***rounding numbers***. The latter can be a minor source of data bias, while abuse of the former can be a major source of misrepresentation and reader irritation. In simple terms, significant numbers refer to their perceived numerical accuracy and analytical utility. Never use more significance for a raw data value than is warranted. If the temperature meter can be read to the nearest 0.1°C, do not record a temperature to the nearest 0.01°C. It is important to understand that 25.3°C and 25.30°C may be the same temperature, but they *do not* have the same analytical meaning. The latter temperature, with four significant numbers, gives meaning (and implies confidence of measurement) to a hundredth of a degree, while the former only to a tenth of a degree. As another example, assume for illustration that three dissolved oxygen values were read and recorded as 8, 7.2, and 3.14 mg/L. Recording different levels of precision may occur between researchers or with older nondigital meters with nonlinear and/or variable scales. Because of differences in significant numbers among the recorded values, the sum may equal 18.34 mg/L but should be reported as 18 mg/L. The sum (or average) cannot be reported with greater confidence than the nearest mg/L since confidence of the first value (8 mg/L) was only to the nearest mg/L.

As a general rule three significant numbers are sufficient. Exceeding three significant numbers usually provides unnecessary numbers and may misrepresent analytical accuracy. For example, a phytoplankton count of 137,593 cells/L (i.e., six significant numbers) is better represented and more easily read as 1.38×10^5 (three significant numbers). Rarely will any scientific conclusion be based on values of greater than three significant numbers. Two common examples of analytical misrepresentation are found in water chemistry. Because some digital pH meters give measurements to the nearest 0.001 pH unit, some researchers feel obligated to report pH values to that same level of precision. Although precise, if the last two digits never stabilize, measurements are certainly not accurate to 0.001 pH units. Similarly, spectrophotometric standard curves (i.e., linear regressions) can give precise nutrient concentration calculations down to parts per trillion if so desired. However, most aquaculture water chemistry laboratories use cuvettes with 1 cm pathlengths, and minimal levels of detection are usually around 0.01 mg/L (Knud-Hansen, 1992b).

Rounding numbers should be done only *after* any and all intermediate calculations are completed. Most calculators and spreadsheet software packages have rounding functions, but they can introduce a subtle bias if they always round the number 5 either up or down. A simple way to eliminate this bias is the odd-even rule, in which rounding is always to the nearest even number (EPA, 1973). For example, 2.5 becomes 2, 3.5 becomes 4, 4.5 becomes 4, 5.5 becomes 6, and so on. Rounding should be done in one step, so if 5.451 were rounded to the nearest 0.1, the answer would 5.5 and not 5.4 (which would be the case if 5.451 were first rounded to 5.45 and then to 5.4). Rounding biases may be only a minor source of experimental error/variation, but a potential source, nonetheless.

Data Scrutiny

Data scrutiny is a quality control mechanism, a source of scientific inspiration, and an essential initial part of any data analysis (for a good review on data scrutiny, see Finney, 1988a). Data scrutiny involves the three steps of (1) a validity check on the actual data, (2) a quick examination of all possible relationships between sets of data, and (3) a check on data distributions.

The validity check requires examining the data for sources of error. Just as a gourmet chef would not cook with dirty pots and pans, the careful scientist should not analyze a "dirty" set of data. First look for extreme values (maximum-minimum spreadsheet functions make this an easy task). Misrecorded data may be discovered (e.g., a temperature of 276°C instead of 27.6°C, or a pH of 0.78 instead of 7.8), or inexplicably extreme values may appear. Do not, however, throw out extreme measurements unless you have an *independent* reason for suspecting their invalidity (e.g., the high ammonia value corresponds to the same beaker/sample in which a fly was found floating during analysis). Extreme values may reflect actual variation of your system or could also provide valuable insights into other nontested sources of variation.

The validity check also involves issues of data recording and data entry. Were there any mistakes? Were appropriate levels of precision given? Were there any changes in precision over time, which could happen if different recorders read the same instrument? For example, one person may try to read a mercury maximum-minimum thermometer to the nearest °C, while another may read it to nearest 0.5 or 0.2°C. Were there any differences due to different laboratory or field technicians, mid-experiment changes in instrumentation, or nonsystematic use of instruments (e.g., not standardizing field instruments before every use)?

The second step of data scrutiny is to examine critically all possible relationships by means of frequency distributions and scatter plots. Examine measured variables against each other and over time. This process may reveal previously unconsidered relationships, develop new hypotheses, or highlight possible experimental error. An unpublished example from a field laboratory in Thailand best illustrates the last consideration. During an experiment, diel measurements of total alkalinity in ponds consistently rose during the night and returned to lower levels during the day. This seemed peculiar since total alkalinity tends to be conservative in oxygenated freshwaters with calcium concentrations below saturation (Wetzel, 1983). At that time alkalinity was measured using a methyl orange indicator (APHA, 1985). As it turned out, this color change was harder to detect at night when the field laboratory was not as well lit; the laboratory technicians, therefore, apparently titrated with a bit more acid to make sure of reaching the endpoint. This conclusion was verified when alkalinity was subsequently measured potentiometrically (using a pH endpoint instead of a visual color change; APHA, 1985), and all diel variation (i.e., higher nighttime values) of alkalinity disappeared. Changing methods also decreased variability between replicate subsamples as well as between laboratory technicians.

The third and last data scrutiny requirement before diving into ANOVAs is to check data for distributional assumptions. This involves verifying normality of your data, or identifying other distributions and testing whether subsequent transformations adequately normalized non-normal data. At this point you are ready to conduct selected ANOVAs or perhaps resort to nonparametric analyses. It would also be wise, however, to examine residual plots from all ANOVAs to help identify possible analytical sources of variation.

CONCLUSION

Publication of Research

A logical conclusion to this chapter is a few words on writing up experimental results for publication. It seems ironic that publications represent the cornerstone of most scientific careers, yet writing papers is often the most difficult task of the overall research process. Not surprisingly, the difficulty in writing is inversely related to how well the research proposal and experimental designs were thought out prior to collecting data. Poorly focused designs, inadequate literature research, and unfounded assumptions (particularly if the researcher wants to "prove" something) generally produce papers of scientific quality comparable to the researcher's efforts.

Everyone has his or her own style and approach to writing scientific papers. Those who find writing a frustrating experience may wish to try the following approach. First, write down the two or three most significant research conclusions you wish the reader to remember a week after reading your paper (i.e., the "take home message"). Write your conclusion based on these points. As we all know, if you can remember more than two points from a research article, that paper was probably outstanding. Second, write your discussion, focusing only on your stated conclusions. How you logically present your conclusions should serve as the conceptual outline for your discussion, results, materials and methods, introduction, and abstract, so give sufficient thought to how the conclusion is written and organized. Third, write the results, presenting only the data that were discussed in the discussion and used to formulate your conclusions. If the data have no value in the discussion, they have no value in the results. Fourth, write the materials and methods, describing only those methods used to collect the data presented in the results. Fifth, write the introduction, giving necessary background information (i.e., setting the stage for your research). The introduction should include a discussion of the research problem, of prior relevant research highlighting current understanding by the scientific community, and of how the following research was designed to address one or more information gaps in this research area. A clear statement of objectives and tested null hypotheses mirroring the conclusion is a logical end to the introduction. Finally, write a clear and concise abstract, focusing on the scientific problem addressed by your research, objectives/hypotheses, and the "take home messages," along with a few notable supporting results/data. Avoid including statements detailing methods (unless methods are the focus of the paper) or statements that give no information beyond "so and so is discussed." Abstracts are read far more frequently than entire articles and should be informational. The end result of this process is a clearly focused and efficient scientific contribution.

An underlying theme of this chapter is that carefully designed research requires some serious effort up front, but this effort pays dividends when it is time to write up the results. At the time of the first measurement, the researcher already has the literature researched for the introduction and much of the discussion, has all the information for materials and methods, has the objectives and null hypotheses clearly articulated, and knows how all the data being collected will be analyzed to address these stated objectives and null hypotheses. Much of the paper could be written while the experiments are being conducted. When you have a clear vision of where you are going, it is easier to get there, which is why writing the conclusion first makes sense. It also may be useful to review one or more of the available guides to scientific writing (e.g., Gopen and Swan, 1990).

Collaborative Research

A final word should be mentioned regarding collaborative research involving two or more people. Research projects can be separated into four phases: experimental design and proposal, execution of experiment(s), data analysis, and manuscript writing. Duties and responsibilities of each participant should be clearly delineated from the outset so the partnership goes smoothly. If opportunities for routine communication are not available (e.g., researchers are based in different countries), then scheduled meetings should be written into the research protocol.

Often the most difficult decision collaborators must make are the names and order of names on subsequent publications. Preferably this decision should be made at the first organizational meeting. A general guideline requires all listed authors to have contributed substantially to at least two of the four research phases (Jackson and Prados, 1983). The lead author certainly has the discretion to add names of those who may have contributed to only one phase of the research (e.g., laboratory technicians). Adding honorary authors (often incapable of explaining or defending the paper's published results), however, is a disservice to everyone.

REFERENCES

APHA (American Public Health Association), *Standard Methods for Examination of Water and Wastewater,* 16th ed., American Public Health Association, American Water Works Association, and Water Pollution Control Federation, Washington, D.C., 1985, 1268 pp.

Baily, N. J., *Statistical Methods in Biology,* Edward Arnold, London, 1981, 216 pp.

Bird, D. F. and Duarte, C. M., Bacteria-organic matter relationship in sediments: a case of spurious correlation, *Can. J. Fish. Aquat. Sci.,* 46, 904–908, 1989.

Boyd, C. E., Water Quality in Ponds for Aquaculture, Auburn University, Auburn, AL, 1990, 482 pp.

Carmer, S. G. and Walker, W. M., Baby bear's dilemma: a statistical tale, *Agron. J.,* 74, 122–124, 1982.

Chew, V., Uses and abuses of Duncan's multiple Range test, *Proc. Fl. State Hort. Soc.,* 89, 251–253, 1976.

Cohen, J., *Statistical Power Analysis for the Behavioral Sciences,* 2nd ed., Erlbaum Associates, Hillsdale, N.J., 1988, 567 pp.

Day, R. W. and Quinn, G. P., Comparisons of treatments after an analysis of variance in ecology, *Ecol. Monogr.,* 59(4), 433–463, 1989.

Draper, N. R. and Smith, H., *Applied Regression Analysis,* 2nd ed., Wiley Interscience, New York, 1981.

Edwards, P., Integrated fish farming, *INFOFISH Int.,* 5/91, 45–52, 1991.

EPA (U.S. Environmental Protection Agency), Biological Field and Laboratory Methods For Measuring the Quality of Surface Waters and Effluents; Chapter Biometrics, Weber, C. I., Ed., Office of Research and Development, U.S. Environmental Protection Agency, Cincinnati, OH, 1973, 27 pp.

Finney, J., Was this in your statistics textbook? II. Data handling, *Expl. Agric.,* 24, 343–353, 1988a.

Finney, J., Was this in your statistics textbook? III. Design and Analysis, *Expl. Agric.,* 24, 421–432, 1988b.

Gopen, G. D. and Swan, J. A., The science of scientific writing, *Am. Sci.,* 78, 550–558, 1990.

Green, B. W. and Teichert-Coddington, D. R., Comparison of two samplers used with an automated data acquisition system in whole-pond, community metabolism studies, *Progr. Fish-Cult.,* 53, 236–242, 1991.

Heath, O. V. S., *Investigation by Experiment, No. 23 in Studies in Biology,* The Camelot Press, Ltd., London, 1970, 74 pp.

Holm-Hansen, O. and Riemann, B., Chlorophyll a determination: improvements in methodology, *Oikos,* 30, 438–447, 1978.

Hopkins, K. and Yakupitiyage, A., Bias in seine sampling of tilapia, *J. World Aquaculture Soc.,* 22(4), 260–262, 1991.

Jackson, C. I. and Prados, J. W., Honor in science, *Am. Sci.,* 71, 462–464, 1983.

Kamphake, L. J., Hannah, S. A., and Cohen, J. M., Automated analysis for nitrate by hydrazine reduction, *Water Res.,* 1, 205–216, 1967.

Knud-Hansen, C. F., Pond history as a source of error in fish culture experiments: a quantitative assessment using covariate analysis, *Aquaculture,* 105, 21–36, 1992a.

Knud-Hansen, C. F., Analyzing standard curves in water chemistry, International Center for Aquatic Living Resource Management (ICLARM), *NAGA ICLARM Q.,* January, 16–19, 1992b.

Knud-Hansen, C. F. and Batterson, T. R., Effect of fertilization frequency on the production of Nile tilapia (*Oreochromis niloticus*), *Aquaculture,* 123, 271–280, 1994.

Knud-Hansen, C. F., Batterson, T. R., and McNabb, C. D., The role of chicken manure in the production of Nile tilapia (*Oreochromis niloticus*), *Aquaculture Fish. Manage.,* 24, 483–493, 1993.

Likens, G. E., Importance of perspective in limnology, in *An Ecosystem Approach to Aquatic Ecology,* Springer-Verlag, New York, 1985, 84–88.

Meade, J. W., Allowable ammonia for fish culture, *Progr. Fish-Cult.,* 47(3), 135–145, 1985.

Parker, R. E., *Introductory Statistics for Biology, No. 43 in Studies in Biology,* The Camelot Press, Ltd., London, 1979, 122 pp.

Peterman, R. M., Application of statistical power analysis to the Oregon coho salmon (*Oncorhynchus kisutch*) problem, *Can. J. Fish. Aquat. Sci.,* 46, 1183–1187, 1989.

Peterson, R. G., Use and misuse of multiple range comparison procedures, *Agron. J.,* 69, 205–208, 1977.

Searcy-Bernal, R., Statistical power and aquacultural research, *Aquaculture,* 127, 371–388, 1994.

Steel, R. G. D. and Torrie, J. H., *Principles and Procedures of Statistics,* 2nd ed., McGraw-Hill, New York, 1980, 633 pp.

Strickland, J. D. H. and Parsons, T. R., A Practical Handbook of Seawater Analysis, Fisheries Research Board of Canada, Ottawa, 1972, 310 pp.

Van Dam, A. A., Multiple regression analysis of accumulated data from aquaculture experiments: a rice-fish culture example, *Aquaculture Fish. Manage.,* 21, 1–15, 1990.

Wetzel, R. G., *Limnology,* 2nd ed., Saunders Publishing, Philadelphia, 1983, 767 pp.

Wetzel, R. G. and Likens, G. E., *Limnological Analyses,* Saunders Publishing, Philadelphia, 1979, 357 pp.

Yoccoz, N. G., Use, overuse, and misuse of significance tests in evolutionary biology and ecology, *Bull. Ecol. Soc. Am.,* 72(2), 106–111, 1991.

15   ECONOMIC CONSIDERATIONS

Carole R. Engle, Revathi Balakrishnan, Terry R. Hanson, and Joseph J. Molnar

INTRODUCTION

An important benefit of the PD/A CRSP is the positive economic and social impact that results from farmers adopting new technologies. CRSP researchers in collaboration with host-country institutions have generated knowledge that increases the understanding of the dynamics of fish ponds. This knowledge is then used to develop new technologies that can be applied by producers (Rogers, 1962). The technology or knowledge diffusion process varies across different countries and regions but often involves an extension component that serves to link research functions and output with the production sector.

The principal focus of CRSP research since its inception has been the study of the biological and chemical dynamics of pond aquaculture systems. One end purpose of this effort is to contribute to economic development through aquaculture production. However, for this to happen, the recommendations must be consistent with the economic and social environment within which these technologies are to be adopted.

In the early years of the PD/A CRSP, primarily pond dynamics studies were conducted. As the overall research effort developed, additional emphasis began to be placed on conducting some sociological and economic studies related to the technologies being developed. As has been the case in many other aquaculture development projects, social scientists in the PD/A CRSP had originally been called upon on a short-term basis to support the aquaculture research program. This type of approach led to a limited scope of economics and social science involvement in the PD/A CRSP prior to 1992.

Involvement of social scientists in the PD/A CRSP has increased gradually over time. However, the limited nature of the economic and social science research that has been conducted to date precludes definitive answers to the following questions: (1) Should aquaculture be promoted to generate income or serve as a nutritional subsidy? (2) What are the key factors that determine economic feasibility of aquaculture that are common to all regions across the world? Yet the key findings from these studies still shed some light on these questions and provide some evidence that contributes to the attempt to provide answers. This chapter summarizes and compares what has been done to date.

Assessment of the economic viability of aquaculture projects requires an interdisciplinary understanding of physical, social, and economic relationships. Quantitative or qualitative estimates of many relevant variables are not readily available because aquaculture is in early stages of development in most areas of the world.

Incorporating economic analysis into aquacultural experiments ensures that unprofitable technologies are not pursued. While a given technology may maximize yields, other technologies may yield higher profits. Market and social constraints vary from region to region and determine product acceptability. Production research that targets those products with greatest market and social

0-56670-274-7/97/$0.00+$.50
© 1997 by CRC Press LLC

acceptance will result in more rapid growth and development of aquaculture. This will increase the overall economic and social impact of the PD/A CRSP.

Technology adoption occurs at many levels, from the researchers' technical decisions on what to investigate down to the micro- or farm-level. However, once a technology is developed, a farmer's decision to adopt a new technology will depend upon many factors that range from simple costs and returns to market factors to complex interactions between the new technology and the farming system practiced by the farmer.

The focus of this chapter is to review the economic and sociological research that has been completed by the PD/A CRSP in Honduras, Rwandan, and Thailand sites. It is not the intent to review the entire literature on the economic and social considerations of aquaculture but rather to focus on the contributions of the social sciences to the PD/A CRSP effort.

TOOLS OF ECONOMIC ANALYSIS

The economics literature includes a host of tools that can be used to address a multitude of economic questions and issues. Which tool is selected is often determined by the type, scope, and quality of data that can be collected in a reasonable period of time. This is particularly true in many developing countries. The following summarizes two such tools that have been used in economics research related to the PD/A CRSP effort.

The Enterprise Budget

The enterprise budget has been the cornerstone of much aquaculture production analysis. An enterprise budget is an estimate of all income and expenditures of a specific enterprise and an estimate of the enterprise's profitability at a single point in time (Kay, 1986). It is an important first step in the analysis of the production economics of any farm product.

Enterprise budgets are often used to assess overall profitability of a new enterprise. For any one given farm, they can be used to compare the relative profitability or relative cost components of different enterprises on that farm. If reliable survey data are available, enterprise budgets can be based on actual farm costs and returns. In the absence of surveys, economic engineering techniques, based on the design of synthetic farms and using research-based production recommendations, are used to develop estimates of costs and returns. The PD/A CRSP Data Analysis and Synthesis Team has developed the PONDCLASS© expert system and POND©, a decision support system, based on data collected from the global experiments. One product of this modeling effort (POND©) includes quick economic profiles of varying tilapia production schemes.

In enterprise budgeting, prices and quantities of the output produced and production inputs used are fixed at a selected point in time. The profitability of the enterprise may change as these prices change over time. Economists use sensitivity analysis to simulate the impact of price changes for important input and output variables of enterprise budgets. Enterprise budgets also do not assess the entire farm operation but only the one enterprise under consideration.

Mathematical Programming Models

Many of the economics issues critical to aquaculture production involve more factors than can be captured in a simple enterprise budget. Whole-farm analysis using mathematical programming techniques (Hazell and Norton, 1986) can incorporate factors such as market, labor, and capital constraints. Conflicts of labor or capital availability across the whole farm can be analyzed within this type of analysis.

Many techniques exist to incorporate risk elements into these models. Mathematical programming models do not analyze market demand and supply factors. Changes in the factors underlying demand and supply often are not accounted for in mathematical programming analysis.

ECONOMICS OF TILAPIA PRODUCTION

Honduras

Green et al. (1994) developed a series of 41 enterprise budgets (using economic engineering techniques and experimental PD/A CRSP data) to compare the relative profitability of various PD/A CRSP nutrient regimes for tilapia production in Honduras. Nutrient input regimes evaluated included (1) inorganic fertilization, (2) organic fertilization only, (3) organic plus inorganic fertilization, (4) organic fertilization plus supplemental feed, and (5) feed only (Table 1). Within each nutrient input regime, treatments included varying rates of feeding and fertilization and varying stocking rates of tilapia.

Net returns were positive for 12 nutrient regimes for tilapia production in Honduras, but the profitability varied with the production technologies used (Table 1). None of the "inorganic fertilization only" treatments had positive net returns. For the other nutrient input regimes analyzed, only those with stocking rates of at least 2 male tilapia/m² were profitable. All lower stocking rates (with the exception of the chicken manure (CM) 1000 kg/ha/wk treatment), regardless of nutrient input levels, were not profitable at assumed parameter costs.

Only two of the "organic fertilization only" treatments had positive net returns. Both of these were treatments at the higher fertilization rates of either 750 or 1000 kg/ha/wk of chicken manure. There may be a threshold of pond productivity based on natural food availability that results in greater marginal returns of fish as compared to slightly higher costs of additional manure. The 1000 kg CM/ha/wk was the only profitable treatment stocked at the lower rate, while this same fertilization rate was unprofitable at the higher stocking rate. This difference is likely due to yield variations caused by seasonal differences.

The C6:N1 treatment within the "organic plus inorganic fertilization" group had the highest net returns. In this treatment weekly chicken litter applications of 750 kg dry matter/ha were supplemented with inorganic nitrogen (30 kg urea/ha) to decrease the C:N ration from 11:1 to 6:1. Results were very similar in a subsequent experiment where weekly chicken litter input was reduced to 500 kg/ha and supplemental urea was increased to 35 kg/ha (Teichert-Coddington et al., 1993). All treatments including the control (C11:N1) were run during the same time. The warm temperatures affected all equally. In fact, the weather was not "exceptionally" warm. Rather, the experiment was run entirely during the warm season instead of bridging the cool season, as in previous years. All C:N ratios were stocked at the profitable 2 tilapia/m² rate, and all had positive net returns.

In the "organic fertilization plus supplemental feed" category, the CM60 per feed (chicken manure for first 60 d at 1000 kg/ha/wk followed by 3% bodyweight feed until harvest) and CM plus feed treatments (chicken manure entire cycle, 3% bodyweight feed per day) and aeration treatments of 10 and 30% saturation had positive net returns. The 10 and 30% dissolved oxygen saturation treatments maintained approximately 1.0 and 2.5 ppm dissolved oxygen, respectively, at 25°C water temperature. Nutrient inputs for the aeration treatments included 2 months of CM at a rate of 1000 kg/ha/wk followed by 3 months feeding at 3% of bodyweight per day. A control that had no aeration had negative net returns, while the other two aeration treatments had positive net returns. The 30% saturation treatment had a higher positive net return than the 10% treatment, even with the additional aeration costs included (Table 1).

In the "feed only" treatments, trials conducted over the wet season had positive net returns, while those conducted in the dry season had negative net returns. In the dry season, water temperatures were lowered due to the evaporative effect of winds over the pond water surface. The cloud cover during the wet season kept the environment hot and humid.

Sensitivity analysis of feed prices showed no change in which nutrient regimes were profitable, as feed price increased from $12.04 to $14.4 per bag (45.4-kg bag of feed), although the level of profitability did change. When feed prices were decreased from $12.04 to $9.63 per bag, the no-aeration alternative became profitable. Neither 20% increases nor 20% decreases in prices of urea or in interest rates changed whether or not alternative regimes were profitable.

Table 1   **Honduras CRSP Enterprise Budget Summary Results for Five Nutrient Categories in Honduras, 1983–1991 (in U.S.$)**

Item	Stocking rate (fish/m²)	Income[a] above variable costs ($)	Net returns to land & mgmt. ($/ha)	Break-even price (variable cost) ($)	Break-even price (total cost) ($)
Inorganic fertilizer only					
1. TSP[b]	1	−680	−362	2.97	5.87
2. Urea + TSP	1	38	−229	1.41	2.39
Organic fertilizer only					
1. Cow manure	1	1071	−38	0.77	1.56
2. CM[c] (750 kg/ha/wk)	0.25	−307	−293	1.87	3.48
3. CM (750 kg/ha/wk)	1	1058	−41	0.90	1.57
4. CM (750 kg/ha/wk)	2	1677	74	0.79	1.30
5. 125 kg CM/ha/wk	1	325	−176	1.16	2.25
6. 250 kg CM/ha/wk	1	819	−85	0.90	1.74
7. 500 kg CM/ha/wk	1	982	−20	0.82	1.49
8. 1000 kg CM/ha/wk	1	1508	43	0.79	1.34
9. 1000 kg CM/ha/wk	2	1083	−36	0.94	1.52
10. CM 750 per no urea	1	704	−106	1.01	1.79
Organic plus inorganic fertilizer					
1. CM 750 per urea	1	799	−89	1.00	1.71
2. C:N[d] control	2	2041	141	0.70	1.16
3. C8:N1	2	1918	119	0.73	1.20
4. C6:N1	2	3171	351	0.56	0.91
5. C4:N1	2	1677	74	0.81	1.28
Organic fertilizer plus supplemental feed					
1. CM 500 (0.0% feed)	1	169	−205	1.29	2.37
2. CM 500 (0.5% feed)	1	338	−194	1.19	2.18
3. CM 500 (1.0% feed)	1	494	−165	1.13	1.97
4. CM 500 (2.0% feed)	1	290	−203	1.27	2.04
5. CM 750 1 mo. then feed 3.0%	1	403	−182	1.29	1.88
6. CM750 2 mo. then feed 3.0%	1	307	−200	1.31	1.95
7. CM750 3 mo. then feed 3.0%	1	694	−128	1.14	1.77
8. CM 60 (feed)	2	2799	261	0.82	1.12
9. CM 500 (1.5% feed)	2	2898	280	0.72	1.06
10. No aeration	2	1146	−45	1.10	1.51
11. 10% sat	2	2134	122	0.92	1.28
12. 30% sat	2	2286	150	0.90	1.25

Table 1 (continued)

Item	Stocking rate (fish/m²)	Income[a] above variable costs ($)	Net returns to land & mgmt. ($/ha)	Break-even price (variable cost) ($)	Break-even price (total cost) ($)
Feed only					
1. 10,000/ha/feed[e]	1	836	–82	1.08	1.61
2. Feed only[f]	2	2260	182	0.94	1.23
3. 20,000/ha/feed[e]	2	1024	–47	1.15	1.50
4. 30,000/ha/feed[f]	3	2015	137	1.01	1.28

[a] 1992 tilapia market price = L.7.72.

[b] TSP = triple superphosphate (0-46-0).

[c] CM = chicken manure.

[d] C:N = carbon:nitrogen.

[e] Dry season.

[f] Wet season.

Adapted from Green, B. W., Teichert-Coddington, D. R., and Hanson, T. R., 1994.

As the selling price of tilapia increased, more nutrient regimes became profitable. Increases in price would need to come from increased demand due to increased consumer income, increased population, or changing consumers' tastes and preferences.

Rwanda

Aquaculture in a Subsistence Economy

Rwanda poses a unique opportunity to evaluate the benefits of subsistence aquaculture. In Rwanda, for example, 37% of the population is deficient in caloric intake (World Bank, 1989). In some areas, as high as 82% of the population has caloric deficiencies. Protein deficiencies are found in 64% of the population nationwide but are 85% in some regions of the country (World Bank, 1989).

A preliminary survey was conducted initially to develop cost and returns estimates for fish production at CRSP and non-CRSP sites in Rwanda (Engle et al., 1993). This study was followed later by a comprehensive cost of production study that included economic data on production of other crops (Hishamunda et al., 1993). This study resulted in comparative cost and returns for fish and other crops. These data were later used in a more rigorous programming analysis of Rwandan subsistence farmers' decision-making when faced with extremely limited resources and a need to provide adequate food for household consumption. Key findings from these studies are discussed below.

Returns to resource utilization in fish farming were estimated for Rwanda by Engle et al. (1993) based on a survey conducted in 1989. Fish farming clearly was considered to be a cash crop by the survey respondents, although fish from the ponds were also used to supplement the diet.

In traditional cost and returns (enterprise budget) analysis, all costs are converted to monetary values as a common denominator. In the case of a subsistence economy, such as Rwanda, the marais land that is used for fish production is public, government-owned land that is allocated to individual and group farmers at no charge. In Rwanda, ponds are constructed by hand, not by hired labor, by the individuals or groups who will manage the ponds. The investment in this case is a labor rather than a capital investment. Traditional economic analysis assumes that capital can be directly substituted for labor, but in the Rwandan subsistence economy, in a noncash, nonmonetary context, the validity of assuming perfect substitutability between labor and capital is questionable (Engle et al., 1993).

The primary source of inputs into fish production in Rwanda is compost made from vegetative matter. Given the general lack of cash and money, most farmers cut grass or use weeds pulled from other crops to add to a compost pile located in the pond itself. Operator's and family labor is the primary variable resource used to add nutrients to ponds. Traditional analyses charge interest on operating capital, but in this situation as in the above discussion on fixed cost, there is an extremely low percentage of capital used. In fact, many farmers use no operating capital but supply all production inputs with their own unpaid labor. Labor represented 93% of all resources used in fish production in Rwanda, while noncash capital resources represented 0.2% and cash resources 7%. Average net returns to labor per are were $6.66/are. Given the lack of capital used in subsistence economies and the high amount of labor, it is more useful to evaluate the economics of a productive activity on the basis of labor rather than capital.

The labor efficiency of food fish production was 0.3 kg of fish produced per person-day of labor (Table 2). For fingerling production, the labor efficiency was 7 fingerlings per day of labor. The total value of labor used in fish production was $0.44 per person-day of labor. While the official daily wage rate in Rwanda is $0.69, employment opportunities, in reality, are occasional, few, and limited to areas surrounding urban centers. The value of fish produced represents the average daily returns from spending time to produce fish. Results indicated that labor, the primary resource used in fish farming, yielded a return that was competitive with daily wage rates even though realistic employment opportunities were scarce.

Table 2   Labor Efficiency and Value of Labor Utilized on Fish Farms in Rwanda

Item	Unit	Value
Harvested production		
Food fish harvested	kg/pond	47
	U.S.$	23
Total value of production	U.S.$	67
Labor		
Production	Person/day	127
Harvest	Person/day	25
Total	Person/day	152
Labor efficiency of food fish production	kg fish/d	0.3
Labor efficiency of fingerlings produced	Fingerlings/d	7.3
Total value of labor used in fish production		44

Adapted from Engle, C. R., Brewster, M., and Hitayezu, F., 1993.

Costs and Returns of Fish Compared to Other Crops

The only actual economic survey conducted of farmers using PD/A CRSP-generated technologies was conducted in Rwanda (Hishamunda et al., in press). While other analyses used experimental data to construct enterprise budgets using economic engineering techniques, this analysis used farm data.

This enterprise budget analysis showed that fish production is superior to other alternative crops that can be raised in the marais in Rwanda in terms of cash income per unit of land (Hishamunda et al., 1993). The importance of this finding lies in the fact that prevailing thought considers aquaculture in a subsistence economy to be primarily a source of animal protein for household consumption. In the case of Rwanda, the higher net returns to land, labor, and management from fish production compared to other crops explains why farmers sold over half of the fish produced.

All enterprises, except Irish potatoes, showed positive income above variable costs, and the fish enterprise (if fingerlings produced could also be sold) had the highest net returns to land, labor, and management (Table 3). The cabbage enterprise ranked second. There was virtually no hired

Table 3 Cost, Returns, and Nutritional Benefits of Marais Agricultural Enterprises

Crop	Net returns to land, labor & mgmt. (U.S.$/are)	Net returns labor/mgmt (U.S.$/are)	Protein (kg/are/yr)	Carbohydrates (kg/are/yr)	Energy (kcal/are/yr)
Fish					
Coop.	15.87	−55.88	2.865	0	15,120
Ind.	17.71	−25.24	2.963	0	15,640
Sweet potato					
Coop.	1.39	−39.80	1.845	33.550	139,491
Ind.	4.45	−11.17	2.104	38.246	159,015
Irish potato					
Coop.	−3.24	−46.41	1.007	14.386	48,575
Ind.	1.43	−13.14	1.668	23.830	80,461
Cassava					
Coop.	4.25	−45.34	0.280	20.354	24,788
Ind.	0.89	−11.59	0.299	21.750	26,489
Taro					
Coop.	2.14	−47.10	0.788	14.735	44,772
Ind.	2.48	−11.03	0.888	16.611	50,471
Sorghum					
Coop.	2.12	−31.46	1.960	19.460	83,269
Ind.	1.43	−7.88	1.264	12.549	53,697
Maize					
Coop.	4.48	−31.52	9.356	33.080	150,256
Ind.	2.83	−10.17	3.103	25.952	117,879
Sweet pea					
Ind.	0.68	−2.36	1.095	3.040	16,651
Beans					
Coop.	5.72	−31.32	6.658	5.645	103,064
Ind.	1.59	−8.97	4.506	3.820	69,754
Soybean					
Coop.	2.48	−26.41	8.235	5.301	97,266
Ind.	2.28	−6.96	5.962	3.838	70,424
Peanuts					
Ind.	12.55	−2.41	1.923	2.782	45,591
Rice					
Ind.	6.59	−3.96	2.115	40.810	109,710
Cabbage					
Coop.	12.91	−39.30	1.783	4.754	2,734
Ind.	17.30	8.20	2.339	6.235	3,586

Adapted from Hishamunda, N., et al., 1993.

labor used on Rwandan farms and much evidence for surplus labor. When opportunity costs were charged for family labor in the analysis, the only profitable alternative was cabbage produced by individually owned farms. A detailed description of labor in subsistence settings is discussed later in this chapter.

In terms of carbohydrates, rice (41 kg/are/yr) and sweet potatoes gave the highest yields (44 to 38 kg/are/yr). The highest amount of energy produced by cooperatives was obtained from maize production (150,256 kg/are/yr), while individually managed farms obtained more energy from sweet potatoes. Fish farming yielded the second lowest quantity of energy/are/yr (15,120 to 15,640 kcal/are/yr).

The least expensive source of protein (without considering opportunity costs of family labor) was maize for cooperatives and peanuts produced by individual farmers. With the cost of family labor included, soybeans were the least expensive source of protein. Costs of producing various forms of animal protein other than fish (e.g., poultry, pork) were not included in this analysis.

Resource Allocation on Subsistence Fish Farms

The farm cost of production survey was one of the few done anywhere in Africa to collect data on how fish farming fits into subsistence economies. Given the comprehensive nature of the data set, representative farms could be modeled using mathematical programming techniques.

This study showed that average land holdings in Rwanda were not adequate to meet nutritional requirements of the family. Land holdings ranged from 1 to 16 ares for individually managed farms and from 0 to 44 ares per cooperative member. When the land allocation was increased to 11 ares per family, the model generated a solution in which 9 ares were allocated to soybean production and 2 ares to fish production for a net annual income of $52.23 (U.S.) (Table 4).

When the energy and protein requirements were removed, the cabbage enterprise was selected to maximize profits in the absence of fish. Without family nutritional requirements, net income

**Table 4   Mathematical Programming Results of Optimal Resource Allocation
on Fish Farms in Rwanda**

Scenario	Net income (U.S.$)	Fish (ares)	Soybeans (ares)	Cabbage (ares)	Peanuts (ares)
No restrictions	52.23	2	9	—	—
Nutritional constraints					
w/o Fish	44.11	—	10	1	—
w/o Fish, Energy & Protein	190.34	—	—	11	—
w/o Fish & Protein	190.34	—	—	11	—
w/o Fish & Energy	44.11	—	10	1	—
w/o Energy & Protein	220.38	11	—	—	—
w/o Protein	220.38	11	—	—	—
w/o Energy	52.23	2	9	—	—
Land constraints					—
15 ares	202.52	9	6	—	—
20 ares	355.97	13	2	5	—
25 ares	464.29	12	—	13	—
30 ares	547.14	10	—	20	—
35 ares	629.99	9	—	26	—
40 ares	706.86	7	—	31	1
50 ares	785.67	2	—	30	17

Adapted from Engle, C. R., et al., 1994.

increased to $190 without fish and to $220 with fish. Farmers may be better off selling fish to generate income to purchase food supplements to improve nutritional levels of Rwandan farm families.

These results indicate that fish enters into the farm product mix of Rwandan farmers primarily as a source of cash income. Land area put into fish is increased to maximize profits only after sufficient soybeans have been raised to satisfy nutritional requirements. At very large (relatively, for Rwanda) sizes of land holdings, adequate nutrition could be derived from peanuts in addition to fish, and overall farm income would be higher than raising soybeans.

Thailand

The Thailand Pond Dynamics/Aquaculture CRSP has, over the course of several years, conducted intensive research on the biological and chemical factors that influence freshwater fish production in Thailand.

Cost and Returns from Fish Farming

Four farm management case studies were conducted to construct a data set of representative farm management indicators for analyzing the economics of tilapia production in Thailand (Engle and Skladany, 1992). Enterprise budgets were developed for fish production with inorganic fertilization, with chicken manure collected from chicken coops, and with chicken manure dropped into ponds from chicken coops constructed over the ponds.

Net returns were highest for fish produced with chicken manure in integrated systems, the second highest were for fish produced with collected chicken manure, and the lowest were for fish produced with inorganic fertilizer (Table 5). The lower yields with inorganic fertilizer caused the lowest net returns for the latter scenario, while the higher production cost of collected chicken manure produced lower returns for collected chicken manure systems as compared to integrated chicken-fish systems.

This study showed results similar to those reported in Honduras. In both these studies, production systems using organic fertilizers were more profitable than those using inorganic fertilizers alone.

Table 5   Cost and Returns to Fish Production (1-ha pond) in Thailand

Item	Inorganic fert. (U.S.$)	Collected chicken manure (U.S.$)	Chicken coops over ponds (U.S.$)
Gross revenue	2,230	2,896	2,896
Variable cost	994	1,010	908
Income above VC	1,236	1,886	1,988
Fixed cost	471	471	471
Total cost	1,464	1,481	1,379
Net returns	766	1,416	1,517
Break-even yield	122	123	115
Break-even price	0.32	0.24	0.23
Return to avg. investment in ponds and equipment	72%	132%	142%
Return to average total investment	5%	10%	10%

Adapted from Engle, C. R. and Skladany, M., 1992.

Comparing Economic Returns from Fish Production to Those from Other Livestock Enterprises

Engle and Skladany (1994) compared the economics of fish production with that of other livestock operations in Thailand. The economic indicators utilized shed light on the profitability and return on investment (ROI) of fish compared to other livestock production alternatives. This study provides further evidence of the diversification strategies utilized by small commercial-scale farmers worldwide to reduce both production and financial risk. Poultry and hog production required high levels of operating capital and little investment. Fish ponds, on the other hand, represented a larger capital and land investment but required little operating capital. Once the pond was constructed, for little additional operating capital, attractive profit levels could be obtained.

The greatest total benefits were obtained from the chicken manure alternatives (Table 6). This largely reflects the increased yield of fish. Net benefits for all scenarios were positive, but highest net benefits were from those options locating the livestock enterprises directly on the pond or levee.

Table 6   Resource Requirements and Profits for Fish and Other Livestock Enterprises, Thailand

Enterprise	Total investment capital (U.S.$)	Annual operating capital (U.S.$)	Estimated profit	
			No land cost (U.S.$)	Land cost of baht/ha (U.S.$)
Poultry-layers	1,270	9,441	852	821
Poultry-broilers	1,092	9,222	881	857
Hogs	1,010	8,250	853	834
Fish w/inorganic fertilizer	2,140	1,050	709	−851
Fish w/collected chicken manure	2,140	1,067	1,359	−201
Fish-chicken	3,410	10,400	2,271	711
Fish w/collected hog manure	2,140	1,145	1,049	−511
Fish-hogs	3,150	9,209	2,033	473

Adapted from Engle, C. R. and Skladany, M., 1994.

Investment capital requirements were higher for fish enterprises than for hogs or poultry. Total investment cost was lowest for the hog alternative, followed by poultry-broilers, poultry-layers, fish-inorganic fertilizer, collected hog and chicken manure, and integrated fish-layers and was highest for the integrated fish-hog operation. Annual operating capital was lowest for the fish enterprises, followed by hogs and chickens. Profit levels from fish production were competitive with those of other livestock enterprises.

Philippines

Philippines PD/A CRSP data were used to compare net economic benefits for the following scenarios: triple superphosphate (TSP) trials during the dry and wet seasons, no feeding or feeding with supplemental chicken manure, or feeding with supplemental inorganic fertilizer, triple monophosphate (TMP, 16-20-0). Wet season chicken manure fertilization trials were conducted at rates of 125, 250, 500, and 1,000 kg/ha/wk (Molnar, Hanson, and Lovshin, 1996). No dry season replications of these treatments were conducted. Dry and wet season TSP trials and wet season trials varying chicken manure produced good yields and positive economic benefits. Treatments combining chicken manure with TMP generated positive economic benefits. However, when feed was added to chicken manure, economic benefits were negative.

When TMP was added to a schedule of feed and chicken manure, production exceeded all other nutrient regime treatments. Nonetheless, three out of five treatments resulted in negative net benefits that were caused by the high cost of commercial feeds. The production alternatives that produced highest yields were not the most economically beneficial. Likewise in Honduras, CRSP researchers have found that biologically optimal production is not necessarily the most profitable.

SOCIAL FACTORS IN AQUACULTURE

Social factors play an important role in technology adoption. In this section, attitudes toward fish farming and its sustainability and gender and family issues related to fish farming will be discussed. This discussion primarily focuses on Rwanda as a context for aquacultural development.*

Attitudes Toward Fish Farming and Its Sustainability

Molnar et el. (1991) obtained data from a sample of 186 Rwandan farmers taken from project rolls throughout the nation. The survey responses suggest a relatively uniform expectation for continuing fish culture, even though many of the factors affecting the sustainability of aquaculture as a farm enterprise relate to the provision of infrastructure that is beyond farmer control.

To be self-sustaining, new technologies like fish culture must conform with the environments where they will be used and interact positively with other activities within the farming system. Otherwise, they will fail to be implemented, and an opportunity to improve food security, nutrition, and income will be lost (Cernea, 1985; Francis and Hildebrand, 1990).

A significant aspect of sustainability is the elimination of dependence on government services for continuing the enterprise. In some locales, seedstock is produced by the public sector that oversees its distribution and utilization by fish producers. In such situations, overall success of aquaculture turns on the efficacy and reliability of the national hatchery system. At later stages of development, better farmers may become seedstock suppliers in their local areas.

Table 7 shows the responses to four survey items reflecting the sustainability of fish culture in Rwanda. More than half the farmers had engaged in fingerling sales. Although small fish can also be purchased for consumption, the availability of seedstock from private sources reduces dependence on state-run hatcheries.

Respondents were asked two questions about possible conflicts between fish culture and their other activities. The combined index shows that few perceived any incongruity with other enterprises. Fish culture seems to be established as a nearly autonomous farm activity in Rwanda. About 91% of the respondents planned new ponds, suggesting a positive outlook for fish culture. Nearly as many respondents thought they could do without extension assistance.

Table 7 further shows the frequency distribution of various measures of technical commitment to fish culture. Most operators visited their ponds every day. Most spent about an hour on each visit. Similarly, 58% reported daily feeding. Women and operators with more cash enterprises reported shorter visits to their ponds, but operators with more food enterprises tended to report staying longer on each visit.

---

* The unrest that troubles Rwanda and much of Africa in the 1990s presents obstacles to aquacultural development. Institutional functioning is less than optimal when resources must be diverted to military support. Personal security for research and extension personnel can be a problem in areas experiencing unrest. Heightened awareness of ethnic, class, or political divisions may disrupt extension contacts across antagonistic groups. In general, conflicts disrupt the orderly provision of government services and the overall process of economic growth and social development. Aquaculture is not sheltered from these forces. Despite the obstacles, Rwandan farmers have embraced fish culture as a farm enterprise with sufficient intrinsic merits and market incentives to continue with or without government support. The PD/A CRSP experience can inform other aquacultural operations in Subsaharan Africa, demonstrating the contributions to be made by aquacultural development.

Table 7   Indicators of Sustainability, Technical Commitments, and Extension Perceptions, Rwanda, 1989

Indicator	Number (no.)	Percent (%)
Sustainability		
1. Cash marketing of seedstock	109	59
2. Anticipate expansion of fish culture	169	91
3. Know enough to do without extension assistance	160	86
4. High response to how fish culture fits with other farm activities	111	96
Technical commitment		
1. Visit pond every day	94	50
2. Visit pond several times a week	73	39
3. Spend an hour at pond	96	53
4. Feed every day	107	58
5. Feed two or more different types of feed	170	92
Extension perceptions		
1. A count of three or more helpful aspects of extension assistance	136	74
2. Satisfaction rating of five aspects; no. of individuals who scored it 5 (highest)	144	93
3. Two or more extension visits/month	151	87
4. Extensionist sometimes comes when expected	87	29
5. Extensionist always comes when expected	50	28

Finally, while about a fourth of the farmers mainly used one or two substances as pond inputs, 24% mentioned five or more items as fish feed. Number of feeds used was negatively associated with the frequency of extension visits. The number of cash enterprises was positively associated with feed diversity, suggesting that wealthier farmers had more byproducts to put in their ponds.

Farmers rated various aspects of the technical assistance they received. Some farmers mentioned five or more aspects of fish culture for which the extensionist was helpful (Table 7). Most respondents gave affirmative responses on all five of the specific subjects where *moniteurs* (Rwandan extension agents) were expected to be informed and useful.

Several contextual factors not measured in this study affect the prospects for aquaculture in Rwanda, regardless of its ecological, socioeconomic, or nutritional merits. The commitment of the Rwandan government may shift to other priorities. The financial condition of the country may disrupt the payment of *moniteur* salaries or fail to provide sufficient resources for recruiting and training replacements. It may fail to allocate sufficient travel funds for the extension *moniteurs*. Farmers have little way of knowing or understanding the larger national questions about the direction of agricultural policy or the status of foreign exchange accounts and the need to redirect spending to export crops.

Molnar et al. (1994) further analyzed those farmers who had ceased to raise fish. The segment of farmers that stopped growing fish did so for reasons other than dissatisfaction with the enterprise. They were slightly more involved in other farm enterprises, and their reasons were more concerned with specific household or community circumstances. Farmers who ceased fish farming perceived more time and effort conflicts with other farm enterprises and household work. They were more interested in the cash proceeds of fish culture than the other sample segments and less likely to feel that the pond was the best use of the land it occupied.

Women in groups seemed the most satisfied and productive segment of the study respondents. They had larger harvests, experienced fewer marketing problems, and were more attentive to the general practice of fish culture. They also seemed to get better prices. Women in groups seemed better able to exploit pond bank sales as a marketing channel for tilapia. Friends, relatives, and neighbors are an immediate network of fish consumers who are readily alerted and mobilized to purchase fish at harvest. Women in groups represent an overlay of multiple social networks.

Gender and Family Issues

As is common in Africa, Rwandan women are active participants in labor-intensive subsistence production. In Rwanda, women represent 51% of the population; 98% of the women are agriculturalists contributing to the nation's food supply, although 78% of households are headed by males (Nyirahabimana, 1991). But the women subsistence farmers are also burdened by constraints of high illiteracy, limited access to land, lack of property ownership rights, frequent pregnancy, and excessive demands on their time. Ford (1990) summarizes the division of labor along gender lines described here. Women do most of the work on subsistence crops (cereals, tubers, legumes), while men work with cash crops (bananas, coffee). There are shared gender tasks for land-clearing activities and for crops that require special heavy work at harvest time, such as manioc. Weeding and other crop maintenance are almost totally women's work, as are most post-harvest activities. Women are also responsible for water-related tasks, such as providing drinking water and water for sanitation. Excluding fish farming chores, general water-related tasks can take up to one-third of a rural woman's time each day. Men are responsible for most money transactions outside the home, such as buying supplies and selling banana beer. Women, however, are the primary producers of banana beer, and use the byproduct to fertilize the ponds. General field-level observations indicate that women work side-by-side with other women in all subsistence production activities including aquaculture.

Extending their roles in subsistence production to aquaculture, Rwandan women have contributed to the rapid adoption of fish culture. "Women farmers are most productive and easiest to work with. Extension workers are asked to seek out women trainees on the rationale that women's economic rewards for fish enterprise will be invested in family well-being" (Nyirahabimana, 1989). In 1990, 25% of the fish farmers were women. However, it is estimated that more than half of the fish ponds were actually "managed" by women although these women were not included in the preceding estimate of the percentage of women fish farmers. "Managed" in this context means that the pond belongs to the family but women are in charge of fertilizing the pond and feeding the fish. But constraints on women's full productive participation in aquaculture continue to persist.

A qualitative study was conducted in Rwanda in 1992 to document the impact of aquaculture intervention on the farm households within their ecological, social, and economic milieu and to identify the women's contributions and constraints which hamper aquaculture productivity at the household level (Balakrishnan et al., 1992). The primary respondents were Rwandan women participating in aquaculture activities, extension agents promoting aquaculture, and scientists involved in aquaculture research. Furthermore, experiences and observations of Rwandan professionals in nongovernment organizations and government agencies involved in aquaculture efforts provided additional information.

Aquaculture technology interventions to improve farm-level production affect resource dynamics and foster involvement of community through farmer organizations. Family resource impacts are associated with household labor allocation and family economics variables, such as opportunities for increased income, expanded choice in food variety, and quality of and access to marais land. Aquaculture practices adopted by Rwandan farm families advance renewable resource management practices in marais land. At the community level, aquaculture production groups have created social organizations to share the tasks and benefits. A major benefit is access to common property land to develop fish ponds.

Among the Rwandan farm households, seasonal fluctuations in food availability are common. The goal of the household is to obtain a good harvest and to plan a food reserve for the rainy season or for prolonged drought. Aquaculture intervention is a viable production alternative for marais land, when population growth and accompanying demand on land resources threaten land productivity and household resource sustainability.

Men own tools and perform tasks requiring these in aquaculture. Tasks perceived to be physically difficult, such as digging the pond, pond preparation, pond cleaning, and harvesting are performed by men, with the assistance of women. Even in women-only groups, men are asked to assist in these tasks. Women are exclusively responsible for collecting household waste for feeding the fish. Women participate extensively in collecting compost materials to enrich the pond nutrient level.

While women are either responsible for or assist in most tasks in the production sphere, in the post-harvest sphere women are exclusively responsible for all tasks including cleaning and processing fish.

In Rwanda aquaculture is often combined with agriculture, horticulture, livestock rearing, and rural development as farmers realize that they can supplement their food supplies and income with small-scale fish farms (Vincke, 1988; Brown, 1985; Schwartz et al., 1988; Kutty, 1986; Coche and Demoulin, 1986). The fish pond levee is also cultivated with vegetable crops, such as cabbage, amaranth, collacase, hot pepper, and occasionally eggplant. Aquaculture ponds thus provide additional land area for cultivation of vegetable crops that add variety to the diet. Locally available materials, such as banana stems, bamboo shoots, cassava leaves, sweet potato leaves, and wild plants, are used to fertilize the ponds.

Most women stated that they have taken up fish culture to provide for family food needs. Fish from the pond is perceived as a ready and relatively cheap source of protein. Frequently women state that they took up aquaculture to provide nutritious food for the children. Women apparently learned from donor-assisted "nutrition centers" the importance of fish as a protein source for good nutrition. However, women-specific constraints other than time for effective participation in aquaculture were identified. These included restricted access to land to develop ponds, limited access or lack of availability of modern inputs and farm animals to fertilize the ponds, inadequate extension and local administrative support, and an almost nonexistent link to a formal market.

Fish culture augments food security during the beginning of the rainy season when food is often scarce (Molnar and Rubagumya, 1988). Similarly, fishponds generate significant secondary benefits when they foster irrigated gardening or other types of animal husbandry that have complementary relationships to fish production (Molnar et al., 1987; Castillo et al., 1991).

ECONOMIC AND SOCIAL IMPACTS OF THE PD/A CRSP

Fertilization Technologies

Inorganic fertilization techniques for tilapia production were less profitable in both the Honduran and Thailand sites. In Honduras, the inorganic fertilization regimes were not profitable, whereas, in Thailand, these were less profitable than strategies using organic fertilizer. Inorganic fertilizer options in Rwanda were not evaluated due to the lack of availability of fertilizers. Since they are not generally available, they cannot be considered a viable option in Rwanda at the present time.

The greater profitability of organic fertilizer strategies in tilapia production was due to a combination of the higher tilapia yields obtained and the generally lower cost of the organic fertilizer. While this cost varies from region to region and with the type of fertilizer used, in most cases, some type of organic material can be identified at a low cost that can be used for tilapia production.

Importance of Labor Efficiency When Evaluating
Economics of Subsistence Aquaculture

The Rwandan studies documented the very small amount of capital used in aquaculture production. While Rwanda may be an extreme case, the issue highlights the importance of evaluating effects of new technologies in terms of labor allocation. While subsistence farmers may be cash poor, the diversified nature of their operations may also result in labor shortages. Fish production technologies that do not require constant labor for weeding, hoeing, or harvesting may fit in well with existing crops farming activities. Very little attention has been paid to evaluating aquaculture crops in terms of labor requirements.

Fish as a Cash or Subsistence Crop

The Rwandan studies clearly indicated that fish are perceived primarily as a cash crop. Yet the small amount consumed still constitutes a significant increase in overall household meat consumption. The Thailand study also showed that farmers perceived of fish primarily as a cash crop, a factor that should be taken into consideration in the design of aquaculture development projects. There is no doubt that fish production remains an important source of animal protein for rural populations. However, its dual role in also constituting an important source of cash income should be considered when designing projects for aquaculture development.

Diversification Strategies of Small Commercial-Scale Farmers

The cost of production survey in Rwanda as well as the case studies conducted in Thailand clearly demonstrate the highly diversified nature of small-scale fish farms that has been described as typical of subsistence and near-subsistence farming in developing nations. Rwandan respondents frequently had from 8 to 10 different marais crops in addition to several hill crops. The Thai case study farms had a similar number of different crops on their farms. CRSP technologies will need to be evaluated in terms of how these technologies will mix in or compete with these other crops.

Economic Significance of PD/A CRSP Results

Thailand

The need for social and economic studies is critical in a country such as Thailand. Freshwater fish is the major source of animal protein for the rural poor in the Northeast (Prapertchob, 1989), where the per capita income level is 40% ($400 U.S.) of the national average. In the 1990s, pond-based aquaculture will require more concerted attention by research, extension, and development agencies in terms of improving overall efficiency (Tomich, 1988).

The CRSP technologies utilizing chicken manure increased net returns by 85 and 98% for chicken manure collected for use on fish ponds and for that applied directly from integrated systems, respectively (Engle and Skladany, 1992). Returns on average investment on ponds and equipment increased by 60 to 70%, while returns on average total investment (including land) increased by 5% (from 5 to 10%) by adopting CRSP technologies of chicken manure fertilization regimes.

There are, to date, no reliable estimates of the total number of farmers who have adopted the technology of fertilizing fish ponds with chicken manure. However, observations from this study indicate that these systems have become numerous in Northeast Thailand over the last 2 to 5 years. The increased returns from the improved technology could benefit not only the individual producers but could also multiply throughout the economy of this region in Thailand.

Honduras

Honduras PD/A CRSP researchers have presented various production options to Honduran fish farmers, and a diverse set of nutrient-input regimes has resulted in profitable aquaculture operations. Sixteen treatments from four categories of nutrient management systems resulted in positive net returns. These 16 nutrient management systems provide a range of profitable options to fish farmers in Honduras. On the low technology end of this range, use of chicken litter resulted in profitable tilapia production. On the high technology end of this range, use of formulated feed and aeration resulted in profitable production systems. Such a range of profitable tilapia production systems allows appropriate input choices in poor areas as well as in conditions favoring capital intensification. A wide choice of profitable production intensities and system alternatives offers efficient resource utilization to the farmer. Farmer efficiency leads to a foundation for sustainable aquaculture.

Rwanda

Fish production is a relatively new production alternative in Rwanda. Farmers have rapidly adopted basic fish culture technologies and have begun to adopt PD/A CRSP-generated recommendations. Economic analysis showed that, even when much of the land was used to meet household requirements for food, fish production increased net farm income by 43% (Engle, in review). This analysis assumed mean yields of fish as determined from survey data.

These magnitudes of percentage increases in net farm income have a greater significance when occurring in a subsistence context than when observed in a more developed market economy. Subsistence economies such as that of Rwanda often hover on the brink of catastrophes that result in famine or other consequences. The marginal, or extra, net farm income to a farm family that has nearly no income has a far greater value than the same amount to a family of better means. The extra income from fish production could well mean the difference between starvation and feeding the family or purchasing medicine.

Nearly one-third of fish farmers used revenue from fish production to pay school fees (Hishamunda et al., in press). While it is difficult to estimate the long-term economic impact of increased educational levels, education and economic development are often correlated positively.

Social Impacts on Households

Molnar, Hanson, and Lovshin (1996) conducted personal interviews with tilapia farmers in PD/A CRSP production areas in 1994 to 1995. About 78% of Filipino farmers and 40% of Thai farmers thought that there were points in the annual farm cycle when the pond was too much work. Previous work suggests that Rwandan women are also likely to report these difficulties. About 80% of the respondents from the Philippines, Thailand, and Rwanda felt that tilapia fit well with other farm activities, but only 64% of the Honduran farmers thought so. Three-quarters or more of the respondents in the Philippines, Honduras, and Thailand noted the benefits of additional cash for their households as something associated with the tilapia crop. Only 5% of the Rwandans agreed with this statement, as the limited amount of cash produced by tilapia likely was used mainly by men to purchase beer or rent more land.

Thai farmers were most likely to note problems over water resources emanating from the tilapia crop (57%), an issue noted by only a few of the respondents from the other countries. Filipino operators had few problems with predators eating their fish, but this was an issue for farmers in each of the other countries. Theft was a concern for 44% of the Honduran farmers and for 11% of the Filipino farmers, but only 20% or so of the respondents from Rwanda or Thailand noted this as an issue. Thai farmers were most likely to agree that tilapia were easier to steal, though a third of the Honduran respondents thought so as well.

Most respondents thought their fish pond produced enough to be worth the work they put into it, though Rwandans were slightly more skeptical. A third of the Hondurans questioned the fit of tilapia with the other activities of their farm household. About 60% of the Hondurans thought that tilapia was less profitable than their other activities. Most respondents thought tilapia was the best use of the land it occupied. Hondurans were likely to report themselves as planning to build new ponds (39%). In land-short Rwanda only 11% thought so. Only 54% of the Rwandans were happy with tilapia as a type of fish to grow; they desired a larger, faster-growing fish.* More than 90% of respondents in the other countries were happy with tilapia.

The perceived profitability of tilapia relative to other farm activities was highest in the Philippines, where 90% thought it was more profitable than other crops. Overall, Hondurans were least happy with the returns from tilapia, although Thai farmers were less convinced than other survey respondents that tilapia ponds were the best use of the land. Lowland Thai farmers with irrigation in the far reaches of the Bangkok marketing area have many enterprise choices and marketing opportunities.

Most fish farmers surveyed in Rwanda, Thailand, Philippines, and Honduras felt that the tilapia pond was the best use of the land it occupied on their farm. As the Thailand respondent's pond area increased, a smaller percentage replied that the pond was the best use of the land occupied. All Filipino owners, regardless of pond category, agreed that aquaculture was the best use of the land. Owners of ponds of all sizes in the Philippines felt very positive about aquaculture in relation to other farm activities. In Thailand, owners of small- and medium-size ponds shared a similar high degree of enthusiasm about tilapia culture, but only about half of the large-size pond owners in Thailand agreed that tilapia was more profitable than other farm activities.

Moving the science of aquaculture from field station to farm brings into focus the human environment within which aquaculture technology is used. This environment encompasses farm families and society. Gender roles in the family, resources, power dynamics, and social norms play key roles in technology assessment and adoption. Rwanda PD/A CRSP activities include linkages with an extensive network of farmer-operated experimental aquaculture ponds and communication with extension agents who assist fish farmers.

Aquaculture sector development at the global level has been viewed as a measure to improve food security and as a means of supplementing income for farm families. In many countries, particularly in Africa, the aquaculture sector is based almost entirely on extensive farming practices, primarily for subsistence and barter, with the surplus being sold in rural markets (Nash et al., 1987). Aquaculture, particularly on a small scale, is labor-intensive, energy-efficient, and conserving of natural resources.

REFERENCES

Balakrishnan, R., Veverica, K. L., Nyirahabimana, P., and Rainey, R., An Approach to Integrate Gender Variable in Rwanda Pond Dynamics and Aquaculture Collaborative Research Support Program, Pond Dynamics/Aquaculture Collaborative Research Support Program, Women in International Development, Oregon State University, Corvallis, 1992.

Blumberg, R. L., Making the Case for the Gender Variable: Women and the Well-Being of Nations, Office of Women in Development, United States Agency for International Development, Technical Report No. 1, Washington, D.C., 1989.

Brown, L. R., Fish farming, *Futurist,* 10, 18, 1985.

Castillo, S., Popma, T., Phelps, R., Hatch, U., and Hanson, T., Family Scale Fish Farming in Guatemala, International Center for Aquaculture Research and Development Series 37, Auburn University, Auburn, AL, 1991.

Cernea, M. M., Sociological knowledge for development projects, in Cernea, M., Ed., *Putting People First,* Oxford University Press, New York, 1985, 3.

Coche, A. G. and Demoulin, F., Report of the Workshop on Aquaculture Planning in the Southern African Development Coordination Conference (SADCC) Countries, CIFA Technical Paper 15, Food and Agricultural Organization of the United Nations, Rome, 1986.

Engle, C. R., Optimal Resource Allocation by Fish Farmers in Rwanda, *J. Appl. Aquacult.,* in press.

---

* In Rwanda, high elevations and cool temperatures result in slower growth of tilapia than at other CRSP sites (see Chapters 3 and 10).

Engle, C. R. and Skladany, M., The Economic Benefit of Chicken Manure Utilization in Fish Production in Thailand, CRSP Research Reports 92-45, Title XII Pond Dynamics/Aquaculture Collaborative Research Support Program, Oregon State University, Corvallis, 1992.

Engle, C. R. and Skladany, M., A Comparative Farm Management Analysis of Integrated Aquaculture-Livestock Production in Thailand, Aquaculture/Fisheries Center Working Paper AFC-94-2, University of Arkansas at Pine Bluff, Pine Bluff, AR, 1994.

Engle, C. R., Brewster, M., and Hitayezu, F., An economic analysis of fish production in a subsistence agricultural economy: the case of Rwanda, *J. Aquaculture Tropics*, 8, 151, 1993.

Engle, C. R., Thomas, M., Brown, D., and Hishamunda, N., Optimal Resource Allocation by Fish Farmers in Rwanda, Aquaculture/Fisheries Center Working Paper AFC-94-6, University of Arkansas at Pine Bluff, Pine Bluff, AR, 1994.

Ford, R. E., The dynamics of human-environment interactions in the tropical montane agrosystems of Rwanda: implications for economic development and environmental stability, *Mountain Res. Dev.,* 10, 43, 1990.

Francis, C. A. and Hildebrand, P. E., Farming systems research-extension and the concepts of sustainability, *Farming Syst. Res. Ext. Newsl.,* 3, 6, 1990.

Green, B. W., Teichert-Coddington, D. R., and Hanson, T. R., Aquacultural Research in Honduras: 1983–1993, Research and Development Series, International Center for Aquaculture, Auburn University, Auburn, AL, 1994.

Hazell, P. B. R. and Norton, R. D., *Mathematical Programming for Economic Analysis in Agriculture*, Macmillan, New York, 1986, 400 pp.

Hishamunda, N., Thomas, M., Brown, D., and Engle, C. R., A Comparative Economic Analysis of Small-Scale Fish Culture in Rwanda, PD/A CRSP Technical Report, Title XII Pond Dynamics/Aquaculture Collaborative Research Support Program, Oregon State University, Corvallis, in press.

Kay, R.D., *Farm Management — Planning, Control and Implementation*, McGraw-Hill, New York, 1986, 401 pp.

Kutty, M. N., Aquaculture development and training in Africa, in Huisman, E. A., Ed., *Aquaculture Research in the Africa Region*, Purdoc Wageningen, Netherlands, 1986.

Molnar, J. J., Cox, C. L., Nyirahabimana, P., and Rubagumya, A., Socioeconomic Factors Affecting the Transfer and Sustainability of Aquacultural Technology in Rwanda, Research and Development Series No. 38, International Center for Aquaculture, Auburn University, Auburn, AL, 1994.

Molnar, J. J., Duncan, B., and Hatch, U., Fish in the farming system: the FSR approach to aquacultural development, in Schwarzweller, H., Ed., *Research in Rural Sociology and Rural Development*, JAI Press, Greenwich, CT, 1987, 169.

Molnar, J. J. and Rubagumya, A., Aquaculture and the Marais: Patterns of Organization, Allocation, and Use of Valley Land Under Conditions of Resource Scarcity and Ecological Complexity, ICA Technical Paper, International Center for Aquaculture, Auburn University, Auburn, AL, 1988.

Molnar, J. J., Rubagumya, A., and Adjavon, V., Sustainability of aquaculture as a farm enterprise in Rwanda, *J. Appl. Aquaculture,* 1, 37, 1991.

Nash, C. E., Engle, C. R., and Crossetti, D., Eds., Women in Aquaculture: Proceedings of the ADCP/NORAD Workshop in Aquaculture, ADCP Technical Report ADCP/REP/87/28, Food and Agricultural Organization of the United Nations, Rome, Italy, 1987.

Nyirahabimana, P., in Why worry about crops when fishing is better, Perelez, J., *New York Times*, New York, 1989.

Nyirahabimana, P., Profil socio-economique de la femme Rwandaise, Service d'Appui a la Cooperation Canadienne en Collaboration avec Reseau des Femme oeurant pour le Developpement Rurale, Kigali, Rwanda, 1991.

Prapertchob, P., Analysis of Freshwater Fish Consumption and Marine Product Marketing in NE Thailand, Department of Fisheries, Bangkok, Thailand, 1989.

Rogers, E. M., *Diffusion of Innovations,* 3rd ed., The Free Press, New York, 1962, 453 pp.

Schwartz, N. B., Molnar, J. J., and Lovshin, L. L., Cooperatively managed projects and rapid assessment: suggestions from a Panama case, *Hum. Org.*, 47, 1, 1988.

Teichert-Coddinton, D. R., Green, B. W., and Parkman, R. P., Substitution of Inorganic Nitrogen and Phosphorous for Chicken Litter in Production of Tilapia, PD/A CRSP Tenth Annual Administrative Report, Title XII Pond Dynamics/Aquaculture Collaborative Research Support Program, Oregon State University, Corvallis, 1993.

Tomich, R. J., A Review of Tablefish Production Systems in Northeast Thailand, Institutional Element Report No. 19, Thailand/Canada Northeast Fishery Project, DOF/CIDA (906/11415), Department of Fisheries, Bangkok, Thailand, 1988.

Vincke, M. N. J., The role of extension in village aquaculture development, in King, H. R. and Ibrahim, K. H., Eds., *Village Level Aquaculture Development in Africa,* The Commonwealth Secretariat, London, England, 1988.

World Bank, Rwanda Agricultural Strategy Review, Report No. 4635-RW, The World Bank, New York, 1989.

16

DEVELOPING AND EXTENDING AQUACULTURE TECHNOLOGY FOR PRODUCERS

Karen L. Veverica and Joseph J. Molnar

INTRODUCTION

Although most aspects of agricultural research have enjoyed a long history and the knowledge base is quite extensive, aquaculture is a relatively new science that still stands to gain considerably from research and development efforts (Shell, 1993). Even though the scientific information base dates back only about 25 years, there is much more information available than is being effectively diffused to prospective clients (Shell, 1993).

Collaborative Research Support Programs (CRSPs) focus on removing constraints to production through the development of technology and rely heavily on other agents and vehicles to disseminate the technologies. Clients of information emanating from the PD/A CRSP are farmers, educators, and other researchers, public policy makers, loan officers, and investors. Vehicles for transmitting the information vary with the client. As with most research programs, it is easiest to detect the products of research in the form of publications in scientific journals, theses, annual reports, and other research reports. Therefore, other researchers and, to a lesser extent, educators and policy makers are served.

The purpose of this chapter is to examine the connection between the Pond Dynamics/Aquaculture CRSP and farm-level efforts to communicate research findings and production strategies and to elucidate the influence of farmers on the PD/A CRSP research program. We review PD/A CRSP efforts to participate in the development and extension of aquaculture technologies in the context of existing extension systems and the range of alternative approaches that are typically available to extension programs.

SUBSISTENCE AND COMMERCIAL AQUACULTURE

Subsistence producers purchase few or no inputs, and the production of fish is only a small aspect of a highly diversified operation unique to a farm family. The highly complex nature of subsistence production, complicates the identification of appropriate research topics. Some researchers believe that subsistence farmers need reliable information more than anything else and that most of the research on aquaculture that can benefit subsistence farmers has already been done. Any remaining subjects will be so site- or farmer-specific that it is best left to the farmers themselves to do some trial and error experimenting. Farmers must have access to information from which they can select the practices best suited to their individual circumstances. Research-based information often must undergo extensive transformation to make it useful to the subsistence farmer (Chambers et al., 1989).

Unlike much of agriculture, though, fish is produced by many subsistence farmers as a cash-generating crop first and secondarily as food (Molnar et al., 1995). In this respect, subsistence farmers can often start to imitate practices of small-scale farmers and may even find the opportunity to increase their investment and resultant income.

0-56670-274-7/97/$0.00+$.50
© 1997 by CRC Press LLC

Small-scale commercial or "complementary" farming is best described as somewhere in between the subsistence and the fully commercial and specialized level of production. Farmers may sell most of their production and consume only a small portion. Inputs available on-farm are used, such as animal manures and agricultural byproducts, but some inputs are also purchased, such as chemical fertilizers and feeds. Some labor may be hired out. Fish production may be one of several enterprises practiced at the farm.

Specialized, commercial aquaculture may vary in its level of intensity, but the enterprise is usually concerned solely with aquaculture, even though more than one species may be produced. Inputs such as feeds and fertilizers are purchased; seed may be purchased or seed production may be the sole concern of the enterprise. Criteria for evaluating production schemes are usually economic returns to investment. The complementary character of aquaculture with other practices and reduction of risk should be considered but are secondary to economic gains. It is easiest to research topics of interest to this level of producer, and the returns to research investments in terms of increased production and price reductions to consumers may be the greatest at this level.

This CRSP has done research to benefit producers of all levels. However, the research program necessarily relies on an effective extension program to relay its findings. Extension providers require transformation of the information into practical management recommendations appropriate for each level of producer. Some of the transformation is done within PD/A CRSP; some is accomplished within the organizations actually conducting outreach activities with producers and villages.

EXTENSION SCIENCE FOR AQUACULTURE

The primary purpose of aquaculture extension is to improve production of aquatic organisms for food and wealth for farmers, as well as for the nation as a whole. Extension programs are justified by the fact that economic development resulting from extension activities should more than compensate for the national investment required to provide these services (Engle and Stone, 1989). Extension activities include education, services, and liaison between public agencies, universities, and producers (Binswanger and Ruttan, 1978). The knowledge transmission process is essential, and the provision of services should only be an initial or temporary phase of extension work.

In traditional models, the technology transfer process is, at its core, a series of communication steps that relay new findings and perspectives to technical representatives in government research and extension systems (Zaltman and Duncan, 1977). The communication process linking an experimental pond to farm practice involves several layers of translation and transmission. Many factors interact to affect the extent and degree of impact of CRSP scientists and research programs on national aquacultural institutions and farm practice. In Thailand, for example, an important intermediate target category for the PD/A CRSP includes hatchery managers and others in commercial sector roles that feature regular contact with fish producers. In turn, these intermediaries are expected to facilitate the diffusion of technology by communicating new information to the farm level.

Experimental findings reflect controlled conditions and careful measurement of a focused set of factors. Farm conditions reflect variable physical and management situations that often mitigate the impact of effects identified by repeated experimental trials. That is, experimental findings often must be compiled from many studies and modified in certain ways to generate a robust field recommendation. In essence, an internal process of recognition and acceptance must take place within national research and extension systems before the findings become a farm-level recommendation. Typically, the greater the deviation from conventional understanding, the slower the process of internal diffusion. It is often not well-recognized that institutions require an internal adoption process before new approaches can be communicated to users (Koehler et al., 1981). In turn, some farmers learn of and use technology in advance of public institutions.

Aquacultural extension has been born of the same traditions that have spawned technical assistance and education efforts in agriculture but has occurred in a later stage of history. In some African nations, fish culture has been promoted since the 1950s, but it is only in recent decades that widespread adoption has taken place (Harrison, 1991; Moehl and Molnar, 1995). Aquaculture industries, export markets, and production systems have begun to grow exponentially around the world. This also has been a period when commitment to the notion of extension is wavering and when donor resources and government budgets have become unreliable sources of continuing support for extension programs. Engle and Stone (1989) have documented the highly variable state of aquaculture extension across a multitude of less-developed countries; connections between CRSP research programs and farm-level practices have recently been investigated by Molnar et al. (1995).

In general, extension approaches and notions developed for land-based agriculture are applicable to aquaculture. Two major differences need to be noted because they influence the appropriateness of extension activities. First, aquaculture has a relatively short tradition in most areas. Second, aquaculture often employs focused aquaculture or fishery extension agents who are remotely connected to the agricultural extension services.

Except for parts of Asia and eastern and central Europe, aquaculture is a relatively new enterprise in most countries. There is little tradition of aquaculture knowledge residing in the rural population, as there is with agriculture. Therefore, aquaculture can be considered in the context of a new, nontraditional crop, or where the recent intensification of aquaculture is in practice, as an activity that has little locally developed technology. A research-extension-farmer partnership can greatly enhance the development of the required new technologies (Kaimowitz, 1991).

In many countries, fish culture extension is carried out by specialized agents and not by the agriculture extension agents. There are some advantages to using specialized agents but many disadvantages. One of the main advantages is that the agents can remain focused on one commodity, and their training can be highly specialized. Research institutes that deal in aquaculture would have their specialized extension service to whom the results and recommendations could be directly transferred. Indeed, specialized commodity extension programs have been cited as the most likely to succeed (Roling, 1988; Axinn, 1987). However, a specialized extension program in aquaculture can be expensive to maintain. Many governments cannot justify the expense of highly educated extension agents for a single commodity and are forced to engage agents with less than ideal educational backgrounds. They require transformation services that would put technical information into usable format by extension agents are often lacking. The single-mindedness that often characterizes aquaculture extension agents does not mesh well with the needs of subsistence-level farmers.

Extension approaches are many, reflecting the diverse national traditions and cultural settings in which these programs have been organized. Albrecht et al. (1990a, 1990b) provide one of the better summaries of extension approaches that have been implemented worldwide. Similarly, Rivera and Schram (1987), Van Den Ban and Hawkins (1990), and Roling (1988) provide a conceptual foundation for agricultural extension programs. In recent years, the fiscal crisis of the state in developed and less-developed nations has forced a rethinking of the limits of government support services and the need to foster reliance on the private sector as a basis for sustainable development (Molnar et al., 1991).

There have been many models proposed and promoted for the conduct of extension work (Benor et al., 1984). The paradigms or framing assumptions for these models also are highly variable, as is reflected in the nomenclature for extension work. Extension *promoters* endeavor to organize participation in the production of a commodity, often in the context of colonial or state-sponsored export development efforts (Harrison, 1991). Similarly, the German *beratung* (berator) implies a particular type of advocacy for changing practices or participation in a commodity. Similarly, a French term for extension worker is a *moniteur* (monitor), one charged with instructing and following compliance with a provided set of procedures. The Dutch *voorlichting* literally means path-lighting and implies voluntary choice in the model underlying the process of how to alter target category behavior through demonstration and communication (Van Den Ban and Hawkins, 1990). The *extension* concept links the university to the countryside. Of English origin but American development, this model endeavors to balance inequities in information and understanding through education and persuasion (Chin and Benne, 1976; Rasmussen, 1989). Extension agents often are referred to as "change agents" in this context.

A dimension of growing importance in extension work is the level of client participation in the development of the technology and the way it is made available to potential users (Chambers et al., 1989). Acker (1992) summarizes extension models in relation to the level of farmer participation and guidance, ranging from a one-way flow of information from technology developer (scientist) to user. Farmer-participatory models, in contrast, encourage farmers to learn to manage their own information and input acquisition systems.

Axinn (1987) categorizes the many different extension systems into two types, delivery and acquisition systems. Delivery systems can try to be participatory in nature but if program targets, goals, and objectives are set by governments and strategies are decided centrally, the system is in essence a delivery system. Delivery systems seem to serve the primary policy of providing cheap food to urban populations, although the stated purpose is to improve the quality of life for rural families. Large commercial farms often are targeted, as these are the most likely to produce large quantities of cheap food. Furthermore, the technologies developed often are geared toward farming systems that rely on inputs purchased off the farm and outputs sold wholesale through commercial marketing channels.

Acquisition systems rely on farmers who organize themselves to acquire the information and services they need. While the scale of their operations may vary, their central interest is to improve the quality of their lives. In acquisition systems, farm families control agricultural extension education. However, Roling and Engel (1991) point out that knowledge often accumulates where it is least needed, i.e., where there is already much knowledge. If small and marginal farmers are to be served, some type of affirmative action is required to avoid the pitfall of preferentially serving the resource-rich. Deliberate action is often needed to mobilize and organize marginal farmers to seek out opportunities for increasing their access to information. Donor agencies and nongovernmental organizations (NGOs) often are key players in keeping the rural poor and women included as parts of acquisition systems.

THE RESEARCH-EXTENSION LINK FOR AQUACULTURE

No matter what extension technique is used, there must be some way to generate appropriate information, to evaluate it, and to put it into usable format for the producer. This link often is not apparent upon examination of the extension service organizational chart. Research-based information may be generated by universities under a ministry of education or higher education, or by agricultural research institutes that stem from a ministry of agriculture. Fish culture extension services are based variously at the agriculture, natural resources, or fisheries ministries and may have their own research staff. Information also is generated by farmers themselves. Even though the fish culture tradition is relatively short, much of the information used by fish farmers often comes from other farmers. This farmer-generated information evolves from several years of trial and error and can be very valuable to researchers.

Figures 1 through 4 depict the network of institutions that generate and diffuse aquaculture information in the present PD/A CRSP countries. There often is competition for limited funds between the different research entities, and communication often depends more on personalities than on a mandated transfer process. The diagrams show the institutional location of the PD/A CRSP, the major organizations conducting research, training, and extension activities, as well as some central features of the private sector. Nongovernmental organizations often relay PD/A CRSP information through training programs and village-level outreach. Formal extension activities are not a central aspect of the PD/A CRSP.

EXTENSION ACTIVITIES OF THE PD/A CRSP

Much of the success in transfer of PD/A CRSP technologies has come from informal linkages with counterpart researchers, extension personnel, and end-users as friends and colleagues, and not from formally established programs (Molnar et al., 1995; Gibbons and Schroeder, 1983). PD/A

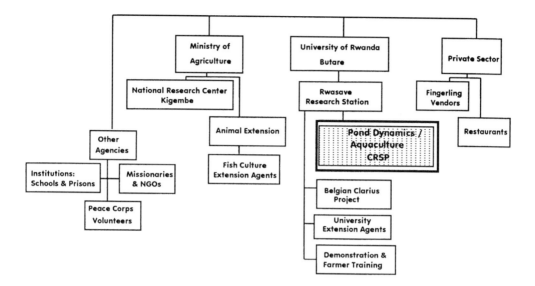

Figure 1.   Institutional networks for tilapia technology in Rwanda.

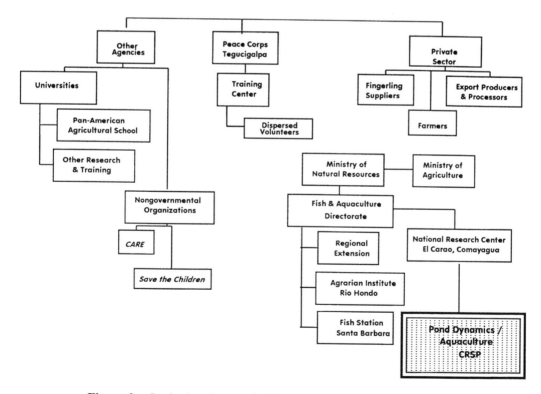

Figure 2.   Institutional networks for tilapia technology in Honduras.

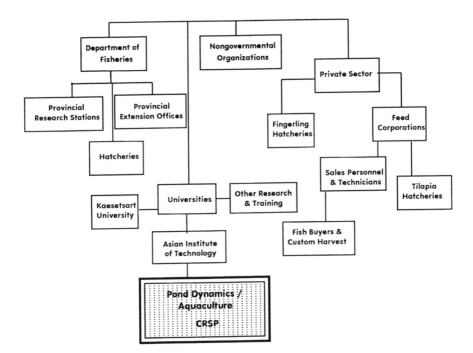

Figure 3.   Institutional networks for tilapia technology in Thailand.

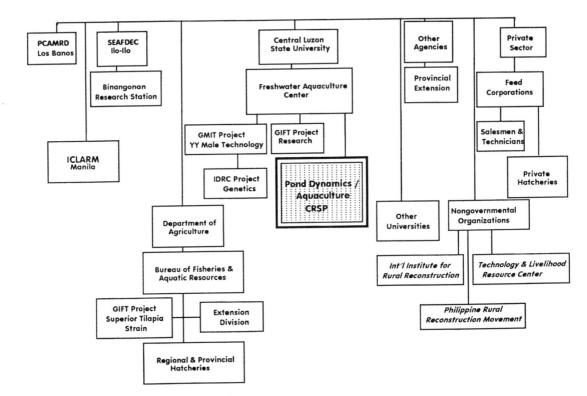

Figure 4.   Institutional networks for tilapia technology in the Philippines.

CRSP scientists interact with research station personnel, sharing insights and perspective on the technology of aquaculture. These influences are then retransmitted to farmers and others who have contacts with station personnel, particularly if the station personnel have training or leadership roles (Brown, 1981). Through interpersonal contacts, PD/A CRSP scientists impart a holistic understanding of pond dynamics and fish behavior that is difficult to obtain through the printed word or other formal means.

Scientists have had direct impacts on farm practice through the various training sessions that research stations have sponsored over the years. The production technologies conveyed directly to farmers through these programs often have had dramatic impact in the operations of receptive individuals with the means and motivation to realize the promise of the enterprise. The effect of personal relationships between host-country personnel and PD/A CRSP scientists should not be underestimated. In turn these individuals influence their neighbors by example and interaction over the proper practice of fish culture.

Another traditional communication channel for PD/A CRSP findings is through written publications. Journal articles and meeting presentations convey research results to the scientific community. Reprints may circulate among some institutional participants in the host country, but rarely do they reach the farm level. Findings of this sort require two sorts of transformation: (1) accumulation and reconstitution to become user-robust recommendations, and (2) translation to the spoken language and level of understanding of the user. Few extension services are organized to facilitate these transformations. Therefore, technologies extended to farmers often are developed by extension service personnel, often duplicating and sometimes contradicting recommendations generated by researchers. Nor are university-based researchers inherently better at producing relevant information than extension service researchers. Many have the disadvantage of physical and social distance from the producers and the resulting difficulty in determining just what research is most useful. Students, many with farm backgrounds, are a continuous source of fresh perspectives for many university-based researchers.

Methods Used in Extending PD/A CRSP Technologies

PD/A CRSP researchers have more to offer farmers and national research systems than simply the results from experimental trials. Most PD/A researchers have many years of experience in aquaculture and in tilapia culture, in particular. Relevant information emanating from PD/A CRSP research is combined with the previous knowledge and experience of the researchers, resulting in a large package of technical information. The extension impact of the PD/A CRSP is centrally manifested in the relationships and communications of the researchers to host-country colleagues, industry actors, and farmers.

The central methods of extension include methods and results demonstrations, from farm visits, field days, meetings, and individual instruction (Sanders, 1966; Lionberger and Gwin, 1983; Albrecht et al., 1990b). Extension methods can be arrayed along a dimension of mass communication to individualized instruction. The following sections examine the way the PD/A CRSP has utilized selected extension methods in furthering aquacultural development.

Communication Strategies

Radio programs, television spots, posters, and pamphlets serve to present general ideas to a large number of people. If the client group is not highly educated, the mass communication methods should be limited to general publicity about fish and fish culture with the information on where to go for help. When farmers have access to one-on-one help from extension agents, books and pamphlets can help reinforce the principles explained by the extension agents. In the Philippines, a widely followed radio farm show occasionally features aquaculture topics and is considered highly influential. Most PD/A CRSP researchers have not utilized mass media approaches in any systematic way.

Newsletters

Quarterly newsletters had been planned for Rwanda, but the program was canceled due to the civil war. Newsletters can provide continuing updates on new technology and constantly reinforce recommendations of extension agents. They also can generate interest and foster solidarity among producers. Farmers and extension agents well-trained also can write in to the newsletter publisher with observations, questions, and news. Newsletters require well trained and highly motivated staff over a long period of time to be successful. Newsletters often are useful as an interorganizational communication strategy, but hold little chance of achieving impact among those who cannot read.

Aquaculture Manuals

Manuals on how to culture fish always should be used in conjunction with an extension service. A good fish culture manual may allow farmers to improve their production with fewer visits from extension agents but rarely replaces the need for on-site technical assistance. All the collaborating countries have tilapia culture manuals. In Honduras, the tilapia culture manual for subsistence-level production was produced by Peace Corps volunteers. No manual yet exists for commercial-level production at the present.

In Rwanda, although less than 50% of farmers can read, many farmer groups have one or two members who can read simple Kinyarwanda. Some of the younger members can read very simple French, but most extension projects try to publish booklets in Kinyarwanda. Translation into local languages can result in the publication of incorrect information. The first fish culture manual printed in Rwanda probably did more harm than good. It was written by individuals who later admitted they knew nothing about fish culture and who used references of questionable quality. A second, improved manual was published, but it also contained bad recommendations. It can be difficult to replace "old information and techniques" by new information simply by publishing a new manual. In countries where anything in print is considered to be true, manuals that promote bad or outdated management practices can seriously inhibit progress.

Technical Publications

The information contained in technical journals is far from the end user in terms of availability, comprehension, and usability. However, a dynamic and qualified extension service often can put bits and pieces of the information obtained from technical journals to practical use. Researchers and extension specialists from other countries can also benefit from this form of communication. Local testing of recommendations is often necessary before translating the information into useable form for farmers. Some shrimp farmers and some intensive-system tilapia farmers may be able to use some of the information directly if they possess an adequate educational background. There is great pressure to publish research results in technical journals, and this may result in ignoring research subjects that may be beneficial to producers but difficult to publish in scientific journals. Publication of information in technical journals directly affects the career development of the researcher but rarely results in direct benefits to farmers. Additional steps are required.

Field Days

In Honduras, most farmers still purchase their fingerlings from the El Carao station. Similarly, large numbers of farmers purchase fingerlings from the government hatchery at Central Luzon State University in the Philippines and from the University of Rwanda's Rwasave Fish Culture Research Station. While visiting the station, a management plan is drawn up with the researcher, using the most current information. Farmers often engage the researchers in conversation about their fish production experiences. PD/A CRSP researchers cite this feedback as a great benefit to the research program.

Many ponds have been constructed within walking distance of the Rwasave station to take advantage of its sales of high-quality fingerlings and the readily available advice. A well developed fish market is also attracting farmers. The station has helped some farmers to sell their fish through the Rwasave station fish market. Farmer training is often conducted at research stations in Rwanda and in Honduras, or at collaborating government stations in Thailand and the Philippines.

Demonstrations and On-Farm Trials

Used in all CRSP countries, on-farm research is a very powerful tool, although somewhat risky to the credibility of researchers if the technology tested is inappropriate. Any technology developed at research stations should be tested on-farm prior to wide dissemination. Following on-farm testing, researchers in Rwanda found that organic fertilizer input rates developed at the research station were probably too high for farm ponds because farm ponds had a lower average depth and because farmers lived too far from their ponds to watch effectively for instances of low dissolved oxygen (Rurangwa et al., 1992).

In Honduras, farmers were offered a choice of experimental treatments (Teichert-Coddington et al., 1993). In Rwanda, farmers' ponds were used as substitute research stations at different elevations, so all farmers were required to apply the same treatments. A standard fertilizer treatment was decided upon after a survey was done to determine the organic inputs available to all farmers in all elevation zones. The Rwanda on-farm experiments were not intended to be an extension tool, but sought to carry out replicated trials at different elevations to provide answers to basic production questions.

The following recommendations were made after on-farm trials were conducted by the PD/A CRSP:

1. It is best if farmers choose the treatments to be assessed. A group meeting prior to the on-farm research allows for the training necessary to carry out the trials. During the meeting, an informal contract should be drawn up to spell out what is expected of the farmers and of the researchers.

2. Compensation for participation in on-farm research is favored by some but is discouraged by most. Thompson and Thompson (1990) recommend that farmers receive some type of compensation for any additional time they must spend managing the trial plots or at least for time it takes to attend organizational meetings. This was the justification used in Rwanda, where farmers often had to wait several hours on prearranged days for pond sampling and the review of their data books. Farmers in Honduras and Thailand usually did not receive compensation, although an initial set of fingerlings was sometimes provided. Compensation can take many forms: in Rwanda, chemical fertilizers were supplied free of charge, which allowed the researchers to accurately weigh the fertilizer for each weekly application. Farmers would not have had the cash available to purchase the chemical fertilizers for the first yield trials, but the additional money they received from fish sales following the yield trials was sufficient for the purchase of fertilizer for the next production cycle.

3. Avoid fostering dependency on the yield verification trials. This was especially difficult in Rwanda, where cash flow was always a problem. Grasses were purchased from farmers to ensure that the amount of grass required for the chosen treatment was indeed applied and to reimburse them for the time spent gathering the grass. Although the purchase price was quite low (farmers made about 50 U.S. cents per week) this was a major source of cash income for many farmers.

4. Encourage farmers to visit each other during the on-farm trials.

5. If a functional extension service is present, extension agents must be aware of the trials and their assistance requested.

6. Discuss the results with the farmers in a conference or other group meeting. Group discussions allow other farmers to directly benefit from the results and to understand the reasons for the differences.

Farm Visits

Researchers often visit farms in conjunction with on-farm trials, and sometimes on special request from farmers. The researcher temporarily serves as the extension agent in this case. Direct farm visits by researchers without the knowledge or participation of the extension service are not cost-effective when the size of the target audience and the opportunity cost of the researchers' time are considered. Farm visits are probably more important for the researcher as a means for maintaining contact with actual producer practices and conditions and may have only a serendipitous effect on technology transfer.

Researchers in all CRSP collaborating countries have used farm visits with the collaboration of the extension services. Not only do the farmers and researchers engage in one-on-one communication, but the researcher can see first-hand the conditions in which the farmer lives, the ponds, and other farm practices. An observant researcher often can judge the farmer's potential for innovation during these visits. It is most productive if the farmer and researcher speak a common language, as inexperienced translators often miss nuances of the producer's experience. Rwanda researchers have found that participating in pond sampling has allowed them to identify one of the major problems in fish culture practices in the country, namely, poor fingerling selection. Farm visits have allowed them to assess the stocks of fish found in a farmers' ponds and show the farmers and their associates how to identify old female fish that were previously passing as fingerlings. Neither the extension agents nor their supervisors were able to identify this problem although farmers continually complained of fish that were too small at harvest.

Extension agents also benefit from farm visits. Farmers often test the veracity of extension agent recommendations by cross-checking with visiting researchers. If the researchers and extension services work in close conjunction, they would give similar answers, and the farmer's confidence in the extension service would be justifiably reinforced. The main drawback to this method is the time it takes. One way to increase the number of people reached in farm visits is to have other farmers accompany the researchers. This method has proven to be of immense benefit in Rwanda. Not only do the farmers gain status by showing their close relationship with the researcher, but they act as some of the best extension agents.

Conferences for Farmers

Farmer-generated information can be transferred successfully to researchers in workshops and meetings where both are present. Workshops that are designed specifically to voice farmers' needs and experiences are excellent avenues of communication. They allow researchers and extension personnel to evaluate the farmers' understanding, and identify research and communication needs. The recently popular idea of participatory rural appraisal seeks to allow the traditional "clients" of extension services to express their needs and concerns. Such conferences help foster the acquisition system of extension education.

A conference for women fish farmers in Rwanda was held at the National Fish Culture Training Center in February 1992, with the intention of creating a forum to hear women's experiences in fish farming (Balakrishnan et al., 1993). The forum was attended by 28 Rwandan women fish farmers, 10 fish culture extension agents, 7 production scientists, 2 development professionals from nongovernmental organizations (NGOs), and 12 policy makers and development program managers from the Ministry of Agriculture, Livestock and Forestry. Women were encouraged to express themselves freely, which they did. The women found it more difficult to get their largely male audience to listen openly to them and to refrain from lecturing them on technical subjects. The women used the opportunity to discuss and verify recommendations the extension agents had been giving and to express what they felt were their greatest constraints to increasing production.

Extension agents should gather feedback from farmers and present it to researchers. Technology packages, once developed with appropriate feedback, can be presented to extension agents, who in turn can teach it to many farmers (Oakley and Garforth, 1985). However, researchers in some CRSP countries have found that the extension services suffer from many constraints and that training extension personnel may not be an effective use of their time.

Expert Systems and Decision Support Systems

Expert systems are decision-making algorithms that could replace an expert. Some CRSP research findings only explain why current practices work and therefore do not merit direct extension to farmers. However, in order to design an expert system, much basic knowledge must be incorporated into the predictions made in the expert system computer programs. The PD/A CRSP expert system POND© requires that the user have a microcomputer and some experience with decision tools and their underlying frameworks. The newest extension approach facilitates self-instruction and learning by simulating the consequences of various combinations of farm inputs, fish characteristics, management approaches, and prices. Learning is then a heuristic process where experience is gained from multiple simulations of a fish cropping cycle. One advantage of decision support systems is that, instead of promoting a single technology package, the program lets the client choose from a wide range of inputs and management levels.

Constraints on Communicating PD/A CRSP Knowledge to Users

Weak or Dysfunctional Extension Institutions

Fish farmers in each of the countries face vastly different institutional support systems. It is generally accepted that Thailand's aquaculture extension service is the most effective of the four considered here. This is probably due to the investment of the Thai government in terms of highly educated and motivated staff. By contrast, the aquacultural extension service in Rwanda had only two university-trained staff members. The extension agents, although specially trained in fish culture, had limited secondary school education. Extension services for fish culture in Honduras and in the Philippines seem to have disappeared, but remain strong in Thailand.

In Thailand, an extensive and well-trained network of Department of Fisheries research stations and extension offices is augmented by a broad set of colleges and universities that provide baccalaureate and post-graduate training in aquaculture. An aggressive private sector makes Thailand the country with the most widespread practice of tilapia production of the PD/A CRSP countries. Producer services in terms of feeds, fingerling supply, and custom harvesting and marketing are the most well-developed here. Several large feed companies and many small processing and fingerling supply firms provide most tilapia farmers with a competitive market for their product and input supply.

Large feed companies also promote tilapia culture in the Philippines, though an extensive private sector fingerling supply has just begun to develop there. Nor is the network of producer services as well developed as in Thailand. Nonetheless, the Philippines has several university programs in aquaculture, but lacks the extensive network of research and extension offices found in Thailand. Central government support for most kinds of extension work has been withdrawn. A few large feed companies are beginning a serious promotion program for tilapia production, albeit with a self-interested formula that emphasizes feeding and neglects pond fertilization.

Tilapia farmers in central Luzon have a variety of publicly supported fingerling producers, while private sector seed suppliers are beginning to increase in number in other parts of the country. The high relative price for tilapia in the Philippines will benefit tilapia producers as long as the fish remains a popular and affordable item in the market place.

Honduras faces fundamental difficulties with an underdeveloped marketing system and an uneven set of consumer preferences for tilapia. The public sector in Honduras is under great financial stress and is widely lacking in public confidence over the ability to deliver services and provide assistance in an effective and reliable manner. One large private university has been a consistent source of training and fingerling supply for producers in central Honduras. Public universities provide some graduates and research support, but financial stress greatly limits the programs that can be offered.

In Rwanda extension agents have about 3 years of post-primary education, usually from an agricultural vocational school. The extension agents in general are hard workers but they do not have the skills to integrate knowledge from various sources and make specific recommendations for particular cases. They were trained to ask for help from supervisors when they encountered a problem they could not solve. The present level of training of the fish culture extension agents was adequate for the initial stages of farmer education, which required much one-on-one discussion and simple demonstration of pond renovation and simple pond management. Now that the basics of fish culture are understood by the majority of fish farmers (Molnar et al., 1995), a different kind of extension is necessary. Although the government has recognized the low level of education to be a constraint on the extension service, the limited financial resources and lack of university-educated supervisors make it impossible to replace the current extension agents with more highly trained individuals. In 1994, plans were to incorporate the fish culture extension service into the newly reorganized agriculture extension service, which would function on the training and visit system. However, no entity was identified to act as transformer and evaluator of the information emanating from research in fish culture. Specialists in the area of aquaculture were not linked with the extension service. Some training in fish culture was anticipated for the agricultural extension agents, and was to be conducted by staff at the university research station and at the National Fish Culture Training Center.

Irrelevant Information

Not all of the information generated by the CRSP is immediately relevant to all users. Much of the information derived from the CRSP "global experiment" using chemical fertilizers is irrelevant for Rwanda and for most of Africa. However, the global experiments are designed to increase our basic understanding of pond fertilization; once this is better understood for inputs such as chemical fertilizers and chicken litter, more general principles can be elaborated, which will include fertilizers more appropriate to the small-scale African fish farmer.

No Mandate for Extension

Acker (1992) makes a case for the difficult situation encountered by CRSPs when it comes to extending research results. CRSPs were not designed to be extension projects. Although efforts to conduct workshops, make surveys, and produce extension manuals are admirable, they were outside the primary mandate of the CRSPs and were done primarily because the researchers were committed to seeing the products of their work put to good use. However, CRSP programs are now being evaluated in terms of benefits to farmers and consumers and not in terms of the collaboration among researchers that they were designed to achieve.

FARMER INFLUENCE ON RESEARCH TOPICS

If the research-extension-farmer links are viewed as a continuum, where each entity learns from the other, the research program will necessarily be influenced by the farmers' needs as perceived by an extension service and directly by the farmers themselves. There is some evidence that farm practices have influenced the research agenda at all of the CRSP sites.

In Rwanda, information on composting rates and methods for fish ponds was lacking, so the extension service had to make fairly conservative recommendations. The only input available in any quantity was grass; manures were only available in small amounts. The PD/A CRSP researchers did a series of experiments to determine efficient composting methods given a limited quantity of inputs. This research was directly influenced by the fish farming situation as perceived by the resident CRSP researcher who had recently transferred from the extension project. Much of the CRSP research in Rwanda was driven by the goal to develop recommendations for efficient use of resources for fish production at the different elevations and resulting water temperatures. Dialogue with farmers and with all levels of the extension service was a habit. The last work plan for the Rwanda site was developed following a meeting that assembled representatives from the extension service and nongovernmental organizations working in fish culture. Attempts to formalize the research-extension-farmer linkage were made during this meeting, when it became apparent that lines of communication were totally dependent on individuals, and these individuals would not remain in the same positions forever.

In Honduras, farmers who visited the fingerling production center (which was also the CRSP research site) often engaged in dialogue with the researchers. However, little of the research conducted by the CRSP at the freshwater site was directly influenced by farmers, the only exception being some special projects on Colossoma polyculture in response to farmers' questions on how many Colossoma to put in a pond with tilapia. The feed and fertilization research done was based on developing economically efficient pond management but was not widely adopted by subsistence-level farmers. Researchers suggest that most of the research of use to subsistence farmers has already been done and that the farmers have often been exposed to most of the knowledge. The reasons why they do not adopt many of the recommended practices are social, stemming mainly from their lack of cash flow, their overall lack of understanding about financial planning, and their desire to minimize risk. Therefore, the most simple systems possible are the most likely to be in operation, and there is little that research can do to make the system more productive and still keep it simple.

Conversely, the brakishwater research site in Honduras is a prime example of farmer-driven research. Shrimp farmers requested the presence of CRSP researchers because they were interested in developing methods for a sustainable shrimp production industry. In other shrimp-producing areas, the effluent from shrimp production ponds has polluted the estuaries from which water is taken to supply the ponds, creating a potential ecological disaster and subsequent ruin of the shrimp culture industry. Many shrimp farmers sought to avoid this problem and asked that the Pond Dynamics/Aquaculture CRSP conduct research to assure the estuarine water quality. There are presently two levels of farmer participation:

1. Estuarine monitoring: Water samples are analyzed free of charge for anyone who wishes to bring them in, provided they do so on a weekly schedule.
2. On-farm experiments: Farmers and the CRSP researcher develop the experimental protocols together. The subjects stem from results of previous research and are often suggested by the CRSP researcher. However, farmers have the final say, as it is they who are responsible for conducting the research. The CRSP researcher mainly functions to assure the scientific quality of the research plan — that the number of replicates are sufficient, that there are not too many variables tested simultaneously, and that the data collection is adequate. Each participating farm runs one replicate of each of the treatments, while the largest farm runs the complete experiment (usually 12 ponds). There is no government experiment station for shrimp culture research.

Even though the research was performed on their own farms, shrimp farmers have been hesitant to change their management based on results from the research. However, once they do accept the changes, their acceptance is long-lasting. Two notable changes have been implemented by farmers following the on-farm trials: ponds are managed differently in wet season and dry season now, and feed conversions have improved.

Farmer-driven research in the shrimp industry in Honduras seems to have avoided some of the disadvantages of other farmer-driven programs. Often, farmers will support only research that enhances the profitability of their operation, to the detriment of the environment. The farmers participating in research in Honduras are interested in the sustainability of their industry, as evidenced by the types of research projects they propose. This may be because the farmers are largely from the local area and hope to hand the operation over to their children. Another potential problem is the marginalization of small farmers. Although the on-farm research is open to all farms, many choose not to participate. Any participating farm must agree that the data from the research will be made public. This seems to discourage some participants.

In Thailand, research on the use of fertilizers to grow tilapia resulted in fish too small to fetch the highest possible market price. This was obvious from the economic analyses performed as part of the research. The resident researcher was aware that slightly larger fish would fetch a much more attractive price and therefore planned research on further grow-out of tilapia using feeds in the later part of the cycle. The outreach program conducted by the Asian Institute of Technology (AIT) involves field testing of pond management recommendations in which farmer consultation plays a part in evaluation of results and recommendations to be tested. The PD/A CRSP has been able to benefit from this program in which on-farm testing of inorganic fertilizers and tilapia stocking strategies provided valuable feedback to the research program (Knud-Hansen et al., 1994).

These are only a few examples of appropriate research topics resulting from the informal communication of needs and ideas from farmers and extension personnel. As part of every 2-year PD/A CRSP work plan, one globally important research topic is maintained, and the remainder of the research has been on topics of regional importance. Special research topics are also permitted. These are not usually part of the work plan and are typically in response to local problems. Indeed, most resident researchers agree that the most important research for the host countries has been done under the "special topics" heading. This level of flexibility has allowed for farmer participation in choice of research topics. Although there is no formal mechanism for choosing topics of interest to farmers or to the extension service, the long-term presence of PD/A CRSP researchers in a country has allowed them to understand local and regional problems and to establish close personal relationships with individuals in the production and extension sectors. More can be done to incorporate extensionists and farmers into the CRSP research planning (Acker, 1992).

MEASURING THE IMPACTS OF RESEARCH AND OUTREACH EFFORTS

Little systematic research has documented the impacts of aquaculture extension programs (Molnar and Duncan, 1989; Cernea and Tepping, 1977). Evaluation tools are necessary for delivery systems. By their nature, pure acquisition systems are continually evaluated by the farmers themselves and changed as needed. The members of the acquisition system must be identified to assure that all have the opportunity to acquire the information they need. The existence of a true acquisition system for aquaculture information in developing countries has not been documented.

Success of a research program can be defined in several ways. Usually, in evaluating the usefulness or impact of agricultural research programs, attempts are made to quantify the rate of adoption of the technology developed. Technologies that are most appropriate should be adopted at a high rate, provided they are disseminated well. One measure of success is an increase in production. This can be misleading, however, because production can increase at no benefit to farmers if the cost of production also increases or if the market price is reduced. Although consumers may benefit in the short run, farmers are not likely to continue an activity that does not benefit their families. A better measure may be the increase in revenue generated from aquaculture or the increase in overall family well-being.

Additionally, many other social and economic constraints that are out of the control of researchers and the extension service can hinder adoption of beneficial technologies. Government price controls on inputs and on produce can prevent adoption of technologies that would otherwise be beneficial in the absence of such controls.

A somewhat more idealistic approach to measure the effectiveness of research/extension programs would be to measure the increase in farmers' knowledge or to measure their perception of the value of the extension service. This can be achieved by direct interviews, by observation of the management practices of farmers, or by witnessing the transfer of information from farmer to farmer. Indeed, many researchers believe that all they should do is get the knowledge to the farmers and let the farmers decide what technologies are most appropriate for their own situation instead of actively promoting a particular technology.

Surveys

To date, there have been surveys conducted in Rwanda, Honduras, Thailand, and Philippines that can estimate the degree to which farmers have adopted some of the technology proposed by CRSP researchers (Molnar et al., 1995). It is more difficult to quantify any increase in income or production due to the adoption of certain techniques because the techniques are difficult to separate out from the ensemble of what farmers are doing. These surveys can be expensive if a sufficiently large sample is to be surveyed to draw statistically supported conclusions. Yet no other method provides quantitative estimates of the needs, characteristics, and farming systems PD/A CRSP research is intended to affect.

Extension Agent Reports

If extension agents produce reliable reports, it is advisable to use the data in their reports to the greatest extent possible. Increases in production along with changes in pond management practices can suggest effective outreach efforts, but conclusive proof will be difficult to assemble. Some bilaterally funded extension projects require considerable reporting (number of people reached, etc.) solely to justify the financial outlays for the program. Extension agents may have several pages of reports to file in this situation and are often encumbered to the point that they are likely to falsify information. Therefore, it is necessary to design a report format that is easy on the reporter while providing the maximum amount of verifiable information. Qualitative reports of farmer perceptions, motivations, and situations are an invaluable part of program management.

ACKNOWLEDGMENTS

Much of the information in this chapter is based upon interviews with the CRSP resident researchers, in Thailand, Dr. C. Kwei Lin, and in Honduras, Dr. David Teichert-Coddington.

REFERENCES

Albrecht, H., Bergmann, H., Diederich, G., Groer, E., Hoffman, V., Keller, P., Payr, G., and Sulzer, R., *Agricultural Extension — Volume 1: Basic Concepts and Methods,* Verlagsgesellschaft mbH fur Technische Zusammenarbeit (GTZ), Eschborn, Germany, 1990a.

Albrecht, H., Bergmann, H., Diederich, G., Groer, E., Hoffman, V., Keller, P., Payr, G., and Sulzer, R., *Agricultural Extension — Volume 2: Examples and Background Materials,* Verlagsgesellschaft mbH fur Technische Zusammenarbeit (GTZ), Eschborn, Germany, 1990b.

Acker, D. G., Developing Effective Researcher-Extension-Farmer Linkages for Technology Transfer, in *Proceedings of the Workshop on Social Science Research and the CRSPs,* INSORTMIL Publication Number 93-3, Department of Rural Sociology, University of Kentucky, Lexington, KY, 1992, 210–231.

Axinn, G. H., The different systems of agricultural education with special attention to Asia and Africa, in Rivera, W. M. and Schram, S. G., Eds., *Agricultural Extension Worldwide,* Croom Helm, London, 1987, 103–114.

Balakrishnan, R., Rwanda Women in Aquaculture: Context, Contributions and Constraints, Office of Women in International Development, Oregon State University, Corvallis, 1993.

Benor, D., Harrison, J. Q., and Baxter, M., Agricultural Extension: The Training and Visit System, World Bank, Washington, D.C., 1984.

Binswarger, H. P. and Ruttan, V. W., *Induced Innovation Technology, Institutions, and Development,* Johns Hopkins University Press, Baltimore, MD, 1978.

Brown, L. A., *Innovation Diffusion: A New Perspective,* Methuen, New York, 1981.

Cernea, M. and Tepping, B. J., A System for Monitoring and Evaluating Agricultural Extension Projects, World Bank Staff Working Paper No. 272, The World Bank, Washington, D.C., 1977.

Chambers, R., Pacey, A., and Thrupp, L., Eds., *Farmer First: Farmer Innovation and Agricultural Research,* Intermediate Technology Publications, London, 1989.

Chin, R. and Benne, K. D., General strategies for effecting change in human systems, in Benne, W. G., Benne, K. D., Chin, R., and Corey, K. E., Eds., *The Planning of Change,* Holt, New York, 1976, chap. 1.2.

Engle, C. R. and Stone, N. M., A Review of Extension Methodologies in Aquaculture, ADCP/REP/89/44, Food and Agriculture Organization of the United Nations, Rome, 1989.

Gibbons, M. J. and Schroeder, R., Agricultural Extension, U.S. Peace Corps, Office of Training and Program Support, Washington, D.C., 1983.

Harrison, E., Aquaculture in Africa: Socioeconomic Dimensions, A Review of the Literature, School of African and Asian Studies, University of Sussex, England, 1991.

Kaimowitz, D., The evolution of links between extension and research in developing countries, in Rivera, W. M. and Gustafson, D. J., Eds., *Agricultural Extension: Worldwide Institutional Evolution and Forces for Change,* Elsevier, Amsterdam, 1991, 101–112.

Knud-Hansen, C. F., Batterson, T. R., Guttman, H., Lin, C. K., and Edwards, P., Field Testing Least Intensive Aquaculture Techniques on Small-Scale Farms in Thailand, in *Eleventh Annual Administrative Report,* Egna, H. S., Bowman, J., and McNamara, M., Eds., Pond Dynamics/Aquaculture Collaborative Research Support Program, 1 September 1992 to 31 August 1993, 1994, 21–45.

Koehler, J. W., Anatol, K. E., and Applebaum, R. L., *Organizational Communication: Behavioral Perspectives,* Holt, New York, 1981.

Lionberger, H. F., *Adoption of New Ideas and Practices,* The Iowa State University Press, Ames, 1960.

Lionberger, H. F. and Gwin, P. H., *Communication Strategies: A Guide for Agricultural Change Agents,* Interstate Printers, Danville, IL, 1983.

Moehl, J. F. and Molnar, J. J., Dilemmas of aquaculture development in Rwanda, in Bailey, C., Jentoff, S., and Sinclair, P., Eds., *Social and Environmental Aspects of Aquaculture Development,* Westview Press, Boulder, CO, 1996.

Molnar, J. J., Adjavon, V., and Rubagumya, A., The sustainability of aquaculture as a farm enterprise in Rwanda, *J. Appl. Aquaculture,* (1)2, 37–62, 1991.

Molnar, J. J. and Duncan, B. L., Monitoring and evaluating aquaculture projects, in Pollnac, R. B., Ed., *Monitoring and Evaluating the Impacts of Small-Scale Fishery Projects,* International Center for Marine Resource Development, Kingston, Rhode Island, 1989, 28–40.

Molnar, J. J., Hanson, T. R., and Lovshin, L. L., The Multiple Identities of Tilapia as a Farm Enterprise: The Growth of a Global Commodity in the Context of Development Assistance, Paper presented at the Annual Meeting of the Rural Sociological Society, Washington, D.C., 1995.

Molnar, J., Hanson, T., and Lovshin, L., Impacts of Aquacultural Research on Tilapia: The Pond Dynamics/Aquaculture CRSP in Rwanda, Honduras, The Philippines, and Thailand, Research and Development Series 40, International Center for Aquaculture and Aquatic Environments, Auburn University, Auburn, AL, 1996.

Oakley, P. and Garforth, C., Guide to Extension Training, Food and Agriculture Organization of the United Nations, Rome, 1985.

Rasmussen, W. D., *Cooperative Extension: Taking the University to the People,* Iowa State University, Ames, 1989.

Rivera, W. M. and Schram, S. G., *Agricultural Extension Worldwide,* Croom Helm, London, 1987, S 544.A59.

Rogers, E. and Shoemaker, F. F., *Communication of Innovations: A Cross-Cultural Approach,* The Free Press, New York, 1971.

Roling, N., *Extension Science: Information Systems for Agricultural Development,* Cambridge University Press, New York, 1988.

Roling, N. and Engel, P., The development of the concept of agricultural knowledge information systems (AKIS): implications for extension, in Rivera, W. M. and Gustafson, D. J., Eds., *Agricultural Extension: Worldwide Institutional Evolution and Forces for Change,* Elsevier, Amsterdam, 1991, 125–137.

Rurangwa, E., Veverica, K. L., Seim, W. K., and Popma, T. J., On-farm production of mixed-sex *Oreochromis niloticus* at different elevations (1370 to 2230 m), in *Ninth Annual Administrative Report,* Pond Dynamics/Aquaculture Collaborative Research Support Program, 1991, Egna, H. S., McNamara, M., and Weidner, N., Eds., Oregon State University, Corvallis, 1992, 35–40.

Sanders, J. B., Ed., *Food and Civilization,* Charles C. Thomas, New York, 1966, TX 345.F3.

Shell, E. W., *The Development of Aquaculture: An Ecosystem Perspective,* Craftmaster Printers, Opelika, AL, 1993, 265 pp.

Teichert-Coddington, D. R., Green, B. W., Rodriquez, M. I., Gomez, R., and Lopez, L. A., On-farm testing of PD/A CRSP Fish Production Systems in Honduras, in *Tenth Annual Administrative Report,* Egna, H. S., McNamara, M., Bowman, J., and Austin, N., Eds., Pond Dynamics/Aquaculture Collaborative Research Support Program, 1 September 1992 to 31 August 1992, 1993, 28–47.

Thompson, R. and Thompson, S., The on-farm research program of practical farmers of Iowa, *Am. J. Alternative Agric.,* 5(4), 163–167, 1990.

Van Den Ban, A. W. and Hawkins, H. S., *Agricultural Extension,* John Wiley and Sons, New York, 1990.

Zaltman, G. and Duncan, R., *Strategies for Planned Change,* John Wiley and Sons, New York, 1977.

INDEX

S